EUROSHOCK –
Drag Reduction by
Passive Shock Control

Edited by
Egon Stanewsky
Jean Délery
John Fulker
and Wolfgang Geißler

W0107365

Notes on Numerical Fluid Mechanics (NNFM)   Volume 56

EUROSHOCK –
Drag Reduction by Passive Shock Control

Results of the Project EUROSHOCK, AER2-CT92-0049 Supported by the European Union, 1993–1995

Edited by
Egon Stanewsky
Jean Délery
John Fulker
and Wolfgang Geißler

Die Deutsche Bibliothek – CIP-Einheitsaufnahme

EUROSHOCK: drag reduction by passive shock control;
results of the project EUROSHOCK, AER 2-CT92-0049 /
ed. by Egon Stanewsky. – Braunschweig; Wiesbaden:
Vieweg, 1997
 (Notes on numerical fluid mechanics; Vol. 56)

ISBN 978-3-322-90713-4   ISBN 978-3-322-90711-0 (eBook)
DOI 10.1007/978-3-322-90711-0

Preface

It has been recognized by the European Union (EU) that the survival of the Aeronautical Industries of Europe in the highly competitive World Aviation Market is strongly dependent on such factors as time-to-market of a new or derivative aircraft and on its manufacturing costs but also on the achievement of a competitive technological advantage by which an increased market share can be gained. Cooperative research is therefore continuously encouraged and co-financed by the European Union in order to strengthen the scientific and technological base of the Aeronautical Industries thus facilitating the future design and manufacture of civil aircraft products and providing the technological edge needed for survival. Targets of research within Area 3, Aeronautics, of the Industrial and Materials Technology Programme (1991 - 1993) have been identified to be aircraft efficiency, cost effectiveness and environmental impact. Concerning aircraft efficiency - relevant to the present research - a reduction in aircraft drag of 10%, a reduction in aircraft fuel consumption of 30% and a reduction in airframe, engine and system weight of 20% are envisaged. Meeting these objectives has, of course, also a strong positive impact on the environment.

In order to further technology, the philosophy is to avoid spreading the available resources too thinly and rather concentrate on the feasibility demonstration of a limited number of technologies of high economic and industrial impact. Examples of such technologies are, for instance, with regard to aircraft efficiency, the application of the hybrid laminar flow technique, advanced aeronautic lightweight structures and engine technologies. In aerodynamics - the focus of the present work - drag reduction technologies in general and means of controlling separation are deemed to be of considerable importance in order to improve not only cruise performance but also to increase the flight envelope of transonic transport aircraft. A prerequisite for technology development is, of course, also the continuous improvement of the theoretical / numerical and experimental tools needed for the understanding of complex viscous compressible flow phenomena such as turbulence, transition, shock boundary layer interaction and separation.

The fundamental research program described here is related to drag reduction and separation control; it is based on the following consideration: Shock waves and their interaction with the boundary layer play a key role in determining the performance of laminar as well as turbulent transonic wings at design and off-design conditions. By controlling the shock strength and the interaction process, it seems therefore possible to not only reduce aircraft cruise drag, hence reduce fuel consumption and detrimental exhaust emissions, but also to shift the drag-rise and buffet boundaries to higher Mach numbers and/or lift coefficients. The specific objective of the research derived from these considerations was thus to study the various aspects of shock boundary layer interaction control (SBLIC), to develop and improve the computational and experimental tools needed to incorporate control concepts into the design of advanced transonic wings and to assess the merits of the SBLIC concept numerically and experimentally up to flight Reynolds numbers.

The work was carried out by five research organizations, viz., Deutsche Forschungsanstalt für Luft- und Raumfahrt e.V. (DLR), Centro Italiano Ricerche Aerospaziali S.C.p.A. (CIRA), Stichting National Lucht-en Ruimtevaartlaboratorium (NLR), Office National d'Etudes et de Recherches Aérospatiales (ONERA) and Defense Evaluation Research Agency (DERA), four universities, viz., the National Technical University of Athens, the University of Cambridge, the University of Karlsruhe and the Universita'di Napoli "Federico II", and Daimler-Benz Aerospace Airbus GmbH (DA) as industrial partner. The program was endorsed by all major airframe manufacturers within the EU.

The present book, containing a thorough description of the work performed, is structured as follows: Firstly, the scientific and economical background leading to this investigation and the approach taken are given. This is succeeded by a comprehensive and critical account of the research and the results obtained - without going into excessive detail. Finally, the individual contributions of the partners are presented in the form of papers giving appropriate details of the fundamental, numerical and experimental research performed.

The editors would like to thank all partners for their contribution to the success of EUROSHOCK (I) and for the effort they put into the preparation of the present book. The work was performed with remarkable enthusiasm and in a very harmonious way which is certainly reflected in the high quality of the results. On behalf of the entire team, we would also like to thank the European Commission for its support and, in particular, for bringing this team together and staying with it during the three-year research effort. Finally, thanks are due to the E.H. Hirschel, the general editor of the „Notes on Numerical Fluid Mechanics" and to the Vieweg Verlag for making this publication possible, although there is a fair amount of experimental work - and not only for validation purposes - involved.

June 1997

Egon Stanewsky   Göttingen
Jean Délery   Paris
John Fulker   Bedford
Wolfgang Geißler   Göttingen

Table of Contents

Table of Contents (continued)   Page

Table of Contents (continued) Page

Table of Contents (continued)   Page

Table of Contents (continued) Page

Table of Contents (continued)   Page

Table of Contents (continued)   Page

Table of Contents (continued)   Page

Table of Contents (continued)   Page

Brief Instruction for Authors

Manuscripts should have well over 100 pages. As they will be reproduced photomechanically they should be produced with utmost care according to the guidelines, which will be supplied on request.
In print, the size will be reduced linearly to approximately 75 per cent. Figures and diagrams should be lettered accordingly so as to produce letters not smaller than 2 mm in print. The same is valid for handwritten formulae. Manuscripts (in English) or proposals should be sent to the general editor, Prof. Dr. E. H. Hirschel, Herzog-Heinrich-Weg 6, D-85604 Zorneding.

A. SYNOPSIS OF THE PROJECT EUROSHOCK

by

E. Stanewsky
DLR, Institute of Fluid Mechanics
Bunsenstraße 10, D-37073 Göttingen

J. Délery
ONERA, Experimental / Fundamental Aerodynamics Branch
29, avenue de la Division Leclerc, 92320 Chatillon, France

John Fulker
DERA, High Speed and Weapon Aerodynamics Department
Bedford, MK41 6AE, UK

Wolfgang Geißler
DLR, Institute of Fluid Mechanics
Bunsenstraße 10, D-37073 Göttingen

1

1 Introduction

The present book is based on research carried out within the EU-funded project EUROSHOCK - European Shock Control Investigation, whose objective was to study, numerically and experimentally, the effect of passive shock boundary layer interaction control (SBLIC) by ventilation on the design and off-design performance of primarily laminar-type airfoils and wings. The considerations leading to this research are outlined in what follows.

When increasing the freestream Mach number or angle of incidence (lift) of a transonic airfoil or wing, a supersonic region develops on the upper surface which will, for a modern turbulent design at cruise conditions, be terminated by an isentropic recompression or a weak shock wave, Figure 1 (T). Further increasing either freestream parameter will generally lead to stronger shocks, hence an increase in wave drag, and finally to separation, thus defining the drag-rise and buffet boundaries. For a laminar wing design, the flow development is inherently characterized by a continuous acceleration of the flow on the upper surface which will cause stronger shock waves even at the design point in order to maximize the reduction in friction drag and, at the same time, maintain high lift (L). Generally, in order to realize the full potential of laminarization, the shock must be placed as far downstream as possible which tends, however, to increase the shock strength and hence wave and viscous drag even further, thus, at some point, offsetting the benefit of laminarization. The latter also causes laminar wings to be more sensitive to changes in the freestream conditions, especially with respect to the performance limits, i.e., the drag-rise and buffet boundaries, which are more rapidly reached. It is, therefore, important for laminar but also for turbulent wings to develop means of reducing the shock strength in order to keep drag low and to postpone the drag-rise and buffet boundaries to higher Mach numbers and/or lift coefficients as demonstrated in Figure 1.

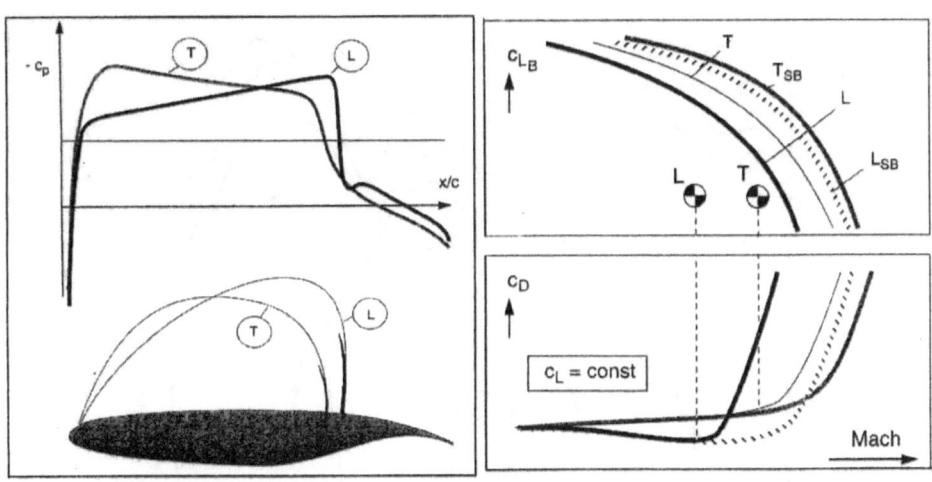

a. Pressure distribution for a turbulent (T) and a laminar (L) airfoil at equal lift

b. Design points and drag-rise and buffet boundaries for a turbulent (T) and a laminar airfoil without (L) and with (SB) shock control

Figure 1: Transonic airfoil flow (schematic)

It has been shown in early experiments that the detrimental effects of strong shock waves and strong rear adverse pressure gradients on the off-design performance of airfoils and wings can be reduced by some means of shock and boundary layer control. Methods that have been investigated with the objective of suppressing shock-induced and/or rear separation included vortex generators, single slot suction and blowing [1.1]. More recently double slots and perforated surfaces with underlying cavities in areas of strong shocks in conjunction with suction have been investigated and positive results have been obtained not only in suppressing shock-induced and rear separation, and hence shifting the buffet boundaries to higher Mach numbers, but also in reducing drag over a considerable range of the flight envelope [1.2, 1.3]. Moreover, it was found, e.g., with the VA-2 airfoil [1.3], that in the case of a perforated surface or a double slot, active suction was not required to achieve these improvements, i.e., this type of shock control constitutes a self-adjusting passive process requiring no additional power. It is shown in Figure 2 that the range parameter L/D is considerably increased by passive shock control which is mainly the result of a reduction in total drag and a small increase in lift due to control. The control mechanism is also indicated in Figure 2: air flows into the cavity in the high pressure region behind the shock and leaves the cavity in the low pressure area upstream of the shock, thus spreading the rapid pressure rise due to the shock reducing shock strength.

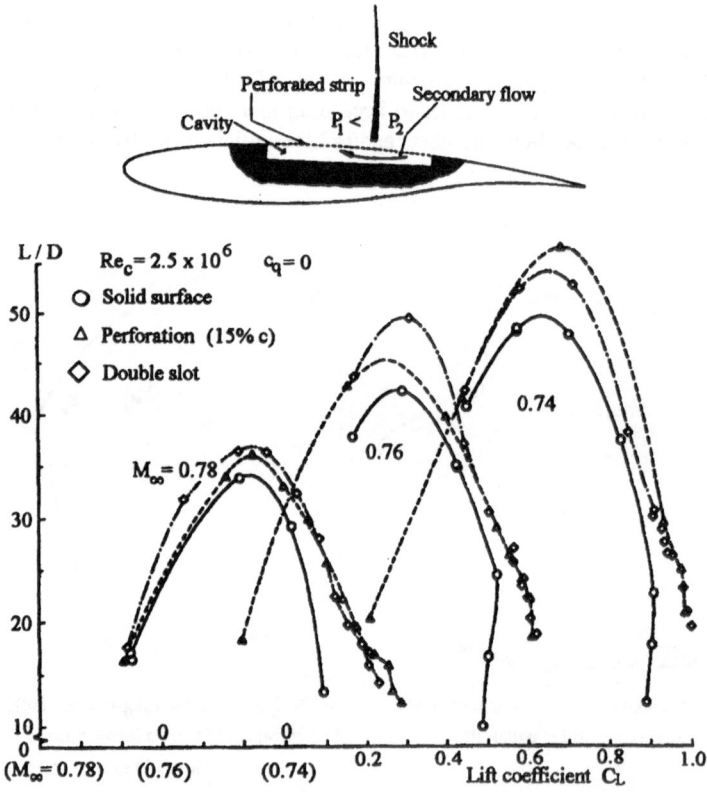

Figure 2: Effect of passive shock control on the range parameter L / D
Airfoil VFW VA-2

The initial success of shock control has initiated activities at various European universities and research institutions concerned with the investigation of the basic physics involved and with first attempts to develop computational methods accounting for shock control effects [1.4-1.9]. Similar research has also been carried out in the US and in Japan [1.10-1.14].

The advance of laminar airfoil technology and the positive effect of shock control on the flow development about transonic airfoils described above led to the present EC-sponsored research program which is mainly concerned with a detailed investigation of **passive** shock control on laminar-type airfoils, previously not considered. The potential for performance improvement at design is much higher for laminar airfoils and, as mentioned above, the flow development in the vicinity of the drag-rise and buffet boundaries is for laminar airfoils much more sensitive to small changes in freestream conditions and shock control can reduce this sensitivity by weakening otherwise existing strong pressure gradients and their inherent effect on (sudden) boundary layer separation.

The specific objective of the research described here was to investigate and devise methods of shock control aiming at the reduction of aircraft drag and the improvement of aircraft off-design performance, and to provide the tools needed for the design of transonic wings with control. This required that one studies first the basic control phenomena and the associated flow developments and the critical parameters involved, such as shock strength, boundary layer conditions and surface geometries. Physical models of the viscous-inviscid interaction had to be established and the rather complex boundary conditions determined. The physical models must be incorporated into computational methods which, in turn, have to be validated by the results of realistic experiments on airfoils. Both, experiments and computations, were employed to assess the merits of the shock control concepts.

The research program was consequently divided into three major research tasks whose interdependence is depicted in Figure 3; the tasks are:

Task 1 Modeling of Sock Boundary Layer Interaction Control (SBLIC) Phenomena with Subtasks 1.1-Basic experiments and 1.2-Physical modeling. Here, basic experiments were carried out on simple configurations allowing a detailed flow analysis of the interaction control process and, based on these experiments, an adaptation and improvement of the physical models to be used in the control region.

Task 2 Transonic Airfoil / Wing Flow Prediction with SBLIC with Subtasks 2.1-Steady flow predictions and 2.2-Unsteady flow predictions. Here, existing numerical methods were extended to predict steady and unsteady flows with shock boundary layer interaction control, employing the results of Task 1; the methods were validated by the results of Task 3 and used for a first parametric analysis of SBLIC-effects.

Task 3 Wind Tunnel Experiments on Airfoils with SBLIC with Subtasks 3.1 Investigation of Reynolds number effects and 3.2-Detailed flow investigation on a large-scale airfoil model. Here, detailed measurements on two laminar airfoil models (later extended to a third turbulent airfoil model) were executed in conventional and cryogenic wind tunnels to relate basic phenomena in the interaction region to improvements in airfoil performance and to study the effect of Reynolds number on control up to flight conditions.

The work was carried out, as already mentioned, by five research organizations, viz., DLR, CIRA, DERA (DRA), NLR and ONERA, four universities, viz., Athens, Cambridge, Karlsruhe and Naples, and by DASA-Airbus as industrial partner; all will be identified within the course of the book and their individual contributions will be presented in Chapters 9 through 23 following a comprehensive and critical account of the research performed and the results obtained - without going, however, into excessive detail.

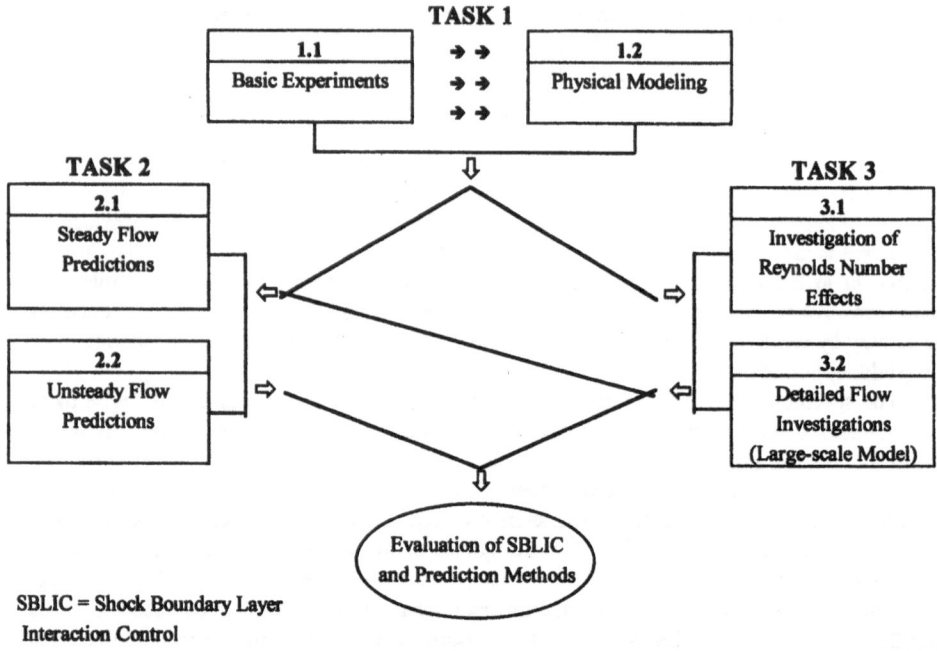

TASK 1

| 1.1 | | 1.2 |
| Basic Experiments | ➔ ➔
➔ ➔
➔ ➔ | Physical Modeling |

TASK 2

| 2.1 |
| Steady Flow
Predictions |

| 2.2 |
| Unsteady Flow
Predictions |

TASK 3

| 3.1 |
| Investigation of
Reynolds Number
Effects |

| 3.2 |
| Detailed Flow
Investigations
(Large-scale Model) |

Evaluation of SBLIC
and Prediction Methods

SBLIC = Shock Boundary Layer
Interaction Control

Figure 3: General task flow chart / interrelation between tasks

2 Basic Experiments and Physical Modeling (Task 1)

The objective of the experimental and theoretical / numerical work performed here can be summarized as follows:

i) Improve the physical understanding of the phenomena involved in a transonic shock wave turbulent boundary layer interaction under passive control conditions.

ii) Assess and, if possible, improve the physical laws associated with passive shock control; these laws are to be introduced into the numerical codes which will be used to predict airfoil / wing performance in the presence a passive control device.

iii) Evaluate the efficiency of a control system by using a simple test arrangement which will allow a rapid modification of the system to investigate essential influence parameters such as cavity depth, perforated plate characteristics, shock wave location.

As also shown in Figure 3, points i) and iii) are linked to Task 3 where optimized control devices shall be incorporated into and tested in conjunction with airfoils, whereas points i) and ii) are obviously closely linked to Task 2.

Three institutions were involved in Task 1, namely, the ONERA Aerodynamics Department, the University of Karlsruhe and the University of Cambridge. These contributors worked in close contact and investigated important and complementary aspects of the problem:

The **ONERA** Aerodynamics Department focused its activity on a detailed analysis of boundary layer flows under passive control conditions, with emphasis placed on flow field topology and turbulence behavior; the **University of Karlsruhe** performed a thorough analysis of the flow

6

through porous plates in order to deduce improved physical laws for the transpiration velocity, this point being vital for predictive methods; **Cambridge University** considered the crucial problem of passive control behavior in three-dimensional flows, which corresponds to the situation met on swept wings. In addition to detailed experiments, all the contributors performed calculations to validate the physical models by careful comparisons with their respective experiments. Details of their research are provided in Chapters 9, 10 and 11.

2.1 Two-dimensional Shock Control

2.1.1 Test Set-up for the basic experiments

Two-dimensional basic experiments were performed by ONERA and the University of Karlsruhe, respectively, utilizing the channel-flow set-ups depicted in Figures 4 and 5.

The ONERA experiments have been executed in a continuous transonic wind tunnel supplied with desiccated atmospheric air, the stagnation conditions being close to ambient conditions. The test set-up, Figure 4, consists of a transonic channel having a height of 100 mm and a span of 120 mm in the test section. The lower wall is flat, the upper wall consists of a contoured profile (nozzle) designed to produce a uniform supersonic flow at a Mach number close to 1.4. An adjustable second throat is placed at the test section outlet to produce a shock wave by choking. Passive control is applied on the flat wall over a length of L = 70 mm with the location selected to reproduce the same ratio L/δ_0 as in reality, with δ_0 being the boundary layer thickness at the origin of the control region. The interacting boundary layer was probed by using a two component LDV system which allowed to measure the mean velocity vector and the Reynolds tensor components in the interacting flows. (For more details on the LDV system, the

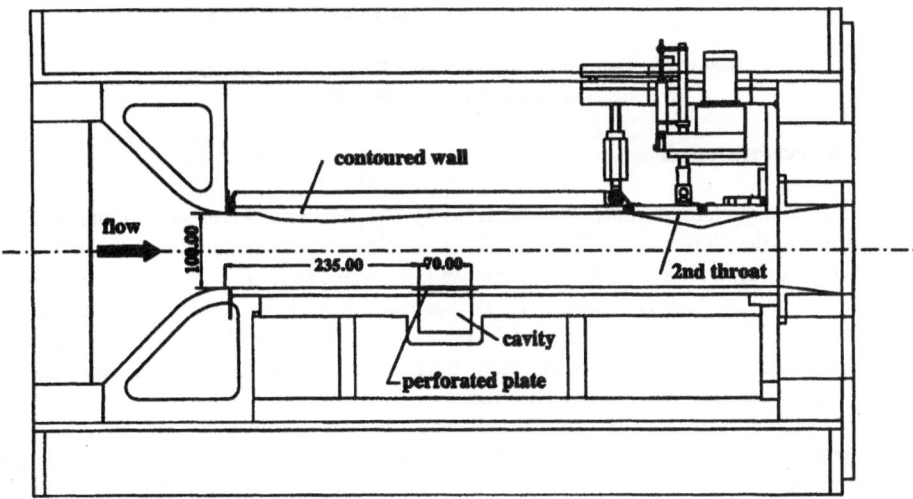

dimensions in mm

Figure 4: Passive control experimental set-up in the ONERA S8 wind tunnel

reader is referred to Chapter 9 and Ref. [2.1].) For the study of passive control, the location of the shock wave in the outer part of the flow was fixed at mid distance between the origin and the end of the control region. This was the same for all configurations tested.

At the University of Karlsruhe, the experiments have been performed in an atmospheric blow-down wind tunnel with a test section having a cross section of about 50 x 200 mm², Figure 5. The test section (Laval-nozzle type) consists of an interchangeable 70 mm long perforated plate allowing the passive control device to be installed in the slightly diverging part of the nozzle. Measurements of the boundary layer profiles were carried out by means of pressure probes. For all plates, measurements have been made for two values of the upstream Mach number: $M = 1.27$ and $M = 1.3$. In the case of $M = 1.27$, the shock is located exactly at the center of the cavity, whereas at $M = 1.3$ it is shifted 10 mm downstream (also see Chapter 10).

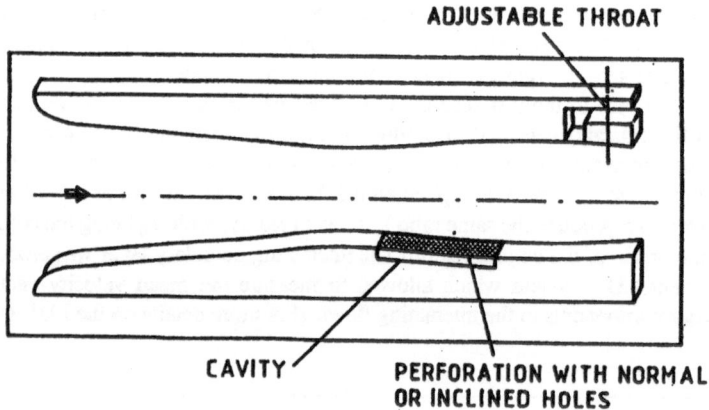

Figure 5: Test arrangement for the Karlsruhe University
shock control experiments

2.1.2 The perforated plates

Two sets of four perforated plates were provided by Daimler Benz Aerospace Airbus (DA) to ONERA and to the University of Karlsruhe. Their nominal characteristics are given in Table I. Here, Plate 0 corresponds to the reference solid wall; for Plates I and III, all holes are normal and for Plates II and IV, they are inclined in the upstream part of the plate and normal in the downstream half. The nominal porosity of the plates was equal to 5.67%. However, as can also be seen in Table I, the quality of the plate manufacturing was poor so that considerable deviations from the nominal geometry occurred.

The actual properties of the perforated plates were determined by the University of Karlsruhe using the experimental facility sketched in Figure 6. The main part of the facility consists of a tube with the porous plate mounting system placed at the inlet. The air is supplied at ambient pressure and temperature conditions. A mass flow rate measurement orifice is located a certain distance downstream allowing an accurate determination of the mass flow rate by using standard procedures. At the downstream side, the tube is connected to a fan sucking the air through the installation. The mass flow rate can be controlled by means of a by-pass valve. The porosity of the plates has been determined by means of hole geometry measurements with a microscope. Here, it was found that the holes are in fact conical with a larger size on the cavity side and that their shape is irregular rather than circular.

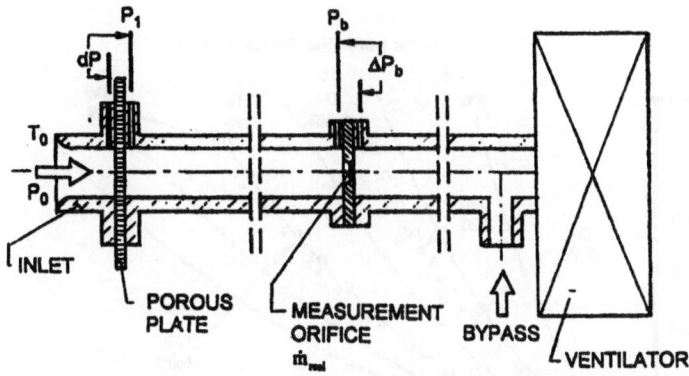

Figure 6: Set-up for plate calibration at Karlsruhe University

Table 1: Nominal and actual characteristics of the plates tested; simulated flow direction is from cavity to main stream

Plate number	Hole diameter	Porosity (%)	Hole inclination
ONERA	Nominal / Actual	Nominal / Actual	Nominal / Actual
0	-	0	-
D-I	0.15 / 0.125 mm	5.67 / 3.55	90°
D-II	0.15 / 0.125 mm	5.67 / 3.30	90° downstream
	0.15 / 0.143 mm	5.67 / 3.30	45° upstream
D-III	0.30 / 0.285 mm	5.67 / 5.00	90°
D-IV	0.30 / 0.285 mm	5.67 / 4.90	90° downstream
	0.30 / 0.305 mm	5.67 / 5.00	45° upstream

Plate number	Porosity (%)	Hole diameter	Hole inclination
Karlsruhe University	Nominal / Actual	Nominal / Actual	Nominal / Actual
0	0	-	-
K-I	5.67 / 4.00	0.15 / 0.125 mm	90°
K-II	5.67 / 3.25	0.15 / 0.125 mm	90° downstream
	5.67 / 3.40	0.15 / 0.140 mm	45° upstream
K-III	5.67 / 5.30	0.30 / 0.30 mm	90°
K-IV	5.67 / 4.94	0.30 / 0.28 mm	90° downstream
	5.67 / 5.14	0.30 / 0.30 mm	45° upstream

Five groups of plates have been tested; however, here we will only consider the ones actually used in the study of passive control by Karlsruhe University and ONERA. The whole set of results and details of the plate - calibration procedure can be found in the Technical Report prepared by the University of Karlsruhe [2.2] and in Chapter 10.

Each plate has been investigated in the facility shown in Figure 6 for blowing and suction, simulating the flow from the cavity to the main stream and from the main stream to the cavity, respectively. Note that, since the holes are conical, somewhat different characteristics prevail at

9

large porosities for blowing and suction. Determined in the investigations were the pressure drop across the plates and the mass flow rates as well as the effective porosity providing the plate characteristics as depicted, e.g., in Figure 7 for the ONERA plates.

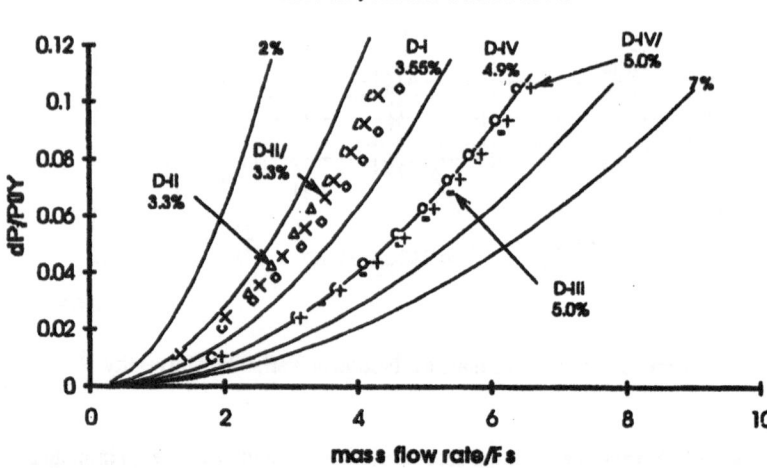

Figure 7: Characteristics of the ONERA plates for blowing (flow from cavity)

From all plate measurements, an empirical relation was derived coupling the effective Mach number downstream of the plate, M_{hole}, to the pressure drop through the plate which is independent of hole size and shape:

$$M_{hole} = 1.2(\frac{dP}{P_0})^{0.55}.$$

With the knowledge of M_{hole}, the velocity downstream of the plate, i.e., the velocity normal to the flow direction in the case of actual shock control, can be determined via the equation

$$v_1 = p_{cor} \frac{M_{hole}\sqrt{\gamma R\, T_0}}{\sqrt{1 + \frac{\gamma - 1}{2} M_{hole}^2}}$$

if the effective porosity, p_{cor}, is known. The equation is derived in Chapter 10 and constitutes the new control law developed at the University of Karlsruhe. Note that the theoretical curves in Figure 7 denoted by integer porosity numbers are based on the equation

$$\frac{dP}{P_0} = \frac{0.063}{p_{cor}^2}(\frac{m_{re}}{F_s})^2$$

where m_{re} is the actual mass flow rate in % and F_s is the total area of the porous plate exposed to the flow. As one sees, there is a fair agreement between the experimental values and the ones given by the formula.

2.1.3 Effect of plate geometry on shock control effectiveness

Passive shock control by a porous plate over a cavity exerts its influence on the flow by a mechanism that causes inflow into the cavity in the high-pressure area downstream of the shock and outflow from the cavity into the low pressure area upstream of the shock. This generally spreads the steep pressure rise due to a shock over the cavity region, as can be seen in the Schlieren photographs taken by ONERA and the Mach-Zehnder interferograms obtained by the University of Karlsruhe in their respective facilities, Figure 8, thus reducing wave drag. The spreading of the shock is also indicated by the numerical simulation of passive shock control, obtained with the University of Karlsruhe standard Navier-Stokes solver [2.2] in conjunction with the above control law.

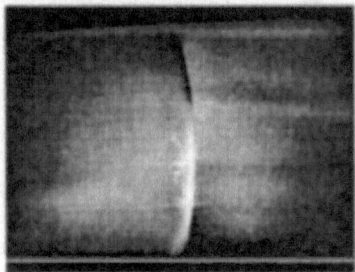 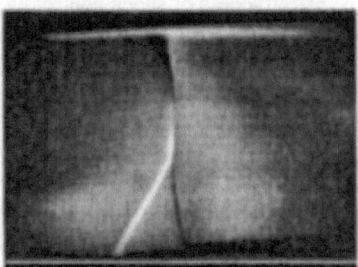

Solid wall   Inclined holes, diameter 0.15 mm

a. ONERA: Schlieren photograph of the flow field w/o and with control

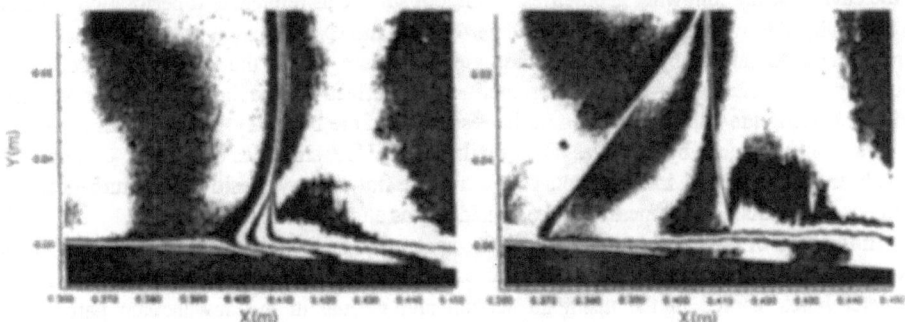

b. Karlsruhe University Mach-Zehnder interferogram;
left solid surface, right porosity 5.3%

Figure 8: Flow field observations obtained by ONERA and Karlsruhe University

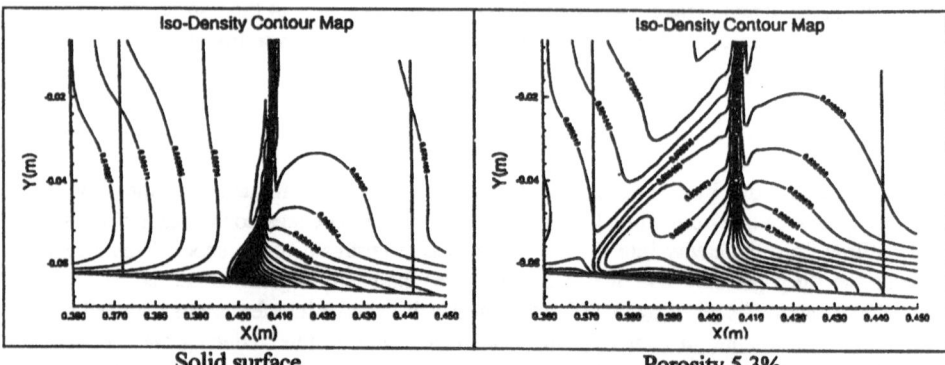

Solid surface   Porosity 5.3%

c. Karlsruhe University Navier-Stokes Simulation

Figure 8: Concluded

The effectiveness of shock control as dependent on the plate geometry was investigated by the University of Karlsruhe for the plate geometries listed in Table 1 and denoted K-I to K-IV at shock-upstream Mach numbers of M = 1.27 and 1.30 where shock control should be effective, utilizing the Laval - type nozzle test set-up of Figure 5; also tested for reference was the smooth surface.

As an example, the effect of passive control on the wall Mach number distribution at a shock -upstream Mach number of M = 1.27 is presented in Figure 9 together with the reference condition. Passive control significantly influences the Mach number distribution along the wall spreading the shock thus reducing its strength. Downstream of the shock, the distributions with passive control indicate higher values of the Mach number compared to the reference case which seems a first indication of flow separation. There is no significant effect of the porous plate geometry on the Mach number distribution. Similar results were obtained at M = 1.30.

In addition to the pressure distribution measurements, boundary layer measurements were carried out upstream of the cavity - undisturbed by the interaction - and somewhat downstream of the cavity for the following geometries: ①smooth surface as reference, ②perforation covered on the cavity side without flow through the perforation and ③ open cavity, i.e., the actual shock control case. As shown by the results presented in Figure 10 for the no-control case, shock boundary layer interaction has an extremely strong effect on the boundary layer velocity profiles, causing the known reduction in the resistance of the boundary layer to separation. The plate roughness, caused by the holes taped on the cavity side, amplifies this effect. Generally, it was observed that at the Mach numbers considered the flow is very sensitive to disturbances and even porosity-induced roughness must be considered as an important factor .

In the case of passive control, Figure 10b, separation of the boundary layer is induced by control with the most detrimental effect being caused by the Plate K-IV, i.e., the plate with large diameter holes slanted in the upstream section and normal in the downstream part of the perforated plate. However, the results presented here do not provide enough evidence to make a final statement about the influence of the hole geometry on passive control. The effect of control on separation will again be addressed when looking at the ONERA results below.

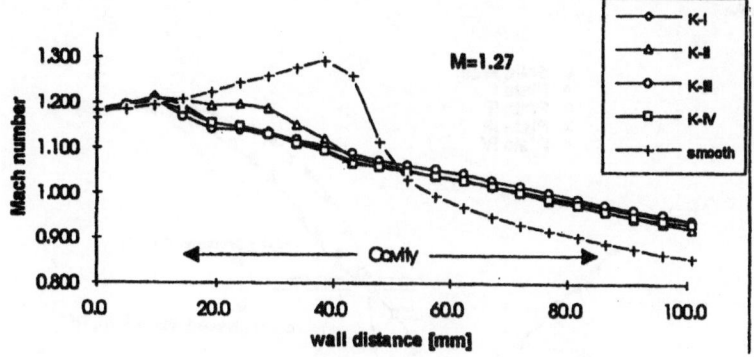

Figure 9: Wall Mach number distributions with and w/o passive control

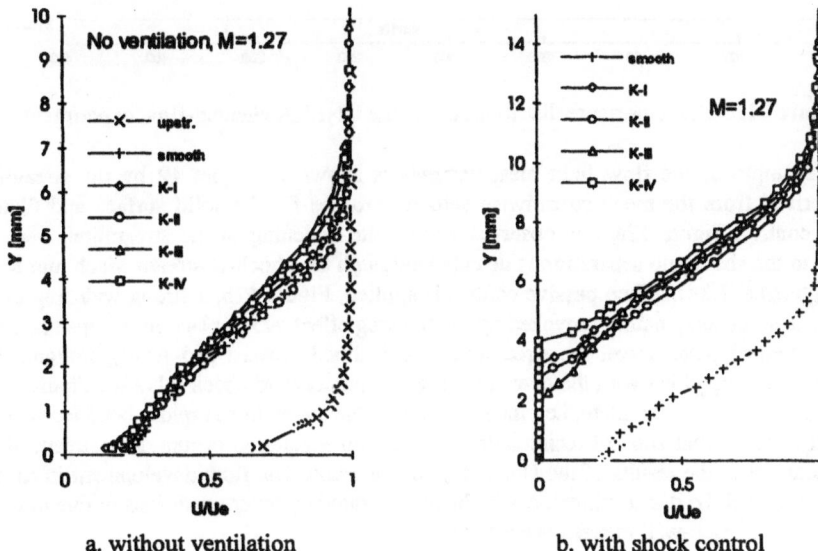

a. without ventilation   b. with shock control

Figure 10: Boundary layer profiles downstream of the cavity with and w/o control; reference condition upstream of the cavity

2.1.4 Boundary layer development over the cavity region

As explained above, the effect of passive shock control is first to spread the pressure rise due to the shock over the cavity region thus reducing wave drag. To explore the overall effect of control, specially on drag, flow field and boundary layer measurements were performed by ONERA utilizing a two-component LDV system. Corresponding surface pressure distributions are plotted in Figure 11. The shock-upstream Mach number without control is M = 1.34. Note that for the plates investigated (see Table 1) there is no difference in the effect of control on the pressure distributions, as already observed in the Karlsruhe investigation.

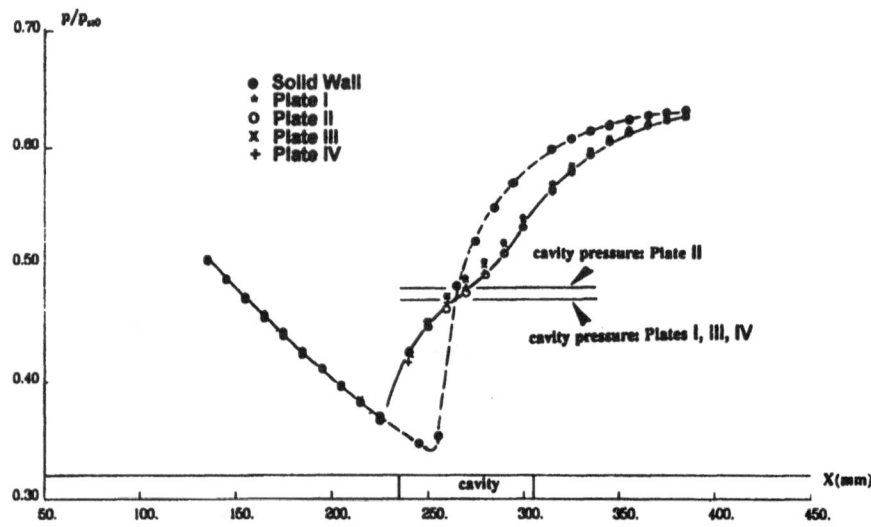

Figure 11: Surface pressure distributions for the ONERA channel-flow experiment

An example of the flow field measurements is shown in Figure 12 by the streamline traces derived from the mean streamwise velocity profiles for the solid surface and Plate I. Without control, Figure 12a, one observes a moderate widening of the streamlines near the wall due to the shock; no separation is detected although the shock-upstream Mach number is fairly high (M = 1.34). When passive control is applied, Figure 12b, a strong widening of the boundary layer occurs, mainly provoked by the blowing effect taking place in the upstream part of the perforated plate. Here, the streamlines are inclined upward and nearly straight, thus producing the ramp effect with the corresponding oblique forward shock (also see Figure 8); in the downstream part of the plate, i.e., the suction part, the streamlines rapidly bend to enter the plate and form a small zone of recirculating flow but no separation occurs downstream of the cavity, contrary to the results of the University of Karlsruhe. The flow development in case of the latter may well be due to differences in the initial boundary layer conditions or due to three-dimensional, i.e., side-wall interference effects.

The global effect of passive control is best illustrated by considering the development of the boundary layer displacement and momentum thickness, δ^* and θ, respectively, in Figure 13 (also shown is the mechanical energy thickness θ^*). By comparison with the reference case, one sees that passive control provokes a much more pronounced thickening of the boundary layer. In particular, the momentum thickness is considerably increased, which indicates a strong amplification of the friction drag. By means of a momentum balance equation, it was possible to determine the total force (total drag) acting on the surface in the control region: a reduction of about 4% was found. This gain is too small to draw an unambiguous conclusion about the potential of passive control to reduce total drag. However, considering the amplifying effect of the rear adverse pressure gradients on the increase in displacement as well as momentum thickness in the case of an airfoil / wing, it is quite possible that the reduction in wave drag may be compensated by an increase in viscous drag, as will be seen later in conjunction with the airfoil investigations. No clear conclusion about the influence of the characteristics of the

14

ONERA plates could be deduced here similar to the conclusion arrived at by Karlsruhe University.

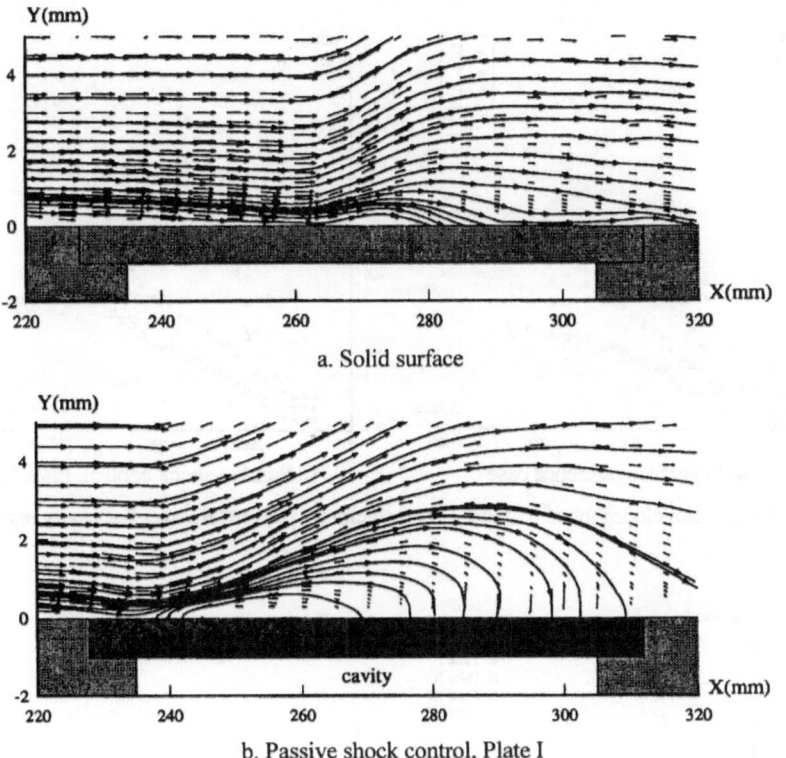

a. Solid surface

b. Passive shock control, Plate I

Figure 12: Mean velocity vector plot with dilatation of the near-wall region

Following the observations described above, a special study was made to determine the drag caused by the holes (excrescence drag) when the perforated plate is covered on the cavity side (no flow through the perforation) and the outer flow is uniform, i.e., without a shock present. As shown by the momentum thickness distributions plotted in Figure 14, the holes produce a considerable increase in drag which can be more than twice the one of the solid wall reference case. The crucial importance of hole roughness was already demonstrated by the results obtained by the University of Karlsruhe. Note that the excrescence drag must also be considered in the numerical treatment of airfoil flow with shock control.

The results of the boundary layer and flow field measurements were also analyzed with respect to turbulence. Generally, passive control tends to increase the turbulence level in the interacting boundary layer with the rise in turbulence intensity linked to the greater destabilization of the boundary layer under the action of passive control. The mechanism is as follows: because of the combined blowing and suction effect, the boundary layer profiles adopt an S-shape with a strong retardation of the flow in the near wall region, especially in the suction part. This distortion of the profiles increases the production of turbulence, as in all interacting

15

flows. The effect reaches a maximum shortly after the control region, with the maximum of shear stress occurring downstream of the maximum of turbulent kinetic energy, which is a usual feature of interacting flows. For more details, please refer to Chapter 9 and Reference [2.2].

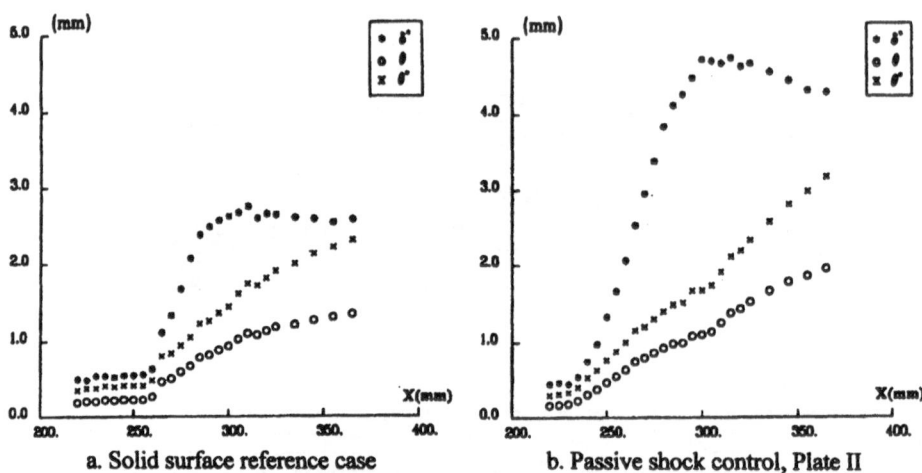

a. Solid surface reference case   b. Passive shock control, Plate II

Figure 13: Development of boundary layer thickness parameters over the cavity region

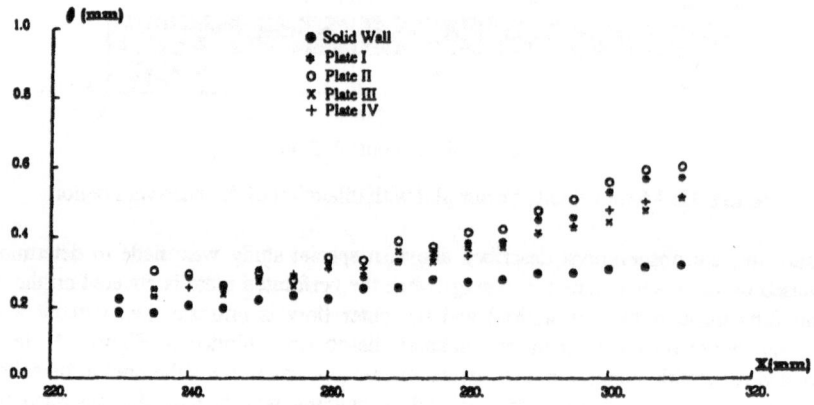

Figure 14: Excrescence drag of perforation / development of momentum thickness

2.1.5 Assessment of turbulence models and shock control laws

A theoretical study was carried out by ONERA to examine the accuracy of the physical laws used in the computation of shock wave/boundary layer interactions under control conditions. Control has a double impact: firstly, it influences the behavior of turbulence, especially in the near wall region; secondly, it introduces a new boundary condition at the wall since the vertical velocity component is no longer zero. These two aspects are difficult to model, so a systematic study was performed employing a boundary layer type approach. Since separation occurs - or is about to occur - in the computed interactions, an inverse mode of solution has been used in which the measured displacement thickness distribution serves as input. The boundary layer code uses a finite difference method and can accommodate various turbulence models. The complete theoretical study is presented in Reference [2.3].

In the present theoretical investigation, the Baldwin-Lomax algebraic turbulence model and the [k, ε] transport equation model were studied in conjunction with shock control; the Algebraic Stress Model (ASM) was only considered in the case of the solid surface; it was rejected in the presence of control because its implementation in computer codes leads to serious numerical difficulties without bringing a substantial improvement compared to the [k, ε] model. Blowing and suction velocity at the wall has been accounted for by using the Cebeci extension of the Van Driest damping function. The problem of representing the non-zero vertical velocity component at the wall in case of shock control has been examined by using as control laws: Darcy's law, two versions of Poll's law, the isentropic law and the law derived at Karlsruhe University (for control laws see Chapter 3.3).

Computed surface pressure distributions are compared to experimental distributions, Figure 15, for the solid wall reference case and for the flow with shock control (Plate III). One observes that agreement with experiment is poor even for the reference case which is mainly due to the deficiencies of the turbulence models used in the presence of shock waves and/or separation, Figure 15a.

There is hardly any effect of the control law applied, except, may be, for Darcy's law, Figure 15b. Furthermore concerning the control laws, they have been directly evaluated by comparing the calculated v_w distribution, using the experimental surface pressure distributions in the control region, to distributions deduced from LDV surveys.

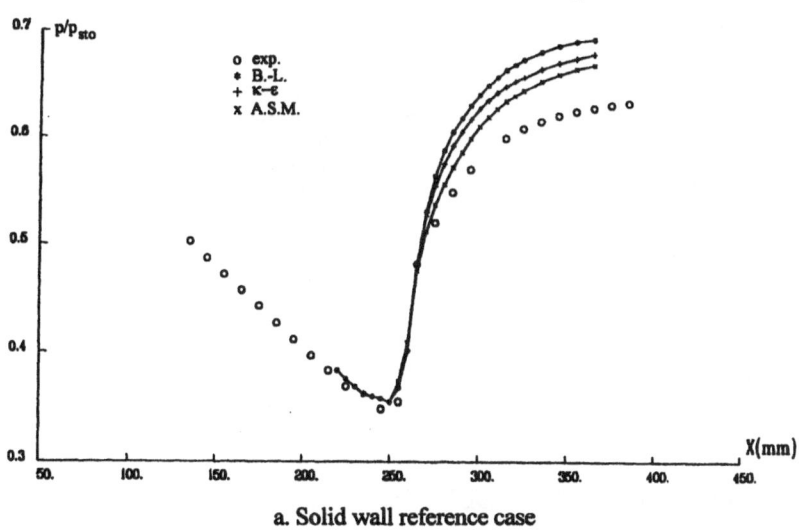

a. Solid wall reference case

Figure 15: Validation of turbulence models by comparison of pressure distributions

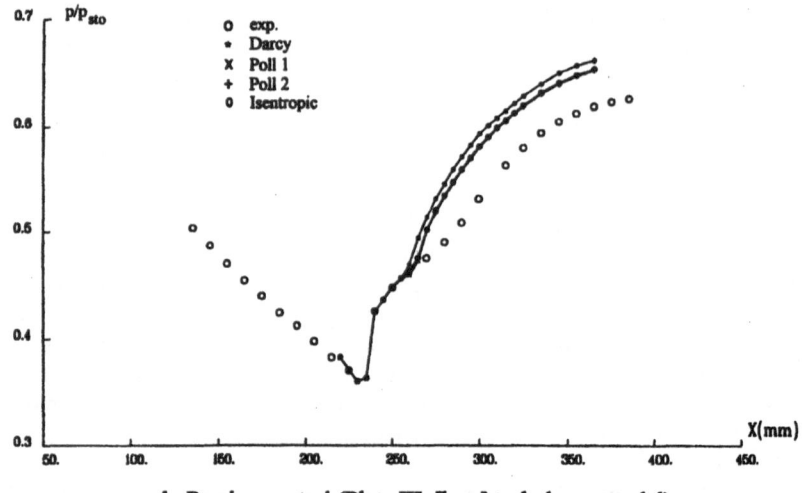

b. Passive control (Plate III, [k, ε] turbulence model)

Figure 15: Concluded

The comparison presented in Figure 16 shows that the laws considered are in fair agreement with the present measurements, with the Karlsruhe law, however, providing the best overall agreement with the measured data. The results obtained with Darcy's law have been omitted here since there is strong disagreement in the blowing region.

Considering Figures 15 and 16, it is justified to state that the discrepancies observed in the surface pressure distributions mainly derive from the baseline turbulence models which are inadequate to predict strongly interacting flows, while the control laws considered seem fairly effective in describing the flow in the control region.

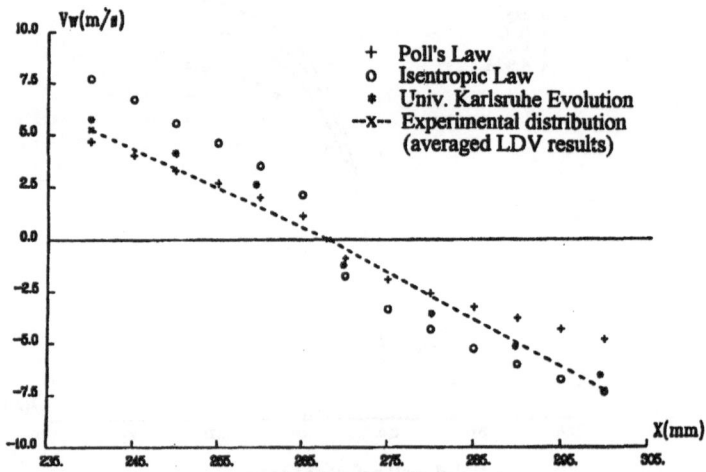

Figure 16: Comparison of wall velocity distributions in the control region

2.2   Three - dimensional Shock Control

2.2.1   Experimental arrangement and procedure

The three-dimensional experiments were carried out in an intermittent blow-down supersonic wind tunnel by the Engineering Department of the University of Cambridge. The tunnel, operating at a maximum pressure of 4.1×10^6 Pa, is equipped with interchangeable nozzle liners on the roof and floor giving a working section with a height of 176 mm and a width of 114 mm. The side walls consist of interchangeable doors equipped with either pressure orifices or 203 mm diameter optical windows for flow visualization. An insert was fitted into the window opening for the present investigation, containing a cavity (plenum) 22 mm deep, 50 mm wide and 190 mm long. The cavity can be covered by solid or perforated plates for reference or shock control experiments, respectively [2.4].

The oblique shocks were generated by a wedge mounted on the roof of the tunnel and positioned such that the shock wave was placed correctly with respect to the perforation and cavity located in the test section side wall, Figure 17; details of the shock generator are presented in Figure 18.

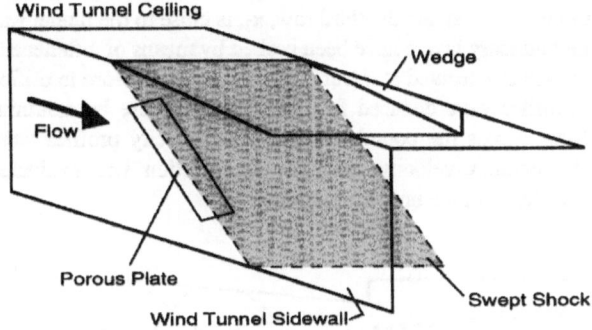

Figure 17: Sketch of the experimental set-up with wedge location and porous-plate side wall position

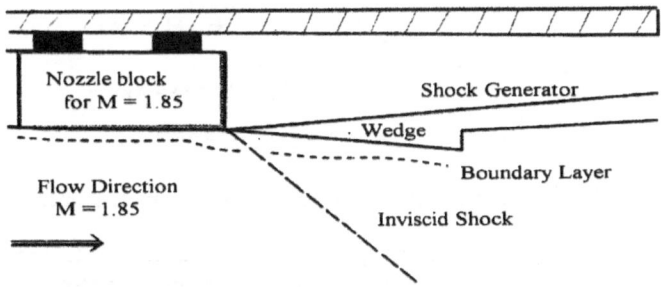

Figure 18: Details of the shock generator

Experiments were executed for the conditions given in Table 2. The choice of these test conditions was motivated by the following reasons:

- Both the pressure rise across the shock, p_2/p_1, the Mach number normal to the shock, M_n, and the unit Reynolds numbers are similar, which allows the effect of sweep to be investigated.

- Without passive control, no shock-induced separation should occur.

- The Mach number normal to the inviscid shock, M_n = 1.14, is common on transonic aircraft wings which often have sweeps of about 30°.

Table 2: General experimental conditions

M_∞	wedge angle$^{(o)}$	shock angle$^{(o)}$	M_n	p_2/p_1	Re/m
1.5	6	49.3	1.14	1.34	3.1×10^7
1.8-1.85	6	38.3	1.15	1.37	3.3×10^7
1.8-1.80	6	39.5	1.14	1.36	3.3×10^7
2.5	6	28.3	1.19	1.51	3.1×10^7

The side wall on which the interactions took place was equipped with static pressure orifices installed along three rows as shown in Figure 19. The first row, x_1, traverses the inception zone of the interaction; the second row, x_2, is at the mid-span location where infinite swept-wing condition is achieved and the third row, x_3, is close to the tunnel wall opposite to the shock generator. The boundary layers have been probed by means of a flattened Pitot probe with the boundary layer traverses executed at about 20 stations located close to orifice rows x_1 and x_2. The Mach number profiles were deduced from the Pitot pressure by assuming the wall static pressure to be constant across the boundary layer. The velocity profiles were determined by using the Crocco temperature-velocity law. The skin-friction was evaluated by fitting the measured velocity profiles with the universal log-law.

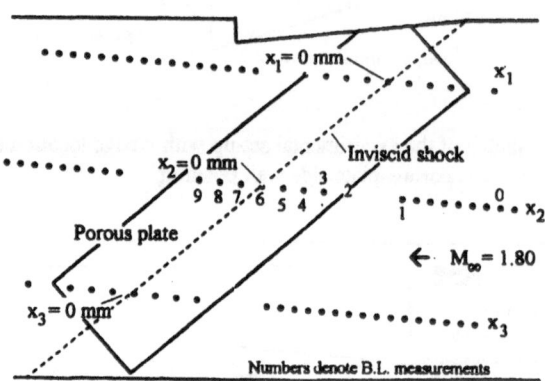

Figure 19: Location of wall pressure and boundary layer measurements (B.L.)

The titanium porous plate used in the experiments had a thickness of 1.2 mm and a porosity of 8%. The laser-drilled 0.0762 mm diameter holes covered an area of 50.8x190.5 mm². The use of such tiny holes in passive control applications has the advantage of physical (and aerodynamic) smoothness and the resulting wall transpiration is expected to have a closer resemblance to the boundary condition of homogeneous normal velocities at the wall, commonly assumed in calculations. The characteristics of the porous plate were determined in a special test rig. In order to minimize the end-wall effects due to the test section constraints, and hence simulate infinite swept wing conditions near the mid-span region, 6 mm thick partition walls were placed inside the plenum chamber (cavity) to eliminate any undesirable spanwise ventilation.

A more complete description of the test conditions and a detailed presentation of results can be found in Chapter 11 and Reference [2.4].

2.2.2 Analysis of experimental results

As an example, the pressure distributions along row x_2 with and without passive control are presented in Figure 20(a) for $M_\infty = 1.80$: The pressure in the case of control shows a continuous rise in both the blowing and suction regions. This reduction of the pressure gradients across the interaction is typical of passive control and is a clear indication of the reduction of the shock strength in the near wall region, a feature that was already discussed in the two-dimensional investigations. Note that at the other freestream Mach numbers investigated, a spreading of the shock is also present, however, it is not always as smooth as for $M_\infty = 1.80$.

The corresponding variation of the boundary layer displacement thickness, δ^*, with and without control is presented in Figure 20(b). It is quite obvious that passive shock control by ventilation increases the boundary layer thickness considerably, as was already shown for two-dimensional interactions. This development, caused by the recirculating flow in the cavity region and the surface roughness introduced by the perforation, leads, in spite of the considerable reduction in wave drag, to only a moderate reduction in total drag, as already pointed out in Paragraph 2.1.4.

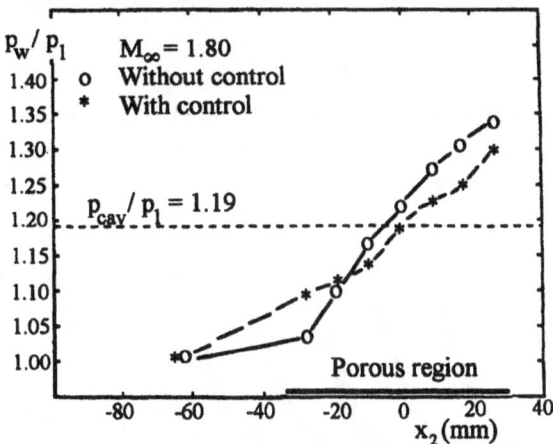

a. Pressure distribution

Figure 20: Effect of passive control on wall pressure and displacement thickness distributions, $M_\infty = 1.80$

21

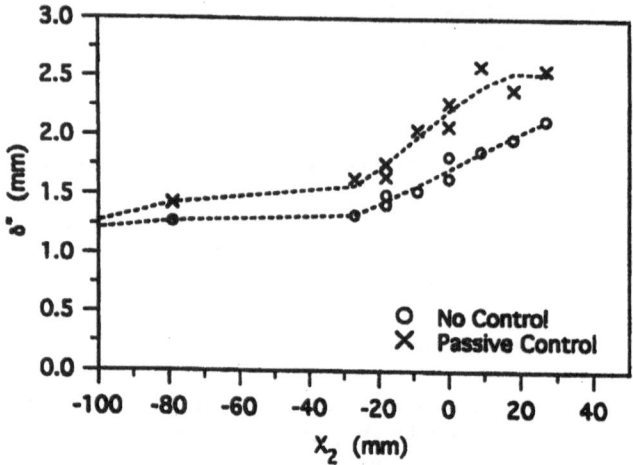

b. Displacement thickness distribution with and w/o shock control

Figure 20: Concluded

The development of the Mach number profiles, here at $M_\infty = 1.85$, with and without control is demonstrated in Figure 21. The effect of blowing on the upstream half of the porous region as well as the gradual recovery of the boundary layer under the action of suction are both evident. However, it can be noted that some deficiencies in the boundary layer profiles still remain at the downstream end of the porous plate (Station 9) due to control, leading, of course, to the larger displacement thickness at this position as shown above.

Another aspect of passive shock control that should be pointed out here is its interactive nature: the extent to which a shock wave boundary layer interaction is influenced by passive control depends largely on the strength of the control, determined essentially by the wall pressure distribution, which in turn is affected by the state and development of the boundary layer.

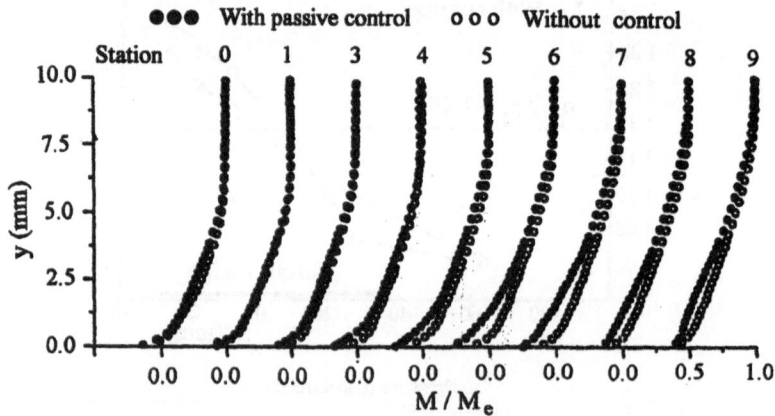

Figure 21: Development of boundary layer profiles with and w/o control, $M_\infty = 1.85$

2.2.3 Numerical simulation of the interaction

The flows investigated experimentally with and without passive control were computed by solving the three-dimensional time-averaged Navier-Stokes equations. The present code uses a finite volume formulation along with an implicit algorithm. Three turbulence models have been considered initially: the Cebeci-Smith and Baldwin-Lomax equilibrium models and the Johnson-King non-equilibrium model. The transpiration effect in the near wall region was taken into account by using the modification of the Van Driest wall damping function proposed by Cebeci. The flow inside the cavity was not calculated. The cavity pressure p_{cav} was assumed constant, its level being determined in an iterative manner by satisfying the condition of global zero mass flux across the porous plate [2.4] (also see Chapter 11).

A comparison of initial computations of the wall pressure distributions and the boundary layer thickness variations with and without control have indicated that the performance of the three turbulence models named above was very similar. However, the need to evaluate the boundary layer thickness in both the Cebeci-Smith and the Johnson-King models rendered these models less efficient than the Boldwin-Lomax model so that the latter was selected for all final computations presented below. The computational mesh used in the final computations is depicted in Figure 22 for the $M_\infty = 1.80$ test case as an example.

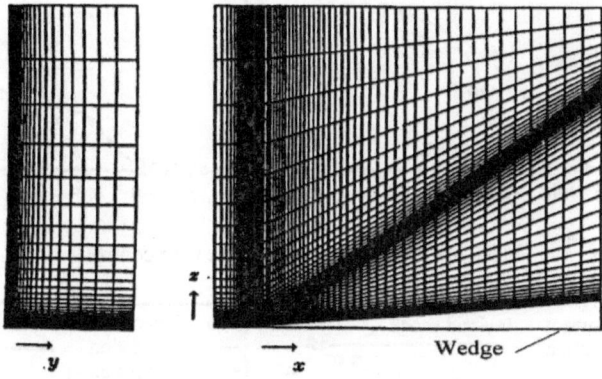

Figure 22: Computational mesh used in the final calculations for $M_\infty = 1.80$

The computed and experimental wall pressure distributions are compared in Figure 23: Even without control, some discrepancies exist between calculation and measurement, although the actual pressure gradients are in excellent agreement. With control there are marked differences in the distributions at Mach numbers of $M_\infty = 1.50$ and 2.5 with the pressure level in the plateau region being higher in the experiments. However, at all Mach numbers the effect of control in moving the start of the interaction forward is correctly predicted as is the position of the plateau. Various attempts were made at $M_\infty = 2.50$ to increase the calculated pressure level in the plateau region and closer agreement with the measured level was found by fixing the cavity pressure at the measured level (shown by the dash-dotted line labeled "calculation -f"). This means, however, that the mass-flow balance into and out of the plenum chamber is not

experiment has still to be determined. One contribution is, of course, likely to be the turbulence model, despite the fact that all models investigated gave similar results, as already indicated in Chapter 2.1.5 for two-dimensional interactions.

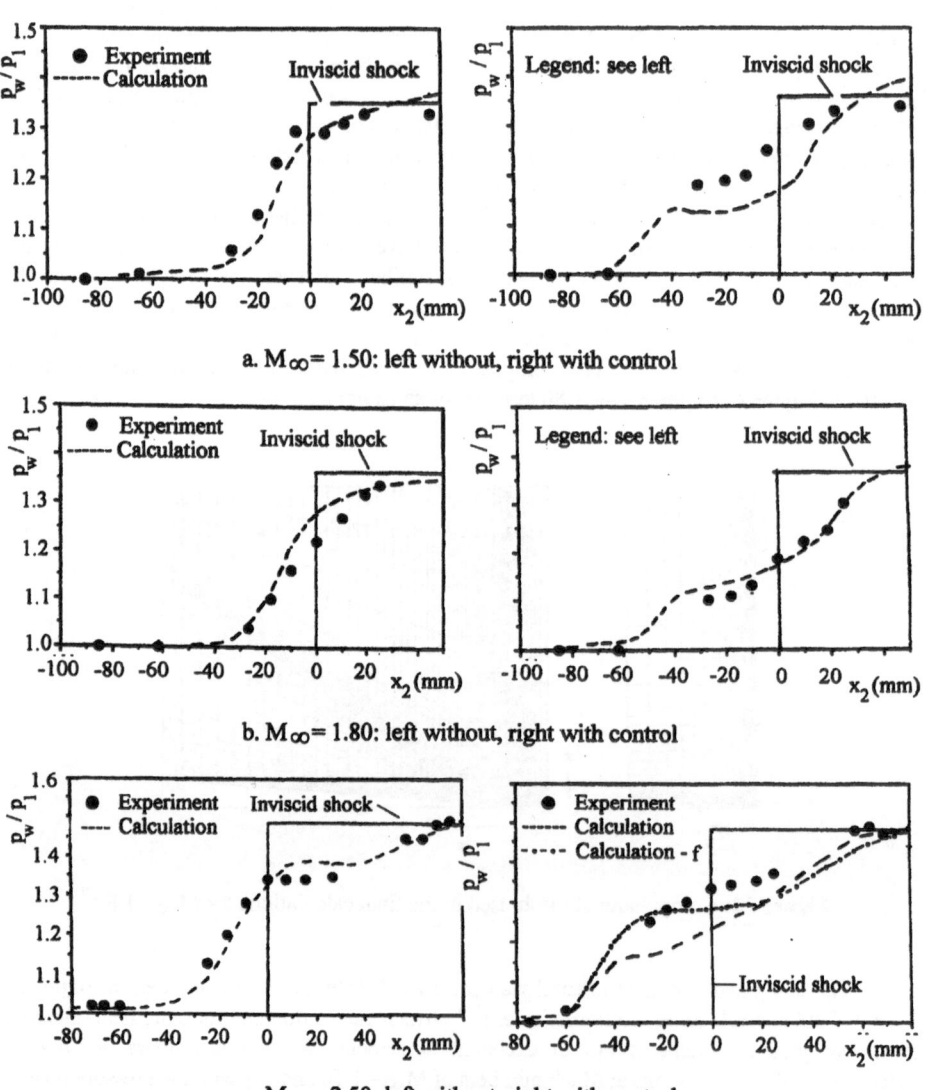

a. $M_\infty = 1.50$: left without, right with control

b. $M_\infty = 1.80$: left without, right with control

c. $M_\infty = 2.50$: left without, right with control

Figure 23: Comparison of measured and calculated pressure distributions

Figure 24 presents the experimental and computed boundary layer displacement thickness. In the computations, the trend in the experimental boundary layer development with and w/o control is well predicted; however, the comparison of the level in the displacement thickness is less satisfactory. Comparing the boundary layer development with and w/o control, one observes

24

that the increase in boundary layer thickness over the interaction region is, as in the two-dimensional interaction, much larger in the presence of shock control indicating a likely increase in total (wing) drag also for three-dimensional shock control.

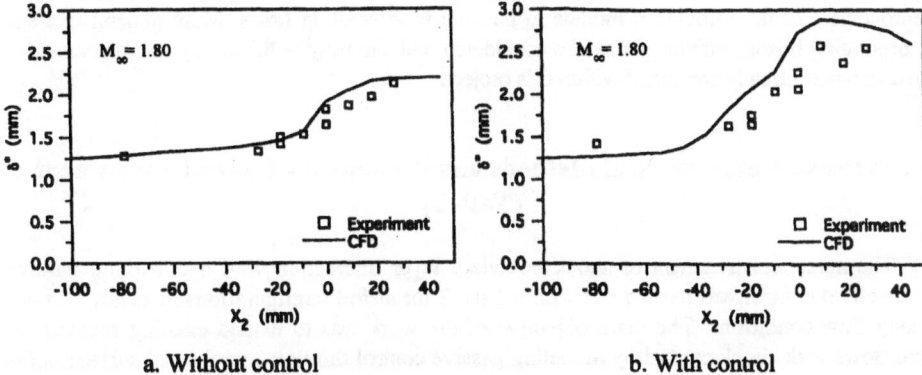

a. Without control   b. With control

Figure 24: Comparison of experimental and computed displacement thickness, $M_\infty = 1.80$

2.3 Conclusion and Future Work

Basic investigations of transonic shock wave boundary layer interaction under passive control conditions have been performed by ONERA, the University of Karlsruhe and Cambridge University, each of these institutions having considered complementary aspects of the problem. Detailed experiments, complemented by a theoretical treatment of the data, have allowed to define some basic physical phenomena involved in the control mechanism. Thus, it is clear that the blowing taking place in the first part of the control region provokes a rapid thickening of the boundary layer with the resulting increase of the displacement thickness being felt by the outer inviscid flow as a ramp with the formation of a lambda shock pattern replacing the single normal shock of the interaction without control. The result is a substantial reduction of the wave drag. However, the combined blowing-suction effect and the hole roughness cause an increase in friction drag which can outbalance the gain in wave drag. The same conclusions are valid both for two-dimensional and three-dimensional interactions. The effect of the porous plate geometry on the effectiveness of shock control seems to be rather weak.

A key problem in the modeling of passive control is the definition of realistic boundary conditions to be prescribed at the wall where a non zero vertical velocity exists in the control region. The studies performed by the Universities of Karlsruhe and Cambridge have permitted the establishment of more accurate transpiration laws taking into account the exact characteristics of the perforated plates. Detailed information on the behavior of turbulence was obtained at ONERA. Turbulence generally increases due to control which can essentially be attributed to the stronger destabilization of the boundary layer by the blowing action.

Numerical simulations of the interactions, both in two-and three-dimensional flows, reveal some weaknesses of the basic turbulence models which often give poor results, even in the reference solid wall case. In spite of these deficiencies, Navier-Stokes calculations with rather simple algebraic turbulence models can be valuable in the investigation of the physics of the

25

flow, to test improved physical models and to execute comparative studies of different control arrangements.

Concerning future work, the transpiration velocity laws should be generalized for the case where a high speed flow is streaming on one side of the cavity. The presence of this flow has certainly an influence on the flow inside the holes. Also, some attention should be paid to the improvement of the turbulence models to be used in interacting flows. More general models, incorporating history effects, should be considered and carefully validated by comparison with the data banks already contained within this project.

3 Extension of Numerical Methods and Preliminary Control Assessment (Task 2)

The numerical treatment of shock boundary layer interaction with and without passive (and active) control was carried out within Task 2 for airfoil configurations at steady and unsteady flow conditions. The main objective of the work was to extend existing methods to treat flows with shock control by modeling passive control through a perforated surface at the position of the shock and implementing the corresponding boundary conditions into the existing codes, i.e., replacing the zero-normal-velocity condition by a ventilation velocity distribution matching the in- and outflow conditions upstream and downstream of the shock, and introducing control laws to determine the appropriate normal (v) velocity distribution. The extended methods were validated by the airfoil experiments of Task 3 and used for a preliminary assessment of the merits of passive shock control. A total of seven contributors have participated in the numerical studies, namely, **CIRA, DASA-Airbus (DA), DLR Göttingen, NLR, ONERA** and the Universities of **Athens** and **Naples**.

3.1   Basic Numerical Methods

The methods employed consisted of Viscous-Inviscid-Interaction (VII) procedures, coupling an outer inviscid flow solution to a boundary layer solution, and solutions of the Navier-Stokes equations. A summary of the basic methods used and their components is given in Table 3. Note that in some instances methods or components of methods were switched during the course of the investigation as also indicated in Table 3. What follows is a brief description of the individual approaches providing the major features of the respective method. More details are given in the individual technical reports referenced below and in Chapters 12 to 18.

CIRA [3.1]: The method employed is based on the iterative coupling of the two-dimensional Euler and the integral boundary layer equations. The inviscid part of the flow is computed using the unsteady 2D-Euler equations which are solved through a standard Jameson's finite volume scheme. The calculation of the laminar boundary layer is based on the method of Cohen-Reshotko, the turbulent boundary layer calculation on the direct and inverse formulation, respectively, of the method of Green. Lighthill's surface source model known as the „method of equivalent sources" is used for the viscous / inviscid interaction. The extension to porous surfaces, i.e., to the treatment of wall transpiration, has been accomplished by modifying the boundary layer equations following the approach suggested by Olling, modified according to results obtained at the University of Naples.

University of Naples [3.2]: The inviscid flow field is determined by solving the Euler equations following the scheme of Jameson, similar to CIRA. One of the features of the present solver is the possibility of handling complex configurations by using a multiblock - structured approach which allows local grid refinement for improved accuracy at low costs. The viscous code is based for the laminar part on the method of Cohen-Reshotko, for the turbulent part on a modification of Green's method. The extension to transpiration was accomplished by a modification of Olling's method.

Daimler-Benz Aerospace Airbus (DA) [3.3]: The method, applicable to airfoil and infinite swept wing flow, is based on a viscous-inviscid coupling of a full potential flow solver with an interactive boundary layer finite difference method. Following the defect formulation concept of Le Balleur, the real viscous flow is split into a viscous defect flow and an equivalent inviscid flow (see below). The latter is computed using a transonic full-potential method for prescribed viscous conditions. The viscous flow solutions are obtained from the compressible boundary layer equation in which the turbulence is given by the Cebeci and Smith eddy-viscosity formulation.

ONERA [3.4]: Two Viscous-Inviscid Interaction (VII) codes, namely, the steady code VIS05c and the unsteady code VIS15, have been extended and utilized to compute airfoil flow with shock control. Both codes have a common methodology and similar coding, including adaptive grids. The approach is based on Le Balleur's „defect-formulation-theory" for the full Navier-Stokes equations that replaces the single-field „Navier-Stokes" domain by a double „viscous-inviscid interaction" field. The steady code VIS05 solves the full potential equation for the inviscid field. The viscous method is a „hybrid field / integral" method solved in a marching thin-layer 2D-numerical technique in direct / inverse modes. (Note that the code VIS05c used in the EUROSHOCK effort is a recent experimental version of VIS05 designed for strongly transonic flows). The time-consistent code VIS15 uses the same viscous methodology as the steady code VIS05c. For the outer inviscid flow the unsteady Transonic Small Perturbation equation (TSP) is solved. A semi-implicit time-consistent coupling algorithm is used in the code. The code is able to discriminate between a steady and an unsteady solution and can, therefore, be used to determine the buffet boundary.

DLR [3.5]: For the present investigation a two-dimensional time-accurate Navier-Stokes code was used. The code is based on the Beam/Warming approximate factorization implicit methodology using central differences in the space coordinates. To represent turbulent flow, the Baldwin/Lomax algebraic turbulence model has been used with a modification to the van Driest damping factor in the case of control.

NLR [3.6]: The computational method ULTRAN-V has been developed at NLR for calculating two-dimensional viscous flow about airfoils in steady or unsteady flow. The method is based on the unsteady Transonic Small Perturbation (TSP) potential equation for the inviscid part of the flow and an integral method for the boundary layer. The latter uses in the laminar region the method of Thwaites, in the turbulent region Green's lag-entrainment method modified for unsteady flow. The viscous layer is coupled with the inviscid flow by two means: a direct method where the velocity of the inviscid flow at the previous step is used for the boundary layer calculation and a „simultaneous" method where the boundary layer equations are solved simultaneously with the inviscid flow which allows the computation of separated flow.

University of Athens [3.7]: For the computation of steady flow an existing integral semi-inverse boundary layer code combined with a time marching Euler solver was extended to treat flow with shock control. For unsteady and steady flows a full unsteady Navier-Stokes solver was tailored to the present requirements. Turbulence was here modeled through either

27

the Baldwin/Lomax model or the two equation [κ-ε] model, though only the latter was employed in the present investigation.

Table 3: Solution techniques employed by contributors

Steady Methods

Inviscid	Transonic Small Perturbation (TSP)	Full Potential (FP)	Euler	Navier-Stokes (NS)
Viscous				
Integral Boundary Layer (IBL)	NLR ONERA 2	ONERA 1	CIRA	
Triple Deck at Shock		DASA-Airbus (1)		
Finite Difference (FD)		DASA-Airbus	Naples Univ.	
Navier-Stokes (NS)				DLR Athens Univ.

(1) Later replaced by Full Potential / Finite Difference method

Unsteady Methods

Inviscid	Transonic Small Perturbation (TSP)	Full Potential (FP)	Euler	Navier-Stokes (NS)
Viscous				
Integral Boundary Layer (IBL)	NLR ONERA 2		Athens Univ. (1)	
Finite Difference (FD)				
Navier-Stokes (NS)				DLR Athens Univ.

(1) Later abandoned

3.2 Pre-computations without Control

Calculations without shock control were first carried out by the various contributors to determine the state-of-the-art of the different numerical codes involved. A set of mandatory as well as optional test cases was defined for both steady and unsteady flow for the airfoils DA LVA-1A, NLR 7301 and NACA 0012 for which steady and, in case of the latter, unsteady experimental data were available.

3.2.1 Steady flow pre-computations for the airfoil DA LVA-1A

Presented here as an example are results of pre-calculations for the transonic airfoil DA LVA-1A for two test cases: $M_\infty = 0.75$, $\alpha = 2°$ ($C_L = 0.53$) and $M_\infty = 0.75$, $\alpha = 3°$ ($C_L = 0.69$); the Reynolds number was $Re_c = 6 \times 10^6$ with transition prescribed at 5% chord on the upper and lower surfaces. Experimental results were obtained in the DLR Transonic Wind Tunnel Braunschweig (TWB), a blow-down tunnel with a slotted test section of 340 mm x 600 mm cross-section [3.8]. To allow a better comparison with the experimental data, the lift coefficients in the computations were adjusted to the experimental values. Table 4 summarizes the measured and calculated coefficients and incidences for the more stringent Test Case 2. Please note the different angles of attack needed in the computations to obtain the same lift coefficient.

Table 4: Numerical and experimental results for the airfoil DA LVA-1A, Test Case 2

Institute	M_∞	$Re_\infty\ 10^6$	α	C_L	C_M	C_D	C_{DW}
Cira	0.749	5.96	2.7580	0.6934	-0.07203	0.0224	0.00509
DA	0.749	5.96	2.20	0.6888	-0.08680	0.01832	0.00481
DLR	0.75	6.0	2.38	0.6932	-0.09214	0.01863	-
NLR	0.75	6.0	2.461	0.68998	-0.08108	0.02326	0.00647
ONERA	0.749	5.96	2.7312	0.69203	-0.08360	0.02040	0.00710
U.Athens	0.75	6.0	3.011	0.72251	-0.05571	0.02325	-
Naples	0.75	5.96	2.35	0.6859	-0.07905	0.01912	0.00584
Experiment	0.75	5.96	3.011	0.6920	-0.08632	0.02093	-

The pressure distributions corresponding to the two test cases are presented in Figure 25. One observes quite good agreement with the measured data, including the shock location, at the lower lift coefficient; however, larger deviations occur at $C_L = 0.69$ especially in the area of the shock. The differences are most pronounced in case of the methods of the University of Athens and NLR, respectively, and immediately downstream of the shock also in the case of the CIRA method. The deviations can mainly be attributed to the different degrees of resolution of the flow field, i.e., the number of grid points employed, especially in the area of the shock, and to differences in turbulence modeling.

Large differences between methods also occur in the predicted boundary layer development, here exemplified by the momentum thickness distribution on the airfoil upper surface for the two lift coefficients, Figure 26. Differences essentially start at the shock and grow towards the trailing edge, the latter, as will be pointed out later, essentially due to the amplifying effect of the rear adverse pressure gradients on the boundary layer. The scatter of the total drag coefficient about the experimental values is correspondingly large, Table 4, and shows deviations to both higher and lower values of up to 23 counts ($\pm$ 0.0023 in drag coefficient) at a nearly constant lift coefficient of $C_L = 0.69$.

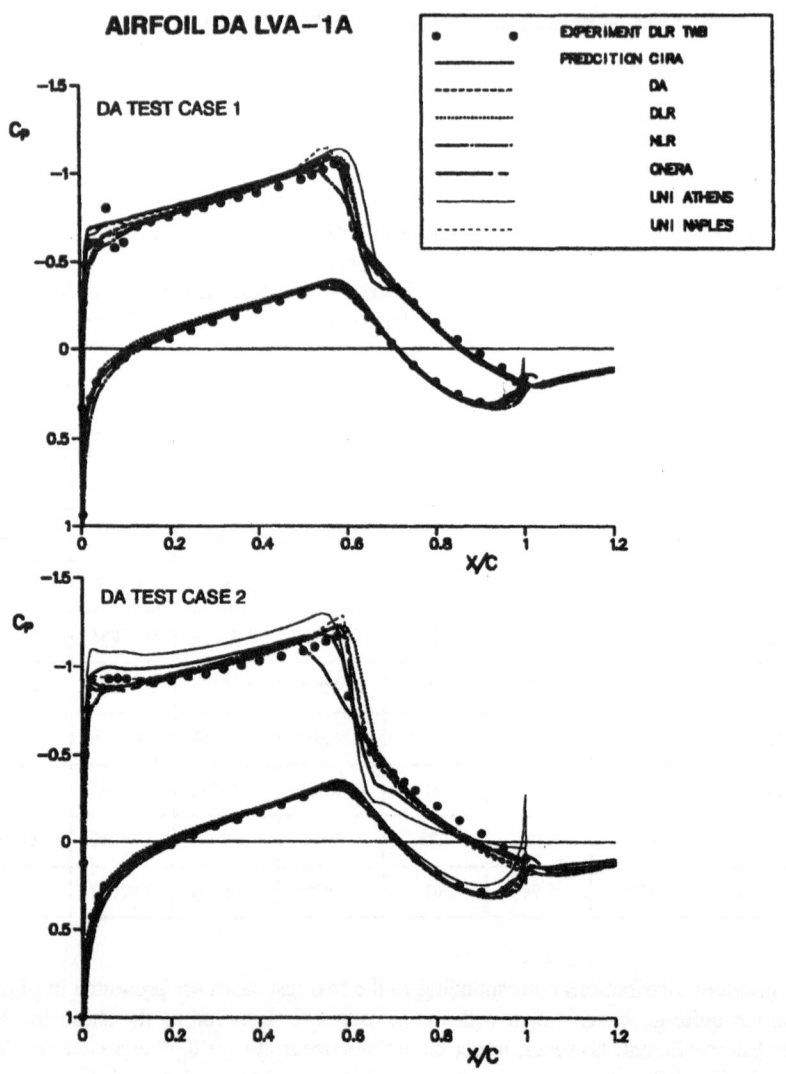

Figure 25: Comparison of experimental and computed pressure distributions for the airfoil DA LVA-1A, $M_\infty = 0.75$, $C_L = 0.53$ (upper plot) and $C_L = 0.69$ (lower plot)

Concerning the pre-calculations for steady flow, it seems that especially the discretization of the flow field at the shock is insufficient and it was pointed out by ONERA that a grid refinement at the shock of $\Delta x = 0.3\,\delta$ is required to sufficiently resolve the physical effects of shock boundary layer interaction. Such a grid spacing was therefore utilized for all shock control computations.

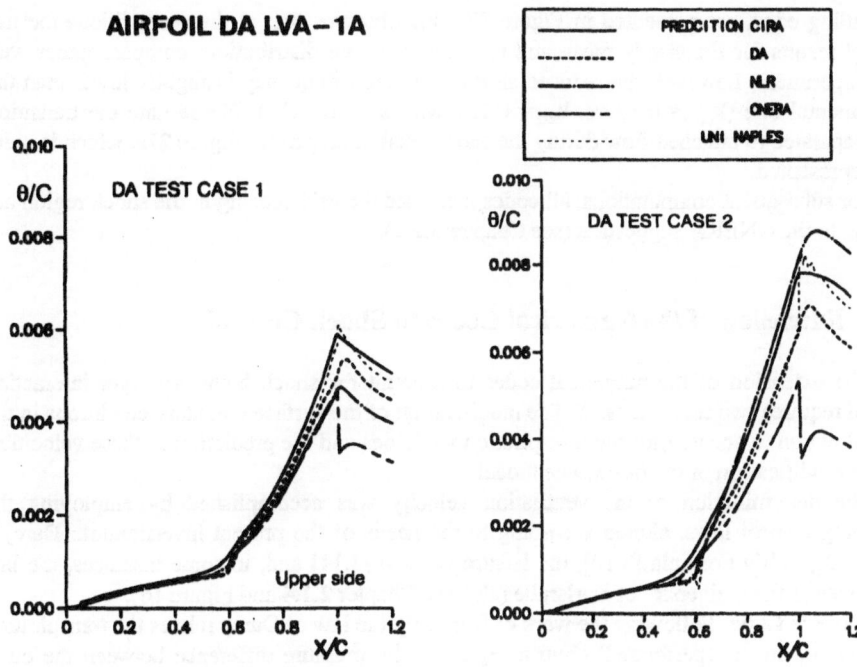

Figure 26: Comparison of momentum thickness distributions; airfoil DA LVA-1A,
$M_\infty = 0.75$, $C_L = 0.53$ (left) and $C_L = 0.69$ (right)

3.2.2 Unsteady flow pre-computations

Similar to the steady pre-calculations without control, unsteady flow cases were defined for the NLR 7301 airfoil, for which experimental data were available from NLR, and for the airfoil NACA 0012. Pre-calculations were carried out by **NLR, ONERA** and **DLR**, the former using time-consistent VII-methods, the latter a time-accurate Navier-Stokes code (see Table 3).

The results have generally shown that, compared to the steady flow predictions, differences between the computations of the various codes and between predicted and experimental results were much more pronounced. Reasons for this are certainly the more stringent requirements at freestream conditions where separation dominates the flow development. This is especially applicable to the number of mesh points needed in regions of strong adverse pressure gradients, such as the shock and the trailing edge regions, since the occurrence and development of separation strongly depend on the pressure gradient, and, for the same reason, to the sophistication of the turbulence model employed. An additional difficulty results, of course, from the fact that the shock oscillates and that, therefore, a high-resolution flow-adaptive grid is required.

Nevertheless, all three time-consistent methods were able to predict steady flows, the transition to unsteady flow, hence buffet onset, and the buffet process. As an example of the latter, ONERA results obtained with the code VIS15 using an H-grid of 160 x100 mesh points with special care of obtaining a sufficient resolution not only in the shock region but also at the trailing edge, are presented in Figure 27. One observes that at these conditions the numerical results for the steady mean and unsteady pressure distributions compare rather well with experiment; however, the reduced shock oscillation frequency is slightly lower than the experimental one (k_{num}= 0.17 vs. k_{exp}= 0.257 with k= $\pi f c / U_{\infty}$). Please note the transition from separated to attached flow during the shock oscillation period, Figure 27a, which is quite well represented.

For subsequent computations, all codes increased the grid spacing in the shock region according to the ONERA suggestion (see Chapter 3.2.1).

3.3 Extension of the Numerical Codes to Shock Control

The extension of the numerical codes to account for shock boundary layer interaction control requires two main steps: ① The modification of the surface boundary conditions in the control region to account for the transpiration velocities and the prediction of these velocities. ② The modification of the turbulence model.

The determination of the ventilation velocity was accomplished by employing the following control laws, altered according to the needs of the present investigation: Darcy's Law [3.9], Poll's Formula [3.10], the Isentropic Law [3.11] and, in some instances, the law developed at the University of Karlsruhe (also see Chapter 2.1.4 and Figure 16).

Darcy's Law: Following the work of Bur [3.9], the law of Darcy relates the transpiration velocity, v_p, in the (perforated) control region to the pressure difference between the outer flow surface pressure, $p(x)$, and the cavity pressure, p_c, with p_c being assumed to be constant:

$$v_p = a \, (\, p_c - p(x) \,)$$

with the proportionality coefficient

$$a = P \, d^2 \, / \, 32 \, e \, \mu$$

where P is the porosity of the porous plate (open area / total area), d the diameter of the holes, e the thickness of the porous plate and μ the molecular viscosity. Experience with this rather simple model was in some cases not very satisfactory, as was also found in the basic channel-flow experiments described in Chapter 2.1.4.

Poll's Formula: This law is based on the experimental investigation of a large number of samples of laser-drilled titanium plates carried out to establish a relation between the mass flow rate and the pressure drop through a porous surface. As a final result this relationship can be expressed by the equation:

$$Y = \frac{1}{K}[40.7X + 1.95X^2]$$

with X proportional to the mass flux $\dot{m}$ and Y proportional to the pressure difference across the porous plate:

$$X = \frac{\dot{m}}{\mu e} \quad \text{with} \quad \dot{m} = \rho \pi \frac{d^2}{4} v$$

$$Y = \frac{(p_c - p(x))d^2}{\rho v^2} \cdot \frac{d^2}{e} \quad .$$

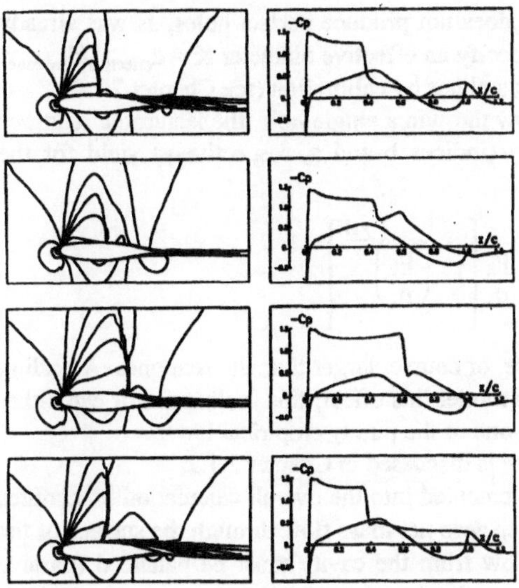

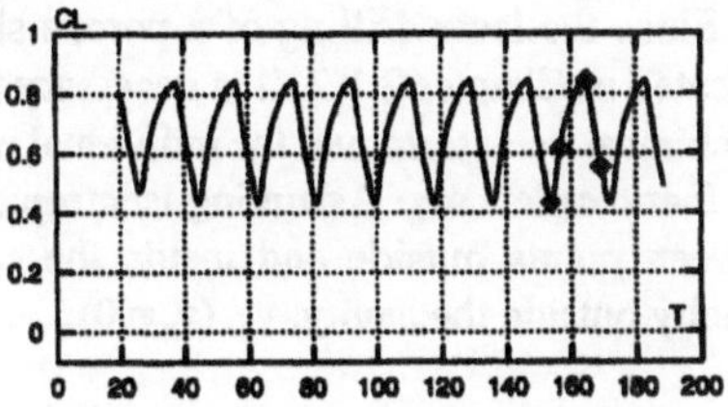

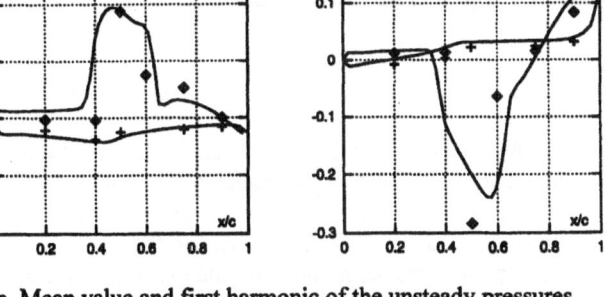

Note: Pressure distributions are for qualitative consideration only

a. Instantaneous Mach field, pressures and lift

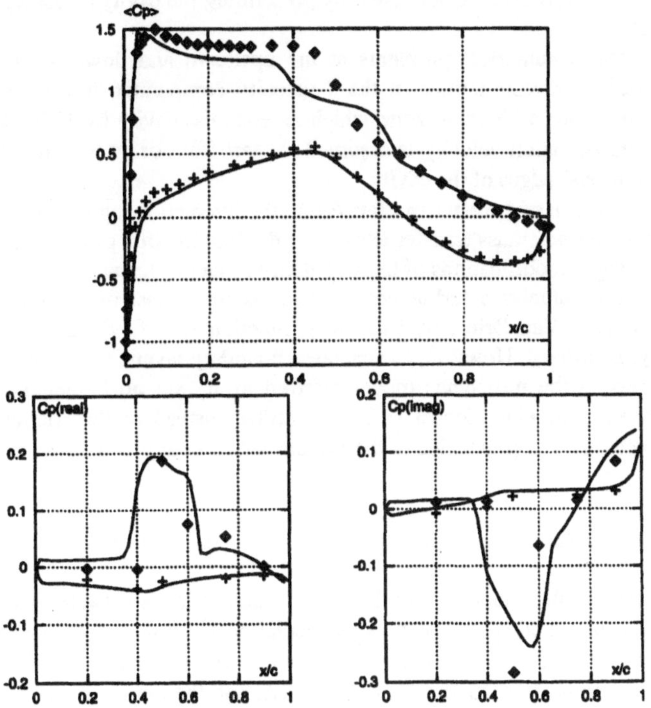

b. Mean value and first harmonic of the unsteady pressures
(symbols denote experiment)

Figure 27: Comparison of time-consistent calculations with experiment; airfoil NLR 7301, $M_\infty = 0.738$, $\alpha = 2.257^\circ$, $Re_c = 11.7 \times 10^6$, $(x/c)_{tr} = 7\%$

Since the laser drilling of a porous sheet does not produce perfect holes, as was already indicated in Chapter 2.1.2, it is necessary to specify an effective diameter $K = d_{effective}/d_{measured}$ which must be determined for individual porous plates by calibration (see Chapter 2.1.2).

Isentropic Law: Assuming isentropic flow through a single hole, the isentropic relations between points outside and inside the cavity (indices b and a, respectively) yield for the velocity outside the cavity, v_b, ($v_a \equiv 0$):

$$v_b^2/2 = \frac{\gamma}{(\gamma - 1)} \cdot \frac{p_a}{\rho_a} \left[1 - \left(\frac{p_b}{p_a} \right)^{\frac{\gamma-1}{\gamma}} \right].$$

Since losses occur, the calculated velocities are, of course, larger than the real ones. A scaling factor must therefore be taken into account to model this effect; this scaling factor can either be determined experimentally or derived from one of the purely empirical laws.

University of Karlsruhe: This control law is discussed in Chapter 2.1.2.

The different control laws have been implemented into the overall calculation procedures of the codes in such a way that at each time step zero net mass flow through the surface of the cavity is achieved, i.e., inflow into and outflow from the cavity must be balanced (passive control). In addition to the passive control, where the final (constant) cavity pressure, p_c, is an output value, active control can be simulated by prescribing the cavity pressure, i.e., p_c is now an input parameter.

In order to avoid numerical problems at the upstream and downstream edges of the control region due to discontinuities in the v-velocity component, it was agreed between contributors to introduce a blending zone which extends the cavity by 10% of its length on both sides. The transpiration velocity is assumed to start from or go to zero, respectively, at the extended (fictitious) edges of the cavity.

Modification of the turbulence model: Along the interaction region the flow is assumed as fully turbulent. Surface mass transfer as well as the longitudinal pressure gradients within this region make some modifications of the turbulence model necessary. In the Cebeci / Smith or Baldwin / Lomax algebraic turbulence models, used in most of the present numerical codes, the generalized Van Driest wall damping function $D = f(y^+, \bar{u}_\tau)$, where $\bar{u}_\tau$ is the friction velocity, is utilized. However, in separated boundary layers, the friction velocity may be zero and terms in the turbulence model referred to $\bar{u}_\tau$ are no longer meaningful. It is therefore proposed, following Cebeci [3.12], to utilize instead of the friction velocity the corresponding velocity $\bar{u}_{rs}$ at the edge of the viscous sublayer at $y^+ = 11.8$.

3.4 Computational Results for Airfoil Flow with Control

Computational results obtained with the participating codes (see Table 3) for three airfoils with and w/o control are compared with experimental data and the performance of the methods employed as well as the merits of shock control are addressed. The airfoils considered are the laminar-type airfoils DRA 2303 and DA LVA-1A, and the turbulent airfoil VA-2. Details of the airfoils investigated, including configurations and experimental facilities, are described in more detail in Chapter 4 (see, e.g., Figures 39, Figure 51 and 58). Note that most of the computational results of the mandatory steady and unsteady test cases selected for these airfoils and the corresponding experimental data will be discussed in Chapters 12 through 23 of this book. Presently we will only consider representative results to allow the assessment mentioned above.

3.4.1 Airfoil DRA 2303 - steady flow conditions

Essential features of the airfoil DRA 2303 are depicted in Figure 51 of Chapter 4; the characteristics of the porous control region, of interest here, are:
- Porosity P = 8%, hole diameter d = 0.076 mm, plate thickness e = 1 mm and the extent of the cavity x/c = 0.5 to 0.6

Various steady test cases with and without control have been selected from the DRA measurements for computations; however, here we will only consider Test Case 2 since the results for all freestream conditions are very similar. Data describing this test case are:
- Mach number M_∞ = 0.68, lift coefficient C_L = 0.81, Re_c = 19x10^6, transition fixed at 5% chord on upper and lower surfaces (Data Points 289 and 1031, respectively). Note that Table 5 includes all measured and computed aerodynamic coefficients for this test case.

Considering first the measured and computed pressure distributions, Figure 28, for this test case, one observes that the overall agreement is fair and that the computed effect of control on the general shock structure is similar to the experimental data: a reduction of the shock strength (spreading of the shock) is obtained due to control. However, essential differences occur between the individual numerical results and, in some instances, between computation and experiment in the shock details: computation frequently shows a change from a single shock to a two-shock system due to control while the experiments indicate a more continuous change in pressure across the interaction region. This discrepancy is possibly a result of a deficiency in the application of the control law employed to determine the transpiration velocity in the cavity region. Nevertheless, a reduction in wave drag due to control, indicated by the spreading of the shock, is shown by both, the computational as well as the experimental results.

Strong differences between computational results can also be observed in the predicted momentum thickness distributions, here considered for the upper surface, Figure 29. The deviations between codes essentially start at the shock and grow considerably towards the trailing edge. Reasons for the differences, which are present with and w/o control, are likely to be found in the grid resolution in the shock region and in the treatment of turbulence within a code. The streamwise increase of the difference in momentum thickness between computations is judged to be in part due to the amplifying effect of the sustained rear adverse pressure gradients on the boundary layer growth.

Concerning the effect of control on the boundary layer development, one can observe, Figure 29b, that the momentum thickness growth downstream of the shock is stronger in the case of control. The reason for this behavior is as follows: the increase in momentum thickness due to control at the downstream end of the cavity caused by the perforation (also see Figure 13) is, as above, amplified by the rear adverse pressure gradients, leading to a considerable increase in viscous drag.

Table 5 shows that an increase in total drag due to control is determined by all calculations in agreement with the experiment. The cause of this drag increase is believed to be the negative influence of control on viscous drag, overcompensating the reduction in wave drag, which is correctly predicted by the various codes. The computed drag levels differ, however, considerably between codes, as is already indicated by the predicted development of the momentum thickness; there is generally also a deviation from the experimental data.

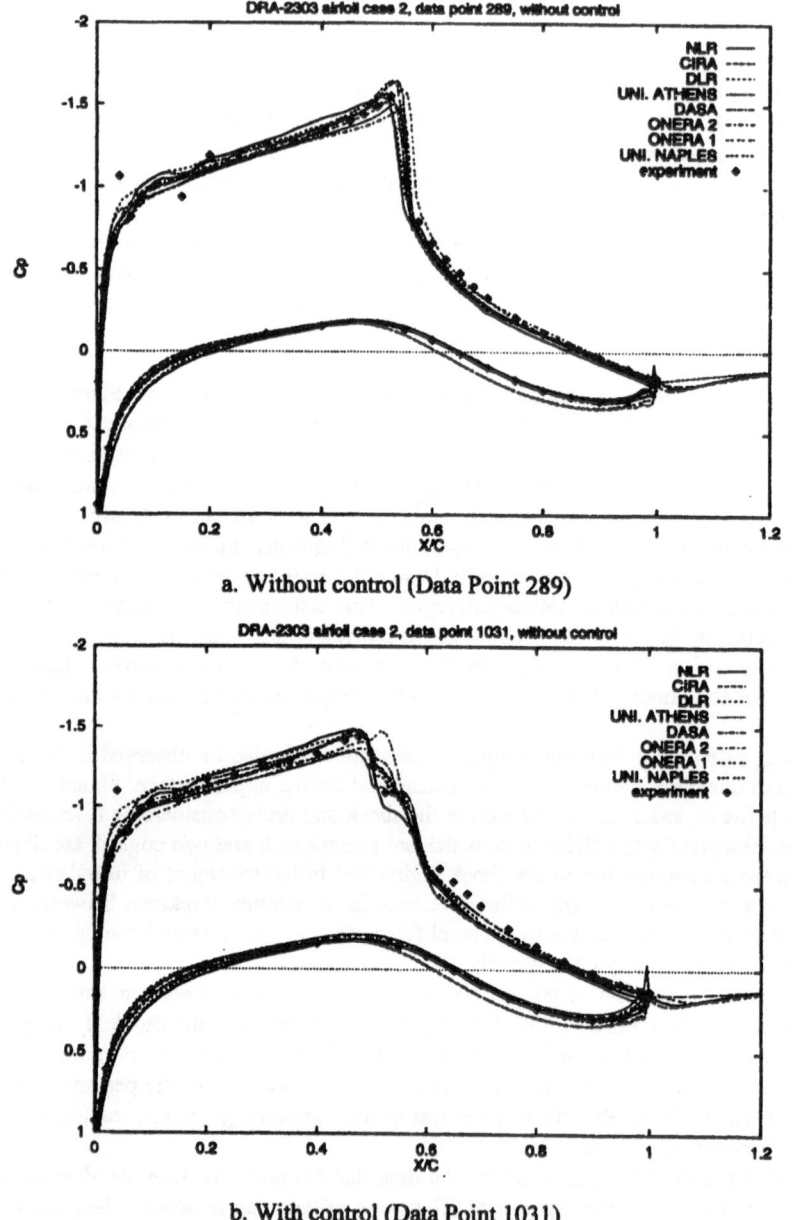

a. Without control (Data Point 289)

b. With control (Data Point 1031)

Figure 28: Computed and measured pressure distributions for the airfoil DRA 2303
$M_\infty = 0.68$, $C_L = 0.81$, $Re_c = 19\times10^6$, $(x/c)_{tr} = 0.05$ (Test Case 2)

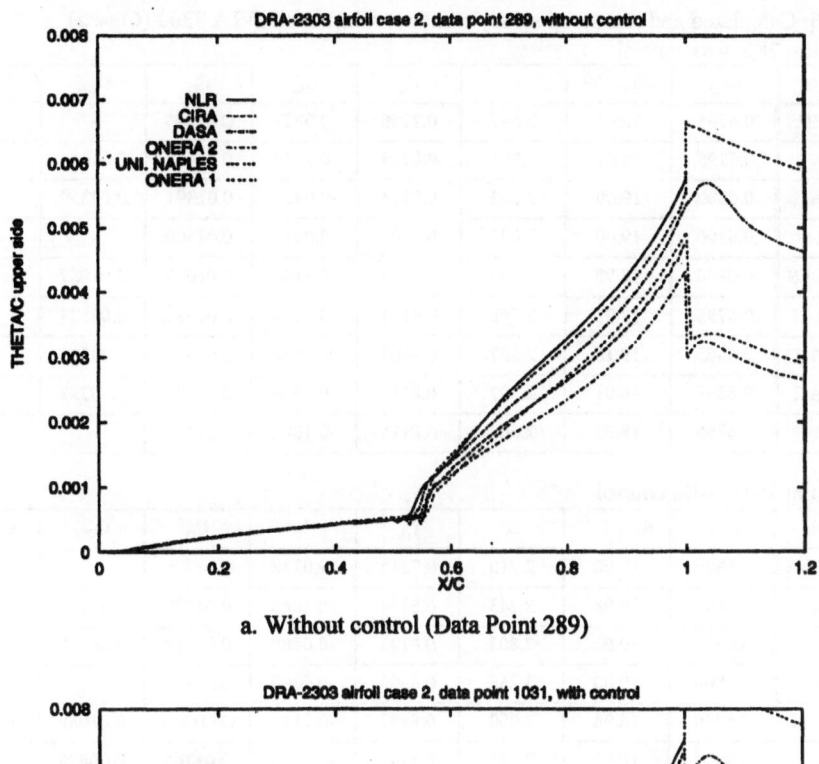

a. Without control (Data Point 289)

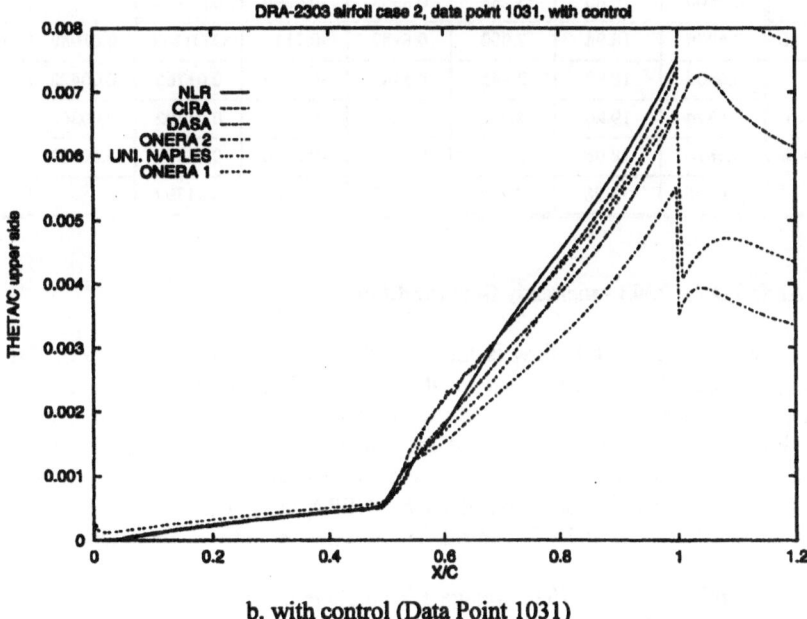

b. with control (Data Point 1031)

Figure 29: Computed momentum thickness distributions for the airfoil DRA 2303
$M_\infty = 0.68$, $C_L = 0.81$, $Re_c = 19 \times 10^6$, $(x/c)_{tr} = 0.05$ (Test Case 2)

37

Table 5: Calculated and measured steady results for the airfoil DRA 2303 (Case 2)
Data point 289, w/o control

Institute	M_∞	$Re_\infty 10^6$	α	C_L	C_M	C_D	C_{DW}	C_{pcav}
U.ATHENS	0.6795	18.91	2.507	0.7756	-0.0884	O.01297	-	-
CIRA	0.6795	18.91	1.791	0.8115	-0.1148	0.01438	0.01014	-
DASA	0.6800	19.00	2.562	0.8153	-0.0954	0.01391	0.00359	-
DLR	0.6800	19.00	3.000	0.8108	-0.0915	0.01500	-	-
U.NAPLES	0.6806	18.98	1.900	0.8292	-0.1148	0.01077	0.01087	-
NLR	0.6795	18.91	2.548	0.8114	-0.0900	0.01557	0.00533	-
ONERA 1	0.6863	19.10	2.507	0.8069	9.9990	0.01622	0.00582	-
ONERA 2	0.6863	18.91	2.507	0.8119	-0.0998	0.01579	0.00358	-
EXP.	0.6795	18.91	2.507	0.8115	-0.1002	0.01458	-	-

Data point 1031, with control

Institute	M_∞	$Re_\infty 10^6$	α	C_L	C_M	C_D	C_{DW}	C_{pcav}
U.ATHENS	0.6806	18.98	2.710	0.7318	-0.0738	0.01306	-	-
CIRA	0.6806	18.98	2.243	0.8134	-0.1045	0.01570	0.01169	-0.976
DASA	0.6800	19.00	2.808	0.8135	-0.0888	0.01519	0.00232	-1.064
DLR	0.6800	19.00	3.266	0.8303	-0.0899	0.01750	-	-1.011
U.NAPLES	0.6806	18.98	2.000	0.8187	-0.1111	O.01144	0.01080	-0.608
NLR	0.6806	18.98	2.845	0.8142	-0.0863	0.01765	0.00420	-0.918
ONERA 1	0.6874	19.20	2.710	0.7632	-0.0954	0.01792	0.00306	-1.058
ONERA 2	0.6874	18.98	2.710	0.8124	-0.0960	0.01797	0.00289	-
EXP.	0.6806	18.98	2.710	0.8142	-0.0984	0.01797	-	-

3.4.2 Airfoil DRA 2303 - unsteady flow conditions

The most complex airfoil flow is the flow at unsteady conditions (buffet). Here, one mandatory test case has been selected for the DRA-2303 airfoil; it has been computed by **DLR, ONERA** and the **University of Athens,** by the latter, however, only for the case w/o control. For the computations, DLR and the University of Athens used a time-accurate Navier-Stokes code [3.5,3.6], while ONERA employed their VII-code VIS15 [3.4] (also see Table 3). Presently, we will only consider the ONERA and DLR computations since they treated both the flow with and w/o control. The freestream and test conditions for this case are:

- Mach number M_∞= 0.702, average lift coefficient: w/o control C_L= 0.67, with control C_L= 0.733, Re_c= 10.5x10^6, $(x/c)_{tr}$= 0.05 on upper and lower surface.

ONERA results: The case without control has been performed for a Mach number of M_∞= 0.709 employing more than 3000 time steps. Transonic buffet has been predicted similar to the experimental results, Figure 30. The mean value of the lift coefficient is rather well predicted by the code; however, the frequency of the computed buffet oscillations, viz., f = 15

Hz, is much smaller than the experimental frequency of f = 36 Hz. The origin of the lower frequency prediction is still not well understood, but may be due to the TSP approximation of the inviscid solver which does not allow to fully treat the leading edge area of the airfoil and hence the motion of the stagnation point may not be predicted correctly.

For the flow with control (M_∞= 0.716), shock control has been initiated after the start of the calculations. An unsteady solution appears initially, but the oscillations are strongly damped and disappear thereafter. Thus, the VIS15 computations predict a suppression of the buffet process due to control in agreement with the experiments conducted by DRA (see Chapter 4). Note that in the computations the zero-mass-flux condition over the porous region was realized at each time step, implying an instantaneous answer of the cavity which is, of course, only an approximation to the actual flow physics not accounting for the lag between outer flow and cavity conditions. The right-hand plot of Figure 30 indicates that the mean pressures are also well predicted by the VIS15 code.

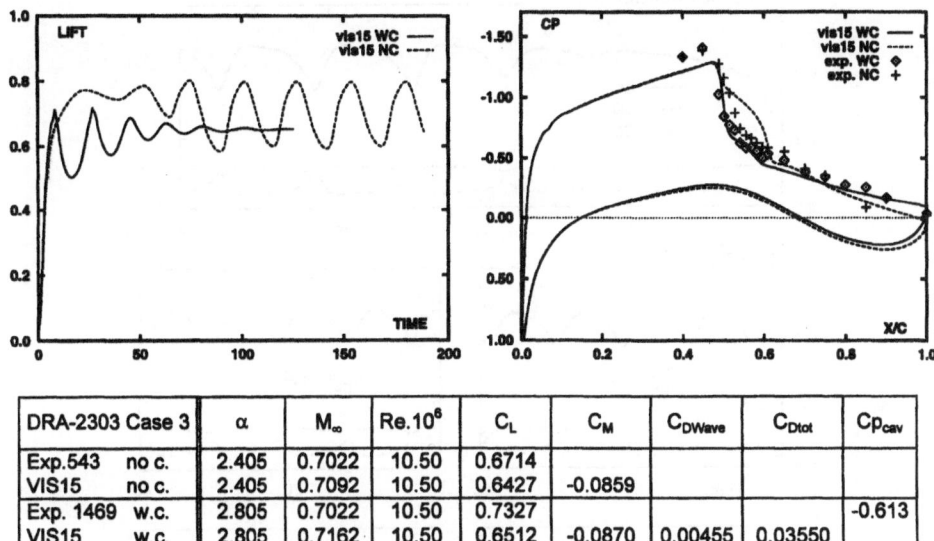

DRA-2303 Case 3		α	M_∞	Re.10^6	C_L	C_M	C_{DWave}	C_{Dtot}	Cp_{cav}
Exp.543	no c.	2.405	0.7022	10.50	0.6714				
VIS15	no c.	2.405	0.7092	10.50	0.6427	-0.0859			
Exp. 1469	w.c.	2.805	0.7022	10.50	0.7327				-0.613
VIS15	w.c.	2.805	0.7162	10.50	0.6512	-0.0870	0.00455	0.03550	

Figure 30: Unsteady calculations (ONERA VIS15-code) for the DRA 2303 airfoil

DLR results: For the DLR Navier-Stokes computations to realize buffet conditions, the Mach number had to be increased from M_∞= 0.702 in the experiments to M_∞= 0.720. This increase seems, at least partly, due to the insufficient grid resolution in the shock boundary layer interaction region, as already discussed in conjunction with the pre-computations, Chapter 3.2.2.

At the higher Mach number, buffet oscillations are observed in the case w/o control as indicated by the corresponding unsteady force and moment coefficients displayed in Figure 31 (a). One observes that it takes considerable time until the coefficients start a periodic dependence with respect to time. After periodicity is established, the frequency of the flow oscillations, taken from the plot, amounts to f = 41.4 Hz which is in reasonable agreement with the 36 Hz determined in the DRA experiments. This good agreement in comparison to the VIS15 computations seems to confirm the deficiency of TSP methods in the airfoil nose region addressed above.

In the experiments the oscillations were strongly damped due to shock control. However, applying in the NS-computations the condition of zero-net-mass-flow in the cavity region at each time step (passive control) did not result in any reduction of the oscillation amplitude which is, of course, also contrary to the VIS15-results. However, when active control, i.e., prescribed cavity pressure, was applied, the buffet oscillations were influenced favorably, Figure 31(b): as shown by the force and moment coefficients, reducing the cavity pressure successively from c_{pc}= -1.3 to -1.4 to -1.5 and finally to c_{pc}= -1.6 resulted first in a reduction of the amplitude (e.g., of lift) and an increase in frequency until finally at c_{pc}= -1.6 steady conditions were reached. The latter is most likely due to the complete suppression of boundary layer separation due to the strong suction in the cavity region.

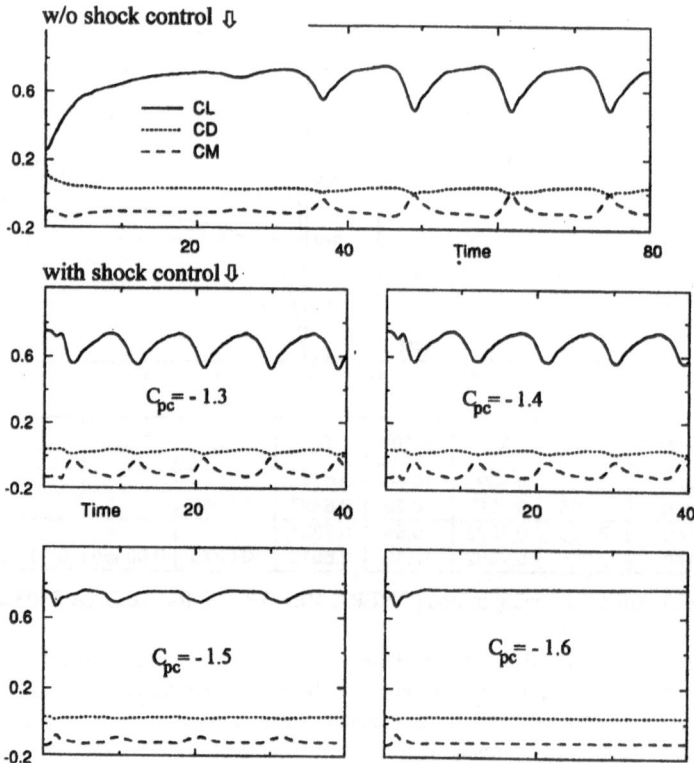

Figure 31: Navier-Stokes computations for the DRA 2303 airfoil at buffet conditions
M_∞= 0.720, α = 3.0°, Re_c= 10.5x10⁶, $(x/c)_{tr}$= 0.05

The numerical results just described have shown that a buffet suppression due to active as well as passive control is being predicted by the computer codes. However, there are certain restrictions one must consider: In the case of VIS15, the physics of the flow in the cavity region are, for passive control, not correctly modeled since the pressure in the cavity is assumed to react without time-lag to changes in the outer flow conditions. The NS-computation did not yield a steady solution for passive control and massive active control had to be applied before a buffet suppression was achieved. It seems, therefore, that in future work a better model of the cavity boundary conditions for oscillating flows has to be established; this is planned by several investigators.

3.4.3 Airfoil DA LVA-1A with and without control

The second laminar-type airfoil investigated was the DA LVA-1A airfoil. Geometrical details and experimental results for this airfoil are presented in Chapter 4. The characteristics of the perforated insert, of interest here, are:
- Porosity P = 5.06%, hole diameter d = 0.1123 mm, plate thickness e = 1.2 mm and the extent of the cavity x/c = 0.58 to 0.695

For this airfoil, one test case with and w/o control has been selected for computation and comparison with corresponding experimental results obtained in the ONERA / CERT T2 - wind tunnel. Data describing this test case are:
- Mach number M_∞= 0.761, lift coefficient w/o control C_L = 0.47, with control C_L = 0.45, Reynolds number Re_c = 4.6x10^6, transition fixed at 48% chord on upper and lower surfaces (Data Points 76 and 73/74). Note that Table 6 includes all measured and computed aerodynamic coefficients for this test case and, in addition, a computed test case with control at a lift coefficient of C_L = 0.47 - equal to the one w/o control - for which experimental data are not available.

Examining first Table 6, one sees that the drag behavior is similar to the one for the laminar DRA 2303 airfoil: at equal lift coefficients, total drag increases due to control while wave drag is reduced. The reason for the former is, as already outlined in previous sections, the dominating effect of control on viscous drag.

Concerning the pressure distributions, we will not consider all computational results since they are generally very similar, but only look at representative data obtained by two VII-codes, viz., the DA-code (DA designed the airfoil) and the ONERA VIS - codes, and the DLR Navier-Stokes method, respectively, Figures 32 to 35. Obviously, the trends observed for the DRA 2303 airfoil are also found for the present configuration: without control, the overall agreement between experiment and computation is quite good; with control both computation and experiment predict a reduction in shock strength (spreading of the shock) due to control, hence a reduction in wave drag. However, while the experimental data exhibit an almost linear decrease in pressure over the cavity region, all codes predict a pressure plateau with weaker shocks at the upstream and downstream faces of the cavity. Moreover, the results of the DA-computations, Figure 32, show that this occurs for both, active (prescribed cavity pressure) as well as passive (zero-net-mass-flow) control. As already pointed out, this discrepancy is most likely a result of a deficiency in the application of the control laws employed to determine the transpiration velocity in the cavity region.

In the ONERA / CERT experiments, Chapters 4 and 20, extensive LDA measurements were performed to probe, for the present test case, flow details in the outer field and the boundary layer development from just upstream of the cavity to the trailing edge. The latter is compared to boundary layer profiles computed by the ONERA VIS05c-code in Figure 35 which indicates that the boundary layer development with and w/o control is well predicted by the code. In particular, both experiment and computation clearly show the stronger growth in boundary layer thickness due to control, which ultimately leads to the overcompensation of the reduction in wave drag by the strongly increasing viscous drag in the presence of control (also see Figures 13, 14, 29).

Table 6: Calculated and measured results for the airfoil DA LVA-1A

Case 1: Run 76, w/o control

Institute	M_∞	$Re_\infty \, 10^6$	α	C_L	C_M	C_D	C_{DW}	C_{pcav}
NLR	0.7613	4.64	0.884	0.4738	-0.0823	0.01227	0.00508	-
CIRA	0.7613	4.64	0.965	0.4736	-0.0840	0.01092	0.00776	-
DLR	0.7613	4.64	1.190	0.4735	-0.0810	0.00976	-	-
U.ATHENS	0.7613	4.64	1.000	0.5059	-0.0933	0.01297	-	-
DASA	0.7613	4.64	0.931	0.4743	-0.0849	0.01130	0.00359	-
ONERA 1	0.7613	4.64	0.890	0.4732	-0.0935	0.01117	0.00492	-
ONERA 2	0.7613	4.64	0.788	0.4736	-0.0877	0.01128	0.00471	-
U.NAPLES	0.7613	4.64	0.854	0.4736	-0.0874	0.01074	0.00282	-
EXP.	0.7698	4.64	1.000	0.4736	-0.0858	0.01172	-	-

Case 1: Run 76, with control

Institute	M_∞	$Re_\infty \, 10^6$	α	C_L	C_M	C_D	C_{DW}	C_{pcav}
NLR	0.7613	4.64	1.261	0.4737	-0.0785	0.01263	0.00201	-0.765
CIRA	0.7613	4.64	1.080	0.4736	-0.0805	0.01206	0.00800	-0.837
DLR	0.7613	4.64	1.190	0.4737	-0.0814	0.00992	-	-0.900
U.ATHENS	0.7613	4.64	1.000	0.5123	-0.0930	0.01282	-	-
DASA	0.7613	4.64	1.056	0.4784	-0.0822	0.01219	0.00200	-0.821
ONERA 1	0.7613	4.64	0.950	0.4726	-0.0922	0.01210	0.00360	-0.922
ONERA 2	0.7613	4.64	0.940	0.4739	-0.0840	0.01151	0.00296	-0.853

Case 2: Run 73, with control

Institute	M_∞	$Re_\infty \, 10^6$	α	C_L	C_M	C_D	C_{DW}	C_{pcav}
NLR	0.7614	4.64	0.915	0.4514	-0.0766	0.01209	0.00179	-0.765
CIRA	0.7614	4.64	0.972	0.4536	-0.0795	0.01140	0.00737	-0.822
DLR	0.7613	4.64	1.080	0.4538	-0.0805	0.00912	-	-0.874
U.ATHENS	0.7614	4.64	0.900	0.4880	-0.0903	0.01186	-	-
DASA	0.7613	4.64	0.948	0.4587	-0.0813	0.01151	0.00173	-0.818
ONERA 1	0.7614	4.61	0.850	0.4538	-0.0909	0.01154	0.00323	-0.910
ONERA 2	0.7613	4.64	0.840	0.4538	-0.0828	0.01073	0.00259	-0.842
U.NAPLES	0.7614	4.64	0.949	0.4537	-0.0801	0.01036	0.00167	-
EXP.(R 74)	0.7692	4.66	1.000	0.4556	-0.0833	0.01233	-	-

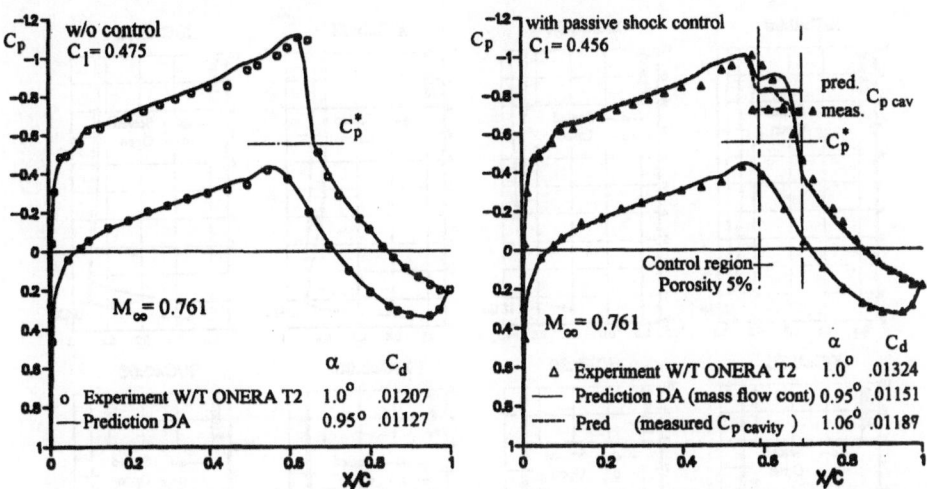

Figure 32: Computed and experimental results for the DA LVA-1A airfoil; DA computation

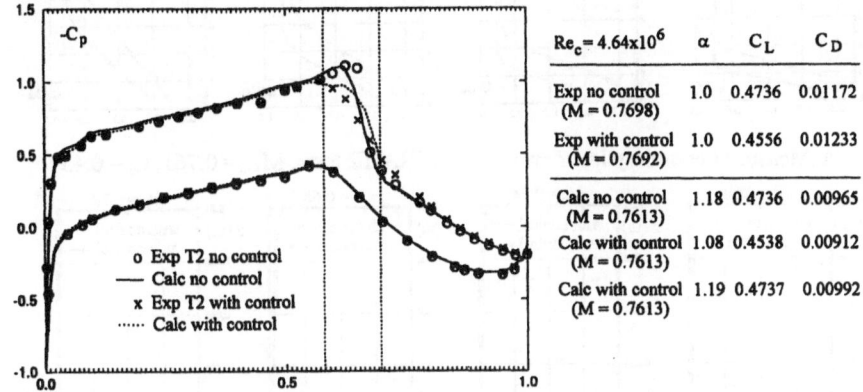

Figure 33: Computed and experimental results for the DA LVA-1A airfoil; DLR NS - code

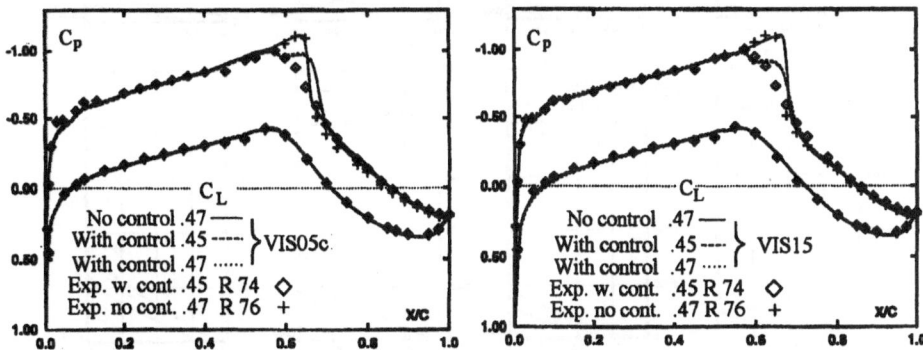

Figure 34: Computed and experimental results for the DA LVA-1A airfoil
ONERA-computation, $M_\infty = 0.761$

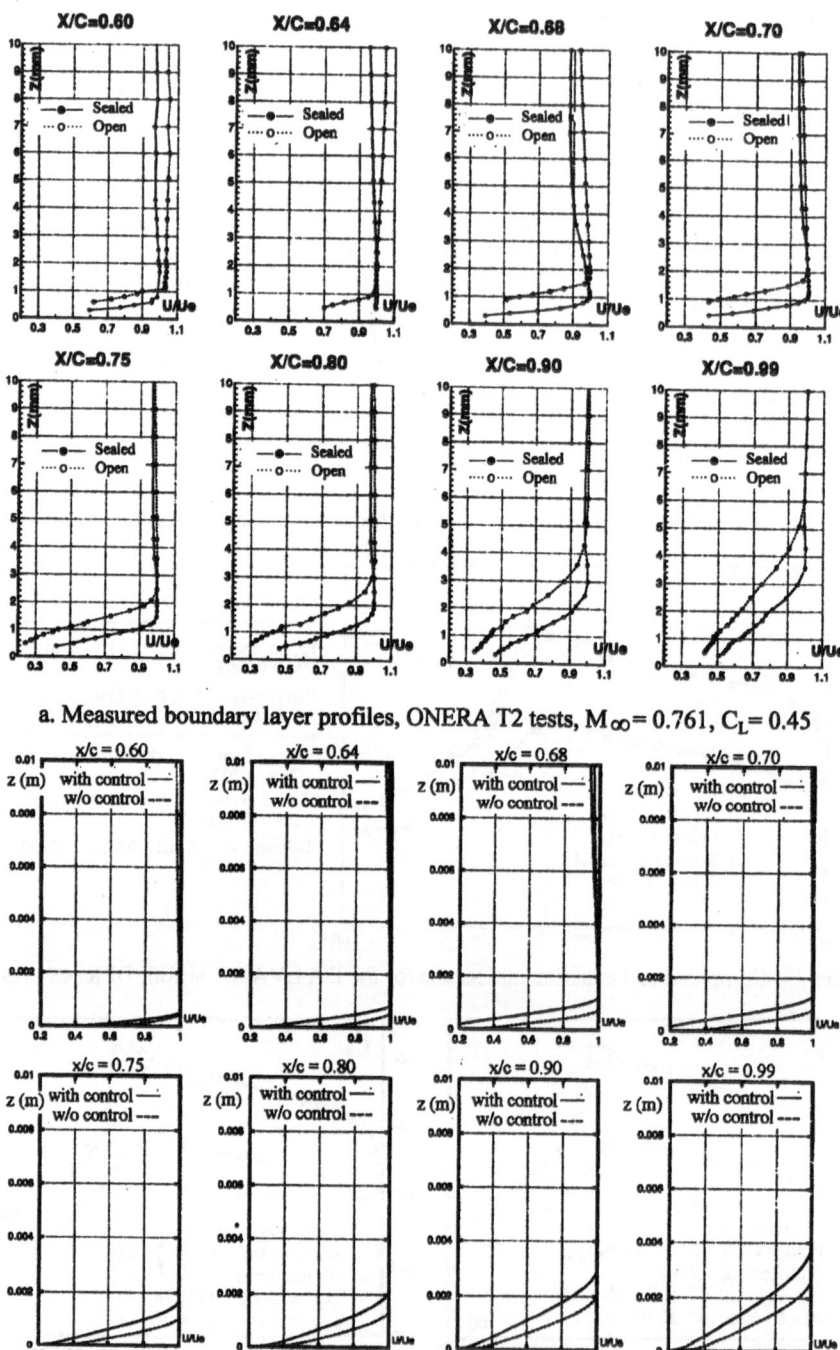

a. Measured boundary layer profiles, ONERA T2 tests, $M_\infty = 0.761$, $C_L = 0.45$

b. Boundary layer profiles computed by ONERA (VIS05c), $M_\infty = 0.761$, $C_L = 0.45$

Figure 35: Measured and computed boundary layer profiles, airfoil DA LVA-1A

3.4.4 Airfoil VA-2 with passive and active control

For the laminar-type airfoils DRA 2303 and DA LVA-1A, originally the only airfoils to be considered within the EUROSHOCK research program, it was found, as discussed above and further to be elaborated on in Chapter 4, that passive shock control was always associated with an increase in total drag. Early measurements with the turbulent airfoil VA-2 [3], however, had shown that a reduction in total drag could be achieved by passive control over a wide range of freestream conditions (also see Figure 2). In order to confirm or reject these results, the early tests were repeated within the present program. Details of the experiments and their results will be given in Chapter 4. Here, we will only consider representative computational results for this airfoil.

Geometrical details for the airfoil VA-2 can also be found in Chapter 4. The characteristics of the perforated insert, of interest in conjunction with the present analysis, are:
- Porosity P = 12.6%, hole diameter d = 0.284 mm, plate thickness e = 1.2 mm and the extent of the cavity x/c = 0.495 to 0.645

Three different flow cases at the same nominal Mach number have been selected as mandatory test cases, viz.,
- M_∞= 0.74 at the three lift coefficients of C_L= 0.65, C_L= 0.90 and C_L= 0.80 (Test Cases 1, 2 and 3); the Reynolds number was Re_c = 2.4x10^6 with transition fixed at 30% chord on the upper and 25% chord on the lower surface.

We will first consider the pressure distributions corresponding to Test Case 1, Figure 36, in comparison to pressure distributions for laminar-type airfoils, here for the airfoil DA LVA-1A, Figure 34, in order to point out essential differences between these two types of airfoils. The laminar airfoil exhibits on the critical upper surface a strong acceleration of the flow from the leading edge down to the shock location which is, of course, needed to keep the flow laminar up to this position. The turbulent airfoil VA-2 is characterized by a nearly constant pressure extending down to the shock - after an initial strong acceleration in the immediate leading edge region. The different shock-upstream pressure gradients will naturally have an effect on the boundary layer development and hence on the characteristics of the boundary layer entering the cavity region in the case of control.

Considering the effect of control on the VA-2 flow development, the pressure distributions, Figure 36 (b), show that the single shock is, as in the case of the laminar airfoils, essentially replaced by a weaker double-shock system which results in the known reduction in wave drag, the latter being indicated in Table 7 which shows the aerodynamic coefficients for this test case. Concerning the important issue of total drag behavior, the table indicates that for the VA-2 at the present freestream conditions (C_L= 0.65), a marginal reduction in total drag is predicted by the majority of the codes. Note that at the highest lift coefficient considered (C_L= 0.9, Test Case 2), three codes predict a decrease in total drag and three codes an increase while at the intermediate lift coefficient of C_L= 0.8 all methods show an increase in drag.

The experiments, Chapter 4, do not allow a direct comparison with the computations since wind-tunnel wall interference did not provide matching freestream conditions, especially not in the effective freestream Mach number on which the efficiency of control is critically dependent. However, it can generally be stated that the experiments have identified, in confirmation of earlier results and similar to the computations, flow conditions where passive shock control reduces total drag and conditions where total drag is increased. This issue will further be discussed below.

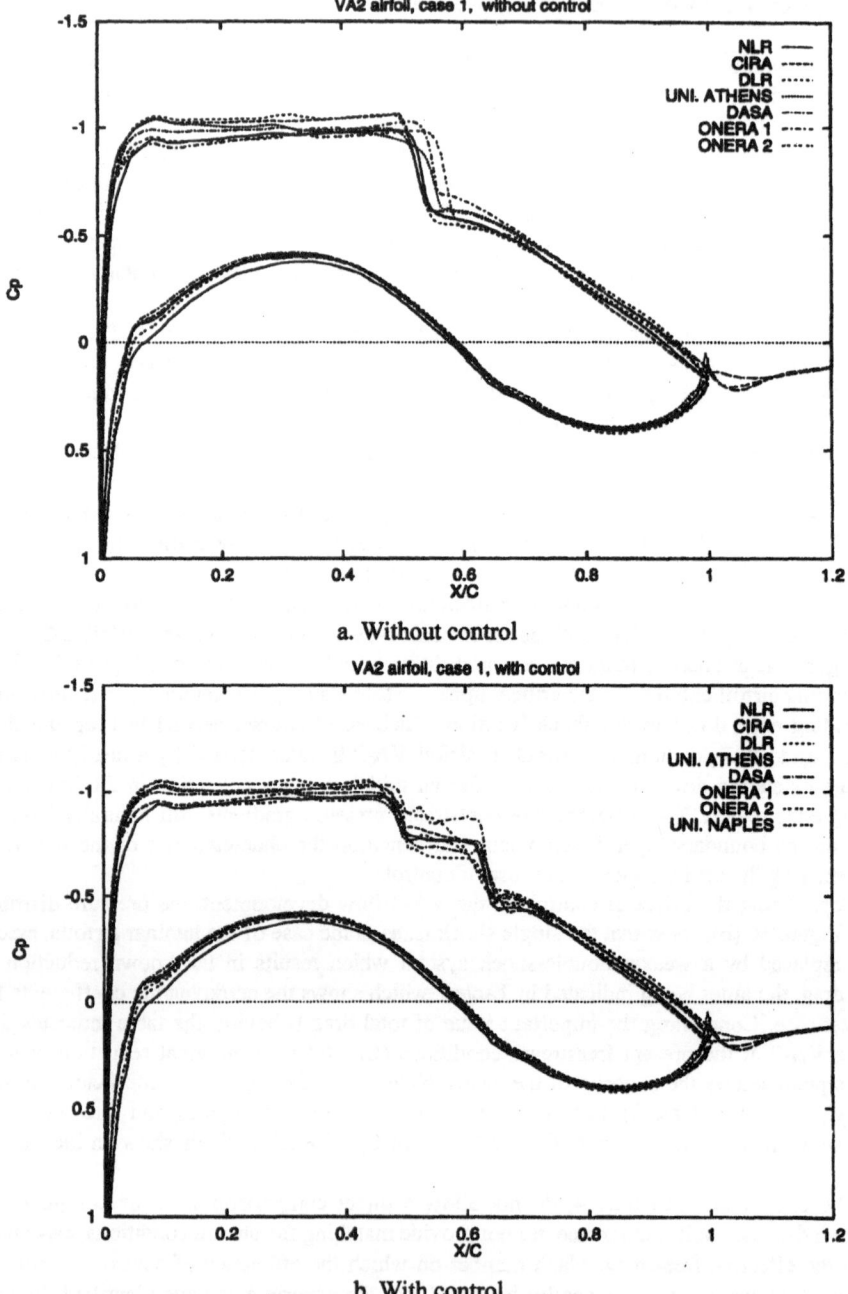

a. Without control

b. With control

Figure 36: Computed pressure distributions for the airfoil VA-2 (Test Case 1)
$M_\infty = 0.74$, $C_L = 0.65$, $Re_c = 2.4 \times 10^6$, $(x/c)_{tr} = 0.30$ upper, 0.25 lower surface

Table 7: Computed aerodynamic coefficients for the airfoil VA-2 (Test Case 1)
Without control

Institute	M_∞	$Re_\infty\ 10^6$	α	C_L	C_M	C_D	C_{DW}	C_{pcav}
NLR	0.7400	2.50	1.128	0.6499	-0.1296	0.00979	0.00084	-
CIRA	0.7400	2.50	1.234	0.6501	-0.1225	0.01250	0.00800	-
DLR	0.7400	2.50	1.350	0.6543	-0.1266	0.01350	-	-
U.ATHENS	0.7400	2.50	1.050	0.6599	-0.1312	0.01281	-	-
DASA	0.7400	2.50	0.801	0.6405	-0.1348	0.01105	0.00076	-
ONERA 1	0.7400	2.50	0.940	0.6516	-0.1419	0.01399	0.00163	-
ONERA 2	0.7474	2.50	0.890	0.6580	-0.1403	0.01218	0.00176	-

With control

Institute	M_∞	$Re_\infty\ 10^6$	α	C_L	C_M	C_D	C_{DW}	C_{pcav}
NLR	0.7400	2.50	1.064	0.6499	-0.1315	0.00902	0.00028	-0.751
CIRA	0.7400	2.50	1.075	0.6506	-0.1281	0.01248	0.00745	-0.737
DLR	0.7400	2.50	1.350	0.6494	-0.1261	0.01360	-	-0.679
U.ATHENS	0.7400	2.50	1.040	0.6523	-0.1305	0.01272	-	-
DASA	0.7400	2.50	0.827	0.6403	-0.1337	0.01096	0.00025	-0.722
ONERA 1	0.7400	2.50	1.090	0.6787	-0.1445	0.01165	0.00154	-0.862
ONERA 2	0.7474	2.50	0.890	0.6530	-0.1402	0.01516	0.00146	-0.802
U.NAPLES	0.7400	2.50	1.143	0.6498	-0.1228	0.01063	0.00096	-

In spite of remaining uncertainties, it seems that for a turbulent airfoil the effect of passive shock control on viscous drag is less severe than for a laminar one which leads, at certain conditions, to a decrease in total drag for the former. Reasons for this behavior are likely to include differences in the boundary layer condition upstream of the cavity due to the different boundary layer developments mentioned above which will, for instance, result in a higher skin friction coefficient upstream of the shock for the laminar airfoil and hence in a higher drag of the holes of the perforation. Furthermore, the relative increase in boundary layer thickness parameters due to the perforation seems to be less for the thicker turbulent boundary layer so that the amplification effect due to the rear adverse pressure gradients on, say, the momentum thickness is less severe, Figure 37, reducing the contribution of viscous drag to the total drag. An additional favorable effect in that regard is caused by the reduced shock strength for the turbulent airfoil when considering equal lift coefficients.

Finally a remark concerning the computer code performance: In the pressure distributions for the three VA-2 flow cases treated considerable differences occur in the predicted height of the pressure plateau on the upper surface. In addition, the predicted shock positions in the case w/o control and the changes in the pressure distributions due to control show much larger deviations between codes than in the computations for the laminar airfoils considered. Particularly at a lift coefficient of $C_L = 0.9$, the differences in the computations with control are severe. It is judged that this is mainly due to the closeness of the flow to maximum lift, hence total separation, where small deviations in the computational (as well as experimental) environment, i.e., in freestream conditions, grid resolution, turbulence models, control laws and their implementation, lead to large differences in the results.

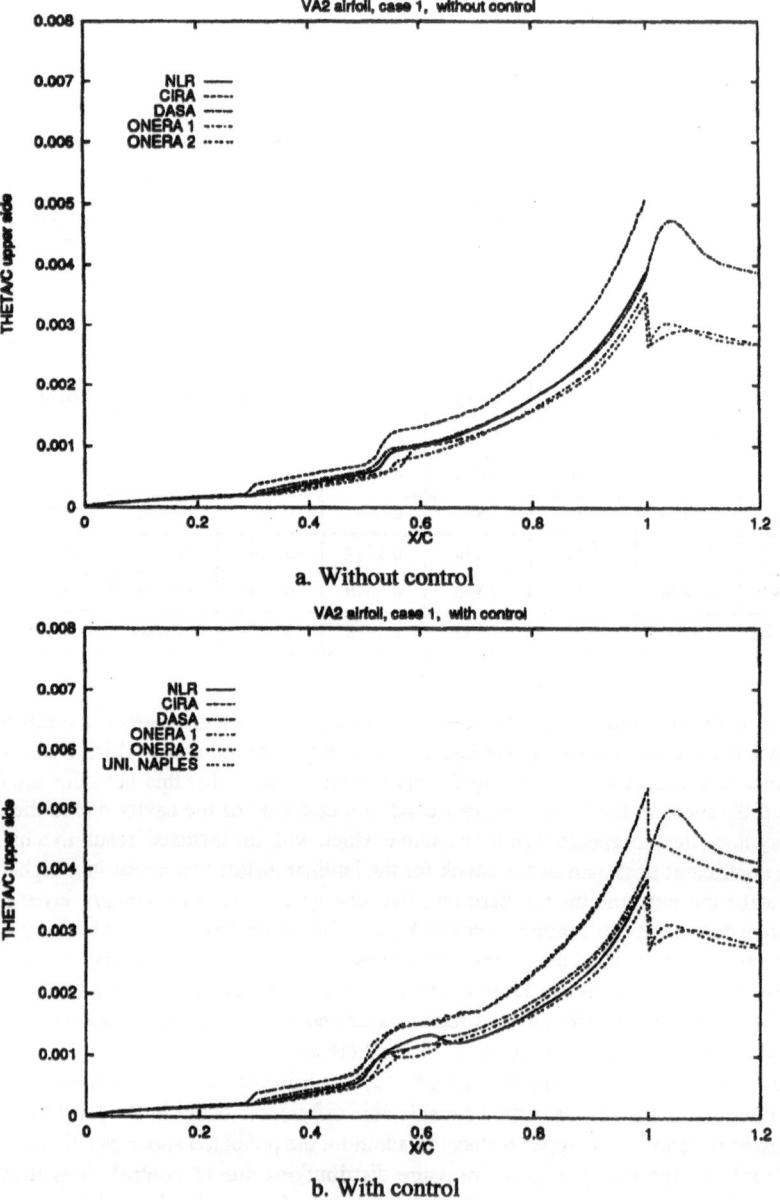

a. Without control

b. With control

Figure 37: Computed momentum thickness distributions for the airfoil VA-2
$M_\infty = 0.74$, $C_L = 0.65$, $Re_c = 2.4 \times 10^6$, $(x/c)_{tr} = 0.30$ upper, 0.25 lower surface

3.5 Conclusion and Future Work

The objective of the present numerical work was the improvement and extension of numerical methods to treat flows with passive shock boundary layer interaction control and to perform a first parametric study to assess the effect of control on transonic airfoil flow.

Concerning the performance of the numerical codes, general agreement has been achieved with most of the experimental data. However, the absolute values of the total drag coefficients show considerable deviations between the different calculations and the corresponding experimental data, even in the case without shock control, although the pressure distributions, including shock location and strength, are in reasonable agreement.

The codes were extended to treat shock control by introducing various control laws for the determination of the velocity distribution in the cavity region. However, in the calculation of the ventilation velocities by the various codes, considerable scatter was observed which can only partly be attributed to the use of different control laws but seems also to be dependent on the numerical method employed and on the implementation of the control law within a code. This scatter in the ventilation velocities seems to be the origin of strong deviations in the boundary layer properties within the control region which leads to the differences observed in the local pressure distributions and in the subsequent flow development. Concerning the control laws, it was found that Poll's formula, derived from the calibration of laser-drilled plates, gives the most reliable results for the present control cases.

Concerning shock control effectiveness for the laminar airfoils DRA-2303 and DA LVA-1A, all contributors predicted a considerable reduction in the wave drag but also an increase in total drag due to control for all flow cases investigated. The computed boundary layer data clearly show a strong increase in momentum thickness downstream of the shock due to control, indicative of an increase in viscous drag which overcompensates the reduction of the wave drag leading, in agreement with experiment, to the observed increase in total drag associated with control. For the turbulent airfoil VA-2 the computational results show a reduced influence of control on viscous drag which also leads, at certain conditions, to a reduction in total drag, again in agreement with experiment.

The results of the unsteady calculations have shown considerably larger deviations than the steady flow cases. Nevertheless, a strong damping of shock oscillations due to control could be predicted, supporting the experimental results.

The experience gathered throughout the present numerical exercise points out the way to future work needed for the performance improvement of computational methods incorporating shock control. Work should include:

- the improvement of shock control laws and the implementation of the best law in the different codes,
- the development of a procedure for a detailed treatment of the boundary layer flow over the control region,
- the development of turbulence models that take into account the turbulence structure generated by the in- and outflow through the openings of the perforation,
- the development of physically correct control laws for unsteady flows and finally
- the extension of the codes to 3D-applications.

4 Airfoil Tests with and without Control (Task 3)

The aims of the present studies were to investigate, experimentally, the effect of passive control of shock boundary layer interactions on the design and off-design performance of airfoils. The investigation concentrated on airfoils designed to achieve large extents of natural laminar flow on both surfaces, since airfoils of this kind have a larger potential for improvement at design conditions, where shock waves are probably already present, and, in addition, have smaller margins as far as drag-rise and buffet boundaries are concerned when compared to turbulent designs. During the course of the investigation, having perused early results, it was agreed that DLR would retest the original airfoil VA-2 which had demonstrated the benefits of passive control [1.3].

A comprehensive wind tunnel test program was designed to provide an extensive data base for cases with and without control for ① a large Reynolds number range, ② different boundary layer characteristics ahead of the shock wave and ③ different shock strengths. The data base thus established was used to assess the computer codes considered in Chapter 3 and to demonstrate the efficiency of passive control up to conditions typical of flight.

4.1   Experimental Program

The experimental program was divided into two parts: firstly, an investigation of the influence of Reynolds number on the effectiveness of passive control was carried out by DLR using the LVA - 1A airfoil model designed and manufactured by DASA - Airbus. The focus of this work was on the determination and analysis of airfoil performance with and without control and its dependence on Reynolds number, Mach number and angle of attack. The DLR-tests were supplemented by tests at ONERA, employing the same model, with emphasis, however, on detailed LDV flow field and boundary layer measurements above the airfoil upper surface with and without control. In addition, the turbulent MBB VA-2 airfoil was retested over a range of transonic freestream Mach numbers and lift coefficients at a constant Reynolds number by DLR to confirm the results obtained earlier.

The second part of the investigation comprised detailed steady pressure measurements carried out by DRA and complementary unsteady measurements performed by DLR on the same large-scale model. The model of the DRA 2303 airfoil, designed and manufactured by DERA, was tested at Reynolds numbers up to those approaching full scale and a variety of Mach numbers at conditions up to and exceeding buffet onset. The measurements allowed the flow development in the control region to be related to the overall airfoil performance. The dynamic pressure measurements allowed the dynamic aspects of control and the mechanism by which buffet is affected by control to be studied in detail.

4.2   Experiments with the Airfoil DA LVA-1A

4.2.1   Airfoil characteristics and wind tunnel model

The basic LVA - 1A airfoil is a transonic laminar flow-type airfoil with a thickness of 12% chord. It was designed by DASA-Airbus (see [4.1] and Chapters 14 and 19) to have natural laminar flow on the upper and lower surfaces up to 50% of the chord at a Mach number of $M_\infty = 0.73$, a lift coefficient of $C_L = 0.4$ and a Reynolds number of $Re_c = 20\times10^6$. The pressure distribution is characterized by a weak shock wave occurring already at design conditions. An essential feature of this airfoil, especially important for shock control, is that with increasing lift coefficient, as the shock grows stronger, the position of the shock remains rather unchanged. The design requires, in order to investigate passive shock control, an exchangeable insert reaching from 55% to 82% of the chord. Typical pressure distributions are, e.g., depicted in Figures 32 and 39.

In order to carry out the experiments in the ONERA T2 and DLR KRG wind tunnels, a cryogenic model has been designed and manufactured by DA with a chord of c = 150 mm and a span of b = 400 mm, Figure 38. The model is equipped with two chordwise pressure plotting sections with a total of 64 pressure orifices and two dynamic pressure transducers (Kulite®) positioned downstream of the shock control region to determine pressure fluctuations, hence buffet behavior. Thermocouples are located on the inside of the model wall to determine transition location.

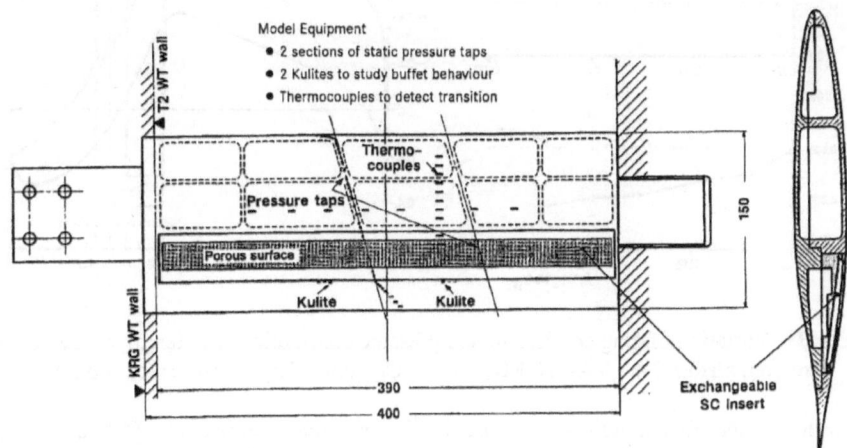

Figure 38: Cryogenic airfoil model DA LVA-1A with instrumentation

The design of the porous surface requires some attention since many parameters may influence the control effectiveness, although, as shown in Chapter 2.1, Figures 9 and 11, there seems essentially no effect of the geometry of the perforation on the pressure distribution, hence wave drag, at least not in the parameter range, e.g., porosity range, considered; however, the development of the momentum thickness, Figure 14, suggests that there might be a strong influence on viscous drag when considering the amplification effect of the airfoil rear adverse pressure gradients.

Geometric parameters that may affect the overall control performance in the case of airfoil flow are the location of the perforation with respect to the shock, the porosity, the hole geometry, i.e., hole diameter, inclination and shape, and the extent of the perforated region. Closely linked to these parameters are the flow conditions prevailing on the airfoil, i.e., essentially the location and movement of the shock and the Reynolds number dependent boundary layer conditions upstream of the control region, determining extent and hole diameter, respectively, of the perforation. Figure 39 depicts, as an example of the effect of the above parameters, the influence of the length of the control region on drag. One observes that wave drag decreases when the extent of the control region increases while viscous drag and total drag increase, very much in agreement with the results described in Chapters 2 and 3.

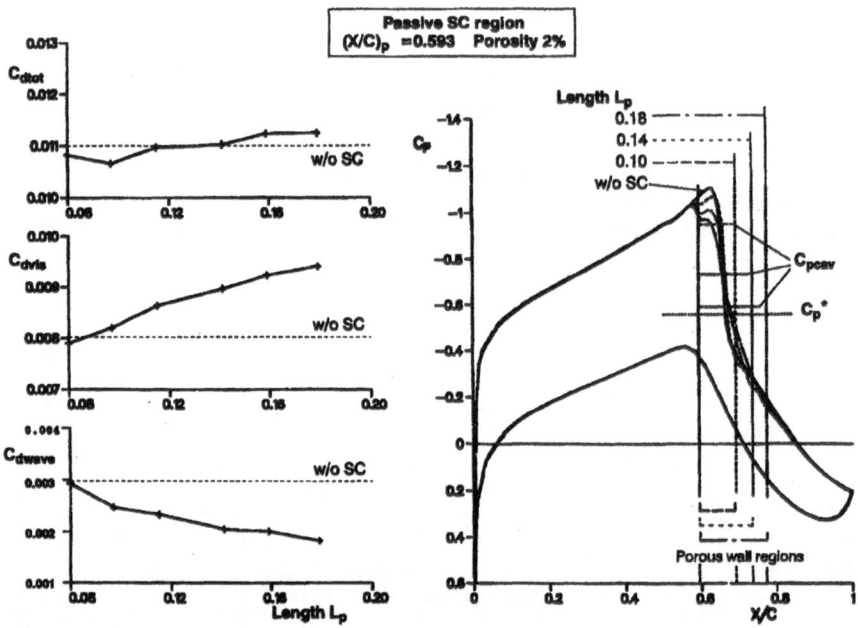

Figure 39: Variation of drag coefficient and pressure distribution with length of the control region; airfoil DA LVA-1A, $M_\infty = 0.76$, $C_L = 0.45$, $Re_c = 6 \times 10^6$, $(x/c)_{tr} = 0.50$

For the present flow conditions, DA has carried out a parametric study, [4.1] and Chapter 14, employing their VII-code, considered in Chapter 3, to determine the perforation characteristics needed here, viz.:

- porosity 3.94%, hole diameter 0.10 mm, perforated plate thickness 1.0 mm, hole spacing 0.48 mm, extent of the perforated region x/c = 0.575 to 0.70.

As already mentioned in Chapter 2, perforations can generally not be (laser-) drilled to specifications and deviations occur which in this case led to the characteristics of the control insert used in the present experiments, viz.,

- porosity 5.1%, hole diameter 0.112 mm, perforated plate thickness 1.0 mm, hole spacing 0.476 mm, extent of the perforated region x/c = 0.58 to 0.695.

In addition, the holes turned out to be conical with the diameter on the cavity side being larger than on the flow side, the latter corresponding to the one given above.

It is furthermore necessary, since the geometry of the perforation cannot be modeled accurately, to carry out calibration tests to establish the characteristics of the perforated plate to be utilized. Such tests were performed by DA for various spanwise sections of the perforated sheet after electron-beam drilling of the holes and before assembly and instrumentation of the insert. Figure 40, depicting the experimental set-up for the ONERA T2-tests and the characteristic pressure drop as function of the effective velocity through the perforation, implies a quite homogeneous distribution of the perforation with essentially equal pressure

52

losses (filled symbols) at these conditions. After instrumentation, assembly and testing in the wind tunnel, a renewed inspection showed larger pressure drops in the vicinity of the pressure plotting sections (P_1 and P_2), indicating a blockage of the holes of the perforation in these areas.

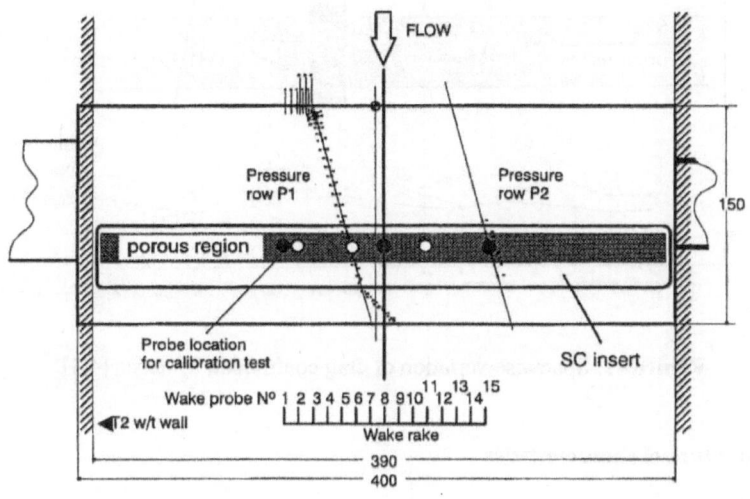

a. Probe locations identifying calibration test locations (T2 test set-up)

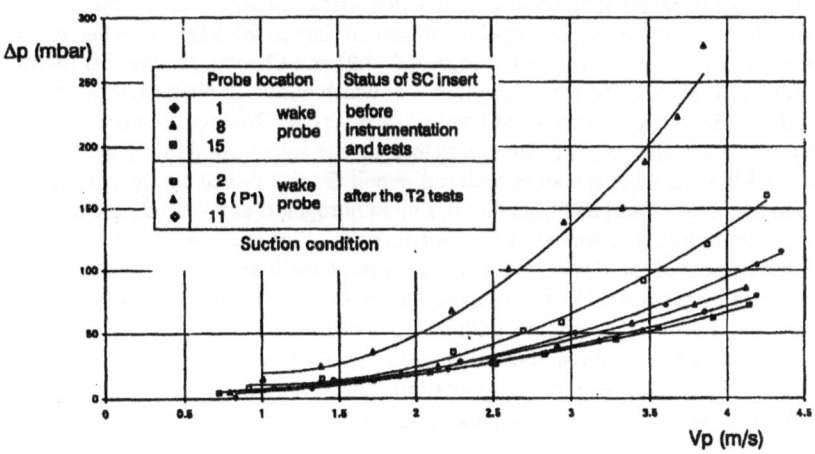

b. Pressure losses of porous control region

Figure 40: Probe locations and calibration tests of the perforated insert
before and after assembly and T2 wind tunnel tests

The blockage was obviously caused by the drilling of the pressure orifices which occurred after manufacturing of the perforated plates. The hole blockage had a considerable effect on the spanwise drag distribution, similar to the effect of locally sealing the perforated surface, Figure 41. Note, however, that for the evaluation of the drag behavior positions of the wake rake located downstream of airfoil sections not affected by hole-blockage were used.

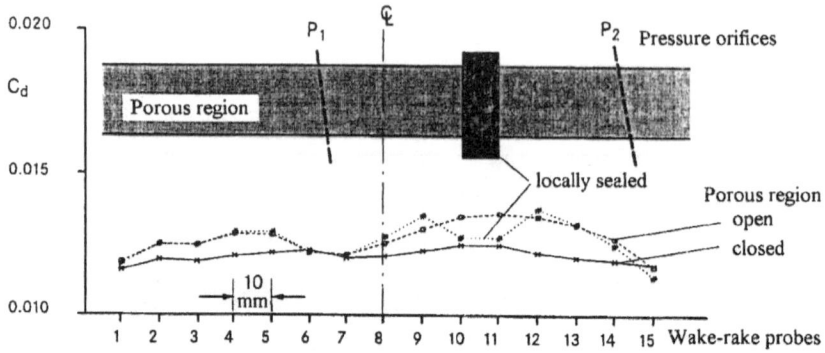

Figure 41: Spanwise variation of drag coefficient, T2-tests, [4.3]

4.2.2 Wind tunnel characteristics

The laminar airfoil DA LVA-1A was investigated in two transonic wind tunnels, the Cryogenic Ludwieg-tube of DLR Göttingen (KRG) and the T2 wind tunnel of ONERA / CERT, Toulouse. The former is a short-duration facility with a test time of up to one second consisting of a 130-meter long tube, a contraction section with a contraction ratio of 3, an adaptive-wall test section, a combination of second throat for Mach number control and quick-opening valve and a dump tank, Figure 42, [4.2] and Chapter 21. The test section has a cross section of 0.40x0.35 m^2 and a length of 2 m which allows to obtain interference-free results for the model chords investigated here, i.e., c = 0.18 m. The performance characteristics of the tunnel, which uses gaseous nitrogen as test gas, are summarized in Figure 42.

The ONERA T2 wind tunnel is a closed circuit facility driven by injecting pressurized dry air into the circuit, [4.3] and Chapter 20. Liquid nitrogen is sprayed into the tunnel to cool the flow and to control the temperature of the mixture of air and gaseous nitrogen which allows the adjustment of the Reynolds number in large steps; fine tuning of the Reynolds number is achieved by pressure. The test section, Figure 43, has a cross section of 0.39 x 0.37 m^2 and a length of 1.4 m; it is equipped with adaptive walls (note the wake rake installation for drag prediction). A second throat located downstream of the test section allows to control the Mach number. Other characteristics are: total pressure variation 1.2 to 3.2 bar, total temperature variation 130K to 300K, Mach number range 0.2 to 0.9 and run time between 20 and 60 seconds.

4.2.3 Discussion of T2 results

The experiments in the ONERA / CERT T2 wind tunnel essentially served to probe the boundary layer development and the flow field associated with shock boundary layer interaction control. Accordingly, thorough LDA-measurements were carried out in addition to regular surface pressure and wake measurements. Since probing the flow field by LDA is very time consuming, only a survey at the most suitable condition with respect to shock location and strength and control performance was conducted.

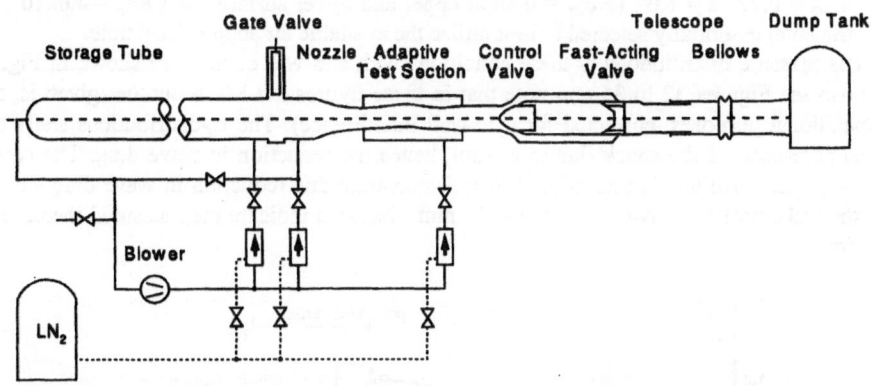

Tube	Diameter	0.8 m	Max. total pressure	10 bar
	Length	130 m	Temp. range	100 - 300 K
	Charge pressure	12.5 bar	Mach range	0.25 - 0.95
Test section	Cross section	0.40x0.35 m^2	Max. Reynolds No.	70x10^6
	Length	2.0 m	Run time	0.6 to 1.0 sec
	Model chord length	0.18 m		

Figure 42: Schematic and characteristics of the Cryogenic Ludwieg-tube (KRG)

Figure 43: ONERA T2 adaptive-wall test section with wake rake

55

To find this condition, initial pressure distribution measurements with and w/o control were carried out at various Mach numbers and angles of attack and the following test case was identified:

- $M_\infty = 0.77$, $\alpha = 1.0°$, $(x/c)_{tr} = 0.48$ at upper and lower surfaces and $Re_c = 4.6 \times 10^6$, the latter essentially selected to best utilize the available air supply / test time.

The pressure distributions at these conditions with and w/o control are shown in Figure 44a (also see Figures 32 to 35 and note that in these figures the Mach number given is, for computational purposes, corrected for side-wall interference). The c_p-distributions show the typical spreading of the shock due to control, hence the reduction in wave drag. The corresponding wake profiles, Figure 44b, clearly demonstrate this reduction in wave drag - compare the wake profiles between $z = 5$ and 10 mm - but also indicate the measured increase in total drag.

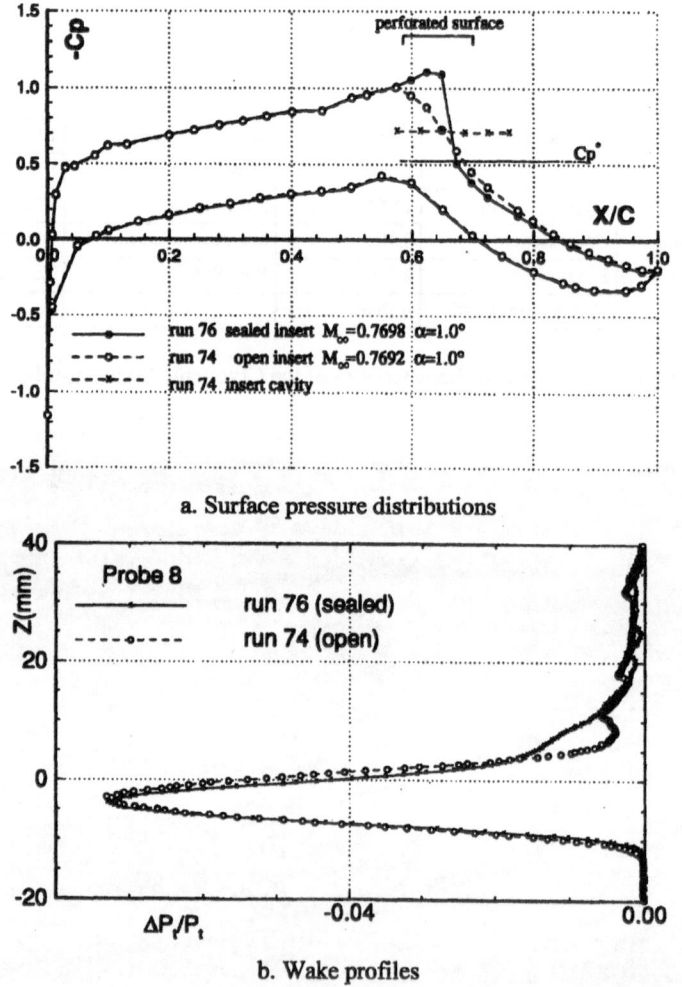

a. Surface pressure distributions

b. Wake profiles

Figure 44: Surface pressure distributions and wake profiles for the airfoil DA LVA-1A $M_\infty = 0.77$, $\alpha = 1.0°$, $(x/c)_{tr} = 0.48$, $Re_c = 4.6 \times 10^6$ (T2-tests)

Typical measured boundary layer profiles corresponding to the above conditions were already shown and discussed in Chapter 3 (see Figure 35a). To repeat here the essential result: the data show clearly that the addition of control leads, as was also demonstrated by the perforated plate measurements of Figure 13, to a considerable thickening of the boundary layer compared to the datum, sealed, airfoil case. This, in turn, causes in conjunction with the sustained rear adverse pressure gradients the large increase in viscous drag which overcompensates the significant reduction in wave drag resulting in the increase in total drag.

Remains to be added that the determination of the fluctuating quantities u', w' and u'w', carried out simultaneously, has shown that also the turbulent stresses increase downstream of the interaction region due to control, Figure 45.

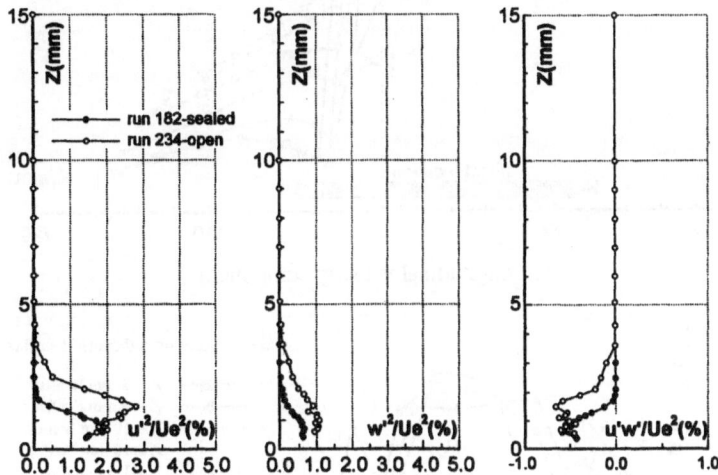

Figure 45: Turbulent stress profiles measured at x/c = 0.80 with and w/o control
$M_\infty = 0.77$, $\alpha = 1.0^\circ$, $(x/c)_{tr} = 0.48$, $Re_c = 4.6 \times 10^6$ (T2-tests)

The second set of LDA-measurements mainly concerned the outer (inviscid) flow field in the shock boundary layer interaction region with and w/o control. Surveys were conducted along lines parallel to the airfoil upper surface ranging from 3 mm to 50 mm above the model.

Considering primarily the near-field up to z = 10 mm, Figure 46, one observes that w/o control the longitudinal velocity component increases sharply up to the shock with decreasing velocity level as one moves away from the surface. With shock control, the velocity decreases gradually due to the compression waves generated by the outflow from the cavity. This is followed by a steeper pressure drop due to the weak remaining shock and by an expansion and/or slow pressure recovery partly caused by the inflow into the cavity in the downstream area. Note that at z = 50 mm, the direct influence of control can still be felt.

The outflow out of the cavity in the forward area is also indicated by the increased absolute normal velocity component, i.e., the upturn of the flow away from the airfoil surface, Figure 46b. Without control the flow follows, of course, the surface curvature with an intermediate upturn, however, due to the interaction of the boundary layer with the shock wave. With control, the flow turns in the downstream region of the cavity back to the surface, very similar to the no-control case (also see Chapter 2, Figure 12).

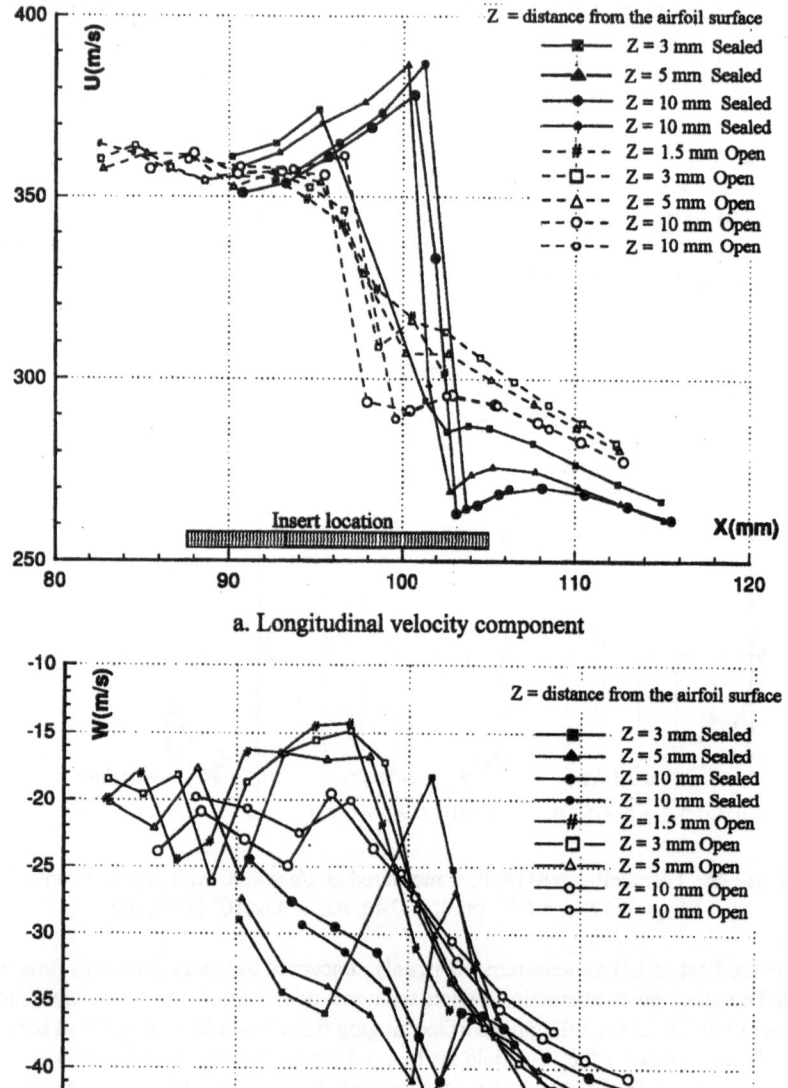

Figure 46: Off-surface velocity components in the shock control region
$M_\infty = 0.77$, $\alpha = 1.0^\circ$, $(x/c)_{tr} = 0.48$, $Re_c = 4.6 \times 10^6$ (T2-tests)

Assessing the results of the boundary layer and flow field measurements one may state that the data have considerably contributed to the understanding of the physics of control and its effect on the overall flow development; most important, the data constitute an excellent data base for computer code validation.

4.2.4 Discussion of KRG results

Experiments were carried out with the DA LVA-1A airfoil in the Cryogenic Ludwieg-tube Göttingen (KRG) mainly at a Mach number of $M_\infty = 0.77$ and angles of attack between $\alpha = -1°$ and 2.5° at Reynolds numbers ranging from Re = 4.6×10^6 to 12×10^6; tests were performed with transition fixed at 48% chord on both surfaces and with free transition, overlapping the test conditions of the T2-experiments. Main objective was the investigation of the control effectiveness as dependent on the Reynolds number or, more general, on the condition of the boundary layer upstream of the shock region.

Of concern was initially a comparison of results from the T2- and KRG-measurements at the same freestream conditions with and w/o control to establish a common basis. Figure 47 indicates that there is a fairly good agreement in the pressure distributions except for a slightly more forward shock position in the KRG-tests which also seems to result in the higher cavity pressure in the case of control. Otherwise, both experiments show the spread of the shock, hence a reduction in wave drag, due to control, but also at the present conditions an increase in total drag as will be shown below.

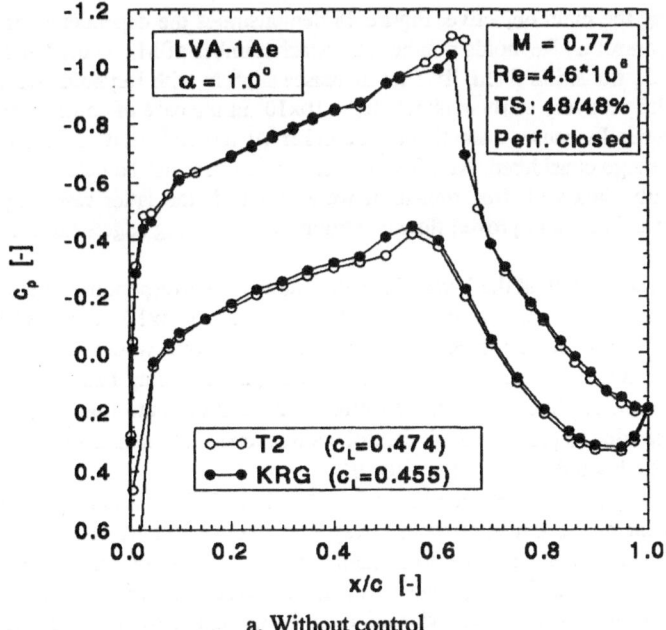

a. Without control

Figure 47: Comparison of T2 and KRG pressure distributions with and w/o control
$M_\infty = 0.77$, $\alpha = 1.0°$, $(x/c)_{tr} = 0.48$, $Re_c = 4.6 \times 10^6$

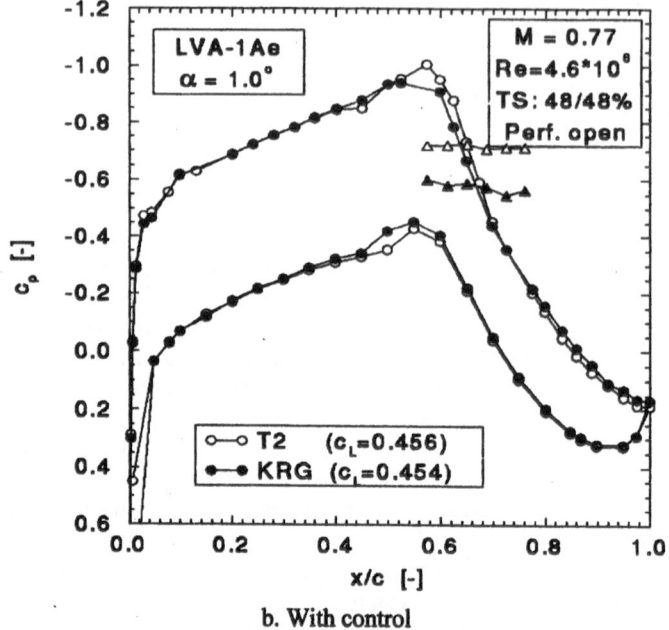

b. With control

Figure 47: Concluded

Considering the main objective, Figure 48 demonstrates the dependence of lift and drag with and w/o control on Reynolds number at a Mach number of $M_\infty = 0.77$ and $\alpha = 1.0°$ and $1.5°$: At $\alpha = 1.0°$, transition fixed, lift first increases slightly with Reynolds number, then decreases, but a lift recovery starts at about $Re_c = 10 \times 10^6$ in the case of control. At these conditions control generally causes a small decrease in lift. At $\alpha = 1.5°$ lift decreases with Reynolds number in the range considered. An effect of control can here not be determined since in the control case only data with free transition were obtained, the latter resulting in a thinner boundary layer and hence improved flow conditions at the trailing edge which tend to increase circulation / lift.

Drag increases slowly at the lower Reynolds numbers, corresponding to the lift decrease, but a strong increase in drag can be observed between $Re_c = 8 \times 10^6$ and 10×10^6. Thereafter drag is reduced with increasing Reynolds number which one would expect at conditions where Reynolds number is the only influence parameter. The drag behavior is similar with and w/o control and for fixed and free transition. The effect of control on drag is such that total drag increases due to passive shock control, as was repeatedly demonstrated; however, this increase is noticeably reduced at $Re_c \geq 10 \times 10^6$.

The rapid drag increase between $Re_c = 8 \times 10^6$ and 10×10^6 is essentially caused by an upstream movement of the transition point away from the 48%-chord location to the leading edge. This causes the boundary layer to thicken considerably resulting in a deterioration of the flow conditions at the trailing edge and in a loss of circulation, hence a stronger decrease in lift, and an increase in friction and viscous drag. In the upper Reynolds number range, the flow is fully turbulent and the above trends are reversed due to the now favorable development of the boundary layer. The movement of the transition point has been confirmed by heat transfer measurements.

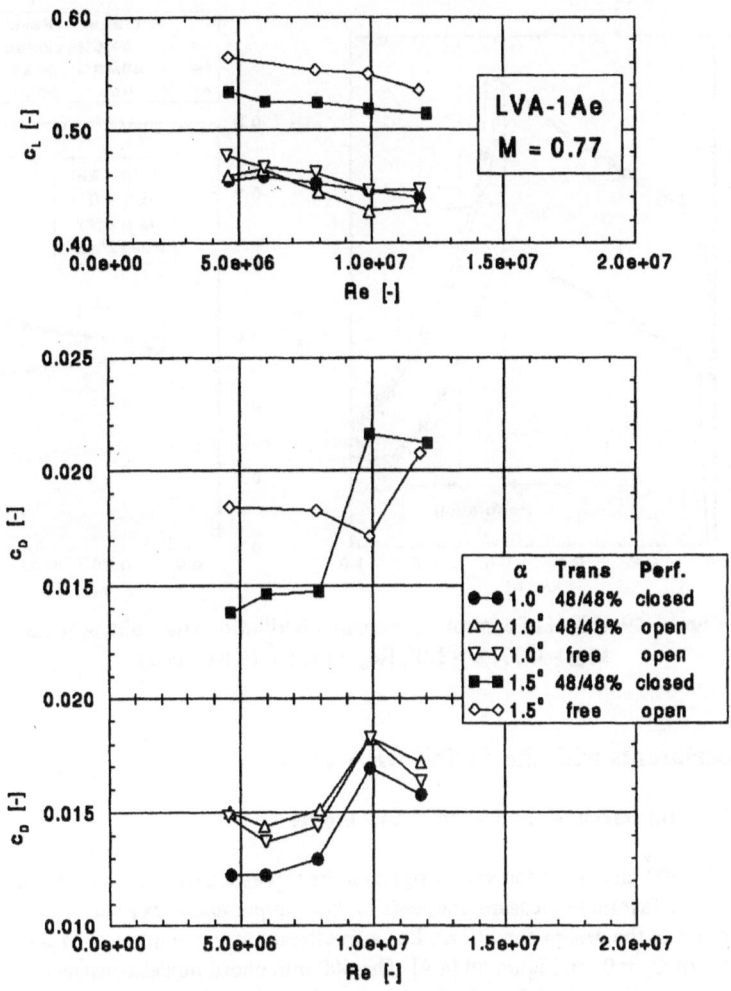

Figure 48: Reynolds number dependence of lift and drag with and w/o control
(KRG-tests)

It is indicated in Figure 48 that the total drag increase due to control is less pronounced at higher Reynolds numbers. The reason for this behavior may be the thicker fully developed turbulent boundary layer approaching the control region. This boundary layer is possibly less susceptible to the additional disturbances emanating from the perforation and, moreover, causes a more pronounced spread of the shock due to control, thereby not only reducing wave drag but also having a less severe influence on viscous drag. The latter is also reflected in the wake development, showing, for instance, much smaller differences in the pressure loss between open and closed perforation, Figure 49.

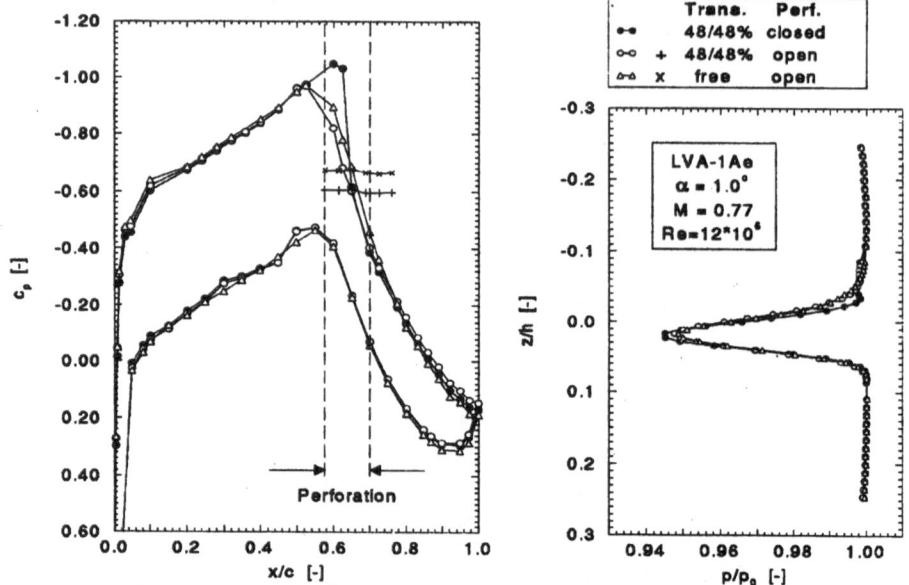

Figure 49: Effect of control on pressure distribution and wake profiles
$M_\infty = 0.77$, $\alpha = 1.0°$, $Re_c = 12\times10^6$ (KRG-tests)

4.3 Experiments with the Airfoil DRA 2303

4.3.1 Airfoil characteristics and wind tunnel model

The DRA 2303 airfoil section was designed to be representative of a laminar-flow section with long runs of favorable pressure gradients on both upper and lower surfaces extending to 50% chord close to the design conditions, i.e., a freestream Mach number of $M_\infty = 0.68$ and a lift coefficient of $C_L = 0.50$, Figure 50 [4.4]. The 600 mm chord model consists of a main spar with detachable leading edge (0 to 0.17 c) and trailing edge (0.7 to 1.0 c) sections; it was manufactured of high-tensile steel. On the upper surface of the main spar there is a removable panel between x/c = 0.39 and 0.69 allowing control systems to be inserted; an insert was also manufactured to form the original profile. The airfoil, including inserts, was equipped with static pressure holes at three spanwise pressure plotting stations, each having 35 orifices on the upper and 22 orifices on the lower surface; five pressure measuring stations were located inside the cavity. The model was also equipped with dynamic pressure transducers to study the control effect on the buffet process.

The passive control insert had a perforated surface between x/c = 0.5 and 0.6 with a porosity of 8% based on the local area; the perforations were formed by laser-drilling with a nominal diameter of 0.076 mm. A calibration of the control surface by the method of Poll et al. [3.10] allowed a value of the so called K-factor (see Chapter 3.3) to be derived which for the perforation employed here was found to be 1.0, which means the hole diameter turned out to be as specified.

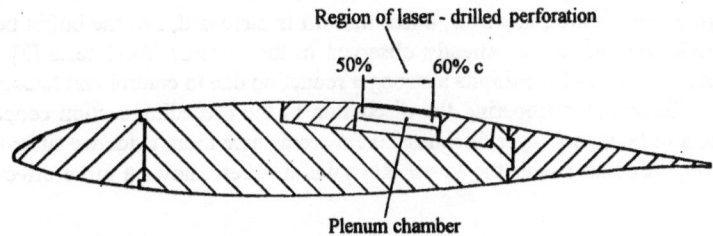

Figure 50: DRA 2303 airfoil model for the study of passive shock control

4.3.2 Wind tunnel characteristics

The DRA-2303 airfoil was tested in the 8Ft x 8Ft Subsonic - Supersonic wind tunnel at DERA Bedford. The tunnel is a variable pressure, closed circuit, continuous running facility with a solid wall working section, therefore appropriate corrections have been applied to the data to account for tunnel-wall interference and blockage, Figure 51 [4.4]. In the figure please note the wake rake used to determine drag.

Figure 51: DRA 2303 airfoil model in the working section of the DRA 8Ft x 8Ft tunnel

4.3.3 Discussion of steady results

The test conditions for the DRA 2303 airfoil investigations were:
- M_∞ = 0.67, 0.68, 0.69 and 0.70 at nominal Reynolds numbers of Re_c = $6x10^6$ and $19x10^6$; transition was fixed at 5% chord on upper and lower surfaces. The angle of attack was varied from about α = -2° to α = 4° dependent on Mach number.

In order to study the effect of control for this airfoil, we consider as example the results at M_∞ = 0.68 and the two Reynolds numbers investigated, starting out with the effect on lift, Figure52. At Re_c = $6x10^6$, the normal-force (lift) coefficient decreases slightly within the

„linear" range due to control, however, maximum lift is increased, i.e., the buffet boundary is raised by shock control, as was already observed in the original VA-2 tests [3]. At Re_c = $19x10^6$, Figure 52b, linear lift exhibits a stronger reduction due to control and maximum lift is basically not affected. Remembering the discussion in the preceding section concerning the favorable effect of boundary layer thickness, there seems here more evidence since the thicker boundary layer, associated with the lower Reynolds number, shows a more effective shock control.

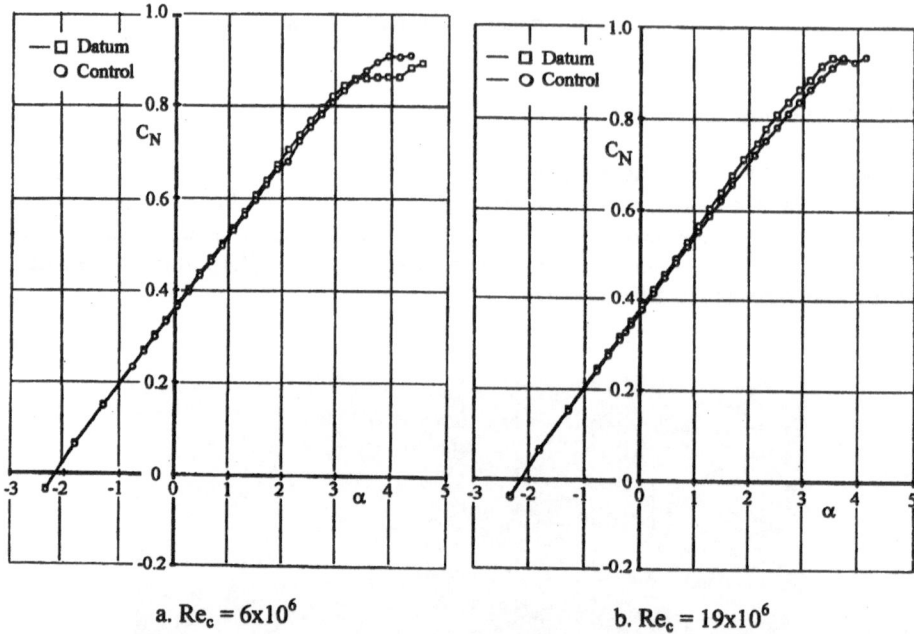

a. Re_c = $6x10^6$ b. Re_c = $19x10^6$

Figure 52: Variation of normal-force coefficient with angle of incidence at M_∞ = 0.68 (DERA-tests)

Considering the corresponding variation of the drag coefficient with normal force (lift), Figure 53, it is obvious that total drag is increased due to control over most of the angle of incidence range investigated. Only at higher normal-force (lift) coefficients, C_N > 0.80, this trend reverses at Re_c = $6x10^6$ and control causes a reduction in total drag. At Re_c = $19x10^6$ such a reversal does not occur which, again, may be caused by the thinner boundary layer present at this Reynolds number.

Representative pressure distributions at transonic conditions, here C_N = 0.70, are presented in Figure 54: As at all conditions, the pressure distributions on the lower surface and those on the upper surface ahead of the shock, when present, are identical with and w/o control. The effect of control is, as has been discussed before, to weaken the shock wave replacing here the single straight shock of the basic airfoil with a multi-shock system which is well known to produce less entropy and thus lower wave drag. The observed increase in total drag can, therefore, only arise from an increase in viscous drag due to the aerodynamic roughness of the porous surface and / or the flow through the surface causing, as was shown in previous sections, an excess thickening of the boundary layer.

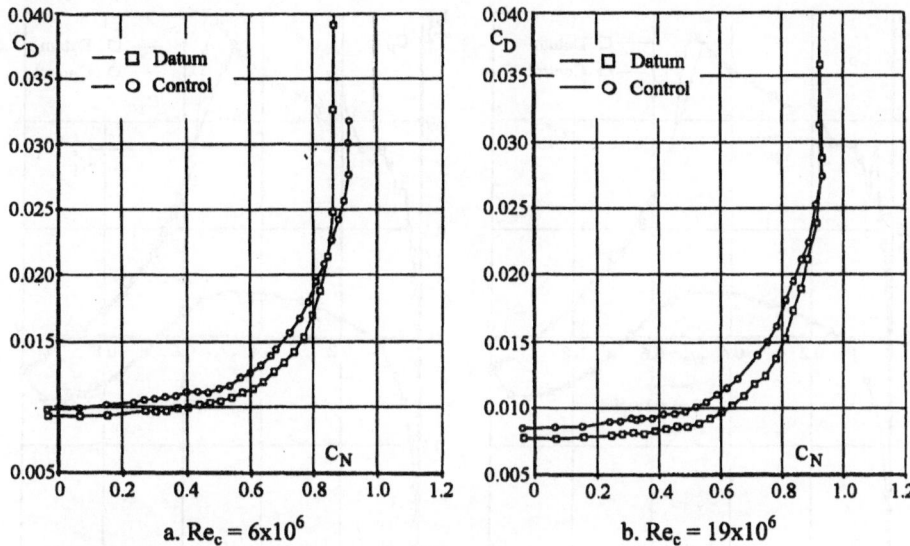

a. $Re_c = 6 \times 10^6$ b. $Re_c = 19 \times 10^6$

Figure 53: Variation of drag coefficient with normal-force coefficient at $M_\infty = 0.68$
(DERA-tests)

It is interesting to note that even without a shock present, say at $M_\infty = 0.68$ and $C_N = 0.10$, there is an increase in drag due to control. There is still a pressure gradient across the control region, Figure 55, $x/c = 0.50$ to 0.60, with the consequence that there still is a secondary flow via the plenum, and one may conclude, since the drag rise is strictly from an increase in viscous drag, that at least part of the increase is due to this secondary flow. Figure 56b, where the local drag distribution in the wake is plotted for the low-lift case, confirms this observation: The area under the curve - representing the overall drag - is greater for the case with control and such that the main contributor to the increase in drag can be identified to be the upper surface.

4.3.4 Control effect on buffet

Parallel with the steady measurements described above, extensive dynamic pressure measurements were made on the DRA 2303 airfoil and analyzed [4.5]. Figure 56 shows for the test conditions $M_\infty = 0.702$, $\alpha = 2.8°$ the effect of the addition of control: the upper half of the figure (a) indicates that without control there are strong unsteady signals from the shock wave to the trailing edge with indications of an unsteady shock movement between $x/c = 0.49$ and 0.54 at a dominant frequency of 34Hz. In the lower half of the figure (b) one observes that control strongly dampens the shock oscillations - with a suggestion that the shock is not moving at all - and that the unsteadiness measured at all transducer locations is essentially zero; this is also in complete agreement with the numerical results presented in Section 3. It suggests, in accordance with the lift behavior shown in Figure 52a, that control delays buffet onset shifting the buffet boundary to higher lift coefficients and / or Mach numbers in agreement with the results for the turbulent airfoil VA-2.

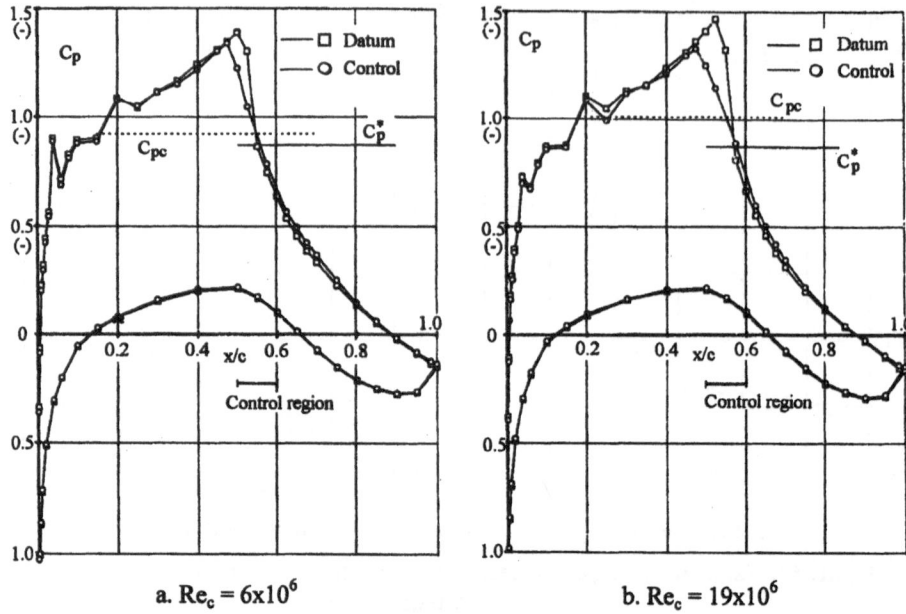

a. $Re_c = 6 \times 10^6$ b. $Re_c = 19 \times 10^6$

Figure 54: Comparison of surface pressure distributions at $M_\infty = 0.68$, $C_N = 0.70$ (DRA-tests)

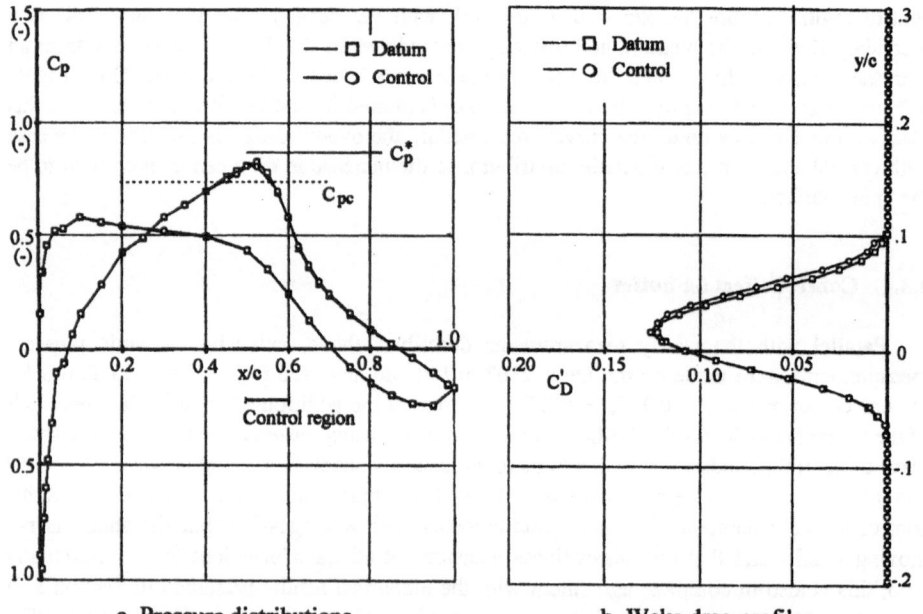

a. Pressure distributions   b. Wake drag profiles

Figure: 55: Comparison of pressure distributions and local drag profiles $M_\infty = 0.68$, $C_N = 0.10$, $Re_c = 6 \times 10^6$ (DRA-tests)

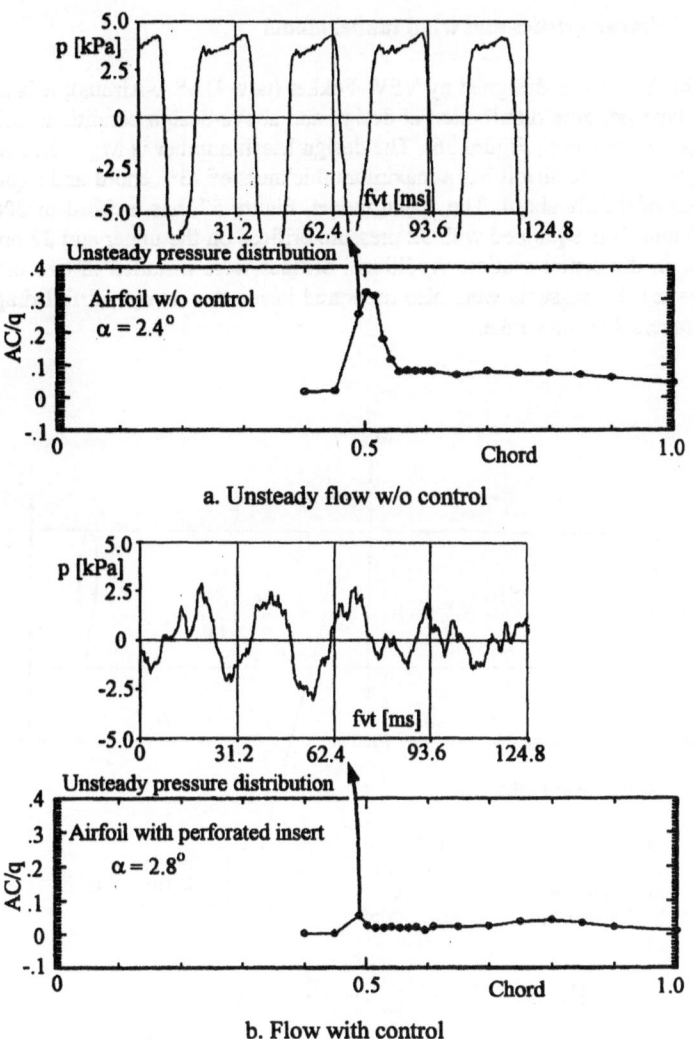

a. Unsteady flow w/o control

b. Flow with control

Figure 56: Effect of control on flow unsteadiness; airfoil DRA 2303, $M_\infty = 0.702$

4.4 Experiments with the Airfoil VA-2

Early measurements with the turbulent airfoil VA-2 had shown, as already mentioned, that through passive shock control by ventilation the drag-rise and buffet boundaries could be improved and that not only wave drag but also total drag could be reduced by passive control [3]. Since the latter is contrary to the results for the laminar-type airfoils described above, the earlier measurements with the airfoil VA-2 were repeated to either confirm or reject the earlier results [4.6].

4.4.1 Airfoil characteristics and wind tunnel model

The airfoil VA-2 was designed by VFW-Fokker (now DASA-Airbus); it is characterized by a plateau-type pressure distribution at design and above-design conditions and by a moderate rear loading (see, e.g., Figure 36). The design Mach number is $M_\infty = 0.73$ at a lift coefficient of $C_L = 0.52$. The airfoil has a maximum thickness of 13% chord and a (blunt) trailing edge thickness of 0.52% chord. The actual model, Figure 57, has a chord of 200 mm and a span of 1000 mm; it is equipped with 32 pressure orifices on the upper and 22 orifices on the lower surface in the center section. Additional orifices were installed in two off-center sections at $y/c = \pm 1.0$. Pressures were also measured inside the cavity. Airfoil drag was determined by a traversable wake rake.

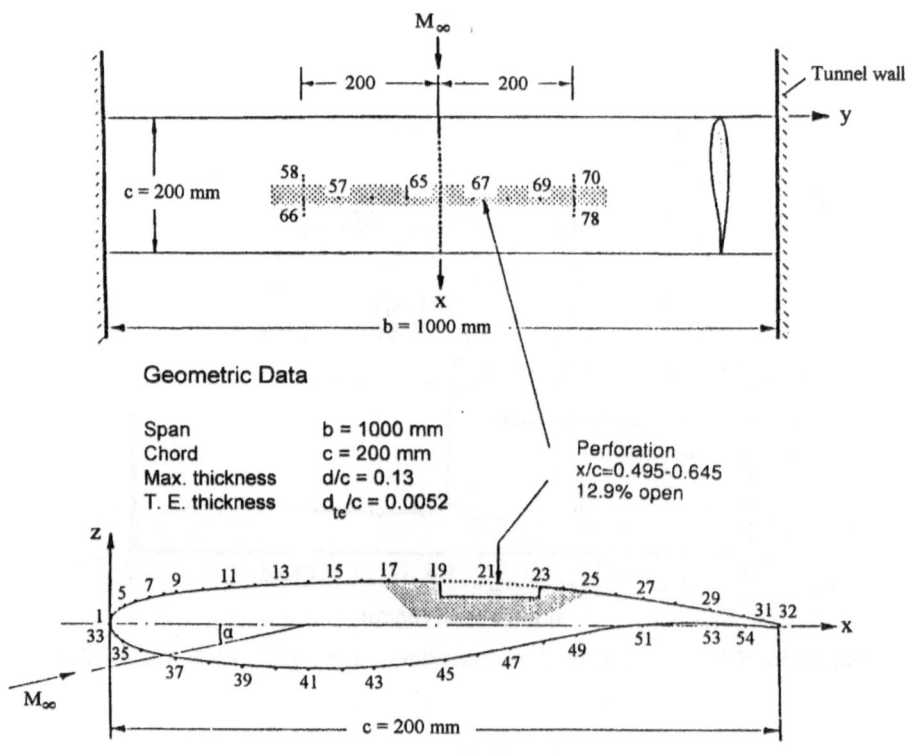

Figure 57: Airfoil model VA-2 with perforation

The perforated insert, Figure 57, had the following measured characteristics:
- Porosity 12.9%, hole diameter 0.284 mm, distance between holes 0.753 mm, plate thickness 1.3 mm, extent of perforation from $x/c = 0.495$ to 0.645.

The holes were electron-beam drilled. The cavity was connected to a vacuum system so that passive as well as active control could be investigated.

4.4.2 Wind tunnel characteristics

The tests were carried out in the 1 x 1 Meter Transonic Wind Tunnel Göttingen of DLR (TWG). The TWG is a closed-circuit continuous tunnel with a cross-section area of 1 x 1 m^2. After a recent modification, there are now three independent test sections available: a perforated test section, 6% open with 60°-slanted holes, for the Mach number range 0.40 to 1.3, an adaptive-wall transonic test section and a supersonic test section for Mach numbers of 1.4 to 2.2. The total pressure can be adjusted between 0.6 and 1.6 bar. The present investigation was performed in the perforated test section of the TWG similar to the original measurements.

4.4.3 Comparison of present and earlier results

In order to judge whether wind tunnel or model inherent changes may cause differences in the flow development about the airfoil VA-2, comparisons are first carried out between the present (TWG 95) and the earlier (TWG 84) results for the datum airfoil and the airfoil with passive shock control at $M_\infty = 0.76$. Corresponding data for lift and drag and representative pressure distributions are given in Figures 58 and 59.

The agreement in lift coefficient, Figure 58a, is generally good with some minor differences only occurring in the maximum lift range where the earlier measurements show slightly higher lift values with a corresponding later break in the lift-curve slope. Drag coefficients are generally somewhat lower for the TWG 84-measurements which is consistent with the slightly better pressure recovery in case of the latter, Figure 58b. The small differences in drag and trailing edge pressure recovery are possibly a result of differences in model roughness including small deviations in the tripping device height and roughness distribution.

A comparison of present and earlier results with passive shock control at $M_\infty = 0.76$, Figure 59, shows that, as in the no-control case, differences are again the later break in the lift-curve slope and the lower total drag in the case of the TWG 84-data. Examining corresponding pressure distributions, one observes that, e.g., at $\alpha = 4.5°$, Figure 59b, prior to the lift-curve break a good agreement in the pressure distributions exists except, again, for the reduced pressure recovery in the present data consistent with the higher drag.

Based on a comparison of all results of the present and the earlier VA-2 measurements, it seems that especially differences in the initial boundary layer development are the cause for the disparities between the two data sets rather than changes in the wind tunnel environment.

4.4.4 Effect of passive and active control

The following configurations were investigated: airfoil model with perforation covered as reference (datum), model with perforation open but cavity disconnected from the vacuum supply (passive control) and cavity connected to the vacuum system utilizing a suction coefficient of $c_q = 5 \times 10^{-4}$ for active control.

Tests were carried out at the following freestream conditions:
- Mach numbers $M_\infty = 0.72, 0.735, 0.755$ and 0.775; angle of attack range $\alpha = 0°$ to $\alpha = 6°$; Reynolds number $Re_c = 2.5 \times 10^6$ with transition fixed at 30% chord on the upper and 25% chord on the lower surface; the rear transition location was chosen to simulate higher (flight) Reynolds number conditions, especially in the shock region.

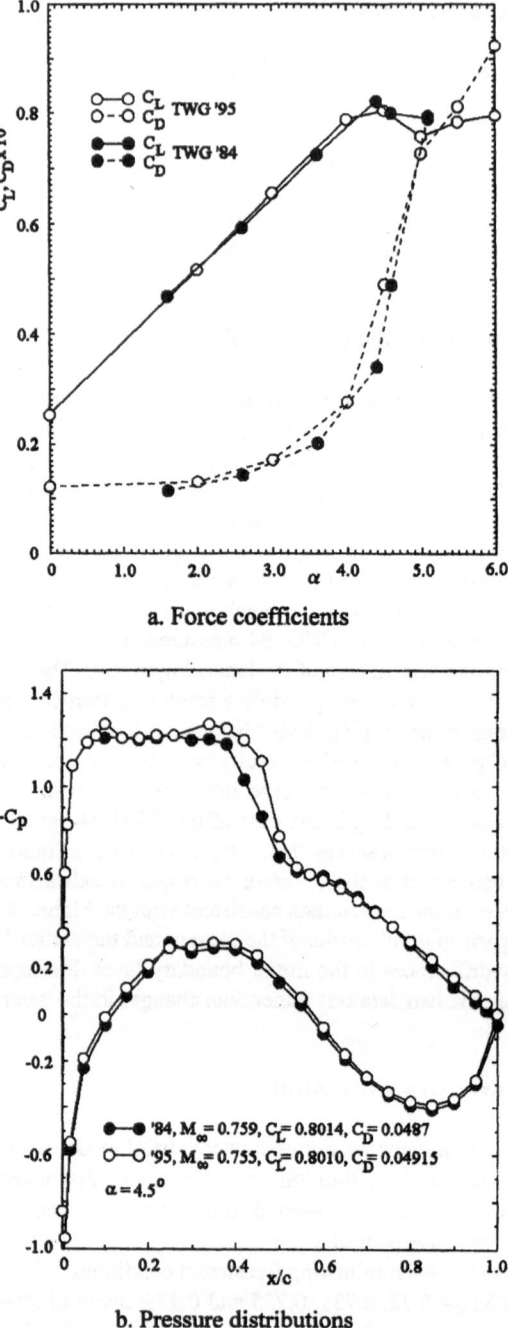

a. Force coefficients

b. Pressure distributions

Figure 58: Airfoil VA-2 comparison between '84 / '95 measurements w/o control
$M_\infty = 0.76$, $Re_c = 2.5 \times 10^6$, $(x/c)_{tr} = 0.3$ upper / 0.2 lower surface

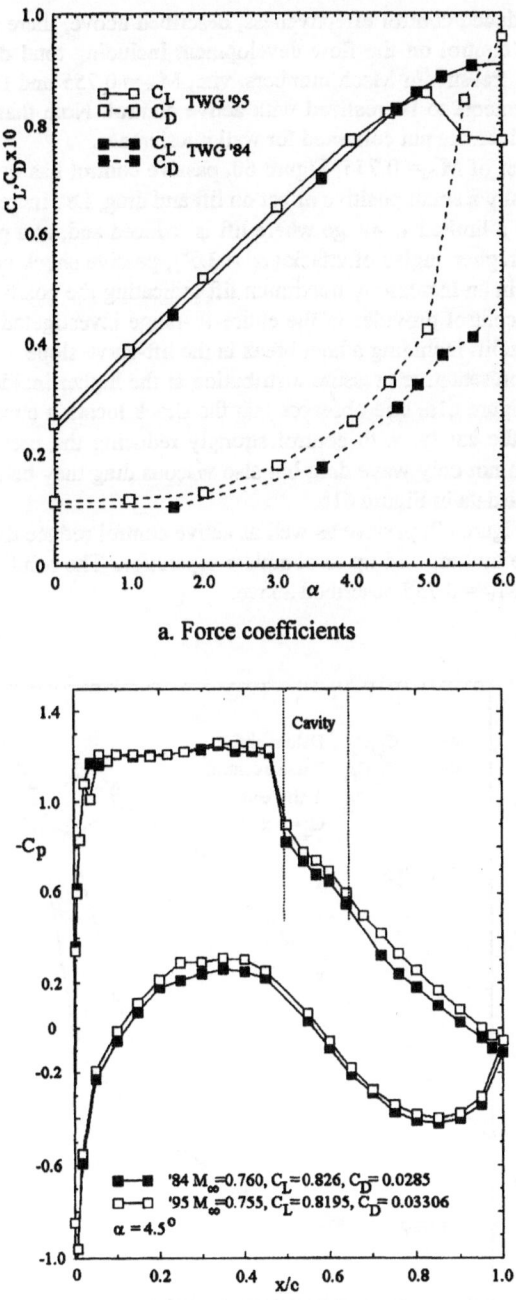

a. Force coefficients

b. Pressure distributions

Figure 59: Airfoil VA-2 comparison between '84 / '95 measurements with control
$M_\infty = 0.76$, $Re_c = 2.5 \times 10^6$, $(x/c)_{tr} = 0.3$ upper / 0.2 lower surface

In spite of the reduced control effectiveness, described above, there still remains a positive effect of passive control on the flow development including total drag. This is demonstrated below for two freestream Mach numbers, viz., $M_\infty = 0.755$ and 0.775. Shown are, in addition, the improvements to be realized with active control. Note that the Mach numbers and angles of attack given are not corrected for wall interference.

At a Mach number of $M_\infty = 0.755$, Figure 60, passive control has in the lower incidence range ($\alpha < 3°$) generally a small positive effect on lift and drag, i.e., increasing lift and lowering drag, followed by a limited α-range where lift is reduced and, at a given lift coefficient, drag is increased. At higher angles of attack ($\alpha > 3.5°$), passive shock control again reduces total drag and results in an increase in maximum lift indicating the positive effect on the buffet boundary. Active control provides in the entire α-range investigated a reduction of total drag and an increase in lift including a later break in the lift-curve slope.

Considering a representative pressure distribution at the higher incidences, here at an incidence of $\alpha = 4°$, Figure 61a, one observes that the shock location coincides here perfectly with the location of the cavity, with control strongly reducing the pressure gradient in the shock region such that not only wave drag but also viscous drag may be reduced; the latter is confirmed by the wake data in Figure 61b.

At $M_\infty = 0.775$, Figure 62, passive as well as active control reduce the total drag over the entire incidence range investigated up to complete separation. The results are otherwise very similar to the ones at $M_\infty = 0.755$ described above.

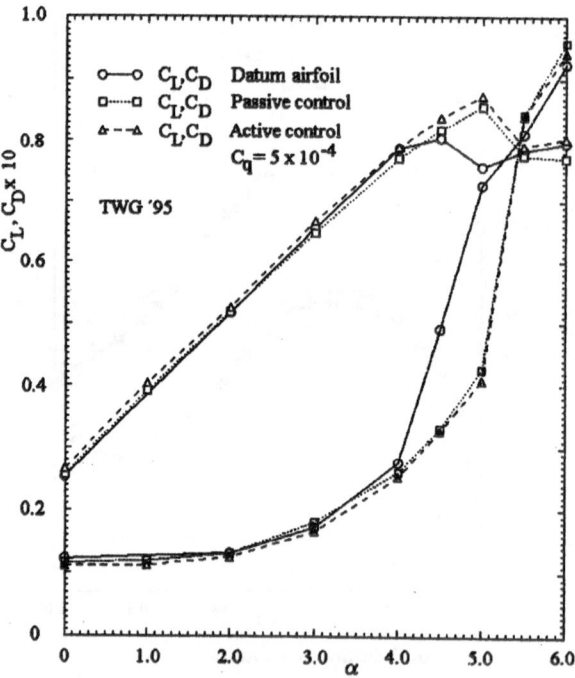

Figure 60: Effect of passive and active control on force coefficients, airfoil VA-2 $M_\infty = 0.755$, $Re_c = 2.5 \times 10^6$, $(x/c)_{tr} = 0.3$ upper / 0.2 lower surface

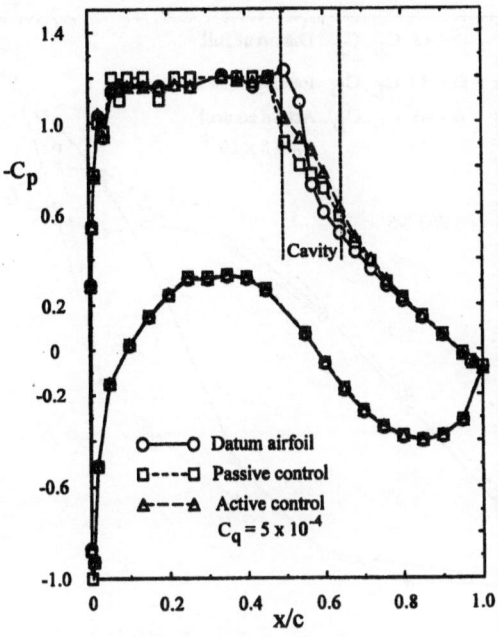

a. Surface pressure distributions

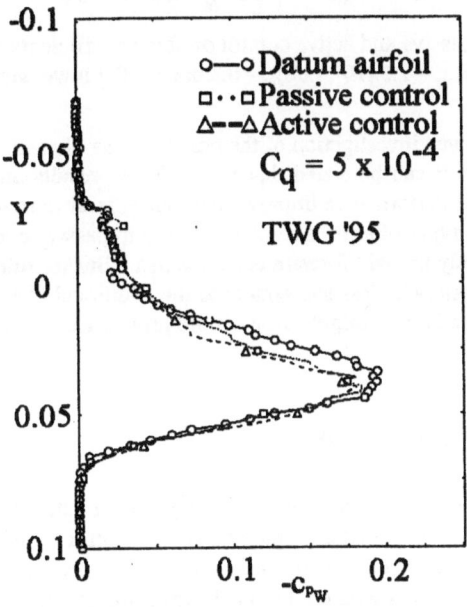

b. Wake profiles with and w/o control

Figure 61: Effect of passive and active control on pressure distributions, airfoil VA-2
$M_\infty = 0.755$, $\alpha = 4°$, $Re_c = 2.5 \times 10^6$, $(x/c)_{tr} = 0.3$ upper / 0.2 lower surface

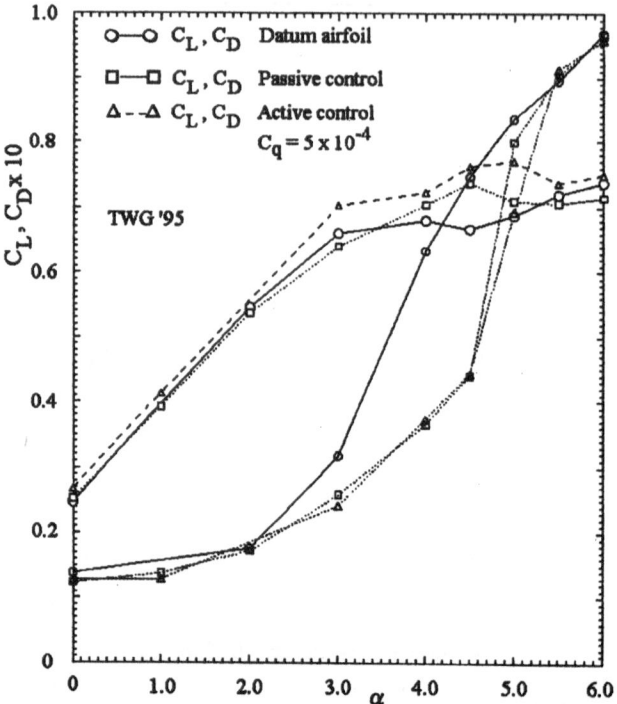

Figure 62: Effect of passive and active control on force coefficients, airfoil VA-2
$M_\infty = 0.775$, $Re_c = 2.5 \times 10^6$, $(x/c)_{tr} = 0.3$ upper / 0.2 lower surface

As a résumé of the renewed investigation of the effectiveness of passive and active shock control for the VA-2 airfoil, it can be stated that the TWG 84 results are being confirmed, although not in magnitude of performance improvement but at least in trend. This then raises the question why do the two types of airfoils react differently to passive control with regard to drag? One explanation, already brought forward above, is that a thicker fully developed turbulent boundary layer may possibly be less susceptible to the additional disturbances emanating from the perforation, and, in addition, might cause a more pronounced spread of the shock due to (passive) control.

4.5 Conclusion and Future Work

The experimental results for airfoils designed for significant extents of laminar flow on the upper surface show that the application of passive shock control is successful in reducing shock strength. The effect of the displacement surface, formed by the injection and suction, is to induce compression waves which weaken the shock wave, thereby reducing wave drag and the likelihood of shock induced separation. However, these investigations have also shown that the drop in wave drag is accompanied by a significant increase in viscous drag. The surface roughness and the recirculating flow in the control region appreciably increase boundary layer momentum thickness causing a strong increase in viscous drag due to the amplification effect of the rear adverse pressure gradients which leads to a considerably thicker boundary layer at the trailing edge than would have been the case in the absence of control. Therefore, although, as shown in Section 2, locally there is a small net reduction in drag due to control, for the full airfoil flow there is an increase in overall drag due to the increase in viscous drag overcompensating the reduction in wave drag.

In the case of an airfoil typical of current design with turbulent flow on the upper surface, a net drag reduction is observed at certain conditions. However, in this case it is not quite clear as to what the important design parameters which give rise to this phenomenon are. The only significant difference lies in the state of the boundary layer approaching the shock wave; in the case of the laminar flow airfoil the boundary layer is growing in a highly favorable pressure gradient and hence the shock wave is interacting with a very robust, thin boundary layer. The turbulent airfoil, on the other hand, has a boundary layer which has grown in an unfavorable or at best equilibrium state. In this case the shock wave interacts with a thick boundary layer approaching separation. It is not clear whether this holds the answer to the successful application of passive control, although there is some evidence that even for laminar airfoils a thicker boundary layer behaves more favorably, at least at the conditions encountered in the present investigation. The final answer can only come from further detailed investigations of the interaction of shock waves with boundary layers in different states of their development.

These investigations might, however, be rather academic since the sensitivity of passive shock control, e.g., to changes in the flow and boundary layer conditions, makes this control device rather impracticable for wing flow control at conditions where drag reduction is of prime interest since even for a turbulent airfoil, the benefits are in that regard marginal. Active control by ventilation, on the other hand, has been shown to be effective in the entire Mach number and incidence range considered. **For this reason, future work should concentrate on this and other means of active control identified during the course of the present investigation.**

As very positive aspect, the airfoil investigations have shown that off-design (buffet) conditions can be noticeably improved by passive shock control.

In summarizing, the wind tunnel tests on various airfoils in several major European wind tunnels have generated the following conclusions:

- Passive shock control via ventilation is very effective at reducing wave drag by the displacement effect of the ventilation.
- For the laminar type airfoils the reduction in wave drag is overcompensated by a large increase in viscous drag due to the increase in boundary layer momentum thickness caused by the control device.
- For the turbulent type airfoil at certain conditions a reduction in overall drag was observed, but the exact reason for this compared to the laminar flow type airfoil is currently the subject of speculation.
- Passive control has been shown to be effective at postponing the onset of buffet; however, further investigation is required to quantify the improvement in the buffet boundary.
- The experimental investigation with the various airfoil models establishes a valuable data base for the validation and improvement of the design capabilities of transonic airfoil codes.

5 Assessment of Shock Control - A Summary

A very thorough investigation of passive shock boundary layer interaction control has been conducted within the EUROSHOCK program consisting of:
- basic experiments and physical modeling of shock control,
- the extension of numerical codes to account for control and a first assessment of control and
- the performance of airfoil experiments to obtain data for the validation of computer codes and - in conjunction with CFD - for the determination of the merits of passive shock control.

The basic investigations, consisting of detailed experiments complemented by a theoretical treatment of the data, were performed by ONERA, the University of Karlsruhe and Cambridge University. The work has allowed some basic phenomena involved in the control mechanism to be defined: it is clear now that the blowing taking place in the upstream part of the control region provokes a rapid thickening of the boundary layer with the resulting increase in displacement thickness being felt by the outer inviscid flow as a ramp with the formation of a λ-shock pattern - or in case of airfoil flow by the formation of an almost continuous (isentropic) compression - replacing the single normal shock present in the interaction w/o control. The result is a substantial reduction in wave drag. However, the combined blowing - suction (secondary flow) effect and the hole roughness cause an increase in viscous drag which, in determining total drag, will oppose the reduction in wave drag. For the perforations investigated it was found that:
- the effect of the geometry of the porous plates on the effectiveness of shock control seems to be rather weak, although there is an effect of geometry on momentum thickness,
- all perforations result in an increase in momentum thickness immediately downstream of the perforation independent of geometry, however,
- there was generally a **reduction in total drag** due to control of approximately 4% over the control region, and
- the same conclusions are valid for both two-dimensional and three-dimensional interactions.

Turning now to airfoil flow and considering first the computational results, with computations having been performed by CIRA, DLR, ONERA, NLR, the Universities of Naples and Athens and DASA-Airbus, the following was determined for the laminar-type airfoils: all codes predicted a considerable reduction in wave drag but also an increase in total drag due to control for all test cases investigated. The computed boundary layer data clearly showed an increase in momentum thickness over the control region - also observed in the basic experiments - here, however, amplified over the rear part of the airfoils due to the strong adverse pressure gradients. This process generates viscous drag which for the present laminar airfoils seems to overcompensate the reduction in wave drag resulting in the observed total drag increase.

The computation of the unsteady test cases showed a strong damping of shock oscillations, hence a positive shift in the buffet boundary.

The results of the airfoil tests carried out by ONERA and DLR - with the model designed and manufactured by DASA-Airbus - and DRA have revealed for the laminar type airfoils that the application of passive control is successful in reducing shock strength, hence wave drag, by the ventilation in the control region already observed in the basic experiments. This drop in wave drag was, however, found to be accompanied by a significant increase in viscous drag caused by surface roughness and recirculating flow in the control region and the amplification of initially small differences in the momentum thickness by the rear adverse pressure gradients. The latter was observed in the boundary layer measurements of ONERA confirming the computations. The increase in viscous drag overcompensates the decrease in wave drag.

For the turbulent airfoil VA-2 investigated experimentally and numerically, a net drag reduction was determined at certain conditions. It is not quite clear what the important design parameters are that cause this phenomenon. The only significant difference in flow parameters lies in the state of the boundary layer approaching the shock wave: in the case of the laminar flow airfoils the shock wave interacts with a very robust, thin boundary layer which, however, may be highly susceptible to the disturbances caused by the perforation. The turbulent airfoil, on the other hand, exhibits the interaction of the shock with a thick boundary layer which may result in a wider spread of the shock wave and a lesser susceptibility to disturbances and, therefore, the influence on viscous drag may be reduced.

Concerning drag it may be concluded that **passive control** can now be **ruled out** as an effective means of reducing drag of laminar wings; even for turbulent wings the sensitivity of the effectiveness of passive control to changes in the flow and boundary layer parameters makes this type of control device rather impracticable at conditions where drag reduction is of prime interest, especially since the benefits are rather marginal.

Nevertheless, there may be applications where **drag reduction** is not the main driver and passive control may still be of use. One immediate application could be in supersonic intakes where shock waves are used to compress and slow the flow at the engine face. One of the main drivers here is to avoid shock induced separation, at which passive control is very efficient by its effective reduction in shock strength, thus avoiding the large adverse pressure gradients which lead to separation.

If aircraft performance is limited by buffet considerations rather than drag, there is evidence that passive control could be of use. The results of the tests with the DRA 2303 airfoil and to some extent those with the VA-2 show that there is an improvement in the C_L for 'lift break' suggesting that the buffet boundary is improved by the addition of control. Also the dynamic pressure measurements on the DRA 2303, confirmed by the unsteady calculations described in Section 3, indicate that passive control is very effective at suppressing the shock unsteadiness and its dynamic effect on the flow up to the trailing edge, which manifests itself in an improvement in buffet characteristics, namely an improvement in the buffet boundary.

Finally, since passive control has been shown to be not a viable method for reducing drag, other control techniques should be considered. For example, a small amount of suction - or active control - in the plenum chamber can certainly increase the efficiency of the system, as demonstrated by the corresponding VA-2 measurements. Also, one can imagine to combine passive control - which is very effective in reducing wave drag and stabilizing the shock wave - with active control downstream of the interaction region as a means to minimize the thickening of the boundary layer and to accelerate its recovery after the interaction. Other methods include contour modifications in the shock region and hybrid control, combining suitable approaches. Note that active boundary layer and shock control is the topic of the EUROSHOCK II project within the 4th Framework Program of the EU.

6 Overall Conclusion and Future Work

Concerning wing flow, it was shown in the preceding sections that passive control is not a viable method for reducing wing drag, but a distinct improvement in the buffet boundary can be achieved. Also other configurations, such as engine inlets, where passive control may be of merit, have been identified The investigations have furthermore shed new light on the complex subject of shock wave boundary layer interaction and have highlighted new active control techniques which will constitute the next steps in the drag reduction efforts to be pursued in the follow-on project EUROSHOCK II.

However, also other conclusions have been drawn from the research work performed during the course of the present project, the latter more concerning the tools used to evaluate shock control rather than control itself, and it is deemed appropriate to also consider these here.

Basic experiments and physical modeling: a key problem in the modeling of passive control was found to be the definition of realistic boundary conditions to be prescribed at the wall where a non-zero vertical velocity exists in the control region. Here, the studies have permitted the establishment of more accurate transpiration laws taking into account the exact characteristics of the perforated plates. These laws should be used in future (active control) studies. It was, furthermore, established that turbulence generally increases due to control which can essentially be attributed to the stronger destabilization of the boundary layer by the blowing action. Numerical simulations of the interaction revealed some weaknesses of the basic turbulence models which often give poor results, even in the solid wall case, in the presence of shocks and eminent separation. In spite of these deficiencies, Navier-Stokes calculations with rather simple algebraic turbulence models can be valuable in the investigation of the physics of the flow, to test improved physical models and to execute comparative studies of different control arrangements to be considered in future work. In that regard attention must also be paid to the following:

- The transpiration velocity laws should be generalized for the case where a high speed flow is streaming on one side of the cavity since the presence of this flow has certainly an influence on the flow inside the perforation.
- Turbulence models to be used in interacting flows must be improved. More general models, incorporating history effects, should be considered and carefully validated by comparison with the data banks already contained within this project.

Extension of numerical codes: the numerical codes achieved generally good qualitative agreement with most of the experimental data; however, the absolute values of the total drag coefficients (lift coefficients were generally prescribed) showed in some instances considerable deviations between computed results and between calculations and the corresponding experimental data even w/o control. Considerable scatter also occurred in the ventilation velocities calculated by the various codes, which can only partly be attributed to the use of different control laws but is also dependent on the numerical method employed and on the implementation of the control law within a code. The scatter in the ventilation velocities seems to be the origin of strong deviations in the boundary layer properties within the control region which contributes to the differences observed in the local pressure distributions and in the subsequent flow development in the case of control. Concerning the control laws, it was found that of the laws employed (not including the new developments mentioned above) Poll's formula, derived from the calibration of laser-drilled plates, gives the most reliable results for the present control cases. Finally, the results of the unsteady calculations have shown considerably larger deviations than the steady flow cases.

The experience gathered throughout the present numerical exercise points out future work needed for the performance improvement of computational methods incorporating shock control. Work should include:

- the improvement of shock control laws and the implementation of the best law in the different codes,
- the development of a procedure for a detailed treatment of the boundary layer flow over the control region,
- the development of turbulence models that take into account the turbulence structure generated by the in- and outflow through the openings of the perforation,
- the development of physically correct control laws for unsteady flows and finally
- the extension of the codes to 3D-applications.

Airfoil flow experiments: Deficits occurred in the manufacturing of the perforation for the DA LVA-1A airfoil model to specifications; it is essential that current techniques be improved. Furthermore, the VA-2 test results could not be compared directly to the computations since wall interference affected the freestream conditions in the experiments. The latter has, however, no effect on the conclusions drawn from these tests. The measurements will, however, be repeated in the adaptive-wall test section of the Transonic Wind Tunnel Göttingen.

7 References

Chapter 1

[1.1]   Délery, J., "Shock Wave / Boundary Layer Interaction and its Control", Progress in Aerospace Science, Vol. 22, 1985.

[1.2]   Stanewsky, E. and Krogmann, P. "Transonic Drag Rise and Drag Reduction by Active / Passive Boundary Layer Control", AGARD Report No. 723, Lecture Series Aircraft Drag Prediction and Reduction, July 1995.

[1.3]   Thiede, P., Krogmann, P. and Stanewsky, E., "Active and passive shock/boundary layer interaction control on supercritical airfoils", AGARD-CP-365, Brussels, 1984.

[1.4]   Savu, G. and Trifu, O., "Porous airfoils in transonic flow", AIAA Journal, Vol. 22, 1984.

[1.5]   Bertelrud, A., "Passive shock modification in flight", EUROMECH Colloquium, Saltsjöbaden, Sweden 1984.

[1.6]   Chanetz, A. and Pot, T., "Expériences fondamentales sur le controle passif de l'interaction onde de choc-couche limite en transsonique", 24ème Colloque d'Aérodynamique Appliquée de l'AAAF, Poitiers, France, 1987.

[1.7]   Raghunathan, S. and Mabey, D.G., "Passive shock wave boundary layer control experiments on a circular arc model. AIAA Paper 86-0285, 1986.

[1.8]   Bohning, R. and Jungbluth, H., "Turbulent shock-boundary layer interaction with control: Theory and experiment", IUTAM-Symposium TRANSSONICUM III, Göttingen, 1988.

[1.9]   Chokani, N. and Squire, L.C., "Passive control of shock/boundary layer interactions: Numerical and experimental studies", IUTAM-Symposium TRANSSONI-CUM III, Göttingen, 1988.

[1.10] Bahi, L., Ross, J.M. and Nagamatsu, H.T., "Passive shock wave/boundary layer control for transonic airfoil drag reduction", AIAA Paper 83-0137, 1983.

[1.11] Chen, C.L., Chow, C.Y., Van Daisem, W.R. and Holst, T.L., "Computation of viscous transonic flow over porous airfoils", AIAA-Paper 87-0359, 1987.

[1.12] Olling, C.R., "Viscous-inviscid interaction in transonic separated flow over solid and porous airfoils and cascades", Ph.D. Thesis, University of Texas at Austin, 1985.

[1.13] Bidlack, T.J., "Passive shock wave-boundary layer control for the Bell FX 69-H-098 airfoil", Ph.D. Thesis, Rensselaer Polytechnic Institute, Troy, New York, 1987.

[1.14] "Aerodynamic Study on a Passively Ventilated Airfoil", Fuji Heavy Industries, Ltd., Aircraft Div., 1989, not published.

Chapter 2

[2.1] Bur, R. and Délery, J., "Study of passive control applied to a transonic shock wave / boundary layer interaction", ONERA final report EUROSHOCK TR AER 2-92-49/1.4 and ONERA RT 107/7078 AY, January 1996 (also see Chapter 9).

[2.2] Bohning, R. and Doerffer, P., "Wind tunnel tests of shock / boundary layer interaction with passive control, porous wall investigation and numerical simulation of the wind tunnel tests", University of Karlsruhe final Report EUROSHOCK TR AER 2-92-49/1.5, December 1995 (also see Chapter 20).

[2.3] Bur, R., "Physical modeling of passive control applied to a transonic shock wave / boundary layer interaction", ONERA intermediate report EUROSHOCK TR AER 2-42-49/1.3 and ONERA RT 102/7078 AY, May 1995.

[2.4] Yeung, A.F.K., Squire, L. C. and Faucher, X., "The passive control of the interaction between swept shocks and boundary layers", University of Cambridge final report EUROSHOCK TR AER 2-92-49/1.2, December 1995 (also see Chapter 11).

Chapter 3

[3.1] Dima, C. and deMatteis, P., P., "Numerical Investigation of the Passive Shock Control Concept for Transonic Airfoils through an Euler / Boundary-Layer Coupling Technique", CIRA final report EUROSHOCK TR AER2-92-49/2.4, December 1995 (also see Chapter 12).

[3.2] DeNicola, C., "University of Naples Contribution to Task 2 of EUROSHOCK", final report EUROSHOCK TR AER2-92-49/2.10, December 1995 (also see Chapter 13).

[3.3] Dargel,G. and Thiede, P., "Passive Shock Control Investigation on Airfoils. Part I: Numerical Flow Simulation, Part II: Model Design and Manufacture", DASA-Airbus final report EUROSHOCK TR AER2-92-49/2.7 and DA Report No. EF 06 / 2.7, February 1996 (also see Chapter 14).

[3.4] LeBalleur, J.C., Girodroux-Lavigne, P. and Gassot, H., "Development of Viscous-Inviscid Interaction Codes for Prediction of Shock Boundary-Layer Interaction Control (SBLIC) and Buffet over Airfoils", ONERA final report EUROSHOCK TR AER2-92-49/2.9, March 1996 (also see Chapter 15).

[3.5] Geißler, W., "Transonic Airfoil Prediction with Shock Boundary Layer Interaction and Control (SBLIC)", DLR final report EUROSHOCK TR AER2-92-49/2.6, December 1995 (also see Chapter 16).

[3.6] Wolles, B., "Computations of Transonic Flows Applying Shock Boundary Layer Interaction Control", NLR final report EUROSHOCK TR AER2-92-49/2.8, December 1995 (also see Chapter 18).

[3.7] Simandirakis, G., Bouras, B. and Papailiou, K.D., "Unsteady Shock/Boundary Layer Interaction Prediction and Control", University of Athens final report EUROSHOCK TR AER2-92-49/2.5, December 1995 (also see Chapter 17).

[3.8] Puffert-Meißner, W., "The Transonic Wind Tunnel at DLR in Braunschweig (Status 1987)", ESA-TT-1114, 1988.

[3.9] Bur, R., "Passive Control of a Shock Wave/Turbulent Boundary Layer Interaction in a Transonic Flow", Rech. Aerosp.-no 1992-6, 1992.

[3.10] Poll, D.I.A., Danks, M. and Humphreys, B.E., "The Aerodynamic Performance of Laser Drilled Sheets", First European Forum on Laminar Flow Technology, Hamburg, March 1992, Paper 92-02-02, 1992.

[3.11] Breitling, T., "Berechnung transsonischer, reibungsbehafteter Kanal- und Profilströmungen mit passiver Beeinflussung (Computation of transonic viscous channel and airfoil flow with passive control)", Ph.D.-Thesis, University of Karlsruhe, Karlsruhe, 1985.

[3.12] Cebeci, T., "Behavior of Turbulent Flow Near a Porous Wall with Pressure Gradient", AIAA-Journal, Vol.8, No. 12, 1970.

Chapter 4

[4.1] Dargel,G., and Thiede, P., "Passive Shock Control Investigation on Airfoils. Part I: Numerical Flow Simulation, Part II: Model Design and Manufacture", DASA-Airbus final report EUROSHOCK TR AER2-92-49/2.7 and DA Report No. EF 06 / 2.7, February 1996 (also see Chapter 19).

[4.2] Rosemann, H., Knauer, A. and Stanewsky, E., "Experimental Investigation of the Transonic Airfoils DA LVA-1A and VA-2 with Shock Control", DLR final report EUROSHOCK TR AER2-92-49/3.5, 1996 and DLR IB 223-96 A 02, 1996 (also see Chapter 21).

[4.3] Archambaud, J.P. and Rodde, A.M., "Qualification by Laser Measurement of the Passive Control on the LVA-1A Airfoil in the T2 Wind Tunnel", ONERA final report EUROSHOCK TR AER2-92-49/3.4, 1996.

[4.4] Fulker, J. L. and Simmons, M. J., "An experimental investigation of passive shock/boundary-layer interaction control on an aerofoil", DRA final report EUROSHOCK TR AER2-92-49/3.2, 1995 and DRA/HWA/CR95216/1, 1995 (also see Chapter 22).

[4.5] Wagner, W., "An experimental investigation of passive shock boundary layer interaction control on an aerofoil: unsteady pressures", DLR final report EUROSHOCK TR AER2-92-49/3.3, 1996 (also see Chapter 23).

B. INDIVIDUAL CONTRIBUTIONS

8 Introduction to the Individual Contributions

E. Stanewsky
DLR, Institute of Fluid Mechanics
Bunsenstraße 10, D-37073 Göttingen

In the preceding chapters an extended summary and résumé of the results obtained during the EUROSHOCK research program was presented. EUROSHOCK is, however, a team effort with the individual work adding up to the final product. The initial summary is therefore followed by a more detailed description of the work performed by the various members of EUROSHOCK, written by the latter. The contributions, designated Chapters 9 through 23, are assembled in the order of the three tasks which comprise the research program. The sequence of presentation and the main contents of the research performed are briefly given below.

Chapter 9 R. Bur, J. Délery and B. Corbel, ONERA: Detailed basic experimental investigation of shock boundary layer interaction with passive control, employing a transonic channel, supplemented by theoretical and numerical studies concerned with the improvement of boundary conditions and turbulence models used in SBLIC.

Chapter 10 R. Bohning and P. Doerffer, University of Karlsruhe: Transonic channel flow experiments with passive shock control through inclined holes and evaluation of loss coefficients for normal and inclined holes; extension of a Navier-Stokes code to treat control and numerical simulation; development of a new control law to determine the ventilation velocity.

Chapter 11 L.C. Squire, A.F.K. Yeung and X. Faucher, University of Cambridge: Basic experimental investigation of a simple swept-back interaction on a flat plate under control conditions supplemented by numerical studies also with respect to the suitability of turbulence models to account for shock control.

The work described in these chapters was performed within Task 1 of EUROSHOCK „Modeling of Shock Boundary Layer Interaction Control Phenomena". The objective was to study characteristic influence parameters involved in shock boundary layer interaction and shock control and to provide optimized control devices and corresponding boundary conditions, physical models and improved control laws.

The following contributions in Chapters 12 through 18 relate to Task 2 „Transonic Airfoil / Wing Flow Prediction with SBLIC". Objectives were to extend existing computer codes to treat shock control, to validate the methods with the experimental data obtained within Task 3 and to assess — together with the corresponding experimental results — the merits of shock control. The methods are summarized in Chapter 3, Table 3. In detail, the contributions are:

Chapter 12 P. de Matteis and C. Dima, CIRA: Extension of a 2D Viscous-Inviscid-Interaction (VII)-method with emphasis on predicting shock-induced separation and shock control effects; performance of validation and control assessment computations.

Chapter 13 C. de Nicola, University of Naples: Extension and validation of a 2D VII-method to account for active / passive SBLIC and performance of assessment computations.

Chapter 14 G. Dargel, DASA-Airbus: Extension and validation of a transonic VII-airfoil-code with SBLIC and performance of parametric studies assessing control.

Chapter 15 J.C. Le Balleur, P. Girodroux-Lavigne and H. Gassot, ONERA: Extension and validation of two VII-methods to account for SBLIC and assessment computations; the basic codes are VIS05 and VIS15, a steady and an unsteady method, respectively, which allowed to determine the effect of shock control on steady as well as time-dependent flows.

The above codes, except for the ONERA VIS15-method, relate to steady flows. Within the domain of unsteady codes, which generally have the advantage to treat steady and unsteady flows, thus being able to determine the boundary between the two conditions (buffet onset), further contributions are:

Chapter 16 W. Geißler, DLR: Improvement and extension of a 2D time-accurate Navier-Stokes code to treat SBLIC and performance of validation and assessment computations at steady and unsteady flow conditions.

Chapter 17 G. Simandirakis, B. Bouras and K.D. Papailiou, University of Athens: Extension of a time-consistent Euler and a steady integral boundary layer method to treat unsteady shock boundary layer interactions including separation, and extension of a Navier-Stokes solver to treat shock control; validation and assessment computations.

Chapter 18 B.A. Wolles, NLR: Extension of the unsteady ULTRAN-V code, comprised of a TSP-method strongly coupled with an integral boundary layer method, to treat shock control and to predict airfoil buffet with and w/o control; performance of validation and assessment computations.

Since all contributors had the same objectives, it is unavoidable that procedures will be described repeatedly within the various contributions, e.g., when discussing shock control laws implemented within the codes. However, there is sufficient variation in the methods that a presentation of the individual approaches by the respective researcher is deemed warranted.

All participants in Task 2 performed computations for specified test cases. These test cases — the ones used in pre-calculations w/o control to assess the state-of-the-art of the basic codes used, and the test cases selected from the experiments of Task 3 with and w/o control — are summarized in Table 8 together with an identification of the airfoil models investigated.

The objective of Task 3 „Wind Tunnel Experiments on Airfoils with SBLIC" was to study the effect of shock control on the flow development about two laminar-type airfoils with respect to drag behavior and the improvement of the drag-rise and buffet boundaries up to flight Reynolds numbers and to provide data for computer code validation. Later during the investigation, after it had turned out that control for the laminar-type airfoils generally resulted in an increase in total drag, the turbulent airfoil VA-2 was retested; the latter had previously shown a favorable drag behavior due to control. Related to Task 3, the following contributions are being presented:

Chapter 19 P. Thiede and G. Dargel, DASA-Airbus: Design and manufacture of the laminar-type airfoil DA LVA-1A, including perforated inserts, for tests in the ONERA-T2 and DLR-KRG wind tunnels; data analysis.

Chapter 20 J.P. Archambaud and A.M. Rodde, ONERA / CERT: Tests in the adaptive-wall cryogenic wind tunnel T2 with the model DA LVA-1A at selected freestream conditions with emphasis on LDV flow field and boundary layer measurements and data analysis and control assessment.

Chapter 21 H. Rosemann, A. Knauer and E. Stanewsky, DLR: Tests in the adaptive-wall cryogenic wind tunnel KRG with the model DA LVA-1A with emphasis on surface pressure and wake measurements up to high Reynolds numbers; tests with the turbulent airfoil model VA-2 with active and passive shock control; data analysis and control assessment.

Chapter 22 J.L. Fulker and M.J. Simmons, DERA: Design, manufacture and testing of the laminar airfoil DRA-2303 in the DRA 8Ft x 8Ft wind tunnel up to high Reynolds numbers with and w/o control and data analysis and control assessment.

Chapter 23 W. Wagner, DLR: Participation in steady and unsteady pressure measurements with the DRA-2303 airfoil in the DRA 8Ft x 8Ft wind tunnel and analysis of the dynamic results and assessment of the effect of shock control on buffet.

Table 8: Test cases for computer code validation and control assessment computations

Airfoil	Type	Test Case	Data Point	M_∞	α°	C_L	$Re_C \times 10^{-6}$	Tested in	Flow condition
Pre-calculations w/o control									
DA LVA-1A[1]	laminar	1	—	0.75	2.0[2]	0.534	5.92	TWB (DLR-BS)	Steady
		2	—	0.75	3.01[2]	0.692	5.96		
NLR 7301	turbulent	2	—	0.738	2.26[2]	3)	11.7	HST (NLR)	Unsteady
		3	—	0.738	0.41[2]	3)	12.6		
NACA 0012	turbulent	1	—	0.75	5[2]	3)	10.0	—	Unsteady
Test cases from the Task 3 experiments with and w/o control									
DRA-2303	laminar	1	271	0.6816	1.068[2]	0.5668	18.97	DRA 8Ftx8Ft (DERA)	Steady w/o control
			1008	0.6807	1.066[2]	0.5529	18.98		Steady with control
		2	289	0.6795	2.507[2]	0.8115	18.91		Steady w/o control
			1031	0.6806	2.71[2]	0.8142	18.98		Steady with control
			290	0.6795	2.71[2]	0.8427	18.91		Steady w/o control[4]
		3	543	0.702	2.405[2]	0.6714	10.5		Unsteady w/o control
			1469	0.702	2.805[2]	0.7327	10.5		Unsteady with control
DA LVA-1A[1]	laminar	1	76	0.7698	1.0[2]	0.4736	4.64	T2 (ONERA)	Steady w/o control
		2	74	0.7692	1.0[2]	0.4556	4.66		Steady with control
VA-2[5]	turbulent	1	—	0.74	3)	0.65	2.5	TWG (DLR-GÖ)	Steady w/o control
			—	0.74	3)	0.65			Steady with control
		2	—	0.74	3)	0.90	2.5		Steady w/o control
			—	0.74	3)	0.90			Steady with control
		3	—	0.743	3)	0.80	2.5		Steady w/o control
			—	0.743	3)	0.80			Steady with control

1) also designated LVA-1Ae 2) experiment 3) varies with contributor 4) same α as with control but different C_L 5) also designated VA2, Va-2

9 BASIC STUDY OF PASSIVE CONTROL APPLIED TO A

TWO-DIMENSIONAL TRANSONIC INTERACTION

R. Bur, J. Délery and B. Corbel

ONERA, Experimental/Fundamental Aerodynamics Branch
29, avenue de la Division Leclerc, 92320 CHATILLON, France

Summary : Passive control applied to a turbulent shock wave/boundary layer interaction has been investigated by considering a two-dimensional channel flow. The field resulting from application of passive control has been probed in detail by using a two-component LDV system to execute mean velocity and turbulence measurements. Four different perforated plates have been considered, as also the solid wall reference case. The performed measurements have shown that passive control deeply modifies the inviscid flowfield structure, the unique strong shock being replaced by a lambda shock system. This fractionning of the compression induces a substantial reduction of the wave drag associated with the interaction. However, the combined injection-suction effect taking place in the control region provokes an important thickening of the boundary layer. There results an increase of the friction drag which nearly outbalances the gain in wave drag. Also, the rugosity of the holes is an important source of drag (excrescence drag) which contributes to compromise the potential benefit of the passive control technique. The most commonly used turbulence models give poor results, even in the solid wall reference case. On the other hand, the wall transpiration velocity distribution in the control region is well represented by usual laws.

9.1 Introduction

One of the objectives of the EUROSHOCK project was to contribute to the understanding and modeling of the physical phenomena involved in a shock wave/boundary layer interaction under control conditions. In this context, the experimental facet of the study performed at ONERA (Bur and Délery, 1996) focuses on a local analysis of the interaction with a view to establishing a detailed picture of the flow field, including both its mean and turbulent properties. These experiments have been executed in a transonic channel flow by considering the interaction between the shock crossing the channel and the boundary layer developing on one of the channel walls. This arrangement allows to work with a boundary layer thick enough to permit accurate definition of its properties during the interaction process.

The precise objective of this study was to define, mainly by means of LDV explorations, the properties of the interaction domain, firstly in the absence of control (solid wall reference case), secondly under passive control conditions, for four perforated plates, having different geometrical characteristics. A method to determine the total drag in the control region is exposed in order to define an optimum passive control configuration. Additional experiments have been executed to determine the excrescence drag of the perforated plates. For this purpose, a uniform flow was established over the plates whose holes were obturated, the drag being deduced from the growth of the boundary layer momentum thickness.

The theoretical facet of the present study focused on a discussion of the validity of the most common models used to represent the turbulence and the wall velocity (or transpiration velocity) in the control region. This discussion was based on a simplified boundary layer type approach and on comparisons of the calculations with the results of the experimental part.

9.2 Experimental Conditions

9.2.1 Test set-up arrangement

These experiments have been executed in the S8 Basic Research Transonic-Supersonic Wind Tunnel of the ONERA Fluid Mechanics Laboratory at Chalais-Meudon. This facility is a continuous wind tunnel supplied with dessicated atmospheric air. The stagnation conditions, nearly constant throughout the tests, had the following average values :
- Stagnation pressure : $p_{sto} = 92\ 000Pa$.
- Stagnation temperature : $T_{sto} = 300K$.

The test set-up used for the present study is shown in Fig. 1. It is constituted by a transonic channel having a height of 100mm and a span of 120mm in the test section itself. The lower wall is rectilinear, the upper wall being made of a contoured profile designed to produce a uniform supersonic flow of nominal Mach number equal to 1.4. The two side walls are equipped with high quality glass windows to allow visualizations and LDV measurements.

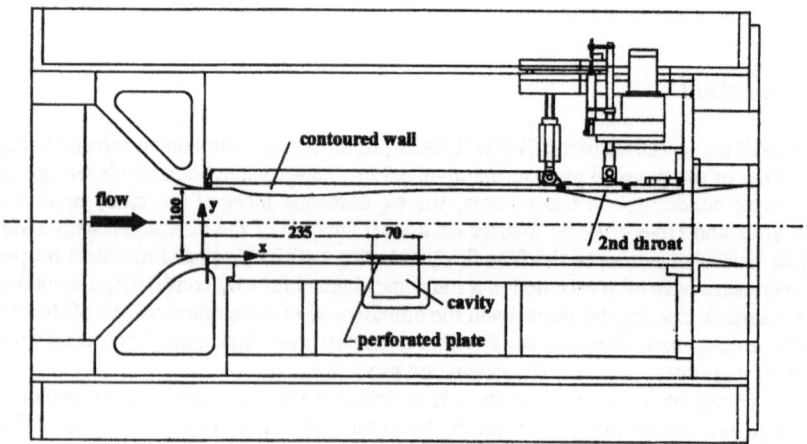

dimensions in mm

Fig. 1 - Passive control experimental set-up in the S8 wind tunnel

A second throat, of adjustable cross section, is placed in the test section outlet, approximately 450mm downstream of the study region, making it possible :

- To produce, by choking effect, a shock wave whose position, and hence intensity, can be adjusted in a continuous and precise manner.

- To isolate the flow field under study from pressure perturbations emanating from downstream ducts. Such a device notably reduces unwanted shock oscillations.

In what follows, X designates the streamwise distance measured from the channel entrance section and Y the distance normal to the study wall (see Fig. 1).

Passive control is applied on the rectilinear lower wall in a region where the outer inviscid flow Mach number is equal to 1.33. The upstream edge of the control cavity, which nearly spans the entire test section, is located at 235mm from the transonic channel entrance. Its length L has been defined in such a way as to approximately reproduce the ratio L/δ_o existing on an airfoil (δ_o is the boundary layer thickness at the cavity origin) ; i.e. $L/\delta_o \sim 18$. Since, in the present experiments $\delta_o \sim 4mm$, the value L = 70mm has been adopted for all the tests. The cavity, or control region, thus extends from X = 235mm to X = 305mm. Its depth h has a fixed value equal to 60mm. A first study (Chanetz and Pot, 1988) having demonstrated that h has a negligible influence on the control mechanism, provided it is not too small (at least 10mm for the present conditions), an unrealistically high value of h has been adopted to facilitate possible LDV measurements inside the cavity. The interchangeable perforated plate is fixed to the wall, upstream and downstream of the cavity, the plate being made flush with the wall. Two lateral narrow beams, of width 2mm, are placed on each side of the cavity to prevent the bending of the plate which could be caused by the pressure difference between the cavity and the outer transonic flow. Air-tightness is insured by a seal placed around the cavity and along the plate.

All the pressure orifices were located in the vertical median plane of the test set-up. The rectilinear study wall was equipped with 10 orifices (diameter : 0.4mm) upstream of the control region and 8 downstream of it. In order to avoid possible perturbation of the LDV measurements, the perforated plates were not equipped with pressure tappings. After completion of these tests, 7 orifices were installed on the plates to determine the pressure distribution in the control region itself. It was checked that this equipment did not affect the flow. Moreover, five pressure taps were located on the bottom part of the cavity.

For the study of the interaction without control (reference case), a rectilinear solid wall extending from the entrance to the end of the channel was manufactured to avoid the perturbations caused by junctions between different parts. This wall was equipped with 24 pressure orifices whose locations are included between X = 135mm and 385mm. The contoured wall, facing the study wall, was equipped with 23 orifices whose indications were used to monitor the shock location.

More details on the experimental conditions, as also on the measurements, can be found in an intermediate report (Bur *et al.*, 1994).

9.2.2 Techniques of investigation and data processing

The Laser Doppler Velocimetry system. The flows under study were qualified by means of: Schlieren visualizations, measurements of wall pressure distributions, probings with a multi-component LDV system, this action constituting the core of the experimental effort.

For the present study, where the flows were nominally two-dimensional, the two-component version of the ONERA three-component LDV system was used (Boutier *et al.*, 1984). The light source is a 15W - Argon laser, used in the present tests at a power of 5W, whose beam is separated into two beams of wave length 0.488μm (blue color) and 0.5145μm (green color) by means of a semi-transparent mirror. The two original beams are first split by classical beam splitters and then traverse Bragg cells to enable the system to detect the velocity direction. The four beams are focused by the emission lens to constitute two fringe patterns inside the measuring volume whose diameter was equal to 0.2mm. The two fringe patterns were rotated at ± 45° with respect to the X-direction, allowing the simultaneous measurement of two velocity components in a vertical plane. The optical adjustments were such that the fringe spacing i had the following values :

    - Green color (λ = 0.5145 μm) : i = 13.82μm.
    - Blue color (λ = 0.488 μm) : i = 13.30μm.

The minimum distance to the wall achievable (safely) by the LDV system, in the two-component version, was equal to 0.3mm.

In the experiments aimed at the determination of the excrescence drag of the perforated plates, a one-component arrangement was employed, since only the X-wise velocity component u was of interest. In this case, it was possible to reduce the probe volume diameter down to 0.1mm and the minimum distance to the wall to 0.1mm.

The LDV system was operated in the forward scattering mode, the green and blue photomultiplier signals being processed by DANTEC type 55 L counters. The emitting and collecting parts are mounted on two separate tables allowing accurate displacements (step of 0.01mm) along three orthogonal axes. At each measurement point, a sample of N couples of the instantaneous values of the velocity components u and v is acquired for further processing. In the present study, where maximum velocities are in the transonic range (around 400m/s), the flow was seeded with droplets of parafine oil injected in the wind tunnel settling chamber.

Mean velocity and Reynolds tensor components determination. The size of the sample was N = 2000 which gives an acceptable statistical uncertainty for the first and second order statistical moments. The mean velocity and Reynolds tensor components were computed by classical formulae (see Bur *et al.*, 1994).

Determination of the mean velocity components allow to compute the local Mach number and define boundary layer global properties : displacement thickness δ^*, momentum thickness θ, mean flow kinetic energy thickness θ^* and incompressible shape parameter H_i.

The two-component version of the LDV system did not allow the determination of the spanwise, or transverse, velocity component w whose mean value is in principle zero (the flow being nominally two-dimensional) but whose fluctuating part is non-zero. The values of the turbulent kinetic energy k were approximated by assuming :

$$k = \frac{1}{2} \ [\overline{u'^2} + \overline{v'^2} + \frac{1}{2}(\overline{u'^2} + \overline{v'^2})] \ .$$

The field quantities were measured with an accuracy depending on uncertainties affecting: 1) the LDV system calibration (uncertainties on the fringe distance, on the Bragg frequency), 2) the determination by the counters of the frequency of the light scattered by the particles, 3) the statistical treatment of the sample.

For the present experiments, the field properties have been determined with an accuracy of:

    - 1% of the maximum velocity modulus for the mean velocity components.
    - $\leq$ to 8% for the normal stress components.
    - $\leq$ to 10% for the turbulent shear stress component.

Domain explored with the LDV system. For the passive control study, the flow fields produced by the various tested configurations, including the reference case (solid wall), have been explored along lines normal to the wall (Y-direction) extending from the surface (Y = 0) to an altitude Y = 22mm and contained in the test section median plane. This extent was chosen to be sure to cover the entire dissipative layer and a part of the outer inviscid flow. Each exploration contained 42 measurement points unevenly distributed in order to refine the probing in the vicinity of the wall (minimum spacing $\Delta Y = 0.1$mm). The interaction domain has been explored along 30 vertical lines whose streamwise locations X were in the range $220 \leq X(mm) \leq 365$ (it is reminded that the perforated plate extends between X = 235mm and X = 305mm) and separated by a distance ΔX = 5mm from X = 220mm to X =325mm and ΔX = 10mm from X = 325mm to X = 365mm.

The excrescence drag of the perforated plates has been deduced from the boundary layer development. To obtain sufficient accuracy in the velocity profile determination, these measurements were performed with a finer mesh close to the wall (minimum spacing ΔY = 0.05mm). The boundary layer has been probed along 17 vertical lines, 5mm apart, whose X locations are within the range $230 \leq X(mm) \leq 310$.

9.2.3 Configurations tested

In addition to the solid wall reference case, four different plates, numbered 1 to 4 and whose nominal characteristics are given in Table I, have been tested. These plates were provided by DASA-Airbus. The holes were drilled by an electron beam technique. The thickness of the plates was equal to 1mm. For Plates 1 and 2, the holes are everywhere normal to the surface. For Plates 3 and 4, the holes are inclined with respect to the surface (in the downstream direction) in the upstream half part of the plate, and normal to it in its downstream half part.

Table I : *Nominal characteristics of the plates tested*

Plate number	Porosity (%)	Hole diameter (mm)	Hole inclination
1	5.67	0.15	90°
2	5.67	0.30	90°
3	5.67	0.15	45° and 90°
4	5.67	0.30	45° and 90°

For the study of passive control, including the reference case (solid wall), the location of the shock wave in the outer part of the flow (i.e., outside the interaction region) was the same for all the configurations tested. This location, which was fixed at mid distance between the origin and the end of the control region, was monitored by considering the pressure distribution on the channel upper wall, the shock position being accurately defined by adjustment of the second throat section. Maintaining a fixed shock location insured an identical pressure rise for all the configurations, which allows a meaningful evaluation of a possible positive effect of passive control.

In the experiments aiming at the estimation of excrescence drag, the second throat was adjusted in such a way that the flow over the plates was uniform with a Mach number equal to 1.36. In these experiments, the holes were obturated by strips of Scotch tape attached to the plate inner face (cavity side).

9.3 Experimental Results

9.3.1 Flow visualizations

Reference case (Solid wall). The Schlieren photograph shown in Fig. 2 reveals the flow structure for the reference case (solid wall). It corresponds to a classical transonic shock wave/boundary layer interaction of moderate strength. One notes the thickening of the boundary layer in the shock foot region and the associated wave pattern forming in the outer inviscid flow. It is comprised of compression waves emanating from the upstream part of the interaction. This front zone is followed by a small triangular region of still supersonic flow terminated by a nearly normal shock, sometimes called the trailing shock. The compression waves and the trailing shock meet in a region from which starts the unique shock crossing the channel. This structure is generally attached to shock-separated flows at transonic speed. In well separated flows, the upstream compression waves rapidly coalesce to constitute the separation or leading shock. In this case, these three shocks meet at a bifurcation, or triple point. The present flow pattern, for which the compression waves do not coalesce, corresponds to a situation nearly coincident with incipient shock-induced separation which occurs for an upstream Mach number M_o of 1.30. The present value $M_o = 1.34$ should lead to the formation of a tiny separation bubble. However, LDV measurements have not detected negative values for the streamwise velocity component. Thus, if it exists, the bubble must be excessively small.

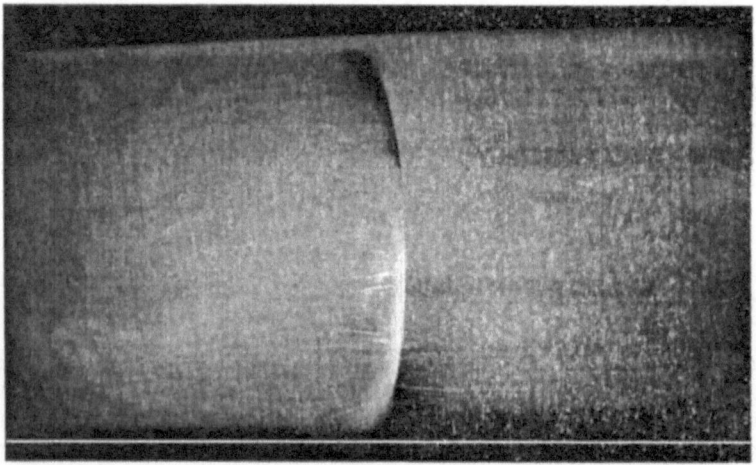

Fig. 2 - Schlieren photograph of the flow field. Reference case (Solid wall)

Flow with passive control. The Schlieren visualization of the flow with passive control is shown in Fig. 3. It corresponds to Plate 1 (normal holes, diameter 0.15mm). Now, the boundary layer starts to thicken suddenly at the cavity origin. Then, it takes a wedge-like shape until it meets with the trailing shock ; afterwards, its thickness remains nearly constant. Hence, one of the passive control effects is to induce an increase of the *friction drag*.

The rapid thickening of the boundary layer, which is provoked by the injection effect taking place in the upstream part of the control region, is felt by the contiguous supersonic flow as a *ramp effect*. There results the formation of an oblique shock wave (C_1), downstream of which the Mach number is still supersonic. This supersonic region is terminated by the trailing shock (C_2) through which the velocity vector changes from an upward direction to become nearly parallel to the wall. The two shocks (C_1) and (C_2) meet at the triple point I, from which starts the unique shock (C_3), to constitute a well defined lambda shape. In the absence of control, such a pattern is associated to strong transonic shock wave/boundary layer interaction with separation.

The triple point I is the origin of a slip line, barely visible on the photograph, which separates the flow having passed through the shock (C_3) from the flow having traversed, successively (C_1) and (C_2). The loss in stagnation pressure is less important through the consecutive shocks (C_1) and (C_2) than across the unique shock (C_3), the upstream Mach numbers and downstream pressure levels being identical in the two flows. Thus, one can conclude that passive control produces a decrease of the *wave drag*, a unique strong shock being replaced by two weaker shocks over a great part of the channel flow. The total entropy production by the shock waves will be less, compared to the reference case where the shock (C_3) occupies nearly the entire channel.

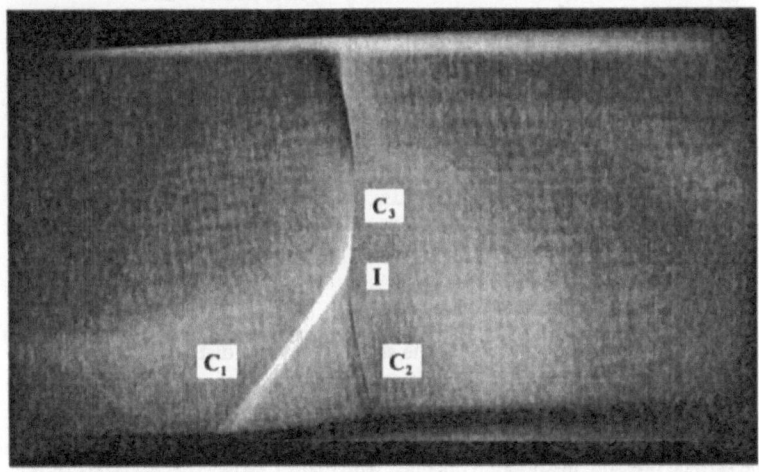

Fig. 3 - Schlieren photograph of the flow field. With passive control (Plate 1)

The Schlieren pictures obtained with Plates 2, 3 and 4 have shown similar modifications on the flow structure in the control region.

9.3.2 Surface pressure distributions

The surface pressure distributions on the channel lower wall for the *reference case* and the *4 control cases* are plotted in Fig. 4 (the local static pressure is referenced to the freestream stagnation pressure p_{sto}).

The decreasing part of the distribution corresponds to the expansion in the supersonic part of the channel. The minimum pressure is reached at $X = 245mm$. If one assumes an isentropic relation, the local value of the Mach number in the inviscid flow, contiguous to the boundary layer, is there equal to 1.33. The interaction with the shock wave produces a rapid rise of the pressure to a downstream nearly constant level $p/p_{sto} = 0.63$, giving an outer flow Mach number equal to 0.84 (here, the isentropic assumption may be less accurate). The general shape of the curve is typical of a transonic interaction without (noticeable) separation.

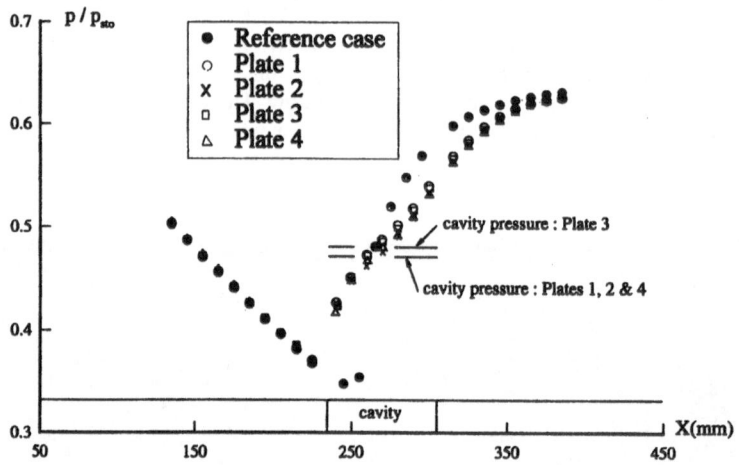

Fig. 4 - Comparison of the surface pressure distributions

The surface pressure distribution obtained with Plate 1 shows that the start of the pressure rise nearly coincides with the cavity forward edge which is located at $X = 235$mm. The local outer flow Mach number just upstream of this position is equal to 1.31 ; i.e., slightly lower than in the reference case. The curve exhibits the three-inflection points typical of shock wave/boundary layer interaction with separation. In this situation, the pressure first undergoes a rapid rise associated with the leading oblique shock (C_1). Then, the slope decreases with the tendency (weakly marked in the present case) towards an intermediate constant level (the plateau of an interaction with extended separation). Thereafter, the pressure gradient increases again, remaining less intense than in the first part of the interaction. Finally, after a third inflection, the pressure tends towards the constant downstream level.

The five pressure taps located in the cavity give the same value $p/p_{sto} = 0.472$ which is indicated in Fig. 4. It practically corresponds to the intermediate inflection point.

The surface pressure distributions measured for the three other plates lead to no obvious differences between the four distributions. It is to be noticed that the cavity pressure is slightly higher for Plate 3 (inclined holes, diameter 0.15mm) than for the other plates ($p/p_{sto} = 0.480$

instead of 0.472). No convincing explanation of this difference has been found.

Differences between the curves corresponding to the perforated plates tested are extremely small. By considering these results, as also the visualizations, it is not possible to define an influence of the hole diameter and/or inclination.

9.3.3 Mean flow field properties

Since the influence of the plate characteristics was found to be weak, only the case of *Plate 1* will be discussed thoroughly and compared with the *reference case*.

The contour lines of the Mach number are traced in Fig. 5 for the reference case, by adopting the same scale (a) for the X-wise and Y-wise distances in order to have an "objective" view of the field structure, and also with an Y-wise dilatation (b) to have a better resolution of the boundary layer region. This tracing shows the system of compression waves generated by the initial thickening of the boundary layer inner region. The apparent thickness of the shock is due to an insufficient refinement of the measurement mesh (X-wise spacing of 5mm) which did not permit a correct "capture" of discontinuities oriented along a normal direction.

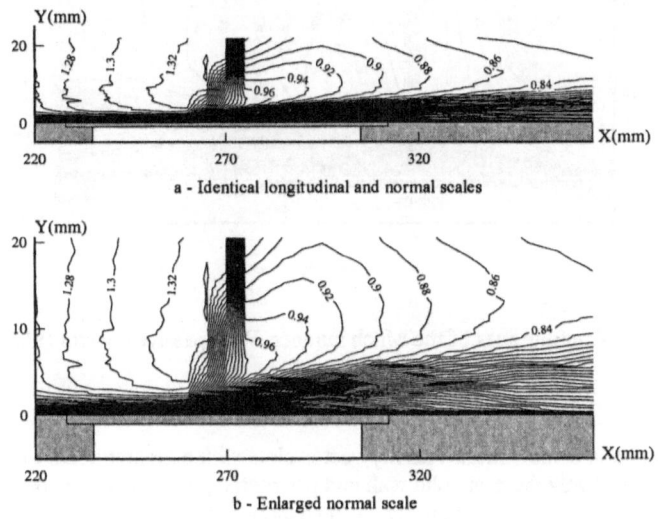

Fig. 5 - Contour lines of the Mach number. Reference case (Solid wall)

The contour lines of the Mach number of the flow under passive control are traced in Fig. 6 with identical Y-and X-scales (a) and with an enlarged Y-scale (b). This figure shows that, in the outer inviscid flow, the leading oblique shock (C_1) provokes a first supersonic compression of the flow from an upstream Mach number of 1.30, to a downstream value of 1.18. In the triangular supersonic region, situated between (C_1) and (C_2), the Mach number varies from 1.18 to 1.08, the flow undergoing an isentropic compression.

The associated pressure rises correspond, respectively, to the quasi-discontinuity observed in the initial pressure rise of the distributions shown in Fig. 4 and to the level of the second inflection point. The Mach number on the downstream face of (C_2) varies between 1.02 in the vicinity of the boundary layer edge and 0.98 at the outer border of the explored domain. A region of supersonic flow, bounded by iso-Mach line M = 1, exists downstream of (C_2), the outer inviscid flow undergoing a nearly isentropic compression to a Mach number of 0.85 at the end of the region explored.

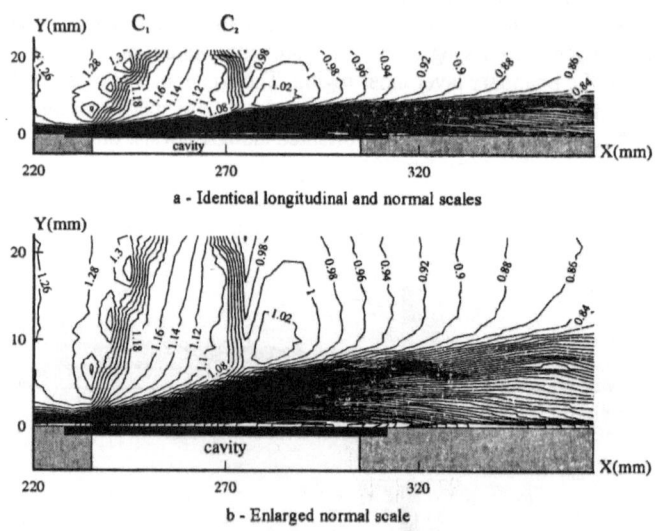

a - Identical longitudinal and normal scales

b - Enlarged normal scale

Fig. 6 - Contour lines of the Mach number. With passive control (Plate 1)

Figures 7 and 8 show mean velocity vector plots with a greater dilatation of the Y-scale emphasizing respectively the near solid wall and perforated plate regions. Some streamlines are superimposed on these plottings to visualize the fluid motion. In the case with passive control, the picture (see Fig. 8) reveals a region where the mean velocity component $\bar{u}$ is negative. Its maximum vertical size is approximately 0.8mm, for a longitudinal extent of 35mm (between X = 270mm and X = 305mm). Thus, the back flow region is extremely flat and contained in a thin zone in contact with the wall. The streamlines reveal the rapid turning of the flow in the downstream part of the control region. On this tracing, the streamlines penetrate into the cavity in the part of the perforated plate beginning at X = 260mm, which indicates a suction effect in a region where the local pressure is still lower than the cavity pressure, according to the pressure distribution shown in Fig. 4. This apparently too early beginning of the suction effect is also observed for the other plates.

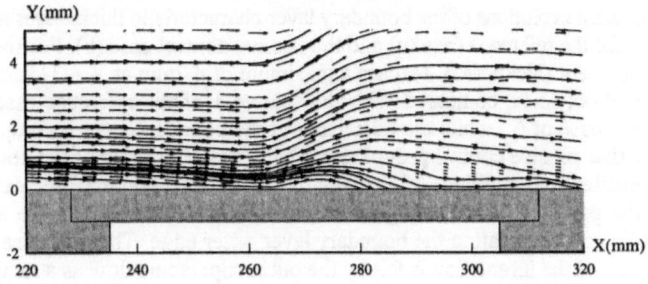

Fig. 7 - Mean velocity vector plot with dilatation of the near wall region. Reference case (Solid wall)

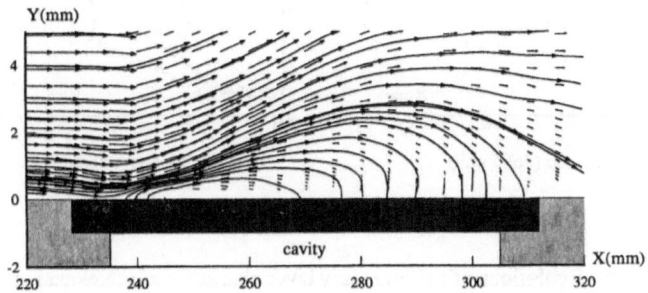

Fig. 8 - Mean velocity vector plot with dilatation of the near wall region. With passive control (Plate 1)

The longitudinal evolutions of the boundary layer characteristic thicknesses are plotted in Fig. 9, respectively for the reference case (a) and the passive control case (b). For the flow under control, the displacement thickness δ^* reaches a maximum of 4.7mm at X ≈ 315mm, leading to a ratio $\delta^*_{max}/\delta^*_o$ = 9.06, to be compared with the value 5.1 of the reference case. This large difference in the increase of δ^*, while the thickening of the boundary layer is comparable in the two cases, proves that passive control provokes a greater destabilization of the boundary layer; i.e., the velocity profiles are less filled. After going through its maximum, δ^* decreases slowly in both cases. For the passive control configuration, downstream values of δ^* are affected by a scatter due to difficulties in locating the boundary layer outer edge. The rapid rise of δ^* taking place in the first part of the interaction is felt by the outer supersonic flow as a ramp effect with the angle : $\varphi = \tan^{-1} (d\delta^*/dx)$. In the passive control case, the value of φ is 3.7°, which is nearly equal to that of the reference case (3.5°). The essential difference is that, in the presence of control, the magnitude of the δ^* rise is much more important.

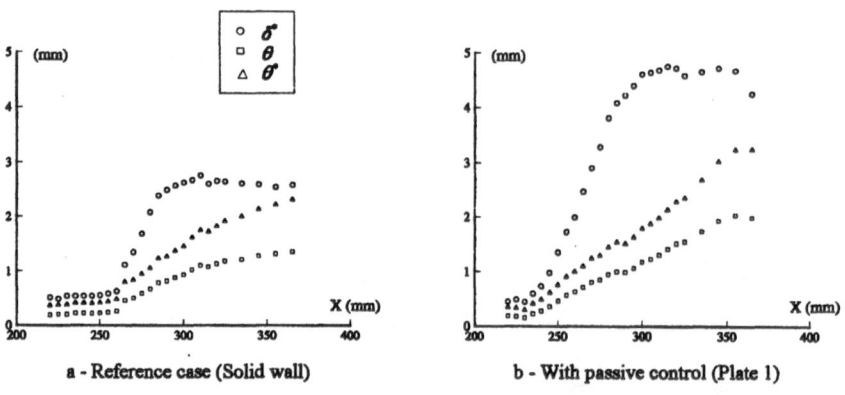

a - Reference case (Solid wall)   b - With passive control (Plate 1)

Fig. 9 - Evolutions of the boundary layer characteristic thicknesses

The evolution of the momentum thickness θ is of special interest since it is a measure of momentum loss undergone by the flow because of dissipative effects. Thus, its rise gives an indication of the friction drag exerted on the surface over the extent of the interaction domain. When passive control is applied, θ reaches a maximum of 2.04mm at the end of the explorated domain, to which corresponds a ratio θ_f/θ_o = 9.27, to be compared with the value 6.14 of the reference case. At this time, it can be concluded that the friction drag is greater when passive control works.

The results relative to the other plates tested lead to no obvious effect of the plate characteristics on the mean flow properties. Table II summarizes typical values of the boundary layer global characteristics for all the tested configurations, by giving the maximum values reached by δ^* and the incompressible shape parameter H_i and the final value of θ. These results do not show any significant effect of the hole diameter or inclination of the plates, the scatter being mainly due to measurement uncertainty.

	$\delta^*_{max}/\delta^*_o$	θ_f/θ_o	$H_{i\,max}$
Solid wall	5.10	6.14	2.50
Plate 1	9.06	9.27	3.50
Plate 2	9.32	9.32	3.57
Plate 3	9.33	9.32	3.65
Plate 4	9.38	8.98	3.70

9.3.4 Turbulent field properties

The turbulent field properties are strongly modified by the presence of passive control in the interaction region. The contour lines of the turbulent shear stress $-\overline{u'v'}$ (normalized by the square of the reference velocity U_o^2) are plotted in Figs. 10 and 11, with an enlarged Y-scale, respectively for the reference case and the control case. It is seen that the maximum in shear stress is almost twice higher under passive control conditions (0.0075 against 0.004 for the reference case) and located farther from the wall, both of them being reached near X = 340mm. The turbulence level downstream of the interaction region remains very high, particularly when passive control occurs (its decreasing after the maximum being rather slow).

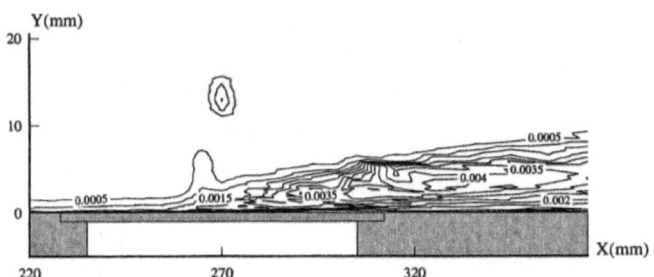

Fig. 10 - Contour lines of the turbulent shear stress.
Reference case (Solid wall)

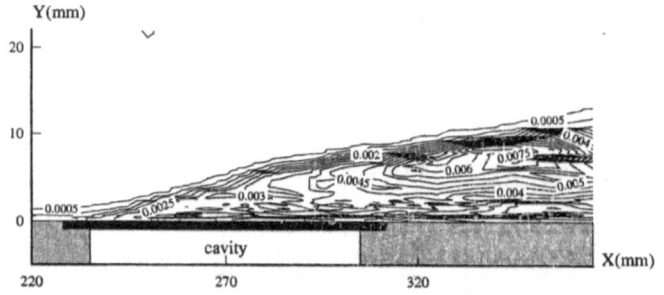

Fig. 11 - Contour lines of the turbulent shear stress.
With passive control (Plate 1)

The contour lines of the (pseudo) turbulent kinetic energy k (normalized by U_o^2) are plotted in Figs. 12 and 13.

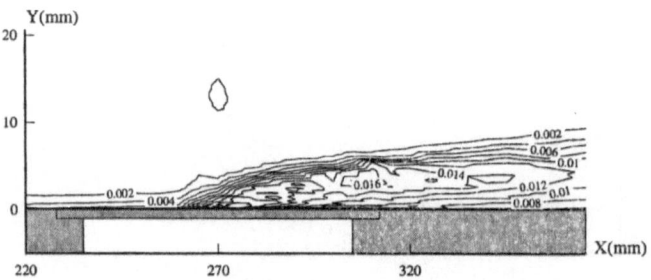

Fig. 12 - Contour lines of the turbulent kinetic energy.
Reference case (Solid wall)

The general rise in turbulence intensity in the passive control case is also observed with k. However, the maxima of k (0.016 for the reference case and 0.022 for the control case) are located around X = 310mm, whereas they occur downstream for the turbulent shear stress (X ~ 340mm). This "lag" of the shear stress is a feature commonly observed in shock wave/boundary layer interactions (Délery, 1983).

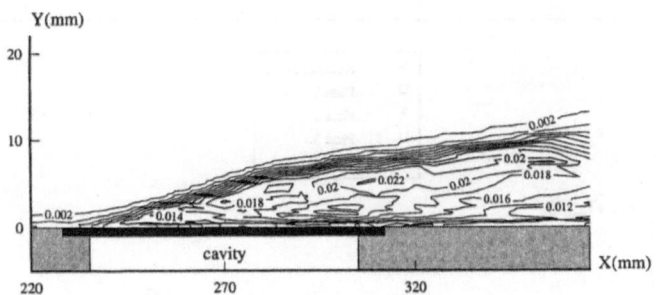

Fig. 13 - Contour lines of the turbulent kinetic energy.
With passive control (Plate 1)

In order to allow a comparison between the reference case and all the plates tested, the X-wise variations of the local maxima of the turbulent kinetic energy and shear stress are shown in Fig. 14. The delayed rise in these quantities for the reference case is due to the fact that the interaction starts at X ~ 265mm on the solid wall, whereas it begins at X = 235mm on the perforated plates. It is clear that passive control produces a higher rise in turbulence levels, which is mainly due to the larger distortion of the velocity profiles. The plottings reveal a somewhat different behaviour between plates with normal holes (Plates 1 and 2) and inclined holes (Plates 3 and 4). The rise in turbulent kinetic energy is more important for the inclined holes, the results for Plates 1 and 2, on the one hand, and Plates 3 and 4, on the other hand, being in good agreement. The effect on the shear stress is less important. The plottings also exhibit the "out-phasing" between k and $-\overline{u'v'}$, the maximum maximorum in $-\overline{u'v'}$ occurring further downstream for each configuration.

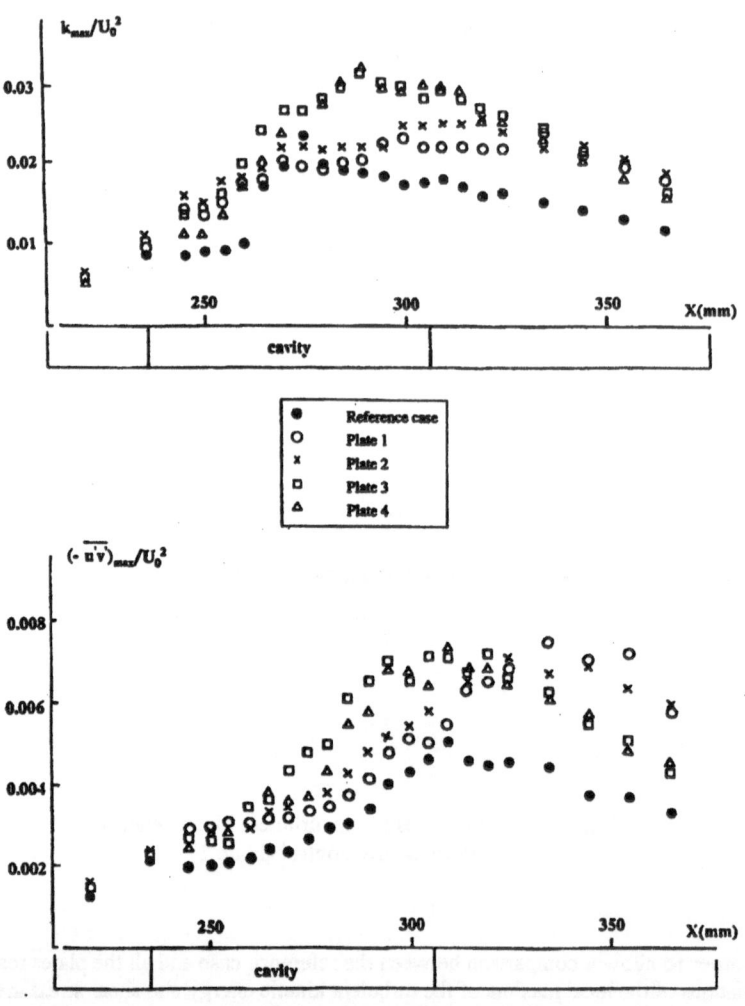

Fig. 14 - Streamwise variations of the turbulent kinetic energy and
shear stress maxima

9.3.5 Total drag coefficient in the control region

One way to evaluate the efficiency of the passive control device is to compute the total
drag force F (and its associated coefficient C_F) acting on the surface where control is applied, for
each tested plate and the solid wall. The force acting on the surface element is determined by
calculating, from the measured quantities, the momentum balance for a control volume which
contains the control region and whose entry and exit sections are perpendicular to the surface and

located at the abscissas X = 220mm and 365mm, respectively.

To evaluate the momentum balance, it is assumed that the pressure in the entry and exit sections is constant along Y which may constitute an approximation. Having seen that the pressure in the cavity is practically constant and that the velocities are quite below those of the external flow, the possibility of the cavity contribution to the drag was not considered.

The associated total drag coefficient is defined as:

$$C_F = \frac{F}{1/2\rho_o U_o^2}$$

where F is the friction force per unit of surface and the reference values ρ_o, U_o (freestream) are the values at the coordinates X = 220mm, Y = δ_o.

Table III gives the values of C_F thus obtained. This table also indicates the corresponding percentage of decrease in C_F, by comparison with the solid wall, defined as:

$$\Delta C_F = \frac{C_{F(solid)} - C_{F(porous)}}{C_{F(solid)}}$$

Table III : *Total drag coefficient*

	Solid wall	Plate 1 normal holes 0.15mm	Plate 2 normal holes 0.30mm	Plate 3 inclined holes 0.15mm	Plate 4 inclined holes 0.30mm
C_F	0.00991	0.00972	· 0.00926	0.00946	0.00952
ΔC_F (%)	-	1.9	6.6	4.5	3.9

The above results show a slight decrease (around 4%) of the perforated plates average coefficient value compared to the solid wall value. This modest gain of drag must be taken with precaution because of measurement uncertainties. Moreover, these results do not give clear indications about differences which could be due to the characteristics of the plates (diameter and inclination of the holes).

As seen in a previous study (Bur, 1992), this shock location (i.e., the shock centered on the cavity) is not the best one to obtain the most important reduction of the drag in the control region ; better results were found for more downstream shock locations.

9.3.6 Perforated plates excrescence drag

As explained above, in these experiments, the shock wave is located downstream of the control region, the expansion part of the channel being followed by a region of uniform flow of Mach number 1.36. The explorations were performed between the control region origin (X = 235mm) and its end (X = 305mm) and executed with the one-component version of the LDV system, special precautions being taken to make measurements as close to the wall as possible. Before calculating the boundary layer integral thicknesses and deducing the excrescence drag, the measured velocity profiles were faired, when necessary, to reduce the scatter of the results.

For a flat plate boundary layer (zero pressure gradient), the von Kàrmàn integral equation reduces to the following relation between the X-derivative of the momentum thickness and the skin friction coefficient :

$$\frac{C_f}{2} = \frac{d\theta}{dX} \ .$$

Thus, for a distance over which the variation of θ can be considered as linear, it is a simple matter to determine the skin friction, which is then constant. By applying this method (a straight line is fitted to the data points by the least square method) to the reference case, one finds: $C_f = 0.00354$, which is in good agreement with classical flat plate boundary layer results.

Table IV gives the values of the skin friction coefficient deduced from the derivative $d\theta/dX$ (assumed constant over the plates) for all the configurations tested. The table also indicates the corresponding percentage of increase in C_f, by comparison with the solid wall, defined as :

$$\Delta C_f = \frac{C_{f\,(holes)} - C_{f\,(solid)}}{C_{f\,(solid)}} \ .$$

Table IV : *Perforated plates excrescence drag coefficient*

	Solid wall	Plate 1 normal holes 0.15mm	Plate 2 normal holes 0.30mm	Plate 3 inclined holes 0.15mm	Plate 4 inclined holes 0.30mm
C_f	0.00354	0.00753	0.00832	0.00900	0.00718
ΔC_f (%)	-	113	135	154	103

The evaluation of the skin friction for the perforated plates leads to an average value which is more than twice the solid wall value. However, these results do not show clear tendencies due to differences in the characteristics of the plates.

9.4 Theoretical Study

9.4.1 Numerical approach

The numerical method used in the present theoretical analysis is based on a conventional *first order boundary layer code*. When there is a risk of separation an *inverse mode* was adopted. Then, the longitudinal distribution of the displacement thickness δ^*, which characterizes the boundary layer development, is prescribed. This low cost numerical approach allowed to test several models of turbulence ; namely, the Baldwin-Lomax algebraic model, the $[k,\epsilon]$ transport equation model and the second-order closure Algebraic Stress Model (ASM).

The boundary layer equations are solved by a finite difference method. The parabolic nature of these equations makes it possible to adopt an implicit upwind scheme to discretize the equations in the main flow direction. In the reversed flow, the parabolic character is maintained by the approximation which consists of replacing the longitudinal component of the main velocity by its absolute value in the X-wise convection term.

The changes due to the presence of passive control in the interaction region occur at two levels :

- Firstly, the main field is directly affected with a non zero (transpiration) velocity distribution at the controlled surface. This distribution is represented by semi-empirical laws relating the velocities through the plate v_w to other flow quantities. The control laws examined are adaptations of the Darcy law and the isentropic law (for details, see Bur, 1991), the Poll calibration law (Poll *et al.*, 1993) and a law proposed by the University of Karlsruhe.

- Secondly, the turbulent field is modified in the region close to the surface. The organization of the turbulent field in this region is modelled by using the wall damping function defined by Van Driest. This function was generalized by Cebeci for flows with surface mass transfer and to represent the effect of a streamwise pressure gradient (Cebeci, 1970).

The full details of the theoretical study can be found in an intermediate report (see Bur, 1995).

9.4.2 Surface pressure distributions

The surface pressure distributions on the channel lower wall, for the *reference case*, are plotted in Fig. 15, the local static pressure being referenced to the freestream stagnation pressure P_{sto}.

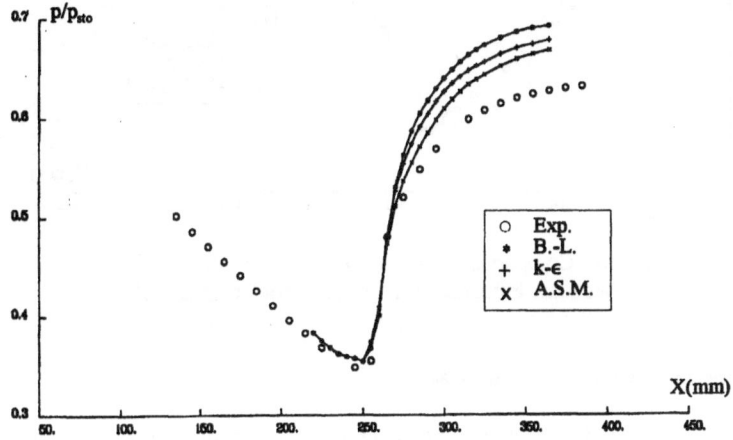

Fig. 15 - Comparison of the surface pressure distributions.
Turbulence model effect. Reference case (Solid wall)

This figure shows distributions calculated with the Baldwin-Lomax model, the [k,ε] model and the ASM, compared to the experimental distribution. The decreasing part of the experimental curve, which corresponds to the expansion in the supersonic part of the channel, is well predicted by calculation, down to the minimum pressure value. The rapid pressure rise in the interaction region is overestimated by all the models. Also, the nearly constant pressure level downstream of

107

the interaction is too high, even with the ASM which gives the best agreement with experiment.

Computed surface pressure distributions are compared in Fig. 16 to experiment in the case of passive control with *Plate 2*. The turbulence model used is the [k,ε] model. To improve the predictions, the original control laws are modified by introducing a corrective coefficient in order to reduce the discrepancy between calculated and experimental surface pressure distributions. However, even if a separation of the flow is predicted by all the calculations, the nearly constant pressure level corresponding to the separated region is always underestimated and the downstream level is too high, particularly for the calculation using the Darcy law. These large discrepancies, already observed in the reference case for all the tested turbulence models, are mainly due to a basic deficiency in turbulence modeling for strongly interacting flows.

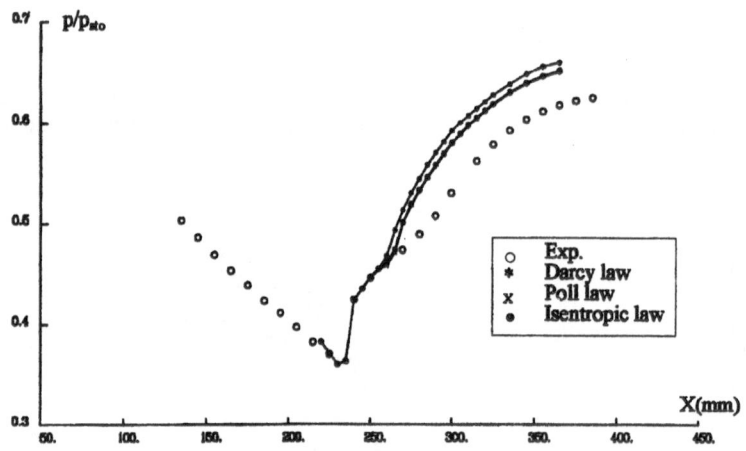

Fig. 16 - Comparison of the surface pressure distributions.
Wall law influence. With passive control (Plate 2)

9.4.3 Transpiration velocity distributions

The present experiments allow to give a representation of the non zero (transpiration) velocity component normal to the surface where control is applied. In the explorations, the minimum distance to the wall achievable (safely) by the two-component LDV system was equal to 0.3mm. The scatter of these measurements has several origins : firstly, there is the limited accuracy of the LDV measurements (the uncertainty on v is of the order of 1% of the outer flow velocity ; i.e., about 3m/s) ; secondly, v_w is not uniformly distributed, injection or suction being performed through discrete holes ; thirdly, slight deformations of the perforated plate due to the pressure difference occur, rendering the exact location of the measurement point relatively to the wall difficult. Moreover, at the altitude retained (0.3mm), the experimental normal velocity v_w distribution is an approximation of the one existing at the surface of the perforated plate.

Nevertheless, one can extrapolate from the LDV measurements an experimental law which is compared to semi-empirical control laws used as wall boundary conditions in numerical approaches. These semi-empirical laws give the transpiration velocities by using the measured surface pressure distribution along the control region. The comparison between the different transpiration velocity distributions is made for *Plate 2* in Fig. 17. All the curves fit rather well with the experimental distribution, the values of injection and suction, respectively at the beginning and the end of the plate, being correctly predicted by semi-empirical laws. However, some discrepancies appear for the near zero normal velocities at the interface between injection and suction. They are due to the quadratic evolution of the semi-empirical laws compared to the linear evolution of the extrapolated experimental law.

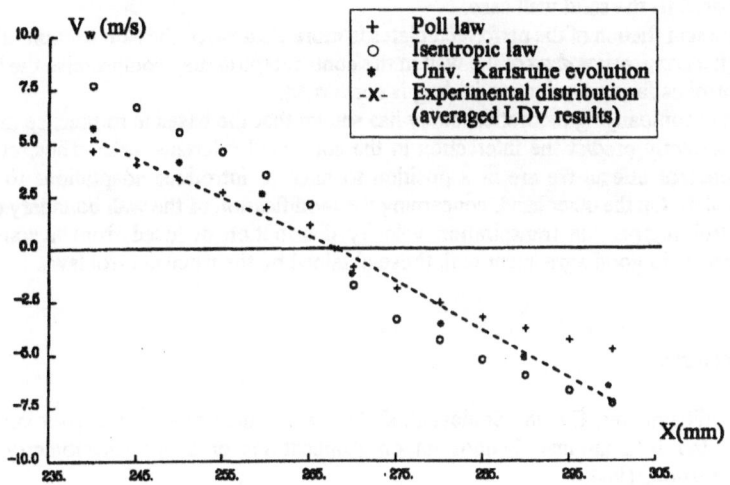

Fig. 17 - Transpiration velocity distributions for the Plate 2.
Comparison of semi-empirical and averaged experimental laws

9.5 Conclusion

A detailed experimental investigation of transonic shock wave/boundary layer interaction under passive control conditions has been performed in a channel type flow in order to work with a thick boundary layer allowing refined explorations. The solid wall reference case and four passive control configurations, with different perforated plates, have been studied by using a two-component LDV system. The results obtained show the following tendencies :

- When passive control is applied, the wall transpiration occurring in the upstream part of the perforated plate induces a viscous ramp effect which provokes the formation of a leading oblique shock wave. The supersonic zone following this shock is terminated by a nearly normal trailing shock, the two shocks meeting at a triple point from which starts the main shock. The net effect of this flow organization is to lead to a substantial reduction of the *wave drag*, the compression being achieved through a two shock system instead of a unique strong shock.

- The boundary layer destabilization is more important in the interactions with passive control, the rise in the boundary layer momentum thickness over the interaction region being 50% greater than in the reference case. This result tends to prove that the *friction drag* is significantly higher when passive control is applied. This negative effect could offset the gain in wave drag.

- As far as the mean flow properties are concerned, there is no conclusive effect of the perforated plates characteristics (hole diameter and inclination).

- Application of passive control provokes a more important rise in the turbulence level of the flow, compared to the reference case. This behaviour is due to the greater distortion of the mean velocity profiles occurring under passive control conditions.

- It was found that passive control induced a modest decrease (around 4%) of the *total drag*, compared to the solid wall case.

- The skin friction of the perforated plates is more than twice the skin friction of the solid wall. Thus, the *excrescence drag* of the wall in the control region may compromise the benefit of passive control as far as a reduction in drag is concerned.

The accompanying theoretical study has shown that the baseline turbulence models are unable to correctly predict the interaction in the solid wall reference case. Thus, in order to represent control effects we are in a position to have to introduce adaptations to basically incorrect models. On the other hand, concerning the modification of the wall boundary condition in the control region, the transpiration velocity distribution deduced from averaged LDV measurements is in good agreement with those obtained by the usual control laws.

9.6 References

Boutier, A. ; d'Humières, Ch. and Soulevant, D. (1984) : "Three-dimensional laser velocimetry: a review". 2nd International Symposium on Applications of Laser Anemometry to Fluid Mechanics, Lisbon, 1984.

Bur, R. (1991) : "Etude fondamentale sur le contrôle passif de l'interaction onde de choc/couche limite turbulente en écoulement transsonique". Ph. D. Thesis, Université Pierre et Marie Curie, Paris, March 1991, ESA English translation available.

Bur, R. (1992) : "Passive control of a shock wave/turbulent boundary layer interaction in a transonic flow". La Recherche Aérospatiale, N° 1992-6, Nov.-Dec. 1992, pp. 11-30.

Bur, R. (1995) : "Physical modelling of passive control applied to a transonic shock wave/ boundary layer interaction. ONERA contribution to Subtask 1.2 of EUROSHOCK". Report TR AER 2-92-49/1.3, May 1995.

Bur, R. and Délery, J. (1996) : "Study of passive control applied to a transonic shock wave/boundary layer interaction. ONERA contribution to Task 1 of EUROSHOCK". Report TR AER 2-92-49/1.4, Jan. 1996.

Bur, R. ; Corbel, B. and Délery, J. (1994) : "Basic experimental study of passive control applied to a transonic shock wave/boundary layer interaction. ONERA contribution to Subtask 1.1 of EUROSHOCK". Report TR AER 2-92-49/1.1, Dec. 1994.

Cebeci, T. (1970) : "Behaviour of turbulent flow near a porous wall with pressure gradient". AIAA Journal, Vol.8, N°12, Dec. 1970, pp. 2152-2156.

Chanetz, B. and Pot, T. (1988) : "Etude fondamentale sur le contrôle passif appliqué à une interaction onde de choc/couche limite en transsonique. Premiers résultats d'essais". ONERA RT N° 75/7078 AN, July 1988.

Délery, J. (1983) : "Experimental investigation of turbulence properties in transonic shock/ boundary layer interactions". AIAA Journal, Vol.21, N°2, Feb. 1983, pp. 180-185.

Poll, D.I.A. ; Danks, M. and Humphreys, B.E. (1993) : "The aerodynamic performance of laser drilled sheets". STAR, Vol.31, N°7, July 1993, pp. 274-277.

10 PASSIVE CONTROL OF SHOCK WAVE - BOUNDARY LAYER INTERACTION AND POROUS PLATE TRANSPIRATION FLOW

R. Bohning and P. Doerffer

University of Karlsruhe, Institut für Strömungslehre und Strömungsmaschinen,
Kaiserstraße 12, 76 128 Karlsruhe, Germany

Summary: The work carried out at Karlsruhe University in the Institut für Strömungslehre und Strömungsmaschinen (ISS) within the frame of the EUROSHOCK project consists of three independent parts: an experimental analysis of a flow through a porous plate, an experimental investigation of passive control effects on shock boundary layer interaction for different kinds of porosity and 2-D numerical simulations of the interaction.

A new model for transpiration flow description has been proposed. The measurements indicated an induction of separation by passive control and no effect of the used porosity kinds. Numerical simulations showed very good agreement with experiments except for some small details and have proven the usefulness of the proposed modelling of porous wall flow.

10.1 Introduction

The first part of the research concerns an experimental investigation of the flow through a porous plate. The considerations presented are based on the results obtained in an experimental facility into which the air is sucked through a porous plate. Therefore the ambient conditions correspond to the stagnation parameters. Such an experiment is very easy to reproduce at any laboratory.

The research involved holes whose diameter varied from 0.085 mm to 0.325 mm. The holes were normal to the plate or inclined by 30° to the normal. Porosity values were in the range of 2%-27%. On the basis of results obtained a new transpiration model has been proposed. A simple function allows to relate pressure drop through a porous plate (normalised by the stagnation pressure) to the Mach number in the holes of a plate. The proposed relation turned out to be independent of hole geometry and quality, providing that the effective (corrected) porosity value of the plate is used. This allows its simple and easy implementation into the existing numerical codes. In order to relate the theoretical approach to the experimental results a corrected porosity of the wall used has to be determined. An appropriate method for the experimental determination of wall porosity has been worked out. Empirical relations obtained on the basis of research carried out allow to determine the value of corrected porosity even from a single measurement of the pressure drop through the plate and the corresponding mass flow rate. The porosity obtained corresponds directly to the porosity used in the model for numerical simulations. Predictions of the transpiration velocity by means of the proposed model for a particular experiment carried out at ONERA agrees very well with the measured velocities by LDV.

The main part of the research carried out concerned, of course, the passive control effect upon a normal shock wave - turbulent boundary layer interaction. Two Mach numbers upstream of the shock wave have been considered, M=1.27 and 1.3. The prescribed cavity length was 70 mm. Originally the investigation concerned four plates which were supposed to have the same porosity of 5.67%. The plate porosity was planned to display two effects:

1. The effect of hole size on passive control. The holes had diameters of 0.125 mm and 0.3 mm. The distance between holes changed accordingly.
2. The effect of hole inclination upstream the shock. There were plates with holes normal to the surface. Other plates had holes inclined by 30° to the plate normal in the upstream half of a plate and normal in the downstream half of a plate.

The effect of passive control is very strong. In all cases investigated it induces separation. At M=1.27 there are weak but noticeable differences between various porous plates without passive control when the cavity side of the perforation was covered. It could be observed that the plates with small holes display fuller profiles than the ones with larger holes. This is apparently caused by the different roughness effect.

The results obtained do not provide enough evidence to make any final statement about the effect of inclined holes upon the boundary layer profile development. The hole inclination effect, noticeable in case of roughness induced disturbance, is too weak to remain noticeable at such strong disturbance to the boundary layer as passive control.

From the results obtained a significant negative effect of passive control on the boundary layer state becomes obvious. The boundary layer shape factor H_{12} is doubled in comparison with the no ventilation case and boundary layer separation sets in for all cases.

The results concerning the effectiveness of passive control will decide about its future implementation into a regular or laminar airfoil technology. The occurrence of separation at the Mach numbers considered indicates that passive control should:

-be implemented at Mach numbers lower than M=1.27,

-be realised by means of a cavity shorter than 70 mm,

-be combined with active control methods.

The last part of the research carried out at Karlsruhe concerned physical modelling. The zonal method, used in the past, for the numerical simulation of the channel flow with shock boundary layer interaction turned out to be too complicated for the implementation of passive control. Therefore the standard code KAPPA of ISS Karlsruhe has been used for the numerical simulation. This is a code solving the Navier-Stokes equation for compressible flows.

The boundary conditions concerning passive control have been formulated using the model developed in Karlsruhe within the present project. The ventilation velocities are dependent on the pressure difference across the porous wall. These are the boundary conditions for the N-S solver. The simulation carried out was 2-D. It has indicated that at regular flow conditions no separation is present for the Mach numbers considered. Application of passive control induces, however, significant separation. Good agreement of the numerical simulation with experiment indicates that the code used is an effective tool for the analysis of passive control effects.

10.2 Porous Plate Flow

10.2.1 Measurements

The test set up for the porous plate flow investigation is sketched in Fig.1. The main part consists of a pipe at the inlet of which was a porous plate mounting arrangement. The plate is placed and fixed with a seal between the inlet part and the pipe. The plate part exposed to the flow is circular, about 40 mm in diameter. Measured parameters are indicated in Fig.1.

On the opposite side the pipe is connected to the ventilator which sucks in the air through the measurement arrangement. The mass flow rate can be controlled by a bypass valve at the ventilator. Ambient pressure and temperature are the stagnation parameters for the flow considered.

In this report all the experimental results carried out at Karlsruhe in recent years that concern porous plate transpiration measurements are used. The porosity of a plate has been determined by means of hole geometry measurements under a microscope. The results obtained are not of high accuracy because irregularities of the hole shapes, due to the manufacturing quality, are difficult to be taken into account. The plates investigated are listed in TABLE 1.

The plates investigated are divided into five groups. The "OLD" set concerns plates used by Braun [2]. These plates have holes of the same diameter, 0.3 mm. The different porosity has been obtained by different spacing of these holes. The "TEST" set concerns plates prepared for porosity tests. It consists of three plates of different hole sizes with holes perpendicular to the plate surface, (plates T-1, T-2, T-3, hole size 0.085 mm, 0.125 mm, 0.185 mm, respectively). The holes of the plates T-4 and T-6 are inclined by 30° to the surface normal (oval openings 0.09/0.12 mm and 0.16/0.18 mm, respectively). "DA" means an additional plate of low porosity (2%) with normal holes of 0.3 mm diameter. The "TUNNEL" set contains the plates actually used in the wind tunnel at ISS Karlsruhe for the investigation of shock boundary layer interaction with passive control (SBLIC) under the EUROSHOCK project. The plates K-I and K-II have a nominal hole size of 0.125 mm and the plates K-III and K-IV have diameter of 0.3 mm. The plates K-II and K-IV have inclined holes in the upstream part of the plate (denoted by "/") and normal holes in the downstream part. The "ONERA" set was used at ONERA for SBLIC investigations and at ISS Karlsruhe for mass flow investigations. They correspond to the "TUNNEL" set and the hole sizes agree well.

10.2.2 Effective flow within a hole and pressure drop through the porous plate

The mass flow measurements reveal a relation between the mass flow rate and the pressure drop through the plates in the form

$$\frac{dP}{P_0} \sim \left(\frac{\dot{m}_{real}}{p}\right)^2, \tag{1}$$

where P_0 means the stagnation pressure and p the porosity.

Furthermore, the measured mass flow rate $\dot{m}_{real}$ allows to determine an average or effective Mach number within a hole of a porous plate. From the formula

$$\dot{m}_{real} = A \frac{M_{hole}}{\left(1 + \frac{\kappa - 1}{2}M_{hole}^2\right)^{\frac{\kappa+1}{2(\kappa-1)}}} P_0 \sqrt{\frac{\kappa}{RT_0}} \tag{2}$$

one may calculate the value of M_{hole}, where "A" is the total open area of a porous plate.

It has been suggested that the effective Mach number within a hole depends on the pressure drop over the porous plate only. This has been confirmed by the same behaviour of all plates used. The similar behaviour of $M_{hole} = f(dP/P_0)$ for all plates without any exception has led to the search of a common function that could represent this relation. The function

$$M_{hole} = 1.2 \left(\frac{dP}{P_0} \right)^{0.55},$$

(3)

fitted very well the average behaviour of distributions measured for all plates. There is a certain scatter for some plates which is not random but shows a parallel shift of experimental curves in respect to function (3). The function is presented in Fig.2 as a thick line. The points marked in the plot will be referred to later.

The aforementioned discrepancy focused the attention on the transpiration flow comparison between plates used. A typical characteristic of a plate is provided by the relation between the mass flow rate and the corresponding pressure drop. Fig.3 presents such characteristics for some of the porous plates. For usefulness a unit area mass flow rate is chosen for presentation. The increase of mass flow rate is inducing an increase of pressure drop through the plate. The curves obtained have the typical parabolic shape described by eq. (1). The important issue is the relative location of curves for plates of different porosity. In case of the same mass flow rate it is natural to expect larger pressure drop through a plate of smaller porosity. Hence, for a smaller porosity one obtains a steeper curve.

In the legend of Fig.3 after the plate name a porosity value is provided. This value is determined under microscope, as described above, and is given in Table 1 as "p". It should be pointed out that for plates of the same measured porosity value the location of obtained curves is quite different (it is for plates K-II/ and K-IV 4.9% and for K-I and K-II 3.94%). For the other two plates (K-III and K-IV) 5.6%) the curves location coincide well. Examples provided show that optical determination of the porosity is insufficient. Hence, instead of the measured porosity an "effective porosity" should define the relative location of curves properly.

The incorrect values of porosity may be responsible for the discrepancy of results in respect to the general function (3). It has been checked how much the porosity of each plate should be changed in order to get the best fit to the function (3). It is not a simple correction of points to a curve but all the results for one plate are corrected by a *single* porosity value only. In spite of that the M_{hole}-distribution obtained for all plates agree with function (3) within the *whole* range of dP, Fig. 2. The effective porosity determined in the described way takes into account not only the geometrical configuration of the holes but, additionally, contains all the flow losses. Due to this for the same hole geometry one obtains different effective porosities for instance for suction and blowing.

The corrections of porosity for each plate are presented in TABLE 1 below. The second column provides the value of the measured porosity by means of a microscope, the third gives the new corrected values and the last column indicates the difference between both porosity values. All porosities are given in [%]. The correction is performed independently for a blowing and a suction case because sometimes the differences are significant. Blowing and suction, respectively, are marked by the last digit in the plate name. 1- means blowing from the cavity into the stream and 2- means suction from the main stream into the cavity. Most of the corrections are very small. Large differences are indicated by plates K-II/ and D-II/ (/ - means that the holes are inclined). A significant difference is also displayed by the plate B-0.6, however, it seems to be very small in relation to the high porosity of this plate. It should be noted that most of the corrections are below 10%. Only in case of the aforementioned two plates are the corrections around 30%. Fig.2 presents the coincidence of measurement results with function (3) when the corrected porosity values are being used.

116

Let us refer again to Fig.3. The plots present only cases with blowing for clarity. In the legend of Fig. 3 the first numbers provide the measured porosity and the second the corrected one. The corrected porosity values are confirmed by the relative location of the curves. It is especially significant for the plates K-II/. The measured porosity 4.91% is in clear disagreement with the other curve's location. The corrected value of 3.4% fits evidently better.

TABLE 1

OLD			
Name	p	p_{cor}	Diff.
B-0.6-1	26.61	24.9	-1.71
B-0.6-2		24.3	-2.31
B-0.8-1	14.97	15.0	0.03
B-0.8-2		14.3	-0.67
B-1.2-1	6.65	6.1	-0.55
B-1.2-2		5.6	-1.05
B-1.7-1	3.31	3.1	-0.21
B-1.7-2		2.9	-0.41

TEST			
Name	p	p_{cor}	Diff.
T-1-1	4.1	4.0	-0.1
T-1-2		4.2	0.1
T-2-1	2.89	2.65	-0.24
T-2-2		2.9	0.01
T-3-1	4.85	5.05	0.2
T-3-2		5.3	0.45
T-4-1	6.12	5.9	-0.22
T-4-2		5.97	-0.15
T-6-1	4.08	4.1	0.02
T-6-2		4.2	0.12

DA			
Name	p	p_{cor}	Diff.
DA-1	2.04	2.27	0.23
DA-2		2.27	0.23

TUNNEL			
Name	p	p_{cor}	Diff.
K-I-1	3.94	4.0	0.06
K-I-2		4.3	0.36
K-II-1	3.94	3.25	-0.69
K-II-2		3.4	-0.54
K-II/-1	4.91	3.4	-1.51
K-II/-2		3.3	-1.61
K-III-1	5.67	5.3	-0.37
K-III-2		5.25	-0.42
K-IV-1	4.94	4.94	0.0
K-IV-2		4.94	0.0
K-IV/-1	5.64	5.14	-0.5
K-IV/-2		4.64	-1.0

ONERA			
Name	p	p_{cor}	Diff.
D-I-1	3.94	3.55	-0.39
D-I-2		3.85	-0.09
D-II-1	3.94	3.3	-0.64
D-II-2		3.25	-0.69
D-II/-1	5.1	3.3	-1.8
D-II/-2		3.25	-1.85
D-III-1	5.12	5.0	-0.12
D-III-2		5.35	0.23
D-IV-1	5.12	4.9	-0.23
D-IV-2		5.35	0.23
D-IV/-1	5.83	5.0	-0.83
D-IV/-2		5.0	-0.83

Based on the corrected porosity values for the porous plates the dependence of dP/P_0 on the unit mass flow rate has been determined. The relation may be described by the following function with the corrected porosity "p_{cor}" as a parameter:

$$\frac{dP}{P_0} = \frac{0.063}{(p_{cor})^2}\left(\frac{\dot{m}_{real}}{F_s}\right)^2. \qquad (4)$$

F_S is the area of the inlet pipe. The porosity is given in %, the mass flow rate in [kg/s] and F_s in [m^2].

In Fig.4 some experimental results are plotted together with the lines calculated by the formula (4). They indicate good agreement for all porous plates.

The function (4) allows one to determine the effective porosity from only one measurement point, but it is suggested to determine the full curve to see if it fits the relation proposed here. If theoretical results are to be compared with experiments it is necessary to make sure that the porosity value used in the calculations is equal to the effective porosity (corrected porosity) of the plate.

10.2.3 Implementation of the porous plate flow model

In reference to the experimental work as basis of the model proposed here a different approach to the case of suction and blowing is necessary. The sketch in Fig.5 illustrates the considerations. The different approach to suction and blowing concerns the choice of stagnation parameters. On the main flow side a turbulent boundary layer develops. Therefore, just at the wall, the stagnation pressure equals the static pressure in the main stream. Stagnation temperature at the wall is equal to the stagnation temperature of the main stream with the accuracy corresponding to the recovery factor.

In the case of suction (from the main stream to the cavity) the stagnation pressure is hence equal to the main stream static pressure (p_s) and the stagnation temperature to T_0. Pressure difference over the plate is $\Delta p = p_s - p_c$.

In the case of blowing the flow is approaching the wall from the cavity side where velocities are very low. Therefore, the stagnation pressure for this case is equal to the cavity pressure p_c. Assuming that the stagnation temperature does not change in the process of suction into the cavity, it may be claimed that stagnation temperature in the case of blowing is equal to the stagnation temperature in the main stream T_0. Pressure difference over the plate is $\Delta p = p_c - p_s$.

For a known pressure drop over a porous plate and appropriate stagnation parameters, one may calculate the Mach number within a single hole independent of the porosity value:

$$M_{hole} = 1.2 \left(\frac{\Delta p}{p_0} \right)^{0.55} . \tag{5}$$

From this Mach number one may calculate the velocity within a hole for given stagnation parameters:

$$v_{hole} = M_{hole} \, a_{hole} = M_{hole} \sqrt{\kappa R T} \tag{6}$$

and, including the temperature relation:

$$v_{hole} = \frac{M_{hole} \sqrt{\kappa R T_0}}{\sqrt{1 + \frac{\kappa - 1}{2} M_{hole}^2}} . \tag{7}$$

Furthermore, one may write a continuity relation for the mass flow rate expressed by the flow parameters within the holes and outside the porous plate where the flow fills up the whole area "A":

$$\dot{m} = A \, p_{cor} \, \rho_{hole} \, v_{hole} = A \, \rho \, v . \tag{8}$$

118

There is the following important question, namely, whether it is necessary to take the changes of density into account. The highest values of $M_{hole}= 0.4$ what corresponds to very little density changes from the stagnation value:

$$\frac{\rho_0}{\rho_{M=0.4}} = 1.08 . \tag{9}$$

In case of neglecting the density changes, the transpiration velocity may be expressed by the following formula:

$$v = p_{cor} \frac{M_{hole} \sqrt{k\,R\,T_0}}{\sqrt{1 + \frac{k-1}{2} M_{hole}^2}} . \tag{10}$$

If one wants to take into account differences in density, one has to distinguish between SUCTION and BLOWING. This is because in the suction case one is interested in the density at the inlet to the plate and in the blowing case one is interested in the density at the outlet from the plate. The formulas then become somewhat more complicated.

10.3 Experimental Investigation of Shock Boundary Layer Interaction with Passive Control

10.3.1 Introduction

The aim of the investigation was to carry out boundary layer profile measurements in order to define the effect of passive control on the boundary layer. The investigation concerned originally four plates which where supposed to have the same porosity of 5.67%. The following effects were to be studied:
1. the effect of hole size on passive control; the nominal hole diameters 0.125 mm for plates K-I and K-II and 0.3 for plates K-III and K-IV, respectively; the distance between holes also changed;
2. the effect of hole inclination upstream of the shock; the plates K-I and K-III have holes normal to the plate surface over the whole length; the plates K-II and K-IV have holes inclined downstream by 30° to the plate normal in the upstream half of a plate and normal holes in the downstream half of a plate; it should be mentioned that originally an angle of inclination of 45° was specified.

The experiments were carried out in the transonic blow-down wind tunnel presented in Fig.6. It works on suction, so the stagnation parameters are very stable. The cross section of the tunnel is 50×200 mm^2. The blow down time is about 10-15 sec. The general layout of the test section is presented in Fig.7. In the supersonic part of the nozzle an appropriate Mach number upstream the shock can be adjusted. In the interaction area a perforated plate with a cavity underneath is located.

The measurement for each flow case consists of:
- flow visualisation by means of Mach-Zehnder interferograms,
- static pressure measurements along the wall and also inside the cavity,
- boundary layer profile measurements.

The measurement of boundary layer profiles was carried out by means of pneumatic probes and therefore restricted to the inlet and outlet profiles, i.e., the profiles upstream and downstream of the porous plates. Measurements have been carried out at two Mach numbers for all plates and cavity cases, namely M=1.27 and M=1.3.

There are two boundary layer profiles that do not depend on the ventilation; these are:
- the boundary layer profile at the beginning of the porous wall; it is relatively far upstream the interaction area and this boundary layer is not affected by the interaction, hence it has been used as a reference one for all test cases,
- the boundary layer profile downstream of a smooth surface (the perforated plate was covered with a very thin adhesive film); this is also a single profile common to all cases, thus being a downstream reference profile.

In the case of M=1.27, the shock is located exactly at the centre of the cavity; at M=1.3 it is shifted 10 mm downstream. For all plates a case without ventilation and with ventilation through the 70 mm long cavity have been investigated.

10.3.2 Mach number at the wall

In order to follow and analyse the boundary layer development along the wall, it is useful to provide the Mach number distribution along the wall. Fig.8 and Fig.9 present the cases without passive control for M=1.27 and M=1.3, respectively. In each of these plots a case of smooth wall, i.e., the wall surface covered by a thin adhesive film, is presented and labelled "smooth". Furthermore, the distributions for all plates are presented with closed cavity but with the holes exposed to the flow. Therefore, the existing effect is caused by the hole-induced roughness. The effect is small but noticeable. Wall roughness causes stretching of the shock wave due to an earlier beginning of the Mach number drop. It is also observed that the curves for large holes K-III and K-IV indicate a stronger deviation. Upstream and downstream the agreement of results for all plates is very good at both Mach numbers.

The wall station "0 mm" is located at the beginning of wall porosity. At this location the upstream boundary layer has been measured. The "100 mm" distance corresponds to the end of wall porosity and the location of the downstream measurement station. At the Mach number M=1.27 the shock is located at the 50 mm wall distance. The shock wave for M=1.3 is located at 60 mm. The 70 mm long cavity never reached so far upstream and downstream, respectively, as the wall porosity would allow.

The effect of the passive control is presented in Fig.10 for M=1.27 and in Fig.11 for M=1.3. In each plot a case with smooth wall is presented as a reference for the other distributions. Cavity size and location is marked by arrows.

Passive control significantly influences the Mach number distribution along the wall. The peak Mach number is not reached due to the shock disturbance by blowing through the porous wall in the upstream part. A significant flattening of the curve takes place. Downstream of the shock the curves for passive control indicate Mach number values higher than the reference case.

In the Mach number distributions no significant effect of the porous plate type can be detected. However, for the large-hole cases K-III and K-IV the distribution is flatter. It means that upstream of the shock the values of Mach number are lower and downstream of the shock these values are higher than for K-I and K-II. There is, however, some noticeable deviation for plate K-II. The difference is qualitatively in the direction of the "smooth" surface without ventilation. It is only present upstream of the shock where the inclined holes are located. Let us point out here that the corrected porosity has indicated a significantly lower value than the other plates (see Table 1, plate K-II/) which is reflected in the different behaviour of the pressure distribution upstream of the shock.

10.3.3 Boundary layer profiles without passive control

The comparison of profiles without passive control allows to observe the change of the profiles between inlet and outlet sections, i.e., the effect of the shock boundary layer interaction on the boundary layer development. For the porous plates sealed on the cavity side (no ventilation), it shows how the porosity induced roughness affects the state of the boundary layer at the outlet section. The corresponding results are presented in Fig.12 and Fig.13 for M=1.27 and M=1.3, respectively.

Shock boundary layer interaction appears to have an extremely strong effect upon the boundary layer profile (comparison of „upstr." and „smooth"). It brings the boundary layer close to separation. The effect is stronger for the higher Mach number.

It turns out that the plate roughness caused by the holes itself with closed cavity displays an effect that is not negligible. At M=1.27 (Fig.12) all profiles are very close to the profile obtained with the smooth surface, however, some differences between porous plates should be indicated. Plates K-I and K-II show nearly identical profiles as in the smooth wall case. These plates have very small holes of 0.125 mm diameter. Plates K-III and K-IV with large holes (0.3 mm) show less full profiles away from the wall, i.e., a stronger disturbance. This effect, although small, is apparent.

An additional effect to be indicated is the difference between cases with normal and inclined holes in the upstream half of the plate. It can be observed that plates with inclined holes show fuller profiles, hence generate smaller disturbances.

At M=1.3 (Fig.13) the profile even in case of the smooth surface is very close to separation. Conditions are so critical that the roughness effect of 5% porosity induces separation for all plates except K-I. Only the profiles of K-III and K-IV confirm the previous conclusion concerning the effect of inclined and normal holes, respectively.

The results obtained show that at the Mach numbers of our investigation the flow is extremely sensitive to disturbances and even porosity induced roughness must be considered as important factor in provoking separation. This observation is very important in the case where the boundary layer state is of great concern. It also means that the Mach number range of interest includes conditions where even relatively weak attempts to control the boundary layer may cause significant negative effects.

10.3.4 Passive control effect

The effect of ventilation is presented in Fig.14 for M=1.27 and in Fig.15 for M=1.3. The profiles are compared with the profile for the smooth wall, the latter being the same as in Figs.12 and 13. The effect of passive control is very strong. In all cases investigated it induces separation. This result could be expected after the observation of the roughness effect. Note that the measurement method used does not allow to determine a reversed flow velocity distribution.

At M=1.27 there are small but noticeable differences between the various porous plates. It can be observed that the plates K-I and K-II display fuller profiles than K-III and K-IV. This is apparently caused by the different roughness effect observed.

At M=1.3 the profiles for K-I, K-II and K-IV are identical, only plate K-II indicates a noticeably better profile. It is rather exceptional and can not be considered as an inclined-hole effect. It is probably the result of the reduced effective porosity in the upstream part of the plate decreasing the disturbance induced by bleeding.

The results presented in Fig.14 and Fig.15 demonstrating the effect of passive control on normal shock wave turbulent boundary layer interaction do not provide enough evidence to make any final statement concerning the effect of inclined holes on the boundary layer development. This is due to the strong disturbance to the boundary layer at passive control.

The effect of ventilation and porosity on the boundary layer integral parameters is presented in the TABLE 2; - listed are: the boundary layer thickness $\delta_{0.995}$, the displacement thickness δ_1 and the boundary layer shape parameter H_{12} as well as the incompressible shape parameter H_i. For each plate and each Mach number the case without ventilation and with ventilation has been included. From TABLE 2 a significant negative effect of passive control on the boundary layer state becomes evident. For instance, the boundary layer shape factor H_{12} is doubled compared to the no-ventilation case, essentially due to separation caused by the ventilation.

TABLE 2

Upstream, undisturbed profile				
Plate	Description	$\delta_{0.995}$	δ_1	H_{12} / H_i
upstream	-----	3.93	0.38	1.71 / 1.21
Downstream profiles				
Plate	Description	$\delta_{0.995}$	δ_1	H_{12} / H_i
M=1.27				
Smooth	no ventilation	7.55	2.40	2.50 / 2.13
K-I	no ventilation	8.03	2.56	2.61 / 2.23
	with ventilation	12.47	6.54	5.71 / 5.02
K-II	no ventilation	7.58	2.28	2.50 / 2.13
	with ventilation	12.18	6.17	5.37 / 4.72
K-III	no ventilation	8.89	2.77	2.44 / 2.08
	with ventilation	12.58	6.52	4.91 / 4.26
K-IV	no ventilation	8.22	2.70	2.55 / 2.18
	with ventilation	12.89	6.99	6.16 / 5.45
M=1.3				
Smooth	no ventilation	7.92	2.85	3.04 / 2.58
K-I	no ventilation	8.56	3.01	3.08 / 2.61
	with ventilation	12.59	7.15	6.65 / 5.84
K-II	no ventilation	8.11	3.31	3.45 / 2.97
	with ventilation	11.83	6.36	6.10 / 5.34
K-III	no ventilation	9.19	3.70	3.97 / 3.45
	with ventilation	13.15	7.29	6.51 / 5.69
K-IV	no ventilation	7.64	3.26	3.87 / 3.36
	with ventilation	12.91	7.26	6.80 / 5.98

10.4 Numerical Simulation of SBLIC

10.4.1 Numerical code

In the past a zonal method has been used [4, 5] for the numerical simulation of the channel flow with SBLI. This method is an effective simulation tool for viscous flows. However, the interactive coupling between domains demanded significant experience and knowledge, especially when ventilation was applied. Besides, it is rather complicated in this method to introduce appropriate turbulence models to describe such strongly disturbed boundary layers properly. Therefore, the results concerning cases with passive control must be considered with limited confidence, most of all those concerning viscous drag.

Due to this reason a new numerical 2D-solution is presented in the present report based on the full Navier-Stokes equations, which describes the complete flow field.

In this study, the standard code of the ISS (University of Karlsruhe), named KAPPA, has been used. This is a Navier-Stokes solver for compressible viscous flows. The code is a cell centred finite-volume scheme with an explicit Runge-Kutta procedure and an implicit LU-SSOR integration in time. In order to increase the convergence rate of the numerical scheme, several convergence techniques, such as local time stepping, implicit residual averaging, multilevel and multigrid, are used. The results obtained were calculated with the Baldwin-Lomax turbulence model.

The boundary conditions concerning passive control have been formulated using the ventilation model described above. The ventilation velocities are dependent on the pressure difference across the porous wall. The pressure inside the cavity, assumed to be constant, is iteratively determined fulfilling the mass conservation law. The resulting normal velocity is used as boundary condition for the Navier-Stokes solution. The grid used for the calculation of the whole channel was based on 290×66 cells.

10.4.2 Results

Figs. 16 and 17 present a comparison between iso-density contour maps obtained numerically and by means of a Mach-Zehnder interferometer for the cases with and without passive control. In the reference case without control (Fig.16a), the shock is centred between two vertical lines which indicate the beginning and the end of the cavity. The corresponding Mach number just upstream of the shock is M=1.3. The shock is very sharply defined in the calculation and a very good prediction of the flow structure can be recognised even in small details. Only just downstream of the shock, in the expansion area at the boundary layer edge a certain smearing of the solution is observed.

In case of passive control the cavity length is 70 mm and the wall porosity is 5.3%. Here too, a very good agreement of prediction and experiment can be observed. The compression propagates up to the upstream edge of the cavity. There is a very strong effect upon the boundary layer downstream of the interaction. The boundary layer is separated due to the interaction, confirming our experimental results. In this case also a smearing of the solution in the expansion region downstream of the shock can be observed.

Figs. 18 and 19 present a comparison of predicted and measured static pressure distributions along the wall in the interaction area. A very good prediction of the measured pressure upstream of the shock is obtained including shock location and the pressure gradient. Downstream of the shock location there is, however, a considerable discrepancy. The possible reasons may be:

- in the experiments the side wall effects are present, which may cause an additional contraction of the channel, accelerating the flow,
- the numerical method itself, i.e., especially the Baldwin-Lomax turbulence model, is not suitable for internal flows with strong interactions.

10.5 References

[1] J. Zierep "Grundzüge der Strömungslehre" Braun Verlag, ISBN 3-7650-2038-9.

[2] W. Braun "Experimentelle Untersuchung der turbulenten Stoß-Grenzschicht-Wechsel-wirkung mit passiver Beeinflussung", PhD thesis, Karlsruhe 1990.

[3] G. Dargel "Flow through a Perforated Surface Used for Shock Control" Deutsche Aerospace Airbus GmbH, Working Paper.

[4] T. Breitling "Berechnung Transsonischer, reibungsbehafteter Kanal- und Profilströmungen mit passiver Beeinflussung", PhD thesis, Karlsruhe 1989.

[5] R. Bohning, P. Thiede, G. Dargel, "Numerical simulation of viscous transonic airfoil flows with passive shock control", Acta Mechanica Suppl. (1994).

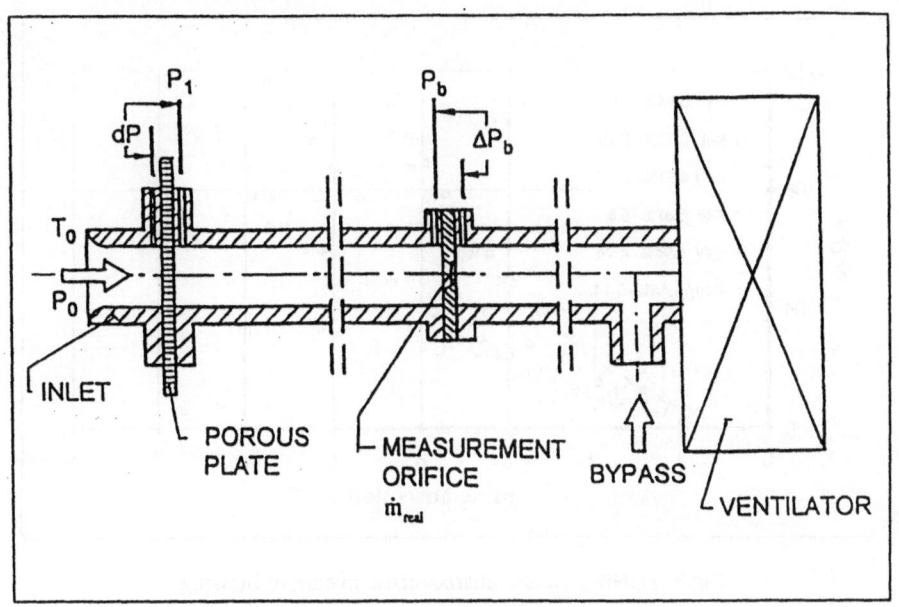

Fig.1 Plate transpiration measurement set up

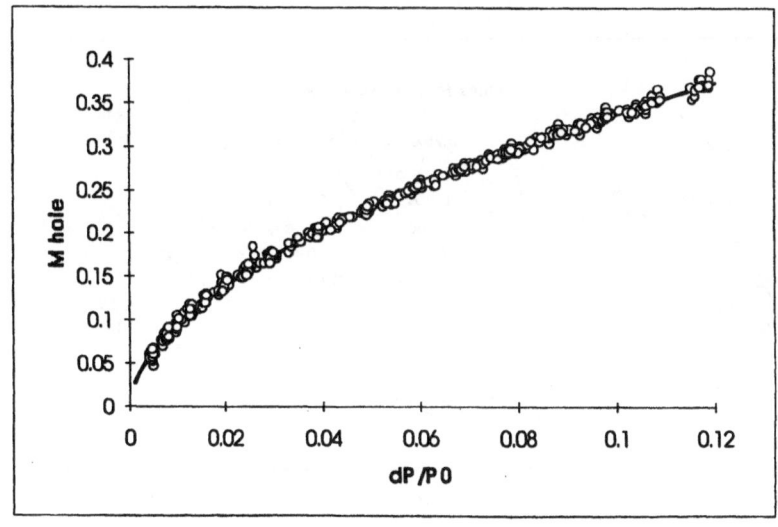

Fig.2 Model function with measurement points for corrected porosity value

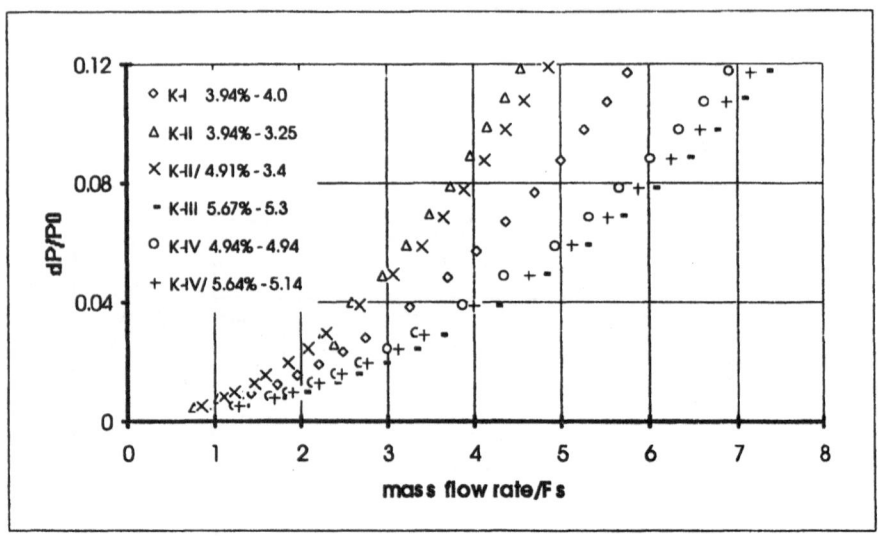

Fig.3 TUNNEL plate. characteristic in case of blowing

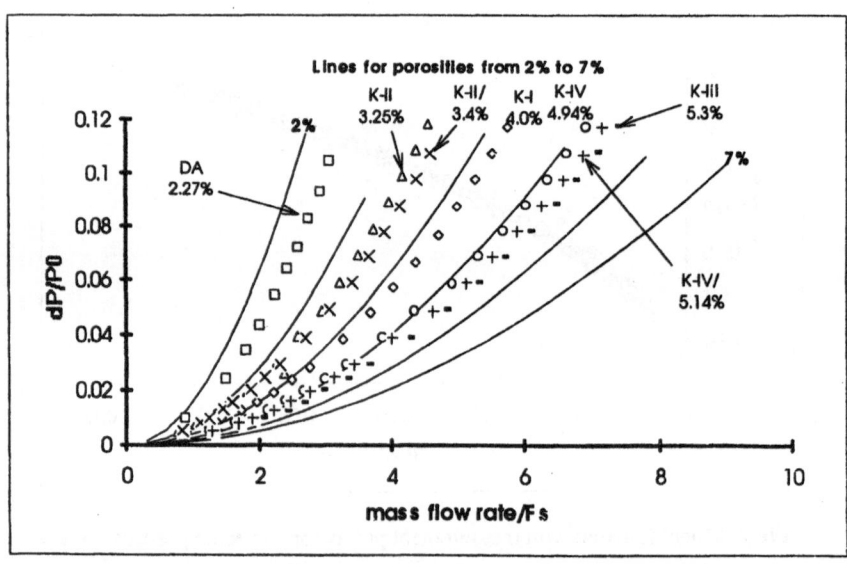

Fig.4 Comparison of TUNNEL and DA plates with general function in case of blowing

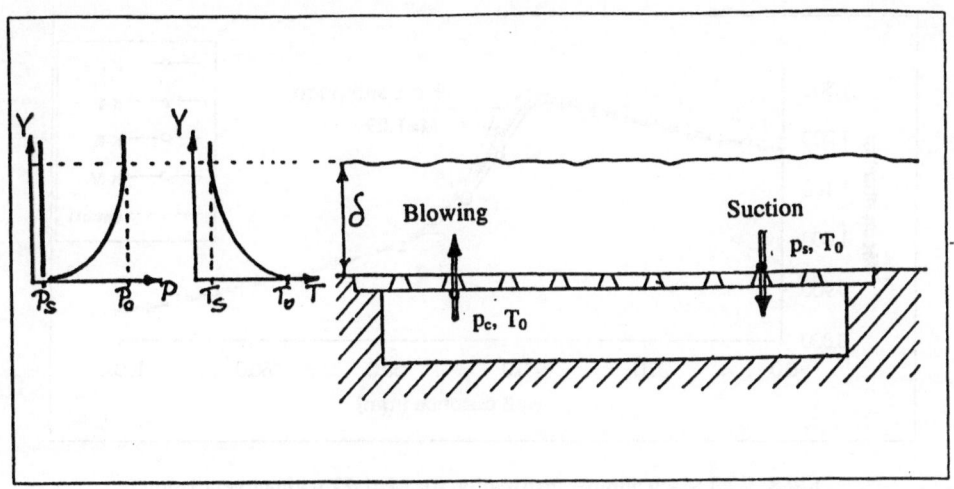

Fig.5 Transpiration flow model

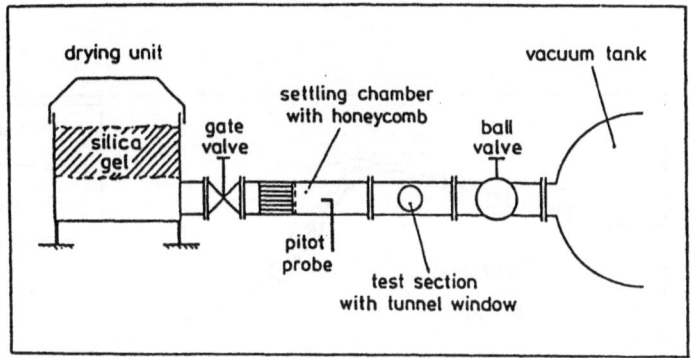

Fig.6 Wind tunnel arrangement

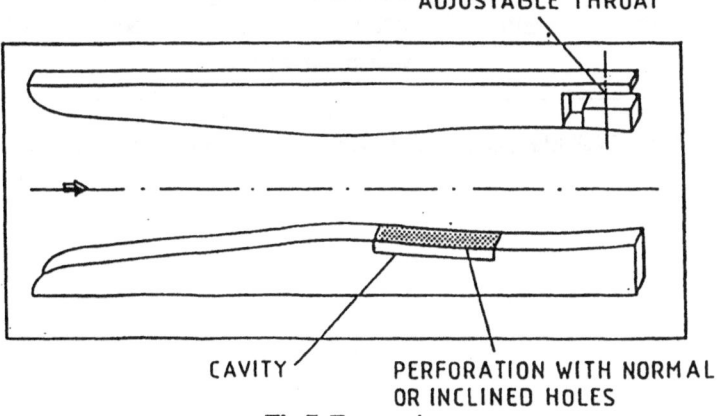

Fig.7 Test section

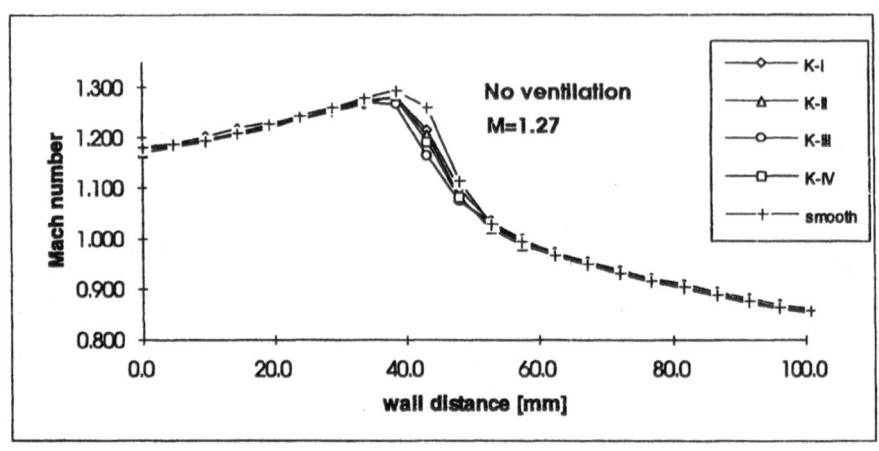

Fig.8 Wall Mach number distribution for M=1.27 without ventilation

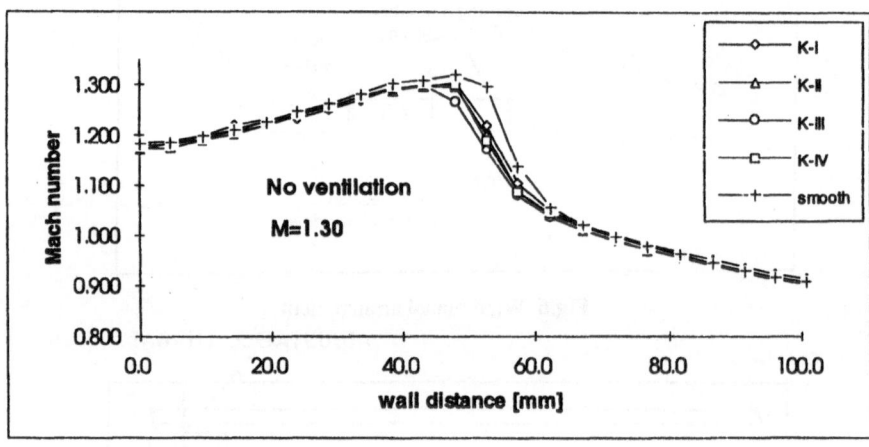

Fig.9 Wall Mach number distribution for M=1.3 without ventilation

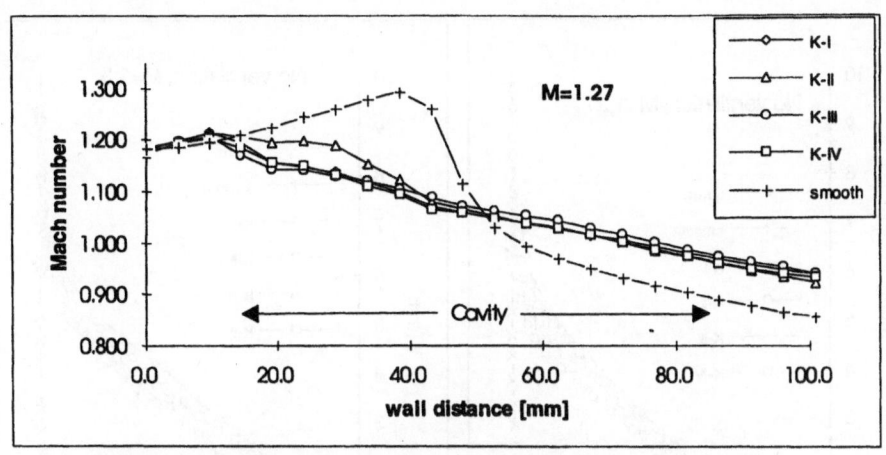

Fig.10 Wall Mach number distribution with passive control, M=1.27

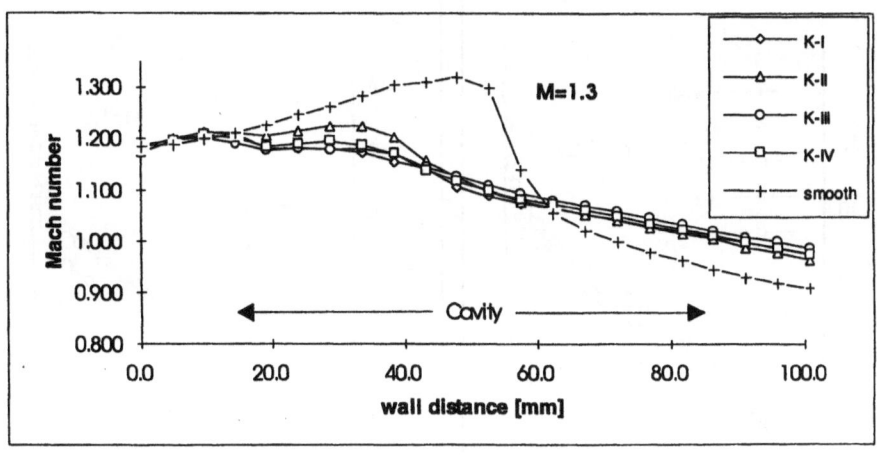

Fig.11 Wall Mach number distribution with passive control, M=1.3

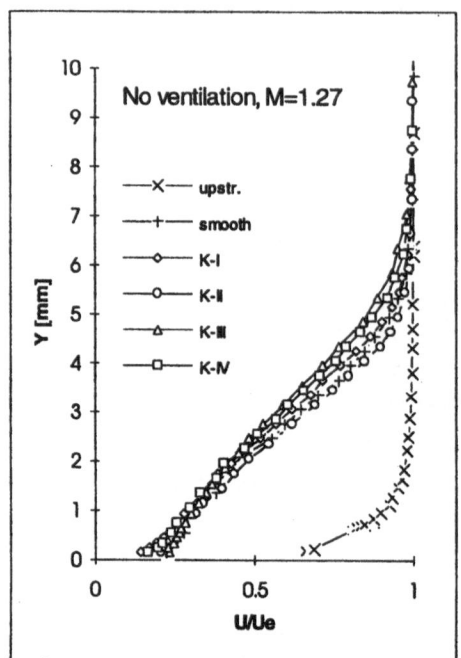

Fig.12 Boundary layer profiles at M=1.27

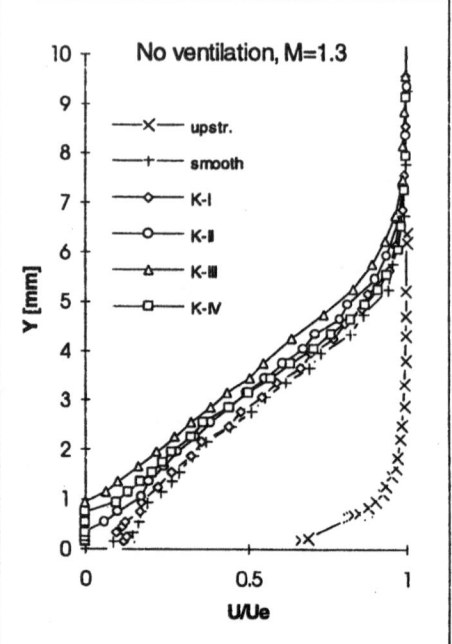

Fig.13 Boundary layer profiles at M=1.3

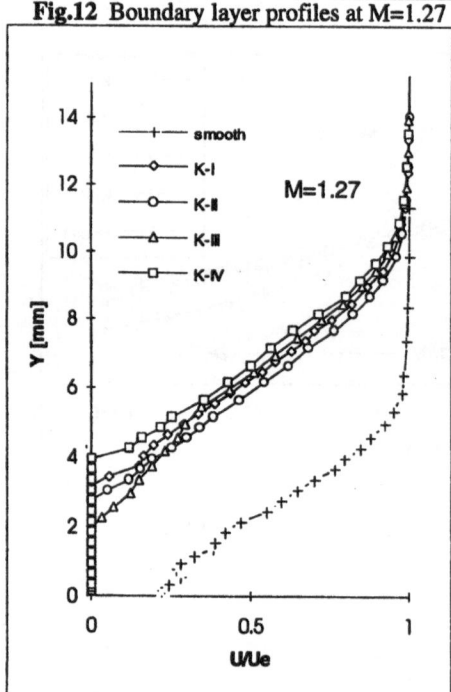

Fig.14 Ventilation effect at M=1.27

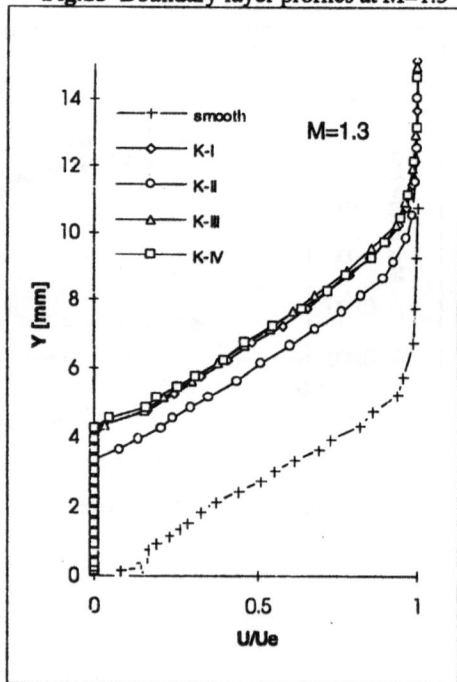

Fig.15 Ventilation effect at M=1.3

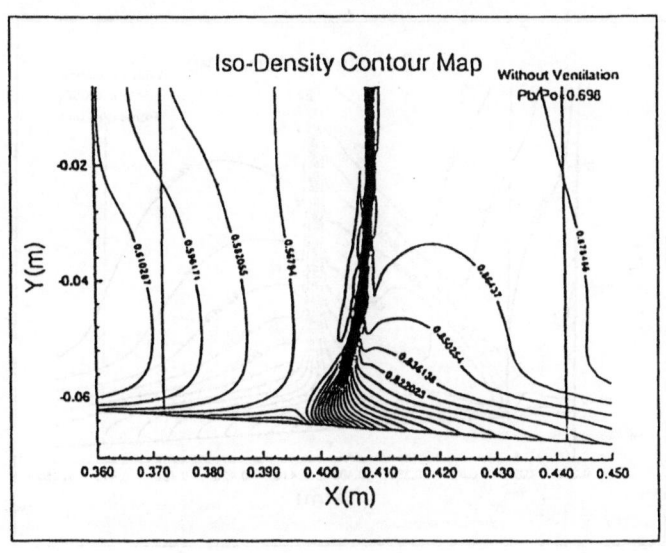

a) Numerical simulation

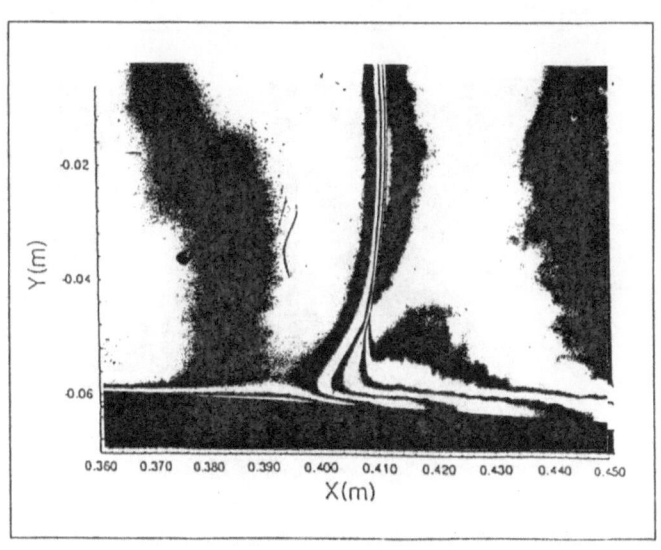

b) Mach-Zehnder interferogram

Fig.16 A comparison between iso-density contours without passive control

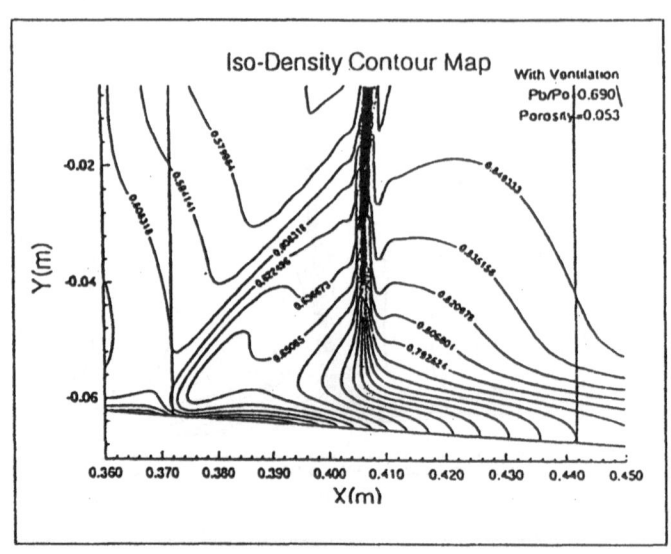

a) Numerical simulation

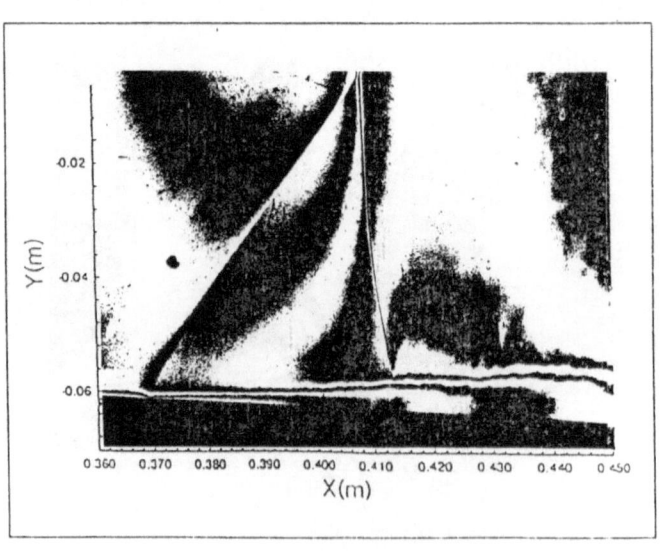

b) Mach-Zehnder interferogram

Fig.17 A comparison between iso-density contours with passive control

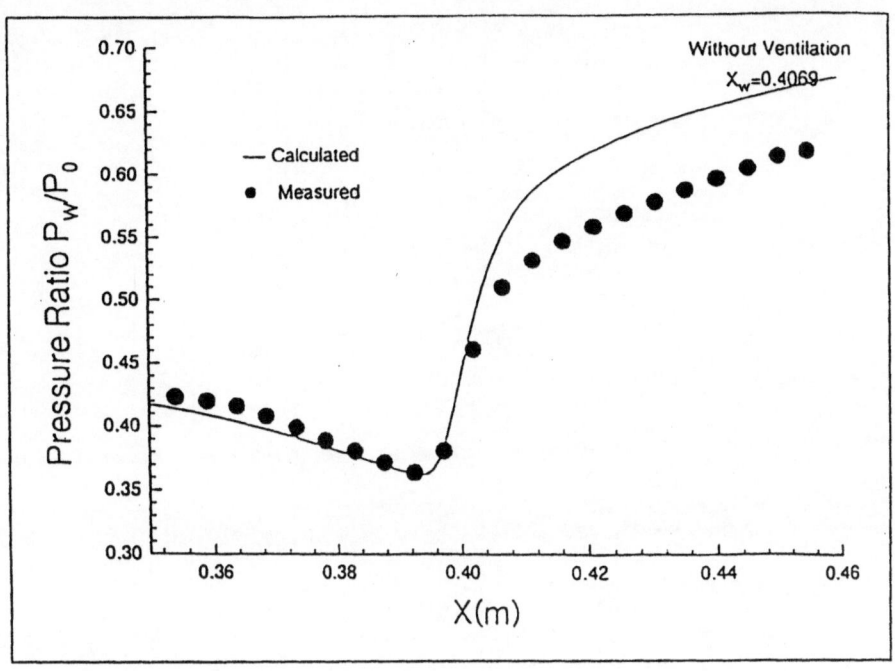

Fig.18 Comparison of pressure distribution along the wall without ventilation

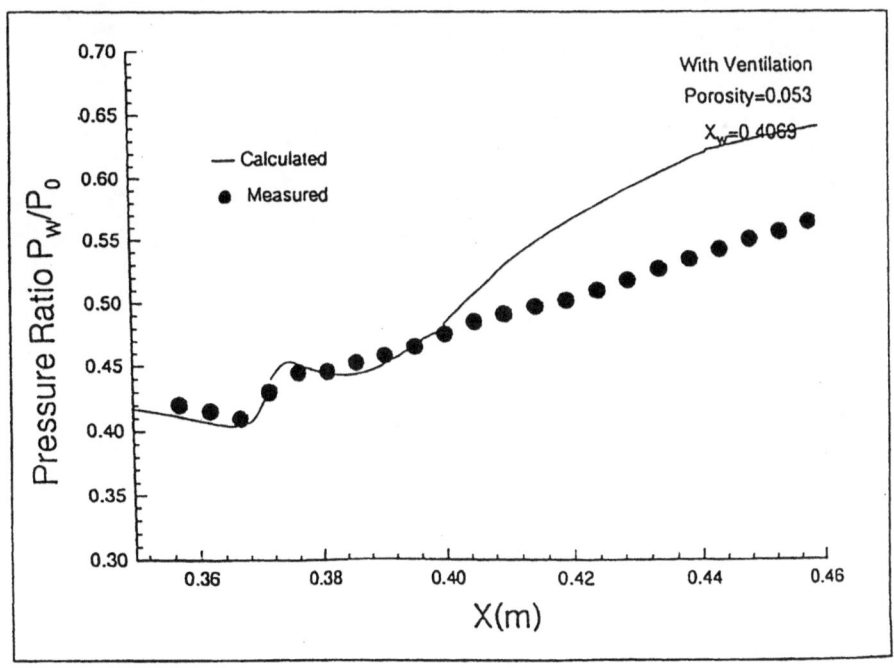

Fig.19 Comparison of pressure distribution along the wall with ventilation

11   AN INVESTIGATION OF PASSIVE CONTROL APPLIED TO SWEPT SHOCK-WAVE / BOUNDARY-LAYER INTERACTIONS

L.C. Squire, A.F.K. Yeung and X. Faucher

University of Cambridge
Engineering Department
Trumpington Street
Cambridge, CB2 1PZ, UK

Summary: The influence of sweep on the effect of passive control of shock-wave / boundary-layer interaction was investigated experimentally and numerically. Passive control was found to have the same effect as observed in other, two-dimensional, investigations. A pressure plateau was formed upstream of the shock and both the displacement and momentum thickness of the boundary-layer downstream of the interaction were increased significantly by the introduction of passive control. Pressure gradients through the interaction were reduced. The variation of sweep was found to cause no significant changes in the effects of passive control. The numerical investigation was also performed using a three-dimensional Navier-Stokes code modelling the wall mass flow through the porous plate with a new empirical relationship. Good agreement was found between the experimental and numerical data with and without passive control.

11.1 Introduction

The work reported in this section is the University of Cambridge contribution to Task 1 of the Euroshock project. In particular the aim is to investigate the effect of passive control applied to a swept shock / boundary-layer interaction of relatively weak strength, i.e. without separation or with weak separation, as commonly found on transonic swept wings.

The concept of passive control has been widely investigated for two-dimensional configurations but, to date, only one experiment has, to the knowledge of the authors, investigated passive control on a swept wing model (Thiede and Krogmann, 1988). They found that drag reductions in the mid-span region could reach values similar to those on two-dimensional aerofoils, but deteriorated towards both the inner and outer wing sections. In the current study an oblique shock-wave, generated by a wedge mounted in the ceiling of a supersonic wind tunnel working section, interacts with the side wall boundary-layer (see Fig. 1). If the interaction region is sufficiently long to minimise end effects, this configuration is a good model for the type of shock-wave / boundary-layer interaction found on a swept wing. By changing the Mach number of the oncoming flow, the shock angle can be varied and thus a similar interaction is formed at a different angle of sweep.

In parallel with the experimental investigation, a numerical study is performed which simulates the same flowfield using a three-dimensional Navier-Stokes flow solver. A variety of turbulence models can be implemented. The porous surface forming the passive control is implemented by using a suitable model for the mass flow through it and ensuring the conservation of mass along the boundary.

11.2 Experimental Apparatus

All experiments were performed in one of the intermittent supersonic blow-down wind-tunnels at the Engineering Department of the University of Cambridge. It has a working section of height 173 mm and width 114 mm, and can be fitted with doors equipped with either pressure tappings or 203 mm diameter optical windows. During a typical run time of 15 to 20 s the stagnation pressure is held constant within ±0.25% and the stagnation temperature variation is below ±1%. Both quantities were assumed constant in this study.

A schematic of the experimental configuration is given in Fig. 1. The wedge angle is 6° in all experiments. A variation of interaction sweep was achieved by changing the oncoming flow Mach number, thus changing the shock-wave angle. As can be seen from the table below, the Mach number normal to the shock front (M_n), the inviscid pressure rise through it (p_2/p_1) and the Reynolds number are relatively constant within the parameters chosen, with the M = 2.5 case giving the strongest interaction.

M_∞	β (°)	p_0 (MN/m^2)	M_n	p_2/p_1	Re (m^{-1})
1.50	49.3	0.20	1.14	1.34	3.1×10^{-7}
1.80	39.5	0.24	1.14	1.36	3.3×10^{-7}
1.85	38.3	0.24	1.15	1.37	3.3×10^{-7}
2.50	28.3	0.30	1.19	1.51	3.1×10^{-7}

Referring to Alvi and Settles' (1991) proposed inviscid separation criterion ($M_n < 1.2$) all test cases would be expected to exhibit attached flow without passive control, with the M = 2.5 case being marginal.

The passive control and the instrumentation are mounted on a rotating plate in order to allow alignment with the shock angle. The inviscid shock location is along the centre of the plate. The porous plate used in the experiments was manufactured from titanium by Aerospace & Technology Ltd, UK. It is 1.2 mm thick and has a porosity of 8% with laser-drilled 0.0762 mm (0.003 inch) holes covering an area of 50.8 mm × 190.5 mm.

The plate was mounted above a plenum chamber which was fitted with two partition walls to prevent spanwise ventilation due to end-effects. Preliminary tests with and without these partitions proved the effectiveness of the arrangement and a number of variations were investigated without any significant differences in the results.

Pressure tappings of 0.3 mm diameter were drilled into the porous plate as well as the rotatable sidewall at intervals of 9 mm or 6 mm along three parallel axes at 45° to the axis of the porous plate as seen in Fig. 2. The direction of these axes, being fixed to the rotatable sidewall, could not be lined up with the initial flow direction and had a misalignment of +5°at $M_\infty = 1.5$, -5° at $M_\infty = 1.8$ and -15° at $M_\infty = 2.5$. Boundary-layer traverses were also performed at about 20 positions along the same three axes. A second wind-tunnel sidewall equipped with over 200 pressure tappings was used for obtaining detailed pressure distributions for the case without passive control. These holes were spaced at quarter inch (6.3 mm) and half inch (12.7 mm) intervals in the streamwise and spanwise directions respectively.

11.3 Computational Method

All computations were performed with a finite-volume implicit algorithm which solves the full three-dimensional Reynolds-averaged Navier-Stokes equations as described by Dawes (1987). The code was originally developed to investigate turbomachinery flows and was later modified by Leung and Squire (1995) for the current configuration. Note that in the present calculations the energy equation is not solved and that total enthalpy is assumed constant. This results in considerable savings of computing time with little effect on the overall accuracy, as discussed by Holmes and Squire (1992). The numerical investigation was carried out in two stages. The first stage involved preliminary calculations with relatively sparse grids to enable modifications and adjustments to the code and the selection of a turbulence model, while the second stage was performed on a larger domain with a finer grid. The number of grid points and equivalent physical dimensions for each computation is given below:

	Preliminary Calculations		Final Calculations	
M_∞	Physical Dimensions ($x \times y \times z$)	Number of Grid Points	Physical Dimensions ($x \times y \times z$)	Number of Grid Points
1.5	$170 \times 76.2 \times 240$	$50 \times 42 \times 70$	$195 \times 76.2 \times 270$	$64 \times 46 \times 80$
1.8	$200 \times 76.2 \times 240$	$50 \times 42 \times 70$	$235 \times 76.2 \times 240$	$64 \times 46 \times 80$
2.5	$310 \times 76.2 \times 240$	$50 \times 42 \times 70$	n/a	n/a

As an example, Fig. 3 shows the computational mesh employed in the final calculations at $M_\infty = 1.8$.

The porous surface was simulated by removing the no-flux condition from the wall-facing side of some of the cells adjacent to the sidewall. Since the shock aligned grid did not overlap the porous plate exactly, all cells cutting the edge of the porous region were treated as part of the porous plate. This increased the effective area of the porous region by less than 1% of the plate size. The flow inside the plenum chamber was not simulated, instead, the static pressure inside the plenum cavity was evaluated to estimate the flow velocity through the surface. This was achieved by using an analytical relationship between the mass-flow through the porous surface and the pressures on either side and iteratively varying the cavity pressure until mass conservation was satisfied, i.e. the net mass-flux was less than 5×10^{-6} kg/s. In order to save computing time, this iteration was performed only once every five time steps, during which the cavity pressure was kept constant.

The functional relationship between mass-flux and static pressures was determined from a number of experiments with a calibration rig, similar to the apparatus described in Chokani and Squire (1993). It was fund that suction and blowing were not strictly symmetrical and the best fit to the experimental data was achieved from the following relationship:

$$\frac{\rho_w v_w}{p_{mean}} = A\left(\frac{p_a - p_b}{p_{mean}}\right)^B , \qquad (1)$$

where $p_{mean} = (p_a + p_b)/2$ and A and B were set to the following values:

	A (sm^{-1})	B
suction	3.15×10^{-4}	0.6052
blowing	2.98×10^{-4}	0.6298

In the general case it appears reasonable that the temperature as well as the tangential wall velocity influence the mass-flux. In the present experiments the temperature of the calibration tests was similar to the wind tunnel test and a temperature dependence is therefore neglected. The effects of tangential velocity however, have not been investigated and are subject to future studies.

The preliminary calculations used three turbulence models, namely Cebeci-Smith, Baldwin-Lomax and Johnson-King. The effect of wall transpiration on turbulence is known to be confined to the inner region of a boundary-layer (Squire 1981). Modifications to the Van Driest damping term to account for wall transpiration effects were implemented into all three turbulence models by Chokani and Squire (1993). To account for the crossflow effects, the T-model proposed by Rotta (1979) for the anisotropic modification was used with a modelling constant of 0.7. A comparison of the wall pressure distributions and the boundary-layer variations, with and without passive control, showed that the performances of the three turbulence models were very similar (see Yeung (1994) for details). However, the need to evaluate the boundary-layer thickness in both the Cebeci-Smith and the Johnson-King models made them less efficient than the Baldwin-Lomax model, and the latter was selected for all final calculations. All results presented here were obtained with the Baldwin-Lomax turbulence model.

11.4 Results and Discussion

11.4.1 Interaction without passive control

Figure 4 shows pressure contours obtained on the wind tunnel sidewall at $M = 1.8$. The theoretical, inviscid, location of the shock generated by the wedge is also indicated. In the mid-span region most of the contours are closely parallel to the inviscid shock line as would be expected in an infinite swept-wing condition. This is confirmed by oil-flow visualisations, as seen in Fig. 5, where the upstream influence line is almost parallel to the inviscid shock line in the mid-span region. The oil-flow visualisations also indicated unseparated flow for all Mach numbers. The pressure contours obtained from the numerical simulation are given in Fig. 6 which also indicate that an infinite swept wing condition has been achieved in the mid-span region of the flow. In general the numerical scheme was found to predict the interaction without control very well. Figure 7 compares the surface pressures measured along the central row (x_2) with the equivalent computational predictions for all three Mach numbers. Figure 8 shows a comparison of boundary-layer displacement thickness distributions (along x_2) obtained from experiment and CFD at $M = 1.8$. As with the pressure data the agreement between CFD and experiment is good.

11.4.2 Interaction with passive control

Figure 9 gives a comparison of the measured surface pressure distributions through the shock / boundary-layer interactions with and without passive control for three Mach numbers. It can be seen that in all cases the onset of the interaction is moved upstream due to passive control. In the M = 1.5 and M = 1.8 cases the rear boundary of the interaction region has also moved further downstream. In these cases a plateau in the pressure distribution is clearly seen for the cases with control, whereas the uncontrolled configurations do not exhibit such a feature. In the M = 2.5 case a plateau is seen even for the flow without passive control, with control the plateau is less distinct, but the interaction is still significantly widened due to the passive control.

Figure 10 gives the displacement thickness distribution throughout the interaction at M = 1.5 with and without control. It can be seen that passive control increases the boundary-layer growth through the interaction and that the onset of boundary-layer growth has moved slightly upstream. The displacement thickness downstream of the interaction is approximately 25% larger than in the solid wall case. The same effects are found for the more highly swept M = 1.8 case, as shown in Fig. 11. The increase in boundary-layer thickness is most likely due to the displacement effect of the upstream blowing. The resulting ramp in the effective boundary-layer edge causes the initial pressure rise upstream of the plateau.

Figure 12 shows a comparison between experimental pressure measurements and the numerical predictions. It can be seen that the agreement is not as good as in the solid wall cases, however, the upstream movement of the interaction as well as the existence of a pressure plateau are well predicted. No consistent conclusion can be drawn from a comparison of the different sweep configurations; at M = 1.8 agreement between experiment and the numerical calculation is relatively good, whereas both the M = 1.5 and the M = 2.5 numerical results significantly underpredict the pressures in the plateau region. An investigation has been undertaken to improve the quality of the numerical predictions and it was found that the pressure in the cavity underneath the porous plate, as obtained from the computation, was below the experimentally measured value. A new computation was performed which used the experimental cavity pressure and it was found that the plateau pressure was increased, however, the mass-flow balance in and out of the cavity was no longer satisfied.

The boundary-layer growth through the interaction with control is also not predicted as well as for the uncontrolled cases, as seen in Fig. 13 which shows a comparison between experiment and CFD for the M = 1.8 case. The displacement thickness downstream of the interaction given by CFD is found to be about 10% above the experimentally determined value.

Finally, Fig. 14 shows a comparison between numerically obtained density contours with and without passive control along a streamwise slice at M = 1.8. It can be seen that the size of the interaction is increased considerably due to passive control and the shock-induced pressure rise is smeared, thus reducing the strength of the shock. The shock-wave structure resembles a typical lambda-foot which explains the plateau observed in the pressure measurements.

From the numerical data the magnitude of wall mass-flux can be obtained and the wall transpiration parameter ($F = \rho_w v_w / \rho_e u_e$) can be calculated. In the present experiments the coefficient F, based on the maximum blowing rate is of the order of 0.01. Most experimental investigations of boundary-layers with distributed injection have concentrated on injection rates F below 0.008, at which value the skin friction is virtually reduced to zero. Thus a value of F = 0.01 represents a very high blowing rate. Since this high blowing rate occurs over only a relatively short distance, followed by suction of similar strength, it does not appear to separate the boundary-layer. However, such a high value of injection must raise doubts as to the applicability of modifications made to turbulence models for blowing and suction, since these modifications are based on results obtained from experiments made on boundary-layers with moderate blowing rates in near equilibrium conditions. In this context it is surprising that the turbulence models used here perform as well as shown.

Considering the effects of sweep, the present results show that when studied in a plane normal to the shock-wave the general effects of passive control are similar to corresponding two-dimensional results. Most of the data shown did not exhibit a significant difference between the three Mach number cases and it appears as if sweep does not change the nature of passive control in any fundamental way. The differences seen in some of the results at the highest Mach number case (M = 2.5) are more likely due to the slightly larger normal Mach number (and hence stronger interaction) than to the difference in sweep. Thus, as in two-dimensional flow, passive control increases the length of the interaction, smears the shock induced pressure rise and so reduces wave drag. On the other hand, the boundary-layer measurements suggest an increase in skin friction due to passive control, as in two-dimensional flow. Thus, the actual effect of passive control on a wing for constant lift is a balance between these two effects and the present tests do not indicate that sweep does anything to significantly influence this balance. Where passive control has been found to be beneficial in two-dimensional tests it should show similar results in a swept configuration.

However, there is one significant effect of sweep which must be considered. With sweep the flow near the surface just ahead of the shock is highly curved and as a result the slight changes in shock position can cause very large changes in flow direction in this region. Thus, holes inclined to match the flow with one shock position may be completely mismatched for the slight changes in shock location which may occur in flight and the potential benefits of inclined holes found in two-dimensional investigations are unlikely to be carried over to swept configurations.

11.5 Conclusions

An experimental and numerical study into the effects of passive control on swept shock-wave / boundary-layer interactions has been performed. Three cases with sweepbacks ranging from 40.7° to 60.7° and normal shock Mach numbers between 1.14 and 1.19 were investigated. It was found that passive control applied to swept interaction shows the same effect as observed in two-dimensional experiments, namely to smear the shock and increase the boundary-layer thickness.

In the presence of passive control a plateau was seen in the wall pressure distributions which is interpreted as a sign of the formation of a lambda-shock foot, induced by the upstream blowing. Boundary-layer displacement thickness measurements confirm this observation and show that he boundary-layer thickness downstream of the interaction is about 25% larger due to passive control. The effect of downstream suction is not as noticeable as the effects of upstream blowing.

Comparison of the experimental results with numerical predictions showed good agreement for flowfields without control and moderate agreement for configurations with passive control. The level of agreement was similar for all of the turbulence models used. The wall transpiration rates obtained from the numerical simulation were found to be relatively large compared to other available experimental data, thus giving rise to doubts over the suitability of turbulence models modified on the basis of previous experiments.

Sweep was found to have little influence on the effect of passive control and it is believed that any difference observed between the various cases is due to variations in the normal shock Mach number. However, swept shock-wave / boundary-layer interactions are subject to a high degree of curvature in streamlines close to the wall underneath the inviscid shock-wave location. Any changes in shock position are likely to change the local flow direction in this region considerably, thus making the usefulness of any alignment of the passive control with the local flow direction doubtful.

11.6 References

Alvi, F.S. and Settles, G.S. (1991), "A Physical Model of the Swept Shock / Boundary-Layer Interaction Flowfield", AIAA Paper 91-1768.

Chokani, N. and Squire, L. C. (1993), "Transonic Shockwave / Turbulent Boundary Layer Interactions on a Porous Surface", *Aeronautical Journal*, May 1993, pp. 163-170.

Dawes, W.N. (1987), "A Numerical Analysis of the Three-Dimensional Viscous Flow in a Transonic Compressor Rotor and Comparison with Experiment", *Trans. ASME, Journal of Turbomachinery*, Vol. 109, pp. 83-90.

Holmes, S.C. and Squire, L.C. (1992), "Numerical Studies of the Supersonic Flow over a Compression Corner", *Aeronautical Journal*, Vol. 96, pp. 87-95.

Leung, A.W.C. and Squire, L.C. (1995), "Reynolds Number Effects in Swept-Shock-Wave / Turbulent-Boundary-Layer Interaction", *AIAA Journal*, Vol. 33, pp. 798-804.

Rotta, J.C. (1979), "A Family of Turbulence Models for Three-Dimensional Boundary Layers", *Turbulent Shear Flows I*, Springer-Verlag, pp. 267-278.

Squire, L.C. (1981), "Turbulent Boundary-Layers with Suction and Injection", *The 1980-81 AROSR-HTIM-Stanford Conference on Complex Turbulent Flows*, Proc. Vol. 1, pp. 112-129.

Thiede, P. and Krogmann, P. (1988), "Passive Control of Transonic Shock / Boundary Layer Interaction", *IUTAM-Symposium Transonicum III, Göttingen*.

Yeung, A.F.K. (1994), "The Passive Control of Swept-Shock / Boundary-Layer Interactions", Ph.D. Thesis, University of Cambridge.

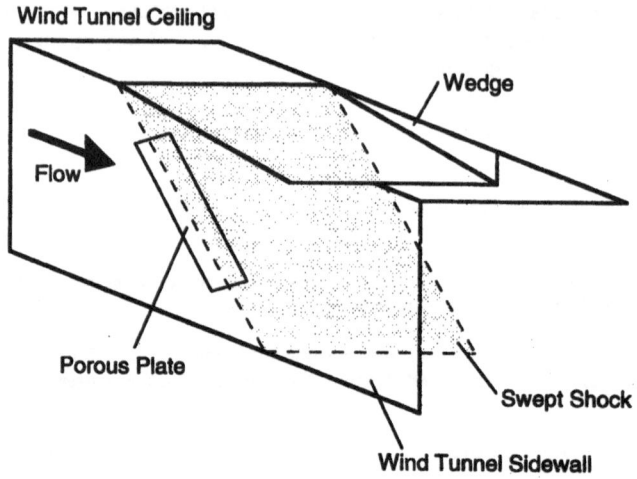

Figure 1   Experimental Configuration

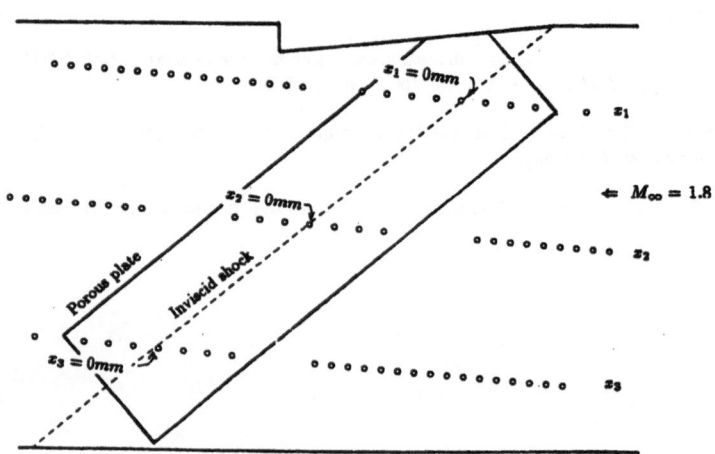

Figure 2   Locations of wall static pressure tapping holes for M = 1.8 tests

142

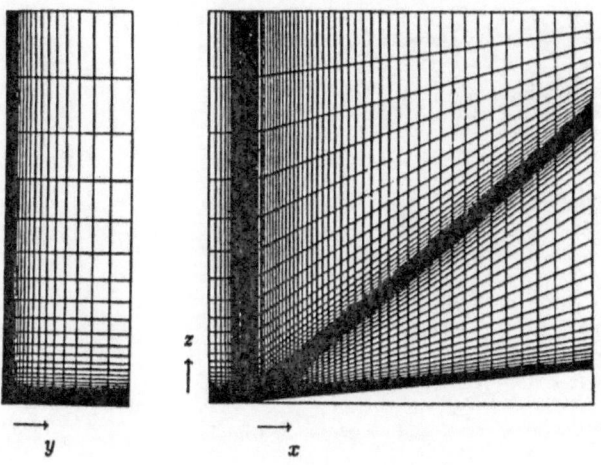

Figure 3   Computational mesh used in final calculations for M = 1.8 case

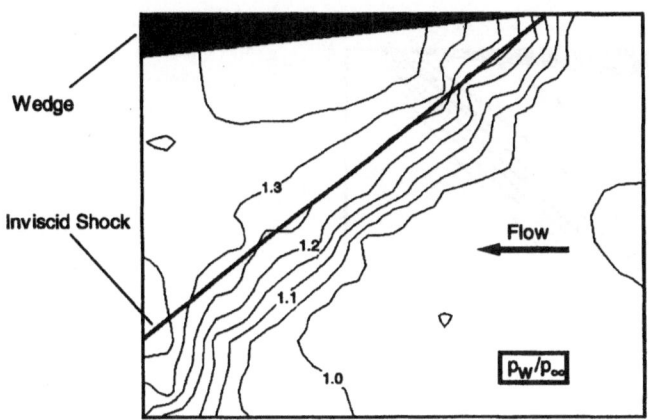

Figure 4   Experimental sidewall pressure contours at M = 1.8 (no control)

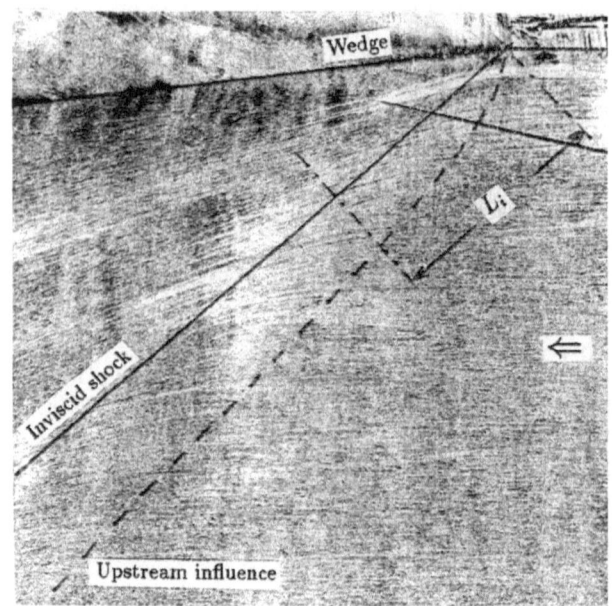

Figure 5   Oil-flow visualisation on sidewall at M = 1.8 (no control)

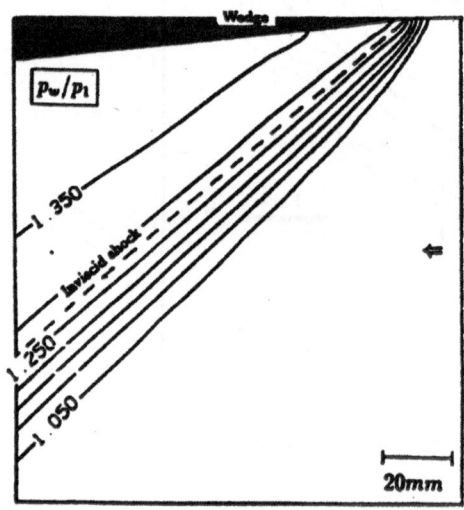

Figure 6   Computed pressure contours on sidewall at M = 1.8 (no control)

144

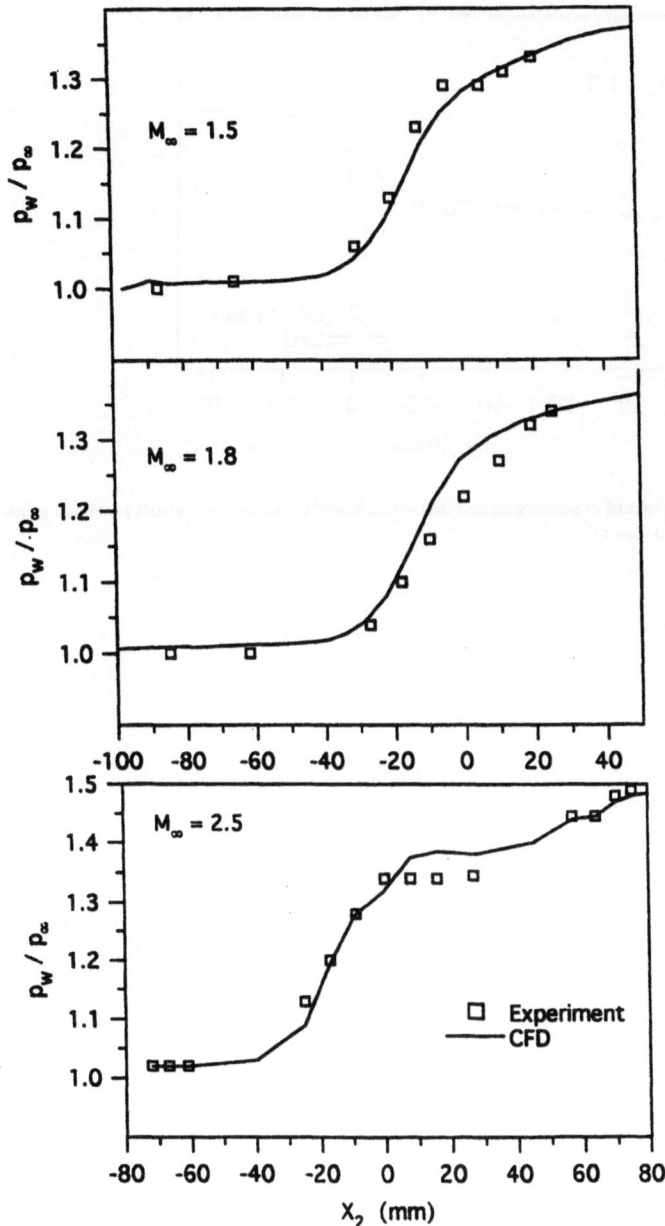

Figure 7   Comparison of experimental and computational wall pressure distributions through the interaction at three Mach numbers (no control)

145

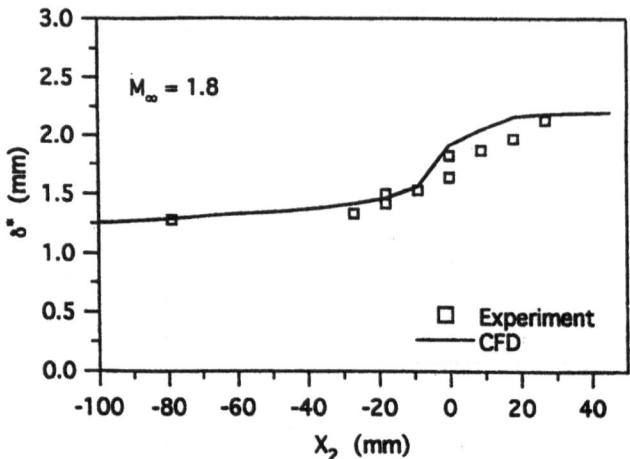

Figure 8 Comparison of experimental and numerical boundary layer displacement thickness distribution through the interaction (no control)

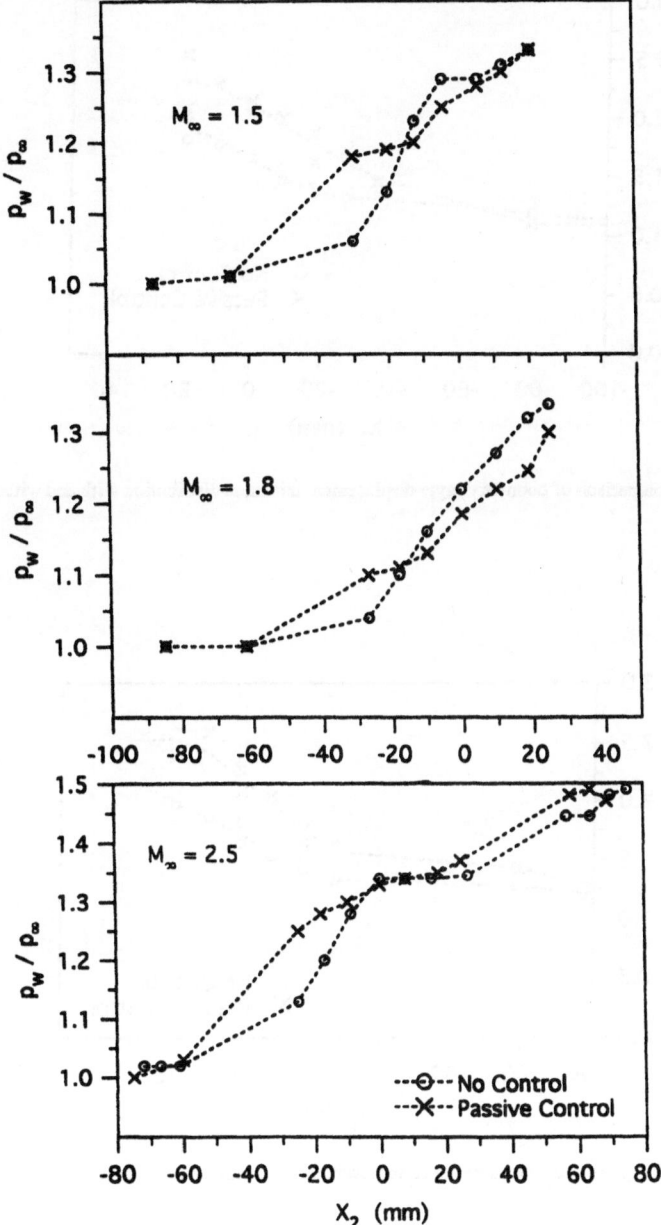

Figure 9   Comparison of wall pressure distributions with and without passive control

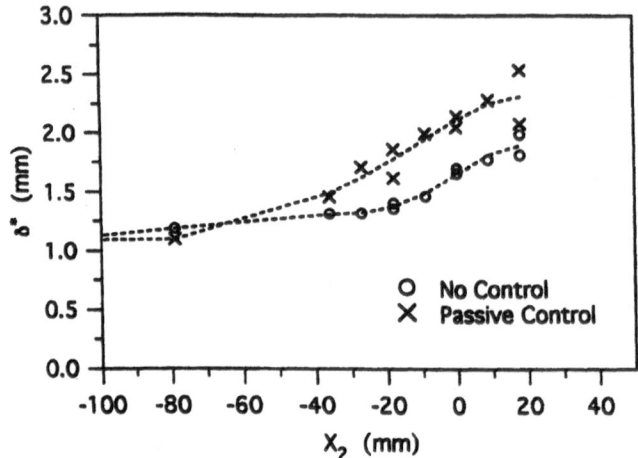

Figure 10   Comparison of boundary-layer displacement thickness distribution with and without control at
M = 1.5

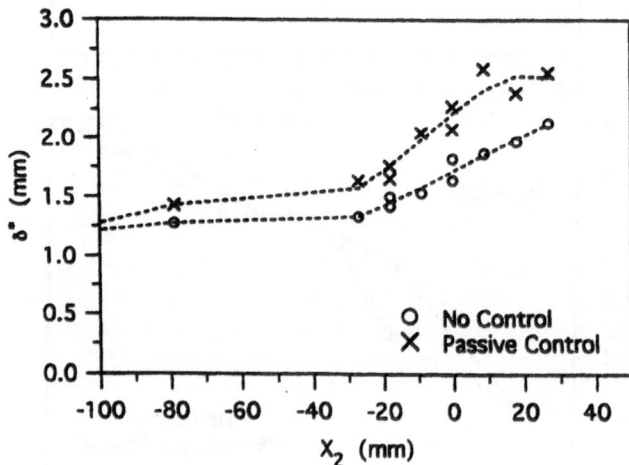

Figure 11   Comparison of boundary-layer displacement thickness distribution with and without control at
M = 1.8

148

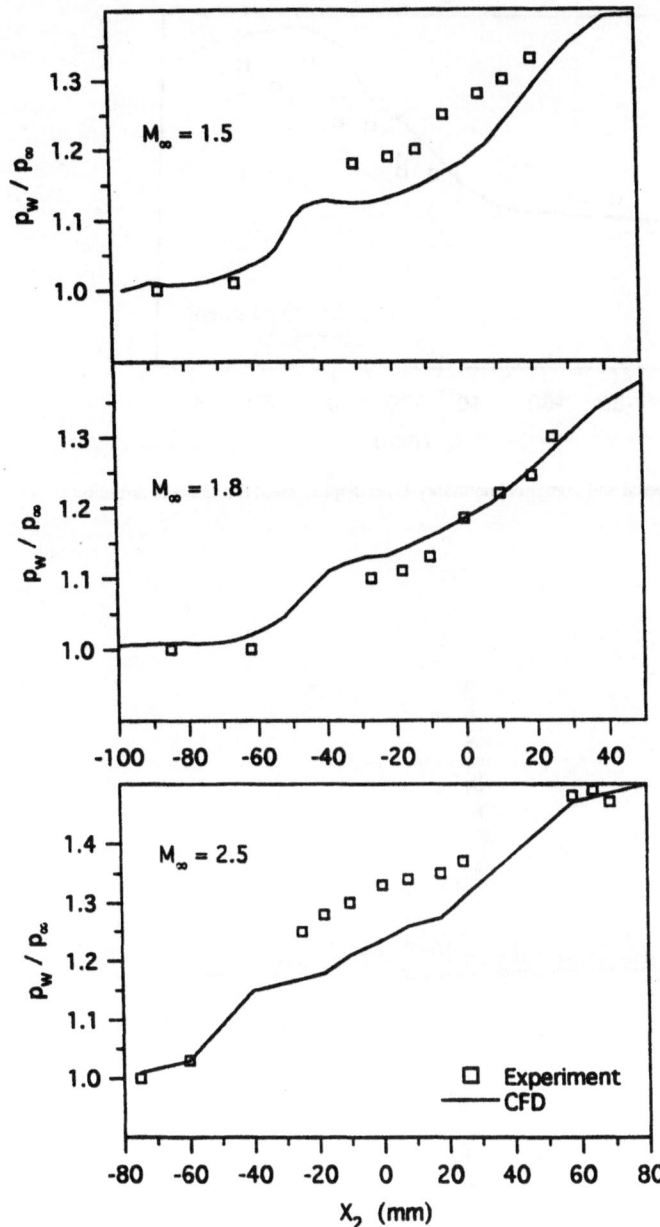

Figure 12 Comparison of experimental and computational wall pressure distributions through the interaction at three Mach numbers (with passive control)

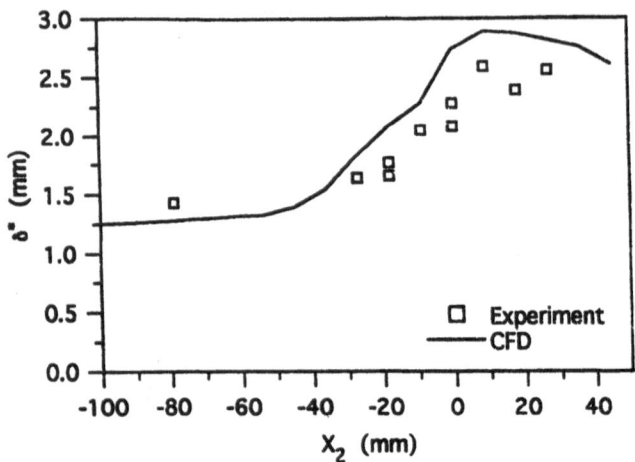

Figure 13 Experimental and numerical boundary-layer displacement thickness distribution at M = 1.8 with passive control

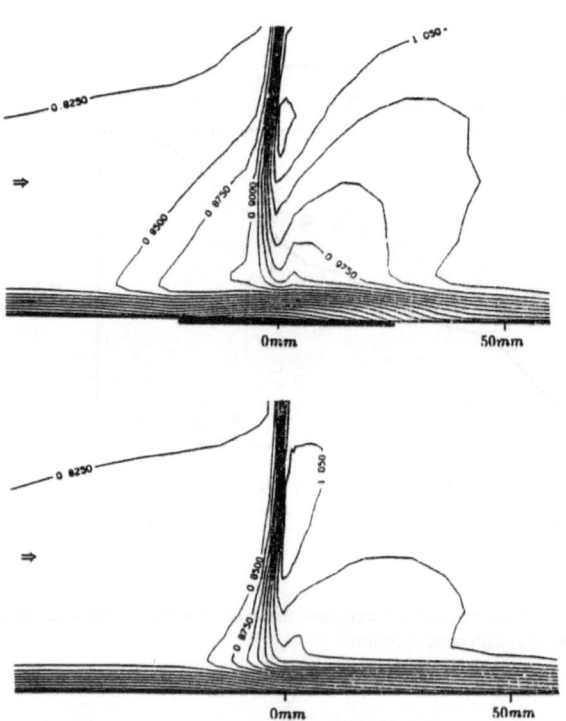

Figure 14 Computational density contours on a plane normal to the inviscid shock at M = 1.8 with passive control (top) and without passive control (bottom)

12 NUMERICAL INVESTIGATION OF THE PASSIVE SHOCK CONTROL ON TRANSONIC AIRFOILS THROUGH AN EULER/BOUNDARY-LAYER COUPLING TECHNIQUE

P. de Matteis, C. Dima

CIRA, Centro Italiano Ricerche Aerospaziali

Via Maiorise, 81043 Capua(CE), Italy

Summary: The present work reports the overall research activity conducted by CIRA within Task 2 of the EUROSHOCK project. The main objective of the activity was to extend the computational capability for the calculation of the transonic viscous flow about airfoils to the simulation of the shock-boundary layer interaction control and to validate it through comparison with the experiments. The method is based on the semi-inverse coupling of Euler equations with integral boundary-layer equations, the latter modified to take the transpiration velocity due to the control into account. In order to simulate the passive control mechanism in the shock region, different models for the calculation of the transpiration velocity in the cavity region of the airfoil surface have been implemented and tested through correlation with experiments performed in Tasks 1 and 3. As far as the effect of passive control on the aerodynamic characteristics of airfoils is concerned, the numerical investigation performed with the present method on the airfoils considered in the project has confirmed the main results of the experiments and of other numerical analyses.

12.1 Introduction

The interaction of shock waves with a turbulent boundary layer is a fundamental aerodynamic aspect of modern transport aircraft, due to its effect on the overall performance at transonic cruise and off-design. The interaction effects on an airfoil are not only limited to the shock region, where complex flow phenomena such as the smearing of the pressure rise across the shock and flow reversal take place, but they also alter significantly the flow properties at the trailing edge possibly resulting in an increase of the base drag. As the Mach number increases, the shock intensity becomes so large to induce separation of the boundary layer. This flow condition represents an upper limit for an aircraft to safely operate, since the unsteady flow phenomena, related to the increasing Mach number or C_l, may have dramatic consequences for the integrity of the airframe.

Thus, the control of the shock-wave boundary layer interaction is not only important to reduce drag but also to prevent or, at most, to delay buffet phenomena. A Shock-Boundary Layer Interaction Control(SBLIC) technique can be applied by using active blowing or suction in the interaction region, or a passive control cavity providing a self-adjusting combination of blowing and suction upstream and downstream of the shock, respectively. Other concepts of SBLIC, which have shown to be effective and inexpensive, make use of bumps [1] or vortex generators [2] positioned upstream of the presumed shock location. Even though some of these concepts for delaying or avoiding shock induced separation and the associated penalties date back to the 50's, they are still of practical interest in current engineering aerodynamic design.

Numerical investigations of SBLIC performed using inviscid flow models, based on the small perturbation theory [3] as well as on Euler equations [4], have dealt only with the wave drag reduction concern. The use of viscous flow models, based on a full-potential/integral boundary-layer coupling [5] or on the solution of the Navier-Stokes equations [6], allows to address the problem of shock control by considering the effect on the overall drag, thus including viscous drag. Indeed, the sequence of experimental tests initiated in 1983 [7] confirmed the importance of the viscous losses increment in the total drag balance as a penalty for the wave drag reduction.

The main goal of the present work was to investigate the numerical treatment of the passive shock control technique on airfoils. To this end, a viscous/inviscid interaction method, EUBL2D, originally developed for transonic flow calculations around airfoils [8], has been modified to include SBLIC simulation [9]. As a first step, the way of adapting Green's integral boundary-layer method to the cases of flows with wall transpiration was investigated on the base of a correlation with experimental data related to a transonic tunnel. As a second step, different models for the calculation of wall transpiration in the control region were implemented into the code and investigated. The comparison with experiments performed on perforated plates was aimed at tuning each law and then at selecting the best transpiration law. Finally, the aerodynamic characteristics of different transonic airfoils(DRA-2303, VA 2, DA LVA-1A) were extensively investigated in comparison with the wind tunnel experiments performed within the same project.

12.2 Extension of the EUBL2D Method to Passive Shock Control Calculations

The EUBL2D method adopted in the present work [8], is based on the semi-inverse iterative coupling of 2D Euler and integral boundary-layer equations and on an adaptive streamline aligned wake model [10,11].

The inviscid part of the flow is computed using the unsteady, two-dimensional Euler equations, which are solved through a standard Jameson's type scheme [12]. The scheme is based on a cell-centred finite volume discretization for the spatial terms and a fourth-order five-stage Runge-Kutta scheme for time integration. Since the transient state is not of interest, the convergence to the steady state is accelerated by using multigrid, enthalpy dumping, residual averaging and the maximum allowed time step in each cell. Riemann invariants are used to calculate boundary conditions at infinity, thus taking the direction of the travelling signals into account.

The calculation of the laminar boundary layer is based on the method of Cohen-Reshotko, an extension of Thwaites' method to the case of compressible flows based on Illingworth-Stewartson transformation [13]. Since the boundary-layer calculation is shifted in the incompressible plane, transition can be predicted using classical criteria for incompressible laminar flows, such as Granville's criterion [14].

The turbulent boundary-layer calculation is based on the direct and inverse formulations of Green's integral method [15]. Also Whitfield's integral method can be used in the inverse part of the boundary layer [16], but its extension to the case of control has not been considered in the present analysis. In both methods, a system of two and three differential equations, respectively, is solved using a fourth order Runge-Kutta integration scheme. In the inverse formulation, the displacement thickness is assigned in place of the inviscid velocity; its distribution is updated at each iteration through Carter's formula [17].

The viscous effects are taken into account following Lighthill's surface source model [18], known as the "equivalent sources" model. In this model, the body thickening due to the boundary layer is accounted for through an equivalent distribution of sources assigned as a wall boundary condition in the inviscid flow field; this results from the integration of the continuity equation across the boundary layer. When a shock control through a perforated plate is applied, a natural or a forced ventilation is present at the wall and the integration yields the following source distribution term:

$$v_e = v_n + v_w \tag{12.1}$$

where

$$\rho_e v_n = \frac{d}{ds} [\rho_e U_e \delta^*] \tag{12.2}$$

is the contribution of the body thickening, with s representing the arc length, ρ_e and U_e respectively the density and the velocity at the edge of the boundary layer, and δ^* the boundary-layer displacement thickness. The velocity v_w represents the ventilation due to control and its modelling will be discussed in the next section.

When the wall transpiration is taken into account, the boundary layer equations and the closure relations have to be modified accordingly. The modifications to be brought to the von Karman equation (integral momentum equation), the lag and lag-entrainment equations, used in Green's integral boundary-layer method, have been introduced following Olling's approach [5]. Firstly, the non-dimensionalized transpiration velocity m_w due to control is introduced; this is defined as:

$$m_w = \frac{\rho_w v_w}{\rho_e U_e} \, . \tag{12.3}$$

The complete set of equations, for both the direct and the inverse formulations, is given below.

- Direct formulation

$$\frac{d\theta}{ds} = 0.5C_f - \left(H + 2 - M^2 \right) \frac{\theta}{U_e} \frac{dU_e}{ds} + m_w \tag{12.4}$$

$$\theta \frac{d\overline{H}}{ds} = \frac{d\overline{H}}{dH_1} \left\{ C_E + m_w - H_1 \left[0.5C_f + m_w - (H+1) \frac{\theta}{U_e} \frac{dU_e}{ds} \right] \right\} \tag{12.5}$$

$$\theta \frac{dC_E}{ds} = C_E \left\{ \frac{2.8}{H + H_1} \left(C_{\tau_{eq}}^{0.5} - C_\tau^{0.5} \right) + \right.$$
$$\left. \left(\frac{\theta}{U_e} \frac{dU_e}{ds} \right)_{eq} - \frac{\theta}{U_e} \frac{dU_e}{ds} \left[1 + 0.075M^2 \frac{1 + 0.2M^2}{1 + 0.1M^2} \right] \right\} \, . \tag{12.6}$$

- Inverse formulation

$$\frac{\theta}{U_e} \frac{dU_e}{ds} = \frac{1}{F_2} \left(\frac{d\delta^*}{ds} - F_1 \right) \tag{12.7}$$

$$\theta \frac{d\overline{H}}{ds} = \frac{d\overline{H}}{dH_1} \left\{ C_E + m_w - H_1 \left[0.5C_f + m_w - (H+1) \frac{\theta}{U_e} \frac{dU_e}{ds} \right] \right\} \tag{12.8}$$

$$\theta \frac{dC_E}{ds} = C_E \left\{ \frac{2.8}{H + H_1} \left(C_{\tau_{eq}}^{0.5} - C_\tau^{0.5} \right) + \right.$$

153

$$\left(\frac{\theta}{U_e}\frac{dU_e}{ds}\right)_{eq} - \frac{\theta}{U_e}\frac{dU_e}{ds}\left[1 + 0.075M^2\frac{1 + 0.2M^2}{1 + 0.1M^2}\right]\Bigg\} \tag{12.9}$$

where

$$F_1 = (1 + 0.2M^2)\frac{d\overline{H}}{dH_1}[C_E + m_w - H(0.5C_f + m_w] +$$

$$H(0.5C_f + m_w) \tag{12.10}$$

$$F_2 = \left(1 + 0.2M^2\right)(H + 1)H_1\frac{d\overline{H}}{dH_1} +$$

$$0.4M^2\left(1 + 0.2M^2\right)\left(\overline{H} + 1\right) - H\left(H + 2 - M^2\right). \tag{12.11}$$

In the above equations, the subscript eq refers to equilibrium flow conditions, M is the Mach number, while $\overline{H}$ and H_1 are defined as

$$\overline{H} = \frac{1}{\theta}\int_0^\delta \frac{\rho}{\rho_e}\left(1 - \frac{U}{U_e}\right)dn \tag{12.12}$$

$$H_1 = \frac{\delta - \delta^*}{\theta}. \tag{12.13}$$

The entrainment coefficient C_E is modified as follows:

$$C_E = \frac{1}{\rho_e U_e}\frac{1}{dx}\int_0^\delta \rho U\,dn - m_w. \tag{12.14}$$

The above set of equations is closed with the following relations:

$$\frac{C_f}{C_{fo}} = 0.9\left(\frac{\overline{H}}{H_o} - 0.4\right)^{-1} - 0.5 \tag{12.15}$$

$$1 - \frac{1}{H_o} = 6.55[0.5C_{fo}\left(1 + 0.04M^2\right)]^{0.5} \tag{12.16}$$

where subscript o refers to the flat plate boundary layer. In the case of flow with transpiration, the flate plate skin friction law is modified according to the experimental work of Kays and Moffat [19]:

$$C_{fo} = C_{fos}\left[\frac{ln(1 + B_f)}{B_f}\right]^{1.25}(1 + B_f)^{0.25} \tag{12.17}$$

$$B_f = 2\frac{m_w}{C_{fo}} \tag{12.18}$$

where the subscript s refers to nonporous surface conditions.

154

The skin friction coefficient C_{fo} is calculated from the above law using a Newton iteration. Because of the presence of the logarithm in the skin friction correlation, the definition domain of C_{fo} is limited in the case with suction.

A preliminary check of the accuracy of Green's method for transpiring turbulent boundary layers was performed by calculating the boundary layer on the wall of the ONERA transonic channel [20]. Olling's original formulation gave a bad agreement with both the experiment and the calculations performed by the University of Naples using a finite-difference boundary-layer solver. A satisfactory agreement was obtained by introducing in the $\overline{H}_o(C_{fo})$ law, defined above, a direct dependence on m_w. The relation

$$1 - \frac{1}{\overline{H}_o} = 6.55[0.5C_{fo}\left(1 + 0.04M^2\right) + m_w]^{0.5} \tag{12.19}$$

is the one that gave a better displacement thickness distribution in the case of the transonic channel, as shown in Fig. 1. This law has then been used in the calculations.

12.3 Calculation of the Transpiration Velocity

The passive control device consists of a cavity and a porous surface underneath the shock. The flow across the cavity reduces the pressure gradient between the regions upstream and downstream of the shock, thus diminishing the intensity of the shock. The mechanism of the passive control consists in a combination of blowing and suction across the porous plate, whose intensity depends on the geometric parameters of the holes as well as on the pressure distribution on the airfoil.

According to different theories concerning the flow through a hole, the following control laws have been investigated by the authors: Poll's law [21], the linear and the nonlinear Hagen-Poiseuille law [22], the isentropic law [23] and Darcy's law [24]. Darcy's law and the Hagen-Poiseuille control law assume a linear dependence of the transpiration velocity on the pressure difference between the extremities of the hole. The other theories introduce quadratic terms whose coefficients are determined through correlation with experimental data. The nonlinear laws are briefly described below.

Poll's law is the calibration law of the laser-drilled plates; it has the following quadratic form:

$$Y = \frac{1}{K}[40.7X + 1.95X^2] \tag{12.20}$$

with

$$Y = \frac{\Delta p}{\rho}\left(\frac{d}{\nu}\right)^2\left(\frac{d}{t}\right)^2 \tag{12.21}$$

$$X = \frac{\dot{m}}{\mu t} \tag{12.22}$$

where Δp is the pressure difference across the surface, d the hole diameter, ρ the density, ν the kinematic viscosity, $\dot{m}$ the mass flow rate per hole, μ the dynamic viscosity and K a calibration coefficient.

The linear Hagen-Poiseuille law has been extended to porous plates with divergent holes by adding a quadratic term, as follows:

$$\Delta p = 0.5\rho v_w{}^2 + 32K_g\mu\frac{L}{d^2}v_w \tag{12.23}$$

$$K_g = 0.25\left(\frac{d}{D}\right)\left(1+\frac{d}{D}\right)\left(1+\left(\frac{d}{D}\right)^2\right) \tag{12.24}$$

where D, d and L are respectively the two diameters, and the length of the divergent hole. Δp is the pressure difference across the hole and v_w is the suction or blowing velocity on the boundary-layer side.

The isentropic control law assumes inviscid flow through the holes; thus, the mass flow rate is computed under isentropic conditions:

$$\frac{\dot{m}_b}{A_h\rho_0 a_0} = \sqrt{\left[\frac{2}{k-1}\left(\frac{\rho_a}{\rho_0}\right)\left(\frac{p_b}{p_0}\right)\left(\frac{p_a}{p_b}\right)^{\frac{-1}{k}}\left(\left(\frac{p_a}{p_b}\right)^{\frac{k-1}{k}}-1\right)\right]} \tag{12.25}$$

where a is the speed of sound, p the pressure, k the ratio of specific heats and A_h the area of the holes. The subscripts a,b and 0 refer to the inner, outer and reservoir conditions, respectively. This inviscid theory is expected to overestimate the transpiration velocity. Thus, in order to account for the viscous losses, the mass flow has to be corrected by a coefficient η_v. In addition, for the whole plate a porosity factor η_P has also to be introduced, leading to the following expression for the transpiration velocity:

$$\frac{v_w}{a_0} = \eta_P\eta_v\frac{\dot{m}_b}{A_h\rho_0 a_0}\frac{1}{\frac{\rho_b}{\rho_0}}. \tag{12.26}$$

The η_v coefficient has to be determined experimentally for each type of porous sheet. Braun [25] correlated this coefficient to the pressure difference across the hole and to the geometry of the hole for the case of metal sheets of conical sections with a porosity of about 0.082. In the case of the laser drilled sheets of the type used in the present project, new correlation curves were built by using Poll's calibration law to compute the experimental value of v_w. The η_v factor was then calculated as the ratio between the transpiration velocity computed with Poll's law and the one resulting from the isentropic law, assuming again that it depends on the local pressure difference at the extremities of the hole. Fig. 2 illustrate the reduction factors corresponding to the DRA-2303 and VA 2 airfoils.

In the case of passive control, the net mass flow through the porous surface must be zero, since it is transpiration due to a natural mechanism. Thus, the following condition has to be satisfied:

$$Q = \int_{s_1}^{s_2}\left(\frac{v_w}{a_0}\frac{\rho_b}{\rho_0}\right)ds = 0 \tag{12.27}$$

where s_1 and s_2 are the coordinates of the extremities of the porous region. The use of this condition allows the iterative calculation of the cavity pressure p_a, supposed to be constant over the porous region, the calculation of the η_v factor from the isentropic transpiration law and, finally, of the transpiration velocity.

12.4 Numerical Results

The numerical tests performed during the project were aimed firstly at validating the viscous/inviscid coupling method in the extended version for ventilated airfoil analysis

156

with particular emphasis on the control models and, successively, at investigating the effectiveness of the passive control concept with a close look at the aerodynamic characteristic improvement. The DRA-2303, DA LVA-1A and VA 2 airfoils, investigated in the wind tunnel experiments within Task 3, have all been considered in the numerical analysis, since they present different characteristics. Indeed, the first two are laminar airfoils while the third one is turbulent.

12.4.1 Validation of the method

Preliminary test cases have been performed on the DRA-2303 airfoil to check the different models for the calculation of the wall transpiration discussed in the previous section. In Fig. 3, the pressure distribution computed with all these models, at Mach = 0.68, Re=19.0×10^6 and at an angle of attack of 0.4°, is shown in comparison with the experiments. The agreement among the different models and the experiments is good except ahead of the shock region. In fact, while Poll's law gives the correct pressure level, the isentropic law overestimates the expansion ahead of the shock with respect to the experiment with some effect on the wave drag. However, it has to be said that in this test the isentropic law has been corrected using the original correlation curves relative to a different type of porous plate. In Fig. 4, results of the transpiration laws are shown. In order to study the effect of the correction of the isentropic transpiration law, a constant reduction has been applied to the inviscid velocity value. The resulting pressure distributions are shown in Fig. 5 for different reduction factors with the best fit with the experiments given by a reduction factor of 80%. As already discussed in the previous section, in the case of the laser-drilled sheets considered in the present project, the correct reduction factor for the isentropic law has resulted from the correlation with Poll's law.

In the test cases with control, some problems related to the numerical stability of the viscous-inviscid interaction scheme were encountered; these problems were essentially due to the way of introducing the control into the code. The best way to avoid such problems consists in introducing it gradually, using a variable relaxation factor with an increase of about 0.3% − 0.5% per iteration step, starting from the first Euler/boundary-layer iteration. It has also been necessary to extend the porous region upstream and downstream with a blending region of 10% of its length.

A grid sensitivity study has also been conducted in order to get the best compromise between accuracy and stability of the calculation. In fact, although it is expected that a grid refinement in the shock region improves the agreement with the experiment as far as the location and the geometry of the shock wave are concerned, it may give origin to numerical instability. As a result, a 170x48 O-type grid has been used for all the three airfoils with the constraint in the control region of a constant step size lower then 1% of the chord length. As an example, the grid around the VA 2 airfoil is shown in Fig.6 .

12.4.2 Numerical assessment of passive control

In all cases, calculations have been performed at a fixed C_l, corresponding to the experimental value. The transition location has been fixed at the same position as in the experiment. The main results of the analysis are next shown separately for the three airfoils.

DRA-2303 airfoil

According to the experimental setup, the presence of a porous plate with 8% porosity has been simulated in the region x/c=0.5 to 0.6. In all calculations, transition has been fixed at x/c=0.05 on both the upper and the lower side. Two test cases are discussed here, referred to as Test Case I and II, each characterized by certain freestream conditions with and without passive control, selected from the wind-tunnel test matrix. The conditions correspond to the same Mach (0.68) and Reynolds (approx. 19×10^6) numbers. In Test I, the experimental lift coefficient is 0.566 without control and 0.553 with control, while in Test II the lift coefficient is higher, being 0.811 without control and 0.814 with control. In Fig. 7 the pressure coefficients calculated for Test Case I at fixed C_l, with and without control, are shown in comparison with the experiment. It can be noticed that the numerical result for the case with control presents a double-shock structure that is not so evident in the experiment. In Table 1 the computed and the experimental aerodynamic coefficients are reported. Although the computed total drag coefficient is lower than the experimental value, both with and without control, the effect of passive control is well predicted by the calculations, at least on a qualitative basis, with control resulting in an increment of the total drag. From a look at the partial contributions to drag, it can be seen that both the pressure and the friction drag increase due to passive control. However, as discussed in [8], an estimate of the entropy rise across the shock wave, based on the local Mach number upstream of the shock, indicated that the wave drag does not increase by the application of passive control. The pressure drag rise may then be attributed to numerical errors (spurious entropy) and, to a minor extent, to the thickening of the boundary layer.

Table 1   DRA-2303 Airfoil - Test Case I

Data point	Re	M	α	C_l	C_m	C_d	C_{dpres}	C_{dfric}
$271 - exp. - n.c.$	18.97	0.6816	1.068	0.5668	-0.0958	0.009428	—	—
$271 - theory - n.c.$	18.97	0.6816	0.468	0.5668	-0.1051	0.008946	0.004695	0.004251
$1008 - exp. - w.c.$	18.984	0.6807	1.066	0.5529	-0.0944	0.010525	—	—
$1008 - theory - w.c.$	18.984	0.6807	0.435	0.5530	-0.1033	0.009297	0.004822	0.004475

In Fig. 8, the experimental and computed pressure distributions for Test Case II are shown. In both cases with and without control, the agreement is excellent. Figs. 9-11 show the development of the boundary-layer parameters. It is clearly evident that the blowing upstream of the shock induces a strong increase in the displacement thickness; from a look at the skin friction coefficient and at the shape factor it results that such blowing is strong enough to generate separation, while in the suction region the separation tendency decreases. As shown in Table 2, the control produces an increase of total drag with respect to the case without control; as far as the partial contributions to the drag rise are concerned, the same conclusions as for Test Case I hold. In the table, a second data point without control is also reported, corresponding to the same angle of attack as in the case with control; in this case the calculations with and without control give quite the same total drag, while in the experiment the total drag with control increases. However, it should be noted that in the calculation-while drag stays constant-lift decreases due to control. As a favourable effect of the control, it can be observed that the tendency of the boundary layer to separate downstream of the shock wave is attenuated.

Table 2　DRA-2303 Airfoil - Test Case II

Data point	Re	M	α	C_l	C_m	C_d	C_{dpres}	C_{dfric}
$289 - exp. - n.c.$	18.907	0.6795	2.507	0.8115	-0.1002	0.014576	–	–
$289 - theory - n.c.$	18.907	0.6795	1.791	0.8115	-0.1148	0.014384	0.010142	0.004242
$1031 - exp. - w.c.$	18.982	0.6806	2.710	0.8142	-0.0984	0.017972	–	–
$1031 - theory - w.c.$	18.982	0.6806	2.243	0.8139	-0.1045	0.015698	0.011690	0.004008
$290 - exp. - n.c.$	18.91	0.6795	2.712	0.8427	-0.1014	0.016557	–	–
$290 - theory - n.c.$	18.91	0.6795	2.135	0.8428	-0.1114	0.015685	0.011444	0.004241

DA-LVA-1A airfoil

In this case the control has been applied in the region $x/c=0.580$ to 0.695 and the porosity of the plate installed on the airfoil model is 5.06%. Transition has been fixed at $x/c=0.48$ on both the upper and lower sides. As shown in Fig. 12, the application of the passive control has split the strong shock wave into a system of two shocks. In Table 3 the computed and measured aerodynamic coefficients are shown; as for the previous airfoil, the control produces an increment of total drag. In this case, the agreement between the computed and measured total drag is good in both cases with and without control. In the case with control the total drag value also includes the excrescence drag, calculated as the friction drag on the clean airfoil in the control region multiplied by a factor of 2.5; this factor resulted from a correlation of experimental data. The value of the excrescence drag is of the order of magnitude $1.0E - 4$; thus, in the present case, its influence on the total drag can be considered as negligible. In Fig. 13, the shape factor computed with and without control is presented; it is clearly evident that the passive transpiration reduces significantly the tendency of the boundary layer to separate at the foot of the shock, the shape factor being reduced from 2.4 to about 1.7, rather than 1.3 as in the case w/o control.

Table 3　DA-LVA Airfoil

Data point	Re	M	α	C_l	C_m	C_d	C_{dpres}	C_{dfric}	$C_{excr.}$
76-theory-n.c.	4.64	0.7613	0.9645	0.4736	-0.08403	0.01092	0.007760	0.003160	–
76-exp.-n.c.	4.64	0.7698	1.0000	0.4736	-0.0858	0.01172	–	–	–
76-theory-w.c.	4.64	0.7613	1.0800	0.4736	-0.0804	0.01206	0.007995	0.003441	0.00063
73-theory-n.c.	4.64	0.7614	0.8695	0.4536	-0.0826	0.01025	0.007128	0.003123	–
73-theory-w.c.	4.64	0.7614	0.9724	0.4536	-0.0799	0.01139	0.007370	0.003408	0.00062
74-exp.-w.c.	4.66	0.7692	1.0000	0.4556	-0.0833	0.01121	–	–	–

VA 2 airfoil

The control region for this turbulent airfoil has been placed at $x/c=0.495$ to 0.645, with a porosity of 12.8%. The effect of control has been investigated at three lift coefficients at constant Reynolds ($2.5x10^6$) and Mach (0.74) numbers. Transition has been fixed at $x/c=0.30$ on the upper side and $x/c=0.25$ on the lower side. The computed pressure and local skin friction coefficients are shown in Figs. 14 and 15 for the lowest C_l. Again, a double-shock structure results from the application of passive control. As shown in Table 4, a reduction of the total drag, though negligible, has been obtained in the presence of passive control for the lowest C_l. This reduction is due to the fact that the pressure drag reduction exceeds the friction drag increment. At $C_l = 0.8$, the overall drag is slightly increased by the application of control, while at the highest lift coefficient $C_l = 0.85$, a more pronounced reduction of total drag has been predicted.

Table 4 VA 2 Airfoil

Data point	Re	M	α	C_l	C_m	C_d	C_{dpres}	C_{dfric}	$C_{excr.}$
Theory-n.c.	2.50	0.74	1.234	0.65010	-0.1225	0.01250	0.00800	0.00450	−
Theory-w.c.	2.50	0.74	1.075	0.65062	-0.1281	0.01248	0.00745	0.00503	−
Theory-n.c.	2.50	0.74	2.0345	0.79998	-0.1334	0.02259	0.01814	0.00444	−
Theory-w.c.	2.50	0.74	1.8257	0.79995	-0.1378	0.02280	0.01674	0.00529	0.000767
Theory-n.c.	2.50	0.74	2.4378	0.84957	-0.1326	0.02859	0.02410	0.00449	−
Theory-w.c.	2.50	0.74	2.1182	0.84988	-0.1418	0.02817	0.02152	0.00591	0.000744

12.5 Conclusions

The control of shock wave/boundary layer interaction on transonic airfoils through porous surfaces has been addressed with a steady Euler/boundary layer semi-inverse coupling method (EUBL2D). The numerical simulation of control required, firstly, the extension of Green's integral boundary-layer equations to take the wall transpiration into account and then the development of a numerical model for the calculation of the transpiration velocity (suction/blowing), produced by the presence of a porous wall with a plenum chamber underneath. The numerical tests, performed for the airfoil geometries investigated in the wind tunnel experiments in Task 3, showed that the above method is capable of reliably simulating the effect of passive control. As far as the assessment of the passive control concept is concerned, the present numerical analysis came to conclusions similar to those of the experimental investigation. They can be summarized as follows:

- the passive control concept does not seem to be effective in reducing the overall drag, the predicted total drag always increasing for laminar airfoils and slightly decreasing in turbulent airfoils in the case of a limited range of C_l;

- the increase of total drag can be attributed to an increase of viscous drag (friction, base and excrescence drag) which is larger than the decrement of wave drag;

- as a favourable effect, the natural blowing/suction mechanism in the shock region reduces the tendency of the boundary layer to separate at the foot of the shock; as a consequence such kind of control is expected to be beneficial to the onset of buffet.

12.6 References

[1] Nagamatsu, H.T., Orozoco, R.D., Ling D.C. "Porosity Effects on Supercritical Airfoil Drag Reduction by Shock Wave/Boundary-Layer Control", AIAA Paper 84-1682, June 1984.

[2] Donaldson, C.P. "Investigation of a Simple Device for Preventing Separation Due to Shock and Boundary-Layer Interaction", NACA RM L50302A, November 1950.

[3] Savu, G., Trifu, O., Dumitrescu, L. Z. "Suppression of Shocks on Transonic Airfoils", Proceeding of the 14th International Symposium on Shock Tubes and Shock Waves, Sidney, 1983.

[4] Hartwich, P.M. "Euler Study on Porous Transonic Airfoils with a View Toward Multipoint Design", J. of Aircraft, March-April 1993, Vol. 30, pp 184-191.

[5] Olling, C.R. "Viscous-Inviscid Interaction in Transonic Separated Flow over Solid and Porous Airfoils and Cascades", Ph.D.Thesis University of Texas (Austin), 1985.

[6] Chen, C.L., Chow, C.Y., Holst, T.L., Van Dalsem, W.R. "Numerical Simulation of Transonic Flow over Solid and Porous Airfoils", AIAA Paper 85-5022, 1985.

[7] Bahi, L., Ross, J.M., Nagamatsu, H.T. "Passive Shock Wave/Boundary-Layer Control for Transonic Airfoil Drag Reduction", AIAA Paper 83-0137, 1983.

[8] Coiro, D., Amato, M., deMatteis, P. "Numerical Prediction of Transonic Viscous Flows Around Airfoils through an Euler/Boundary-Layer Interaction Method", AIAA Paper 90-1537, June 1990.

[9] Dima, C., deMatteis, P. "Extension of a VII Interaction Method to Airfoil Computations with Passive Control- Preliminary Test Cases", CIRA Report CIRA-TR-95-0002, EUROSHOCK Report TR AER2-92-49/2.2, 1995.

[10] deMatteis, P., Dima, C. "On the Modelling of Separated Flows About Airfoils", AIAA Paper 93-3479, August 1993.

[11] Dima, C. "Calculation of Flow with Separation Using a Viscous-Inviscid Interaction Method", Mécanique Appliquée, Sept./Oct. 1989, Vol.34, No.5, pp 501-508.

[12] Jameson,A., Schmidt,W. and Turkel,E. "Numerical solution of the Euler equations by finite volume methods using Runge-Kutta time-stepping schemes", AIAA Paper 81-1259, June 1981.

[13] Stewartson, K. "Correlated Incompressible and Compressible Boundary Layers", Proc. Roy. Soc. (London), Ser. A., Vol. 200, 1949, pp 213-228.

[14] Granville, P.S. "The Prediction of Transition from Laminar to Turbulent Flow in Boundary Layers on Bodies of Revolution", Rep. No. 3900 of the Naval Ship Research and Development Center, Bethesda, Maryland, 1974.

[15] Green, J.E., Weeks,D.J., Broman, J.W.F. "Prediction of Turbulent Boundary Layers and Wakes in Compressible Flow by a Lag- Entrainment Method", R.A.E. TR 72231, 1972.

[16] Whitfield, D.L. "Analytical Description of the Complete Turbulent Boundary Layer Velocity Profile", AIAA Paper 78-1158, 1978.

[17] Carter, J.E. "A New Boundary Layer Inviscid Interaction Technique for Separated Flows", AIAA Paper 79-1450, 1979.

[18] Lighthill, M.J. "On displacement thickness", J. of Fluid Mechanics, Vol 4, Part 4, 1958, pp 661-663.

[19] Kays, W.M., Moffat, R.J. "The Behavior of Transpired Turbulent Boundary Layers", NASA CR-119147, 1985.

[20] Bur, R. "Étude Fondamentale sur le Contrôle Passif de l'Interaction Onde de Choc/Couche Limite Turbulente en Écoulement Transsonique", ONERA, Note Technique 1991-9, 1991.

[21] Poll, D.I.A., Danks, M., Humphreys, B.E. "The Aerodynamic Performance of Laser Drilled Sheets", First European Forum on Laminar Flow Technology", Hamburg, March 1992.

[22] Bieler, H., Preist J. "HLFC for Commercial Aircraft-First ELFIN Test Results", First European Forum on Laminar Flow Technology, Hamburg, March 1992.

[23] Breitling, T. "Berechnung transsonischer, reibungsbehafteter Kanal-und Profilströmungen mit passiver Beeinflussung", Ph.D. Thesis, University of Karlsruhe(TH), Karlsruhe, November 1985.

[24] Darcy, H. "Les Fontaines Publiques de la Ville de Dijon", Paris, 1896.

[25] Braun, W. "Experimentelle Untersuchung der turbulenten Stoss-Grenzschicht-Wechselwirkung mit passiver Beeinflussung", Ph.D. Thesis, University of Karlsruhe(TH), Karlsruhe, October 1990.

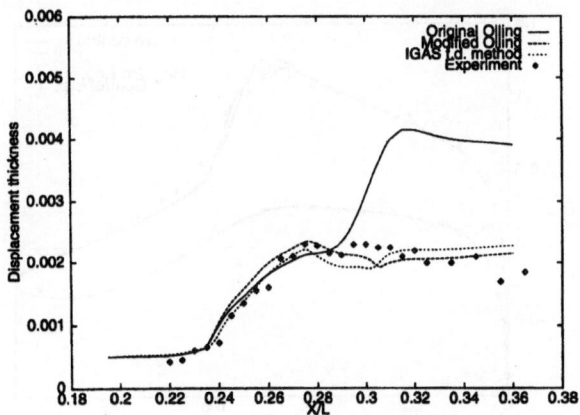

Fig. 1 ONERA transonic channel flow test

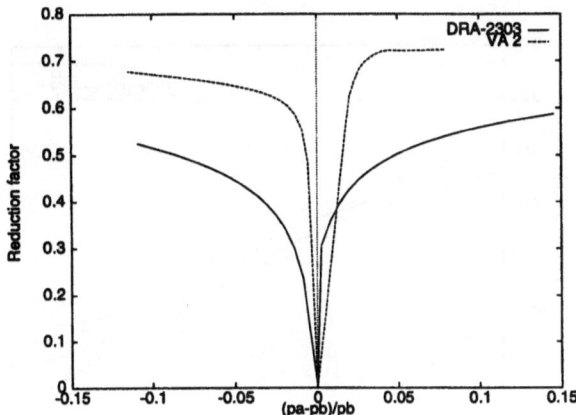

Fig. 2 Transpiration reduction factor for the DRA-2303 and
VA 2 airfoils

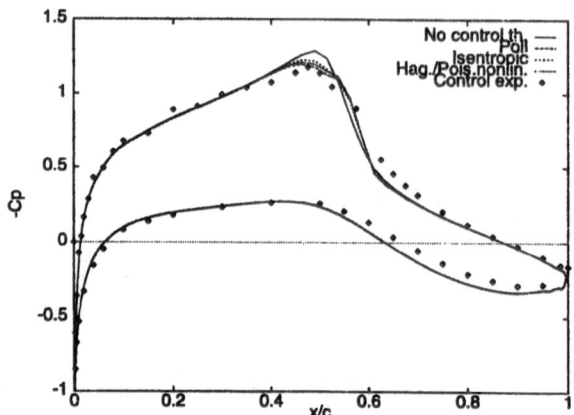

Fig. 3 Control effect on the pressure distribution: DRA-2303, M=0.68, Re=19.0x10^6, α= 0.400

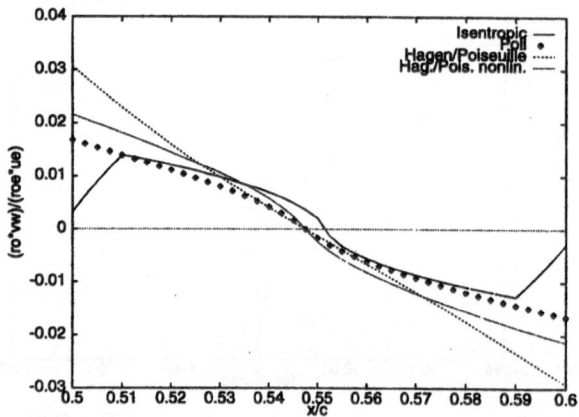

Fig. 4 Transpiration velocity distribution on the DRA-2303

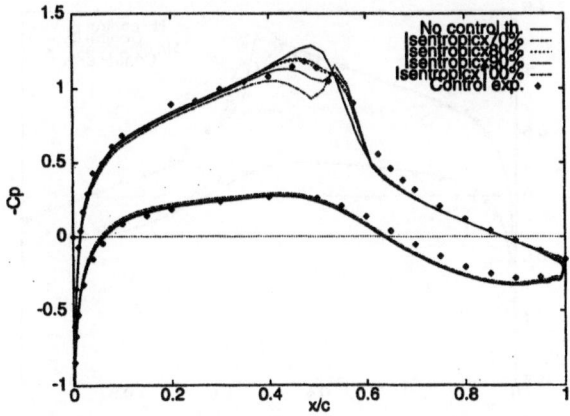

Fig. 5 Transpiration intensity effect on the pressure distribution: DRA-2303, M= 0.68, Re=19.0x10^6, α=0.400

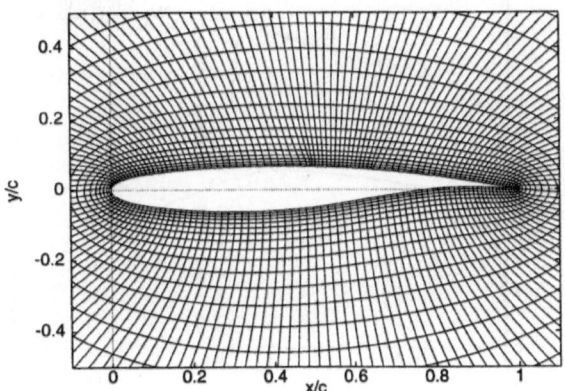

Fig. 6 170x48 grid around the VA 2 airfoil

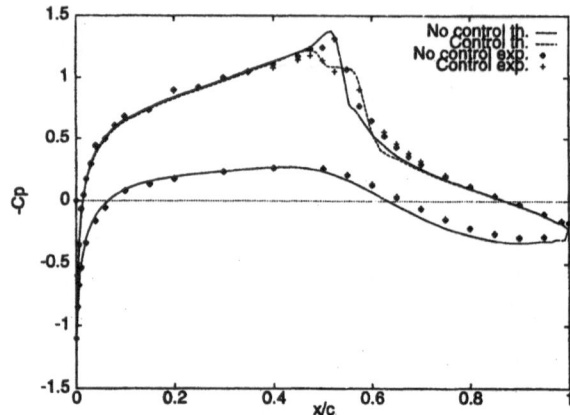

Fig. 7 DRA-2303 Test Case I, M=0.68, Re=19.0x10⁶

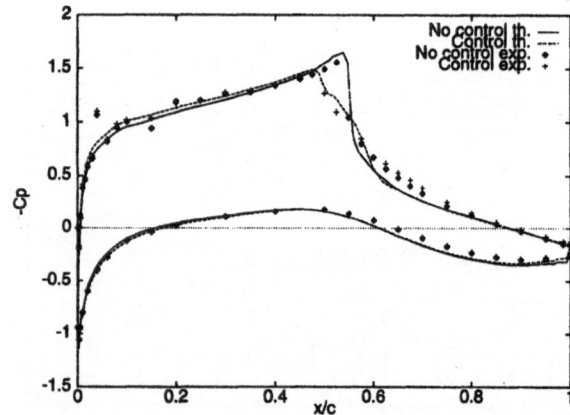

Fig. 8 DRA-2303, Test Case II, M=0.68, Re=19.0x10⁶

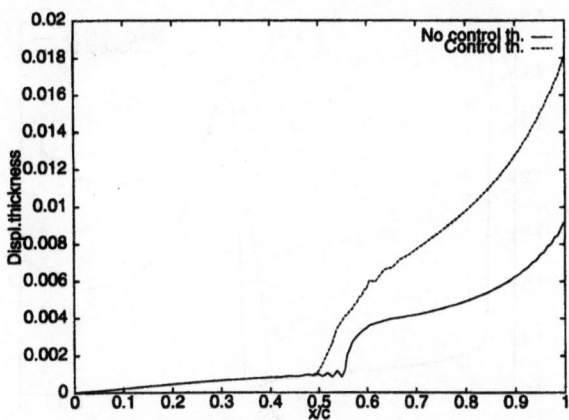

Fig. 9 DRA-2303, Test Case II, M=0.68, Re=19.0x10^6

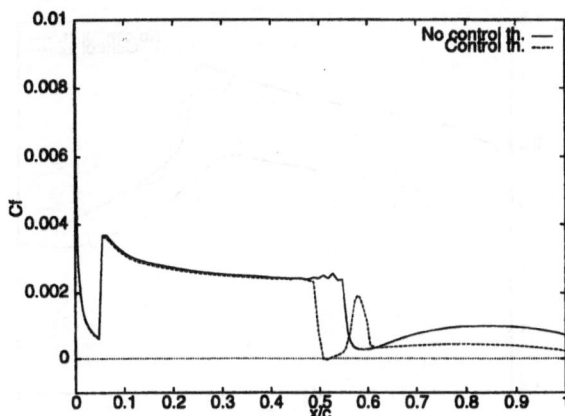

Fig. 10 DRA-2303, Test Case II, M=0.68, Re=19.0x10^6

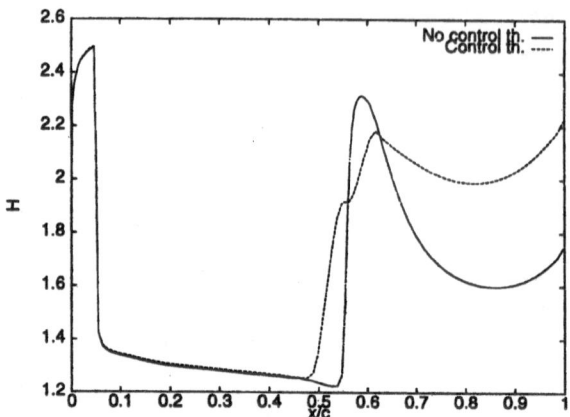

Fig. 11 DRA-2303, Test Case II, M=0.68, Re=19.0x10⁶

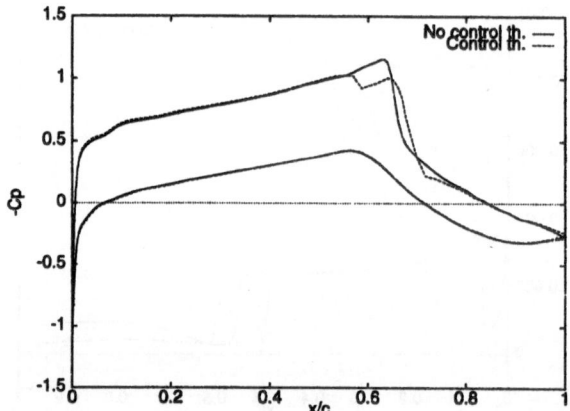

Fig. 12 DA LVA-1A, M=0.7613, Re=4.64x10⁶

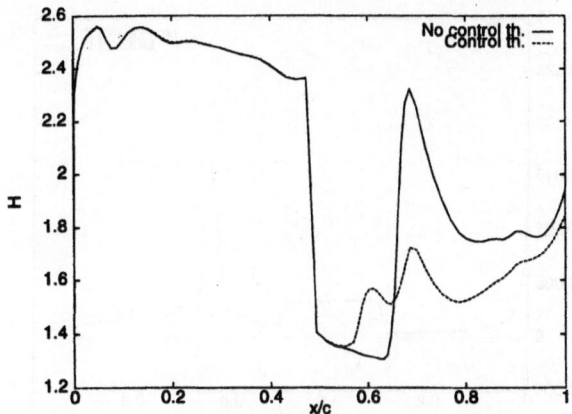

Fig. 13 DA-LVA-1A, M=0.7613, Re=4.64x10^6

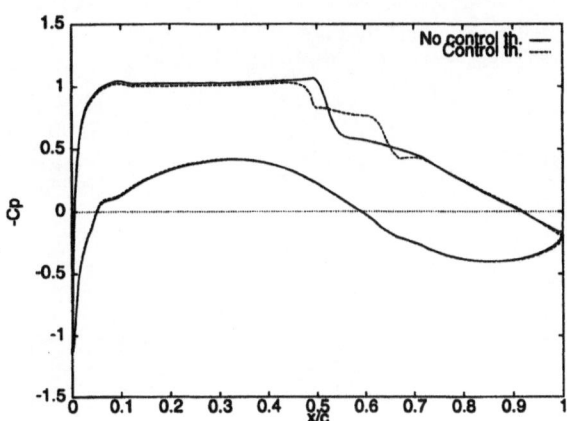

Fig. 14 VA 2, M=0.74, Re=2.5x10^6

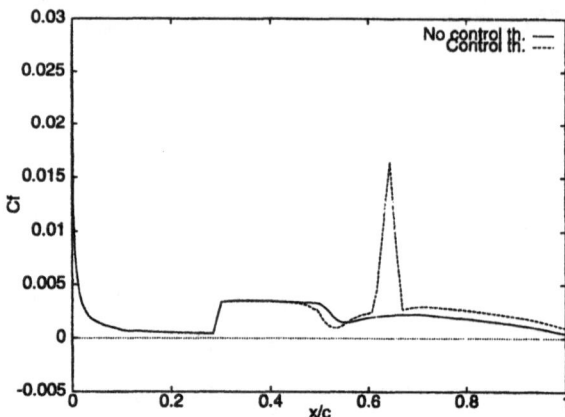

Fig. 15 VA 2, M=0.74, Re=2.5x10⁶

13 INTRODUCTION OF PASSIVE SHOCK CONTROL IN AN INTERACTIVE BOUNDARY LAYER METHOD

C. de Nicola
Gasdynamics Institute
P.le Tecchio, 80 - Naples, Italy

Summary: Objective of this research program was to examine the capability to predict flows around transonic airfoils with passive Shock wave - Boundary Layer Interaction Control (SBLIC), that is performed by using a porous cavity located in the shock region. Its simulation has been obtained by using the inviscid flow/boundary layer interaction model, improved by the introduction of a proper ventilation through the airfoil surface. Concerning the inviscid flow, the Euler model has been considered, while an integral method for the boundary layer has been used for accuracy and computing time reasons.

13.1 Introduction

Three main aims have been pursued to set up a numerical method for the calculation of the Shock Wave Boundary Layer Interaction following a Viscous/Inviscid Interactive (VII) Procedure.

1. The acquisition of an inviscid solver: a more general Euler solver, instead of a potential method, has been preferred for more generality reasons and in view of applications of the VII method in the presence of strong normal shocks. The multiblock approach, coupled to a new Local Grid Refinement technique, has been followed to obtain very accurate solutions with reasonable computational time.
2. The application of a suitable integral boundary layer method: the extension of a 2D boundary layer method to take into account transpiration effects has been performed.
3. The development of an appropriate method for the introduction of the shock wave control into the code: the transpiration mechanism through a cavity in the presence of a normal shock wave has been investigated for both finite difference and integral boundary layer solutions. Different transpiration models have been implemented and compared. A shock wave control mechanism with both assigned transpiration laws and use of a cavity has been implemented into the code.

A number of related problems have been faced and solved, e.g. model of transpiration, design of the cavity location, calculation of total drag, wake effect.

During the research, at the various levels of development and efficiency of the code, the test cases have been computed as shown in the last paragraph, including all mandatory and optional test cases.

13.2 Inviscid Flow Solver

The inviscid flow field is computed by solving the Euler equation on a structured grid. The algorithm is the well known and robust scheme developed by Schmidt, Jameson and Turkel [1].

The method is based on a central space discretization with explicit adaptive artificial dissipation. The artificial viscosity is constructed by blending a 2nd and 4th order contribution. The 4th order background viscosity is necessary to provide stability to the scheme and avoid odd-even decoupling, while 2nd order dissipation is introduced at discontinuities by a shock sensor based on the pressure jump to locally improve the accuracy. Steady state calculations are performed by using a pseudo-transient approach.

The unsteady Euler equation are integrated, until steady state is reached, by using an explicit multi-stage scheme (Runge-Kutta type) with local time stepping. Convergence speed is improved by using different techniques: enthalpy damping, implicit smoothing, multigrid.

One of the features of the present solver is the capability of handling complex configurations by using a Multiblock Structured Approach (MSA). Moreover, MSA allows to locally refine grid in order to improve accuracy without increasing the computational cost (Local Grid Refinement technique, LGR).

In the MSA the flowfield is initially divided into a number of quadrilateral blocks; each block includes a proper region of the flowfield, and the calculation starts with a uniform mesh. Depending on the presence of discontinuities or large gradients of the flow variables (causing a higher local error), the mesh of one or more blocks is automatically refined during the convergence procedure; this leads to a drastic improvement of the quality of the solution, that can reach the same accuracy as a solution obtained by using a uniform finer mesh. So, accurate solutions with a high convergence speed have been obtained, i.e., a computing time reduction of about 50% has been reached for inviscid compressible flows with a fixed level of accuracy. Details of the basic algorithm can be found in [1], while the local grid refinement is discussed in [2].

13.3 Viscous Solver

The integral boundary layer code is based on the Cohen-Reshotko method (Stewartson transformation and Thwaites method) for the laminar part and a modified Green method [3] (equations solved in direct and inverse mode, both adapted to SBLIC calculation) for the turbulent part.

Particular care has been required for the extension of the boundary-layer method to a flow with transpiration. The Olling approach to transpiration [4] provided unsatisfactory results (see also Fig. 3), so a modification has been proposed: by using a finite-difference method, assumed as 'exact', a data-base has been generated to extend the Olling correlation [5].

The same finite-difference method has been used to check the correlation of the Green method and to perform a validation of the resulting code in presence of blowing/suction for the case of a flat plate in supersonic flow.

13.4 Transpiration Model

During the development of the research different transpiration models have been used and their efficiency studied.

First the Darcy law and the isentropic model have been considered. The isentropic model has been preferred to the Darcy law due to more realistic previsions (the order of magnitude of the blowing/suction velocity provided by the two models are different). Only the isentropic model has been used for the preliminary boundary layer calculations with passive control. In any case, the value of the transpiration velocity resulted large enough to likely give separation of the boundary layer at the beginning of the blowing region. So a reduction factor for the strength of the normal velocity (SCF) has been introduced for both model flows initially analysed, i.e., a theoretical normal shock wave impinging on a flat plate and the transonic ONERA channel flow test case.

Furthermore, the effect of the transpiration model for different locations of the shock wave with respect to the cavity has been numerically evaluated.

As alternative to the use of the isentropic law, the calculation of both the transpiration velocity with Poll's law and the excrescence drag coefficient have been implemented into the code. The suction/blowing transpiration velocity distribution has been calculated starting with a known pressure difference Δp between the external flow and the cavity. Since the pressure in the cavity is not fixed a priori, its correct value has been obtained with an iterative calculation by letting the total mass flux vanish inside the cavity. However, it has been observed that, once the convergence has been achieved, the blowing/suction velocity has the same order and trend as the one obtained with the isentropic transpiration model.

13.5 Design of the Cavity Location

A design version of the SBLIC code has been developed with a particular procedure to determine the cavity location on a given transonic airfoil with specified flight conditions. The procedure can be summarised as follows: when the existence of the shock has been detected during the convergence procedure, the code 'creates' a cavity, whose characteristics (e.g., extent, porosity and position with respect to the shock) have been previously fixed by the user; so, to have the optimum control, the shock can be centered or asymmetrical with respect to the cavity. Successively, during the evolution of the solution, the cavity moves together with the shock position up to its final location, when convergence has been reached.

13.6 Calculation of Total Drag

The calculation of the pressure drag coefficient C_d using the 'near-field' approach, e.g., by integrating the pressure on the airfoil surface, requires a very accurate calculation of the pressure field on the airfoil, and, in presence of separation, lack of accuracy can result. The 'far-field' approach [6] can avoid this problem by shifting the drag evaluation from the body to the far-field, i.e., to the border of the computational domain. A first evaluation of the total drag coefficient has been obtained by adding the pressure C_d to the friction C_d.

In addition, the calculation of C_d has been performed by considering a viscous term and a wave term. The first has been calculated using the Squire-Young formula, while the second has been calculated by a line integral along a path surrounding the shock(s). In 2-D this method has been easily implemented, although the exact evaluation of the shock wave positions are needed; this is possible by seeking a jump of the local Mach number at the mesh points where the flow is supersonic.

13.7 Wake Effect

The wake effect has been evaluated in the same manner as the viscosity effect on the airfoil. In this case the solid wall is replaced by the separating stream line of the inviscid flow. For airfoils at low incidence this line can be regarded as coincident with the co-ordinate line of the grid that emanates from the trailing edge.

By extending the boundary-layer calculation on both the upper and lower side of this line, the calculation of the normal velocity jump Δw necessary to complete the boundary condition for the wake effect is possible.

Depending on the wake length, a systematic increase of the drag coefficient has been found. The results do not change when the wake length is greater than 50% of the chord. This is demonstrated in Table 1 referring to the results of the airfoil DRA 2303 without control.

13.8 Results

In this section the main results of the research activity are summarised. The section is comprised as follows:
1) Block decomposition and Local Grid Refinement (LGR) concept.
2) New correlation for a ventilated boundary layer.
3) Test cases for the airfoil DRA 2303 with and without control.
4) Test cases for the airfoil DA-LVA-1A with and without control.
5) Test cases 1 and 2 for the VA-2 airfoil with and without control.
6) Test cases for the VA-2 airfoil at $0.75 < C_l < 0.85$ with and without control.

13.8.1 Block decomposition and LGR concept

In Figs. 1 and 2 the block decomposition of the flowfield and a typical locally refined grid are shown: blocks in the region of the shock and around the leading edge have been highly refined. The computational domain is first divided into twelve blocks. Afterwards, an increase in the local density of the computational grid close to particular regions is possible, so very accurate solutions with minimum computational time (compared to a solution obtained using a uniformly increased computational grid density) have been performed.

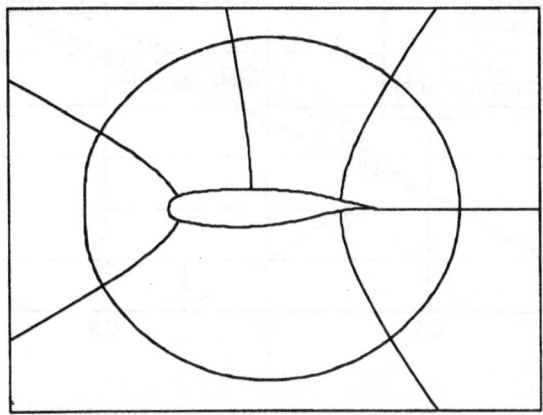

Fig. 1. Block decomposition of the computational domain

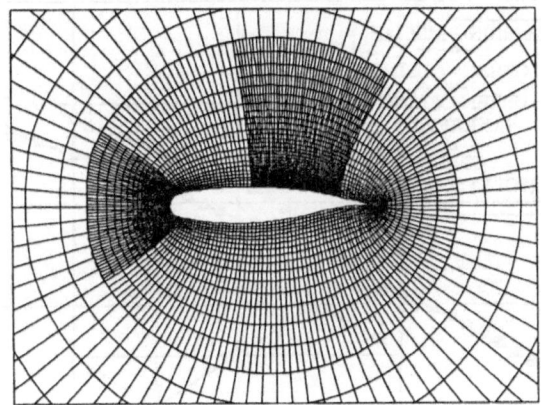

Fig. 2. Locally refined grid

13.8.2 New correlation for a ventilated boundary layer

The Fig. 3 shows the effect of the extension of the Olling correlation to correctly account for the transpiration. In the picture the comparison among different methods of boundary layer calculation for a flat plate with blowing and suction is presented. The finite-difference method solution, assumed as 'exact', has been used as a tool for both the modification and validation of the integral method.

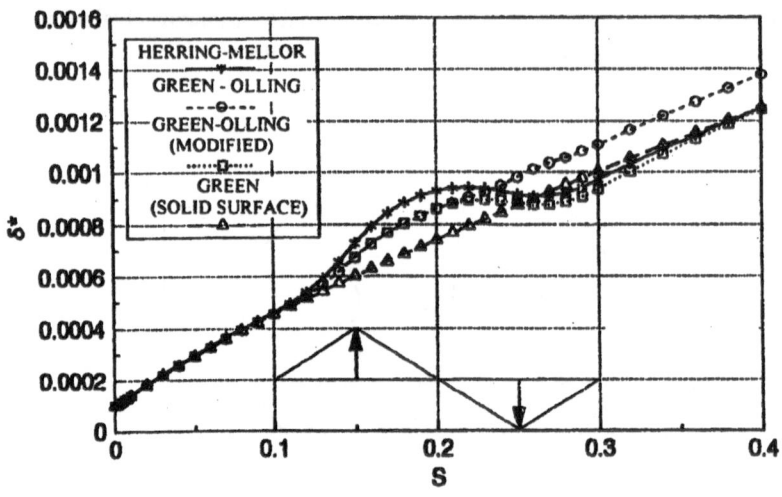

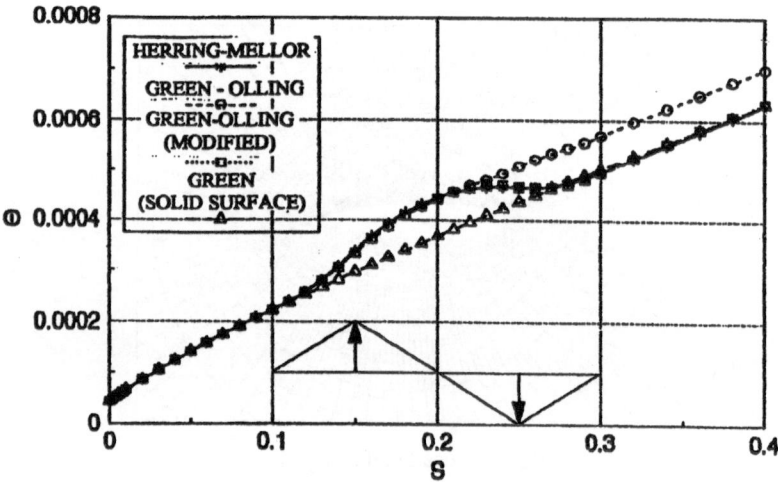

Fig. 3. Comparison of calculated dispacement and momentum thickness for a flat plate with wall transpiration. Mach=1.2, Re=10^7

13.8.3 Test cases for the airfoil DRA 2303 with and without control

In the following figures and tables the results for the DRA 2303 airfoil are reported, showing a good agreement with the experimental data (with/without wake effect).

Calculations with and without control and comparisons with experimental results have been performed for the Test Cases conventionally indicated as 290 (without control) and 1031 (with control).

A locally refined O-Type grid (160x60), Fig. 4, equivalent to a (320x120) grid, has been used. The averaged mesh size on the airfoil surface is about 0.01 respect to the chord, while in the refined region it reduces to 0.005.

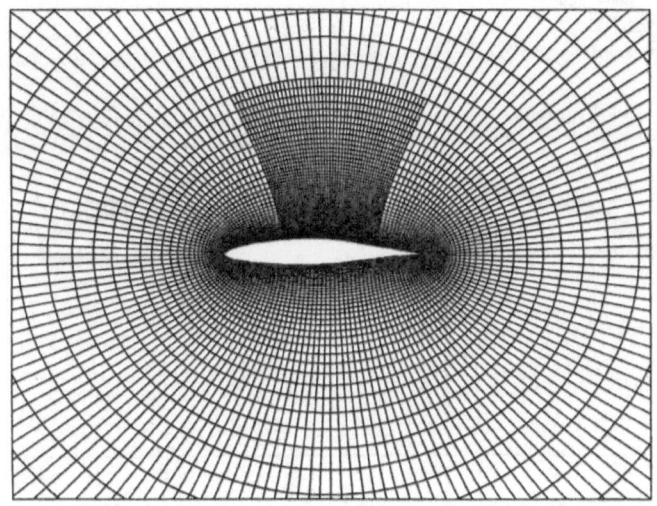

Fig. 4. DRA 2303 AIRFOIL LGR O-Type grid (160x60 - 320x120)

In many cases comparisons between experimental data and numerical results have been done at equal C_l: the angle of attack used for numerical tests is different from the angle of attack of experiments.

No drag reduction has been detected by applying passive control. The experimental values of total drag coefficient are even greater than the calculated ones. With regard to test 290, a good capturing of the shock can be observed together with a slight overestimation of the calculated pressure values behind the shock. In presence of passive control (test 1031) the calculated pressure coefficient distribution shows an underestimation before the shock.

Transition is fixed at x/c=.05 on upper and lower surfaces, as in the experiments.

In Fig.5 pressure coefficient distributions and aerodynamic coefficients are presented: it should be noted that the angle of attack used for computations is lower than the one used for experiments, to fit the location of the shock wave and the pressure distribution in the expansion region. Consequently, the aerodynamic coefficients do not match, moreover a discrepancy in the pressure distribution in the rear part of the airfoil can be observed.

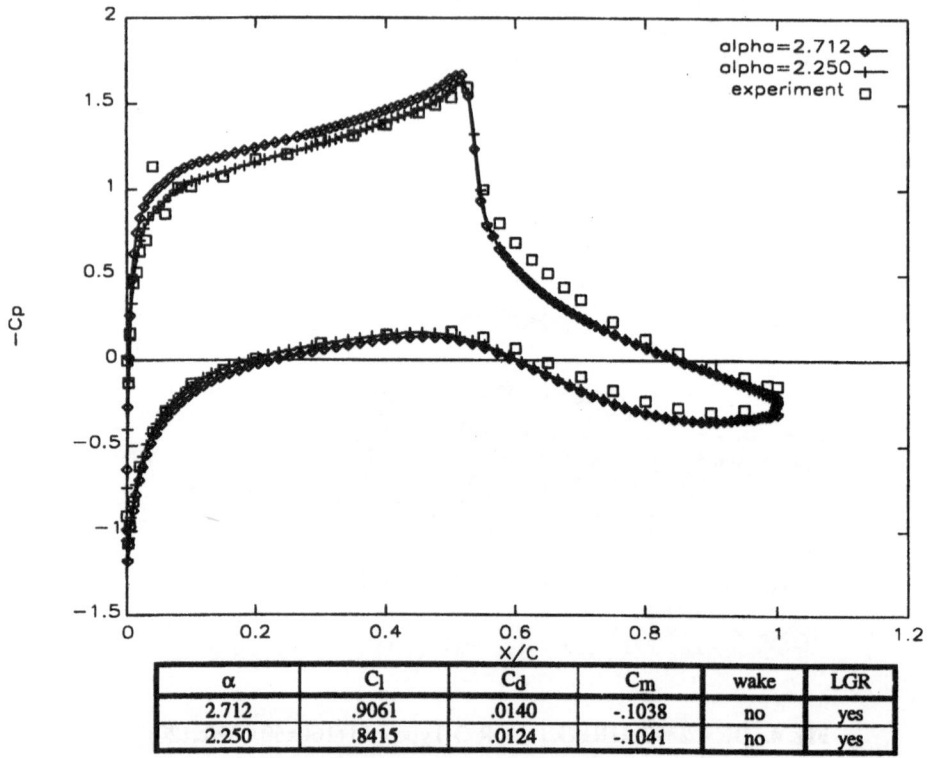

α	C_l	C_d	C_m	wake	LGR
2.712	.9061	.0140	-.1038	no	yes
2.250	.8415	.0124	-.1041	no	yes

Fig. 5. DRA 2303 AIRFOIL - TEST 290, O-type LGR mesh. Comparison between the calculated and experimental pressure distributions.
Exp: Mach =0.795, Re=18,907,000, α=2.712, C_l=0.8427, C_d=0.0165, C_m = -.101

In Fig. 6 the effect of the local grid refinement is presented: for an assigned grid, e.g., a (320x120) grid, the same accuracy has been obtained by halving the mesh except than in one properly chosen block (the block embedding the shock wave). By using the LGR concept, in the present case a CPU time saving of 60% has been obtained.

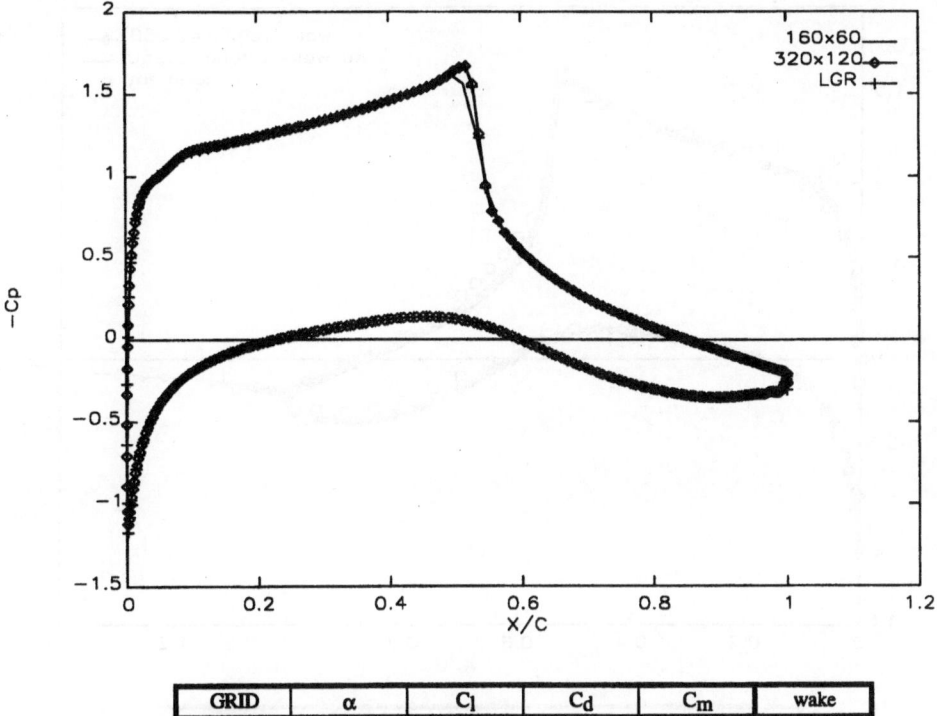

GRID	α	C_l	C_d	C_m	wake
160x60	2.712	0.9058	.0141	-.1051	no
320x120	2.712	0.9079	.0140	-.1043	no
LGR	2.712	0.9061	.0140	-.1038	no

Fig. 6. DRA 2303 AIRFOIL - TEST 290. Comparison among the calculated pressure
distributions with different grids
Exp.: Mach=0.6795, Re=18,907,000, α=2.712, C_l=0.8427, C_d=0.0165, C_m= -.101

In Fig.7 the influence of the presence of the wake in the computational model is shown.
Typically, by adding the wake effects an increase of the lift coefficient and, consequently, of
the drag coefficient is found. Hence, in the present case, the angle of attack has been reduced
to match the lift coefficient: anyway, the drag coefficient has still increased, due to the
downstream shock displacement.

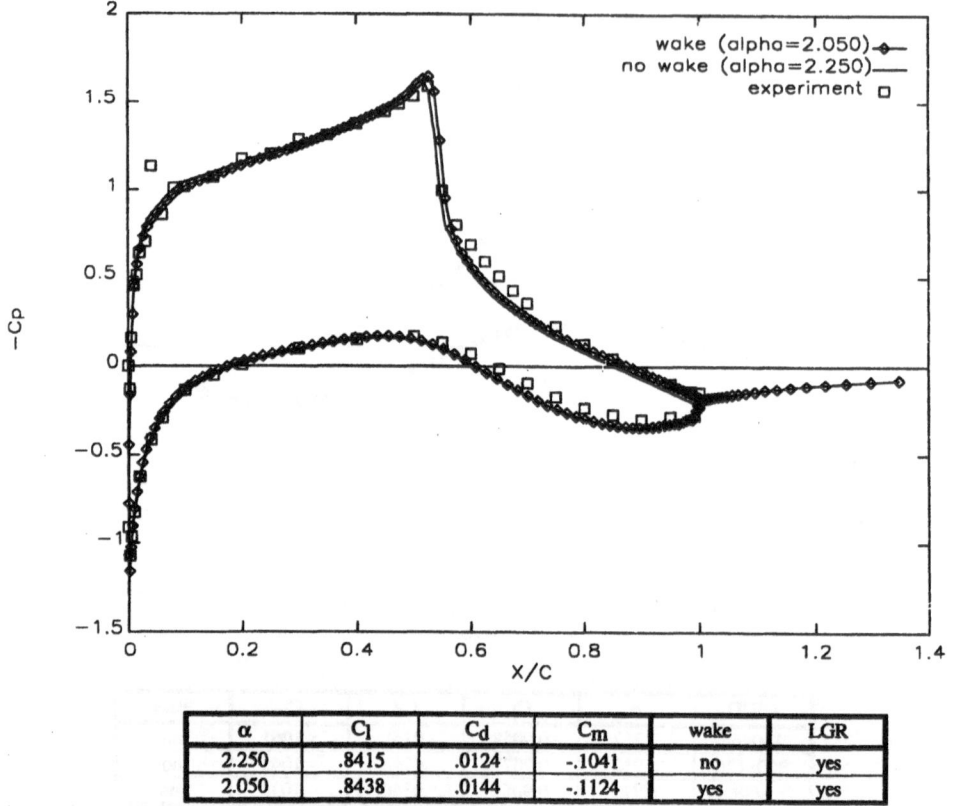

α	C_l	C_d	C_m	wake	LGR
2.250	.8415	.0124	-.1041	no	yes
2.050	.8438	.0144	-.1124	yes	yes

Fig. 7. DRA 2303 AIRFOIL - TEST 290, O-type LGR mesh. Comparison among the calculated and experimental pressure distributions including the wake effect
Exp.: Mach=0.6795, Re=18,907,000, α=2.712, C_l=0.8427, C_d=0.0165, C_m= -0.101

Table 1. DRA 2303 AIRFOIL - TEST 290. Wake effects: aerodynamic coefficients versus wake length. Mach=0.6795 Re=18,907,000 α=2.050

%wake	C_l	C_d	C_{dS-Y}	C_{dw}	C_m
0	.8116	.0109	.0106	.0125	-.1041
10	.8399	.0137	.0108	.0135	-.1112
20	.8433	.0143	.0109	.0136	-.1122
30	.8438	.0144	.0109	.0134	-.1124
40	.8442	.0145	.0109	.0134	-.1126

Legenda:
$C_d = C_d$ (pressure)+C_d(friction); $C_{dS-Y} = C_d$ (Squire - Young); $C_{dw} = C_{dS-Y} + C_d$ wave

Table 1 shows, togheter with the lift coefficient, different terms of total drag coefficient versus wake length considered in the computations: a wake lenght of about 40% of the chord can be sufficient to fully simulate the wake effects.

The effects of the control will be now shown and discussed.

In the present case the cavity is located at ($.50<x/c<.60$), while the porosity is 7.8%. Different transpiration laws have been implemented, in addition to the isentropic law originally used by the Author ('Present' in the pictures). It is evident that a reduction factor for the transpiration velocity distributions is needed in order to avoid a double shock structure not revealed by the experiments (Fig. 8).

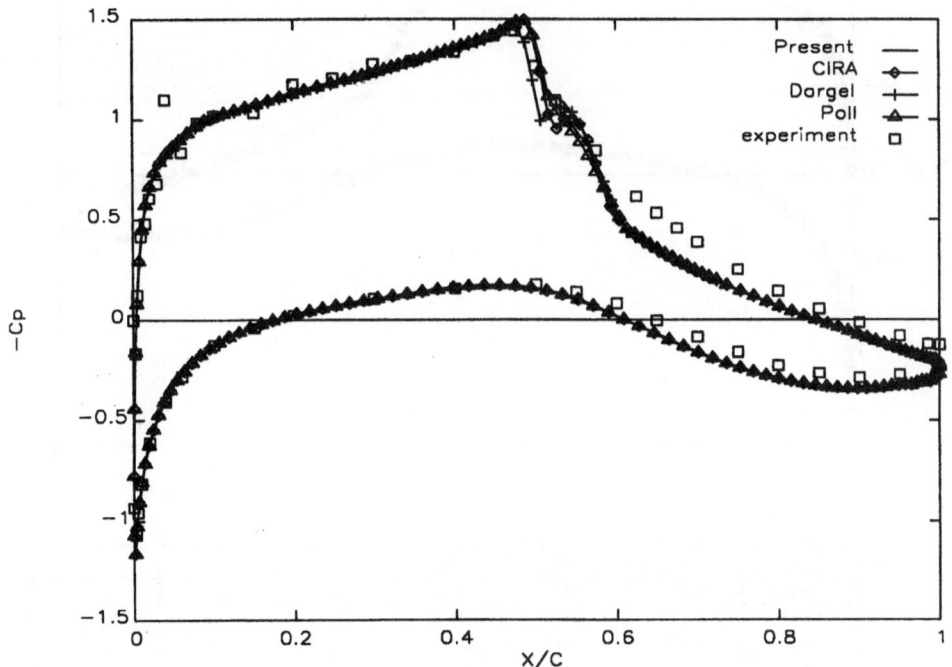

LAW	α	C_l	C_d	C_m	wake	LGR
Present	2.250	.8043	.0108	-.0975	no	yes
CIRA	2.250	.7973	.0107	-.0967	no	yes
Dargel	2.250	.7878	.0108	-.0955	no	yes
Poll	2.250	.7950	.0104	-.0960	no	yes

Fig.8. DRA 2303 AIRFOIL - TEST 1031, FULL LAWS, LGR mesh. Comparison between the calculated and experimental pressure distributions for different transpiration laws
Exp.: Mach=0.6806, Re=18,982,000, α=2.710, C_l=0.8142, C_d=0.0180, C_m=-0.0984
Calc.: Mach=0.6795, Re=18,907,000, α=2.250

When a proper reduction factor (SCF=0.5) is introduced, this problem disappears, Fig. 9.

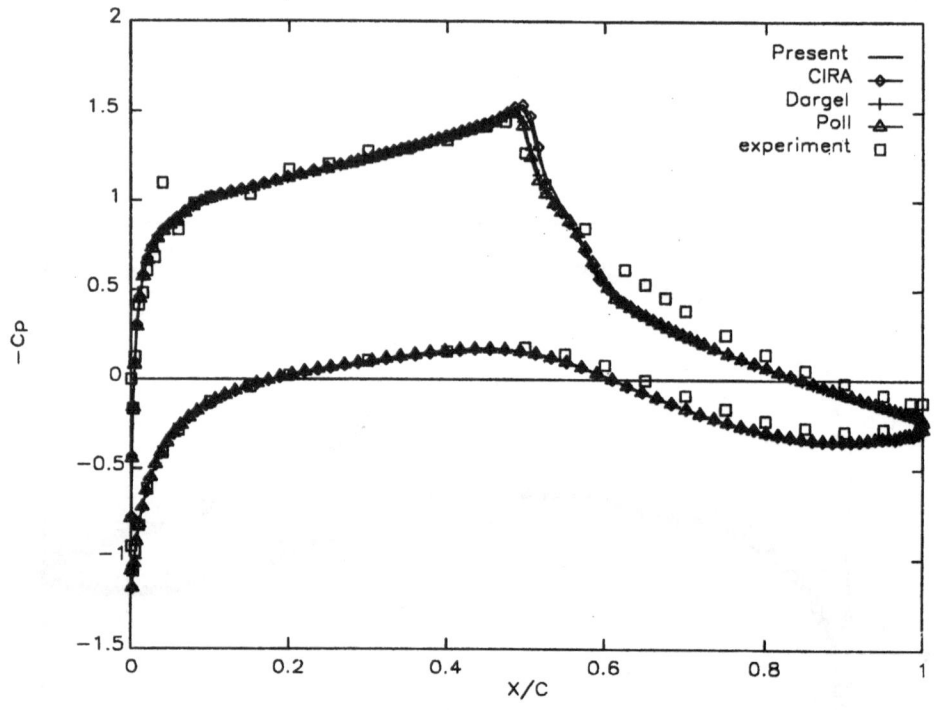

LAW	α	C_l	C_d	C_m	wake	LGR
Present	2.250	.8075	.0107	-.0978	no	yes
CIRA	2.250	.8097	.0108	-.0984	no	yes
Dargel	2.250	.7971	.0104	-.0963	no	yes
Poll	2.250	.7950	.0104	-.0960	no	yes

Fig. 9. DRA 2303 AIRFOIL - TEST 1031, SCALED TRANSPIRATION (SCF=0.5),O type LGR mesh. Comparison between the calculated and experimental pressure distributions for different transpiration laws
Exp.: Mach=0.6806, Re=18,982,000, α=2.710, C_l=0.8142, C_d=0.0180, C_m= -0.0984
Calc.: Mach=0.6795, Re=18,907,000, α=2.250

In Fig. 10 (a)-b) detailed pressure distribution in the shock region are presented, and in Fig. 11 a description of the flow field around the airfoil (isomach lines) is shown.

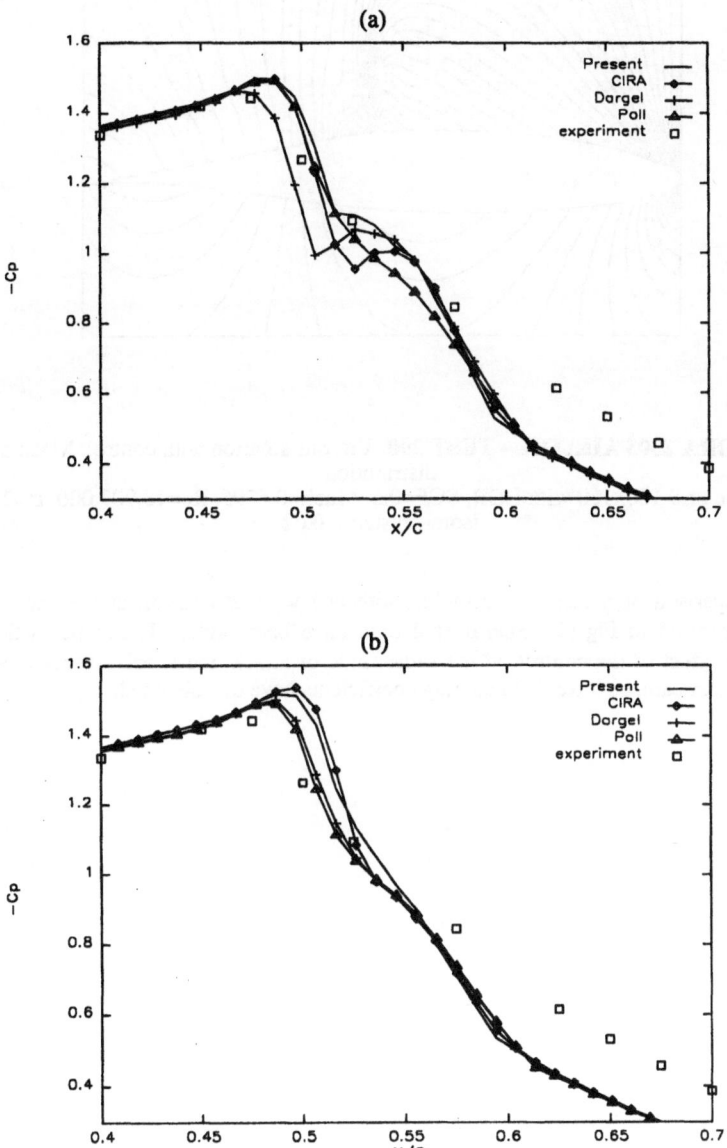

Fig. 10. DRA 2303 AIRFOIL - TEST 1031. Detailed pressure distributions:
(a) **FULL LAWS**, (b) **SCALED TRANSPIRATION (SCF=0.5)**

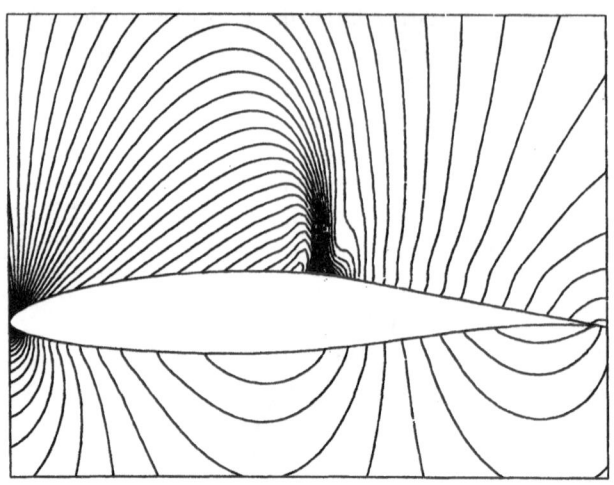

Fig. 11. DRA 2303 AIRFOIL - TEST 290. Viscous solution with control: Mach number distribution
O-type grid 160x160 with LGR, SCF=0.5, Mach=0.6795, Re=18,907,000, α=250
isomach step = 0.02

Comparison of pressure distribution with and without control, at the same angle of attack, is reported in Fig.12; experimental data have been added. It can be evaluated the (theoretical) effect of the control, at the same angle of attack: the shock structure is weaker, and moved upstream, so lower lift and drag coefficients have ben obtained.

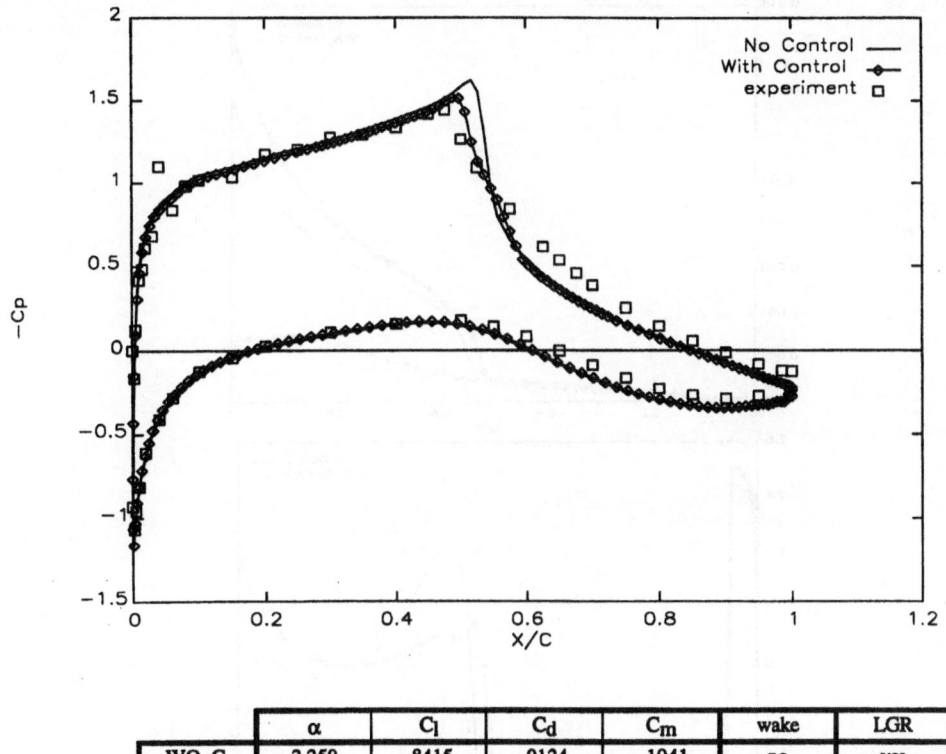

	α	C_l	C_d	C_m	wake	LGR
WO. C.	2.250	.8415	.0124	-.1041	no	yes
W. C.	2.250	.8075	.0107	-.0978	no	yes

Fig.12. DRA 2303 AIRFOIL - TEST 1031, O-type LGR mesh. Comparison among the
calculated and experimental pressure distributions w/wo control
Exp.: Mach=0.6806, Re=18,982,000, α=2.710, C_l=0.8142, C_d=0.0180, C_m=-0.0984
Calc.: Mach=0.6795, Re=18,907,000, α=2.250

Figg. 13 and 14 show the calculated boundary layer characteristics, in presence of
control, with full (Fig. 13) and reduced (Fig.14) transpiration. By using the full transpiration
isentropic law, and the corresponding pressure distribution reported in Figg. 8 and 10 (a), the
double shock structure influences strongly the boundary layer development; obviously, this
influence reduces when the reduced transpiration velocity distribution is used: in both cases a
tendency to separate on the blowing side of the cavity and a large beneficial effect on flow
separation behind the shock can be observed.

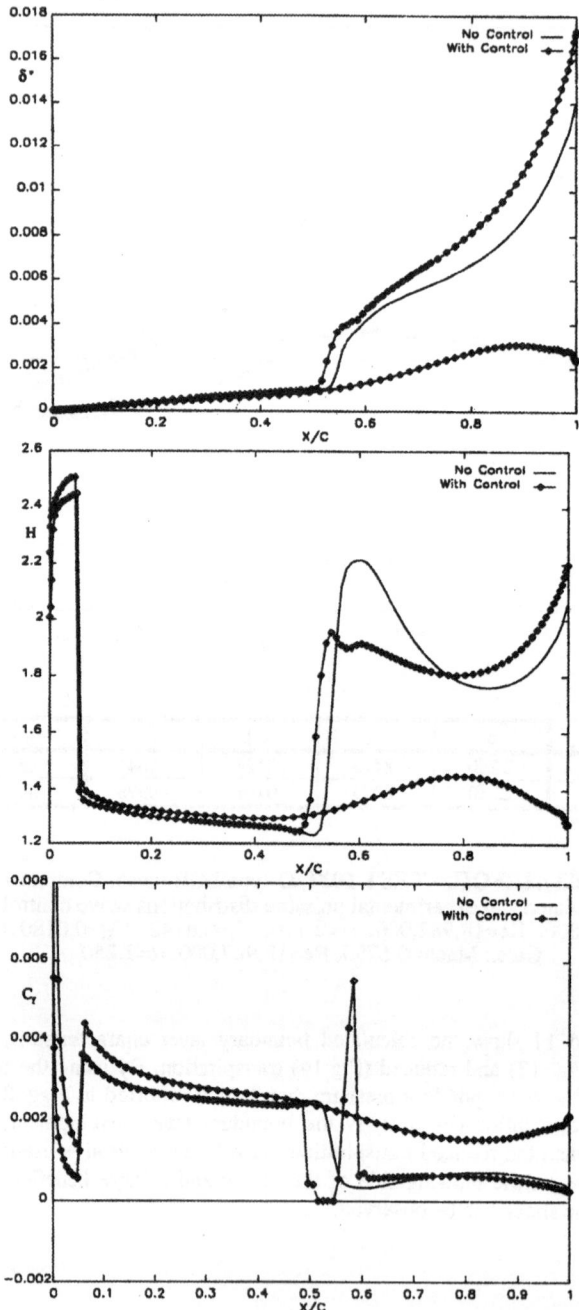

Fig. 13. DRA 2303 AIRFOIL - TEST 1031 (SCF=1.0). Calculated boundary layer parameters δ^*, H, C_f

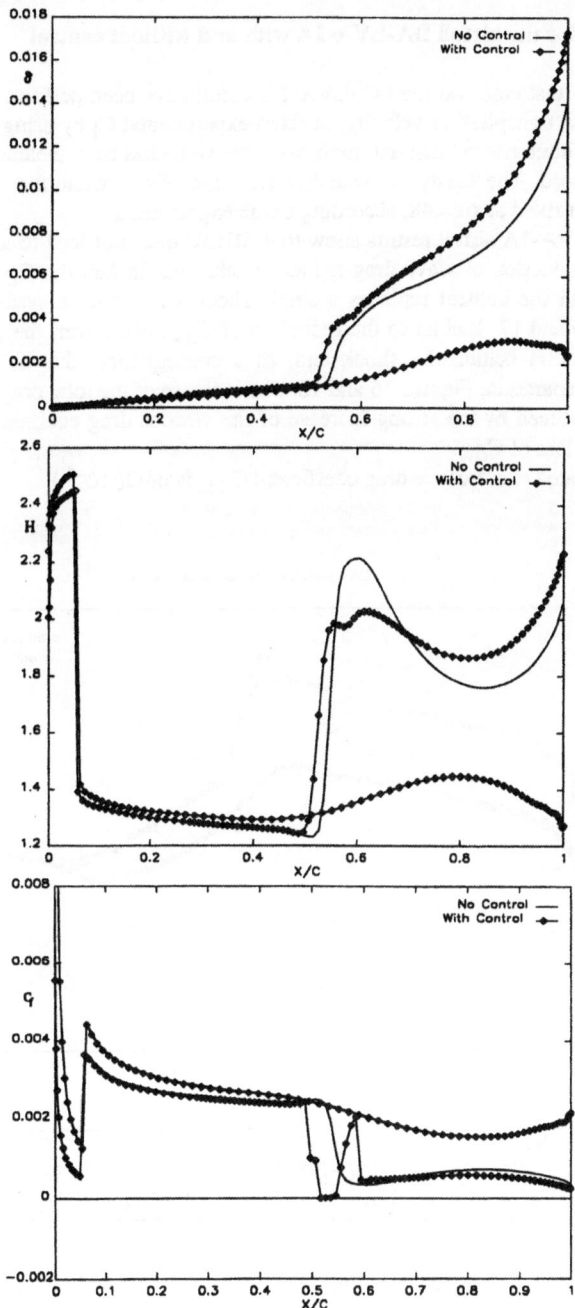

Fig. 14. DRA 2303 AIRFOIL - TEST 1031 (SCF=0.3). Calculated boundary layer parameters δ^*, H, C_f

13.8.4 Test case for the airfoil DA-LVA-1A with and without control

Numerical test cases on the DA-LVA-1A airfoil have been performed using Poll's law for determining the transpiration velocity, at fixed experimental C_l by using an O-type grid of 160x60 points without grid refinement; moreover, the wake has been included up to a distance of 30% of the chord. The cavity is located at $(.58 < x/c < .695)$, while the porosity is 5.06%; transition has been fixed at $x/c = .48$, according to the experiments.

The DA-LVA-1A airfoil results show that SBLIC does not lead to a reduction of total drag; anyway, a reduction of wave drag is indeed achieved. In fact the C_{dw} is related to the shock strength and the control replaces a single shock wave with a system of two weaker waves, Figures 15 and 17, leading to the reduction of C_{dw}. Moreover, the control results in a larger shape factor H behind the shock and, as a consequence, this has a positive effect concerning flow separation, Figures 16 and 18. The increase of the total drag coefficient can be interpreted as produced by the strong increase of the viscous drag coefficient due to control, compensating the loss of C_{dw}.

The calculated excrescence drag coefficient C_{des} is of $O(10^{-4})$.

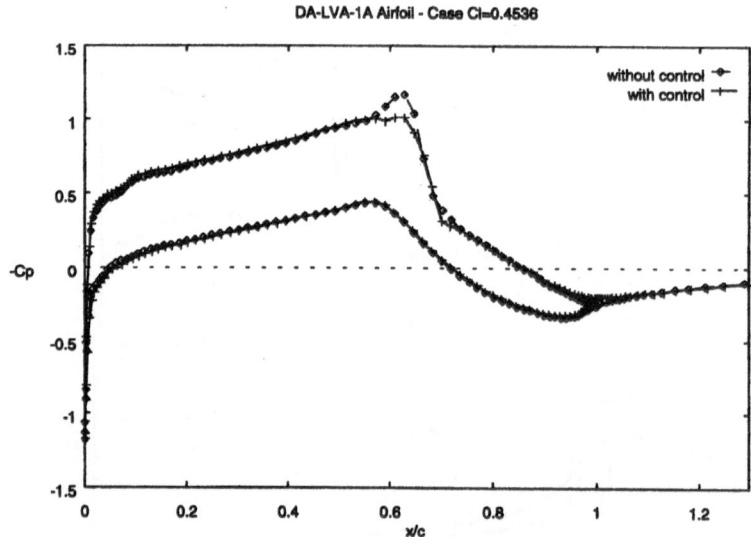

DA-LVA-1A Airfoil - Case Cl=0.4536

contr.	M_∞	Re	α	C_l	C_m	C_d	C_{dw}	C_{des}
off	0.7614	4.64E+6	0.7021	0.4536	-0.08874	0.01073	0.00743	0.00078
on	0.7614	4.64E+6	0.9000	0.4536	-0.08273	0.01326	0.00731	-

Fig.15. Calculated pressure distribution with and without control

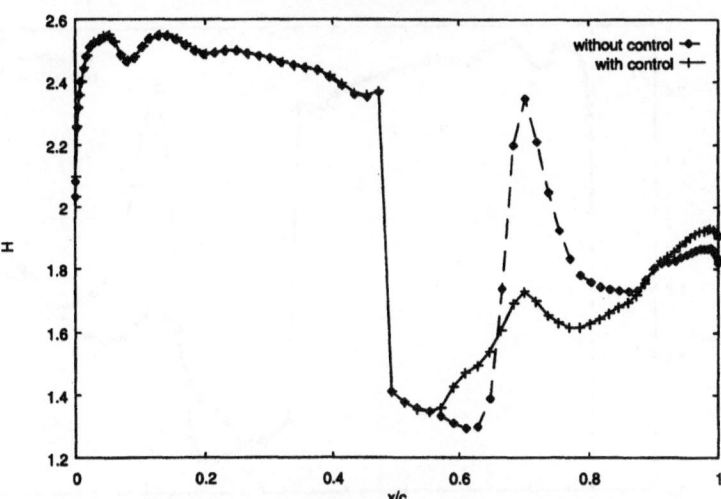

Fig.16. Calculated shape factor (upper surface)

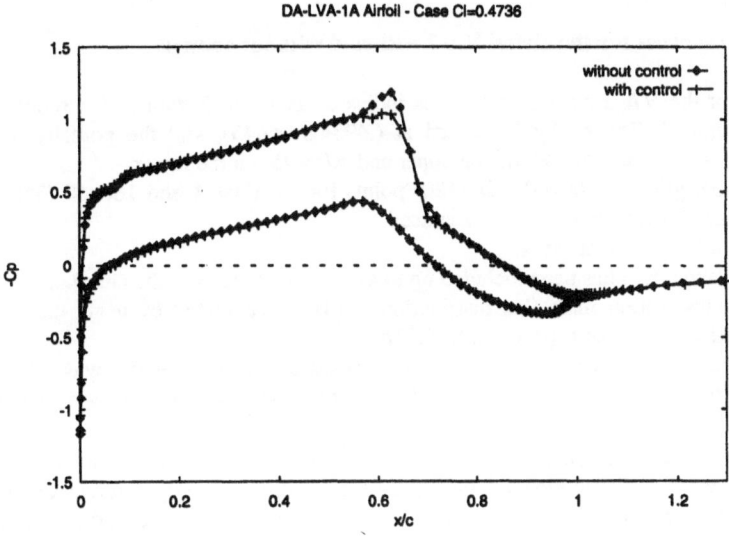

contr.	M_∞	Re	α	C_l	C_m	C_d	C_{dw}	C_{des}
off	0.7613	4.64E+6	0.7947	0.4739	-0.09034	0.01137	0.00748	0.00079
on	0.7613	4.64E+6	1.0040	0.4737	-0.08389	0.01392	0.00735	-

Fig.17. Calculated pressure distribution with and without control

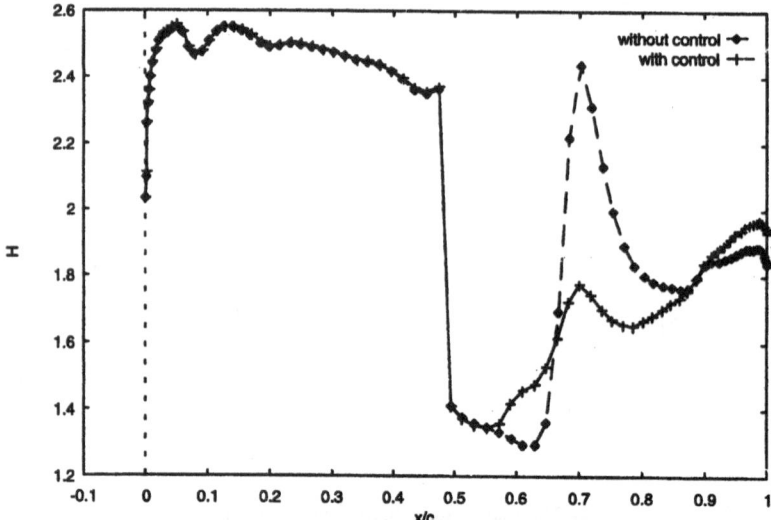

Fig.18. Calculated shape factor (upper surface)

13.8.5 Test cases for the airfoil VA-2 with and without control

For the VA-2 airfoil, Test Cases indicated as 1 and 2 with and without control have been computed. The cavity is located at ($.495<x/c<.645$), and the porosity is 12.8%; the transition is fixed at $x/c=.30$ on the upper and $x/c=.25$ on the lower.

LGR grids of 160x60 (320x120) points for the Case 1 and 180x60 (360x120) points for the Case 2, respectively, have been used.

For both the test cases:
- the wake effect has been included up to a distance of 30% of the chord and
- the transpiration mass flux distribution has been calculated by using the isentropic law scaled by coefficients provided by CIRA.

The introduction of passive control causes a reduction of the wave drag coefficient (14% for the Case 1, 18% for Case 2), while the total drag increases in Case 1 and decreases in Case 2.

Note that the transpiration mass flux values at the first and last point of the cavity influence the C_p distribution on the upper side: without a proper reduction of the values coming from the corrected isentropic law, the ventilation does not lead to a gradual rise of pressure but causes a strong shock on the upstream edge of the cavity (Fig. 20).

The following figures present the results for the VA-2 test cases.

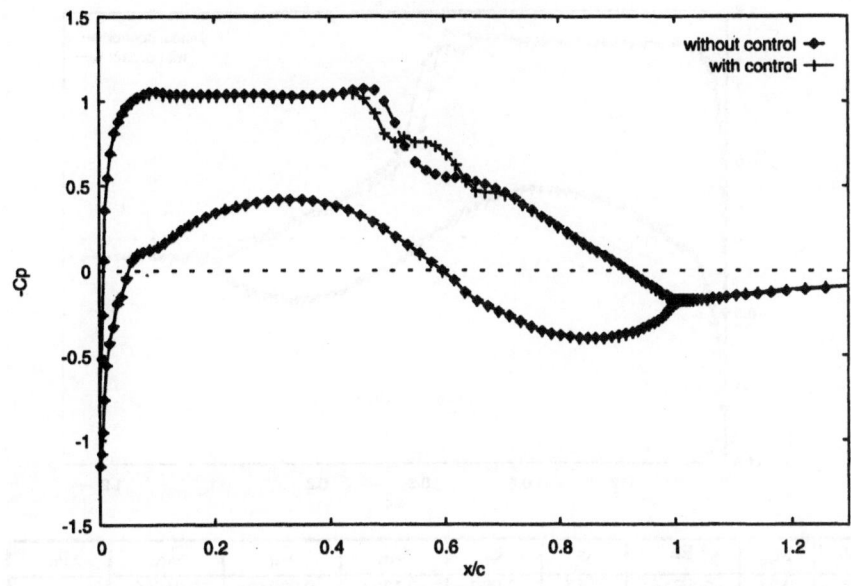

Fig. 19. Calculated pressure distribution, Case 1

contr.	M_∞	Re	α	C_l	C_m	C_d	C_{dw}	p_c
off	0.74	2.5E+6	1.180	0.651	-0.1214	0.01054	0.00112	-
on	0.74	2.5E+6	1.143	0.650	-0.1228	0.01063	0.00096	0.7155

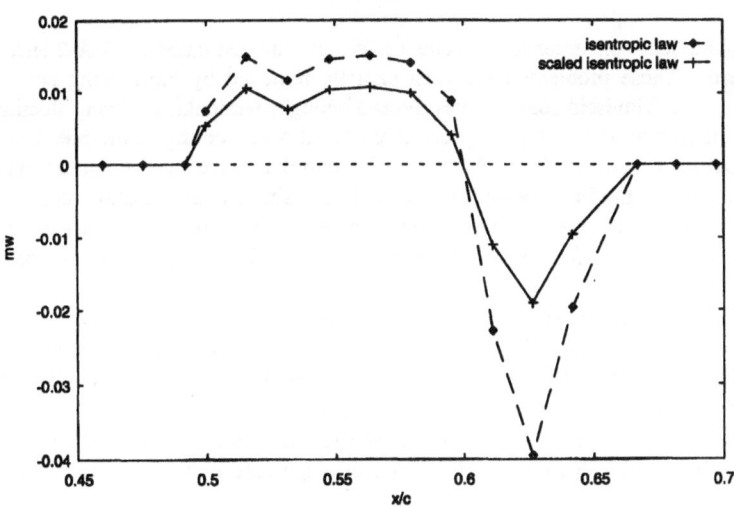

Fig. 20. Non-dimensional ventilation velocity

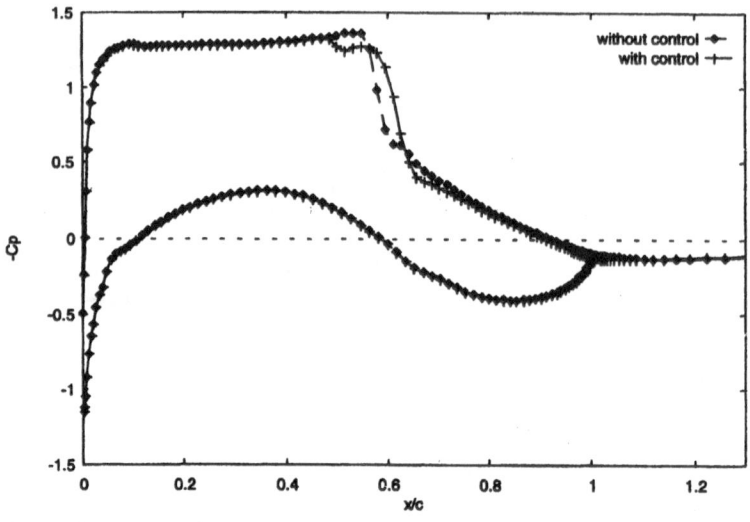

Fig. 21. Calculated pressure distribution, Case 2

contr.	M_∞	Re	α	C_l	C_m	C_d	C_{dw}	p_c
off	0.74	2.5E+6	2.65	0.890	-0.1324	0.02942	0.01080	-
on	0.74	2.5E+6	2.50	0.890	-0.1320	0.02808	0.00879	0.5366

13.8.6 Test case for the VA-2 airfoil at $0.75 < C_l < 0.85$ with and without control: a summary

Numerical convergence problems performing the test cases for VA-2 airfoil have been encountered. These problems have been partially removed by introducing control when the interactive viscid/inviscid solution was relaxed enough; particularly, strong fluctuations of the shock position during the iterative procedure caused overshooting of the pressure distribution, with subsequent numerical problems for the correlations of the boundary layer method. Moreover, control has been introduced gradually by using an amplification factor that increases its value by 0.33% at every iteration. However, by using the present release of the interaction code, to date no satisfactory results have been obtained at $C_l > 0.82$ in the presence of ventilation.

SBLIC has not always had a beneficial effect on the value of the total drag coefficient: in fact, as previously noted, the introduction of passive control causes a reduction of the wave drag coefficient, while the total drag increases. However, in some cases (C_l=0.890 and C_l=0.775), a reduction in total drag has been found.

As for the DA-LVA-1A airfoil, SBLIC has certainly a positive effect on flow separation behind the shock wave. The order of magnitude of the excrescence drag coefficient is 10^{-4}.

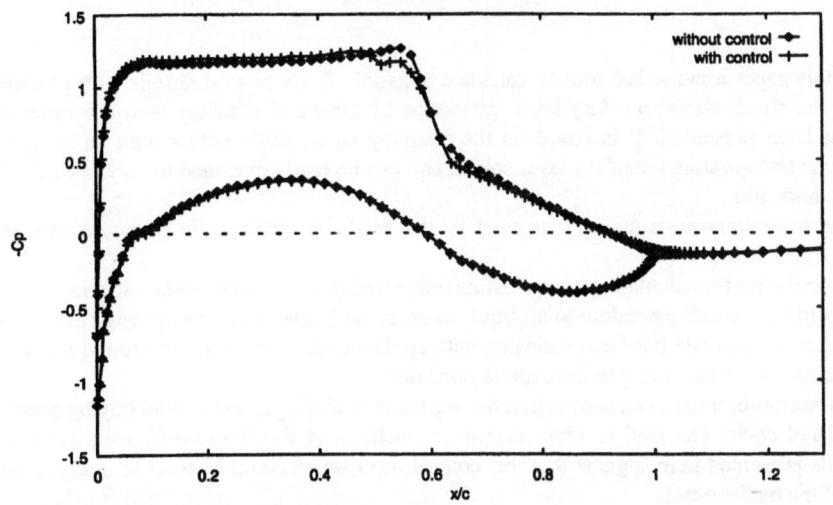

Fig.22. Calculated pressure distribution w/wo control

contr.	M_∞	Re	α	C_l	C_m	C_d	C_{dw}	C_{des}
off	0.743	2.5E+6	1.8172	0.8001	-0.13544	0.01756	0.01234	0.00085
on	0.743	2.5E+6	2.1132	0.8007	-0.12753	0.02351	0.01225	-

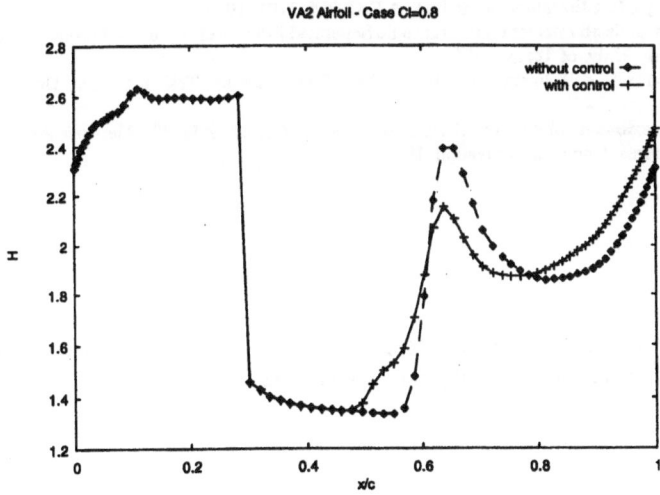

Fig.23. Calculated shape factor

13.9 Conclusion

In this paper a numerical tool to calculate transonic flows around airfoils with passive control of the shock-wave/boundary layer interaction by means of a cavitiy with a ventilated surface has been presented. It is based on the coupling of an Euler solver with an integral direct/inverse transpirating boundary layer solver, and can be easily extended to simulate active control methods too.

The main improvements with respect to the methods available in literature can be summarized as follows:
1) a new correlation for calculating compressible transpirating turbulent boundary layers;
2) a local grid refinement procedure to improve accuracy with reasonable computing cost.

The resulting code has been validated with applications to some flows around airfoils, giving reliable results according to the experimental data.

Furthermore, it has been shown that the application of the passive technology by means of a ventilated cavity can lead to only marginal reductions of the drag coefficient; anyway, some results presented here suggest that this control mechanism could be used to attempt the delaying of the buffet onset.

13.10 References

[1] A. Jameson, W. Schmidt, E. Turkel, 'Numerical Solution of the Euler Equation by Finite Volume Methods Using Runge-Kutta Time-Stepping Schemes', AIAA Paper 81-1259, 1981.
[2] A. Kassier, R. Tognaccini, 'Boundary Conditions for Euler Equations at Internal Block Faces of Multi-Block Domains Using Local Grid Refinement', AIAA Paper 90-1590, Seattle 1990.
[3] J. E. Green, P. J. Weeks, J. W. F. Brooman, 'Prediction of Turbulent Boundary Layer add Wakes in Compressible Flow by a Lag Entrainment Method', A.R.C.&M. 3791, 1977.
[4] C.R. Olling, 'Viscous-Inviscid Interaction in Transonic Separated Flow over Solid and Porous Airfoils and Cascades', Ph.D Thesis, Univ. of Texas, 1985.
[5] M. Lazzaro, 'Sul Controllo dell' Interazione Onda d' Urto-Strato Limite', Tesi di Laurea, Universita' di Napoli, 1992.
[6] M. Bidello, 'Ottimizzazione di Metodi per il Calcolo di Flussi Viscosi su Profili Alari per Applicazioni Avanzate', Tesi di Laurea, Universita' di Napoli, 1993.

Aknowledgements

The author is gratefull to Dr. R. Tognaccini, M. Lazzaro, P. Visingardi, M. Bidello, A. Ragni for the help that they provided in the development of this research.

14 EXTENSION, VALIDATION AND APPLICATION OF THE DA VII TRANSONIC AIRFOIL CODE WITH PASSIVE SHOCK CONTROL

G. Dargel

Daimler–Benz Aerospace Airbus GmbH

Hünefeldstr. 1–5, 28183 Bremen

Summary: The contribution of Dasa Airbus (DA) to the EUROSHOCK project within Task 2.1 includes the extension of the DA viscous inviscid interaction transonic airfoil code to passive shock control (SC) by the incorporation of a SC law procedure for the prediction of the ventilation velocity and by the modification of the boundary conditions and of the algebraic eddy viscosity turbulence model for wall mass flow.

Using the data base of Task 3 the extended SC code has been validated with passive SC test cases of laminar– and turbulent–type supercritical airfoils under consideration of the shock boundary layer interaction with grid adaptation. The numerical results are in agreement with the experimental data concerning the wave drag reduction but also the viscous drag increase. In addition, the SC code has been applied for a parametric study of the main passive SC design parameters to support the design of the DA LVA–1A model for passive SC.

14.1 Introduction

In the last years predictive capabilities for transonic airfoil flows with shock control have been developed allowing the design of passive SC devices. A first approach to compute viscous transonic flow over porous airfoils has been made by Chen et al. [1] on the basis of the thin layer Navier–Stokes equations. The viscous inviscid interaction (VII) transonic airfoil code of Dasa Airbus, Dargel et al. [2], has been extended to passive shock control by incorporating a local triple deck solution of Bohning [3] and Breitling [4]. Furthermore, using this code numerical studies have been carried out showing promising drag rise improvements for a transonic laminar–type airfoil by passive control.

The contribution of Dasa Airbus (DA) to the EUROSHOCK project is devoted to the numerical simulation by a steady airfoil flow prediction method with the objective to improve the numerical approach to this complex flow field and to study control effects on drag and flow separation. The present chapter describes the basic viscous inviscid interaction airfoil prediction method, the extension to passive shock control, the validation with airfoil test cases, measured in Task 3 of the EUROSHOCK project, and some application aspects.

14.2 Basic Airfoil Flow Prediction Method

The basic DA prediction method had been previously developed for the prediction of transonic airfoil and infinite swept wing flows, Dargel [5]. It is based on an inviscid viscous coupling of a full potential flow solver with an interactive boundary layer finite difference method. Following the defect formulation concept of Le Balleur [6], the real viscous flow is split into a viscous defect flow and an equivalent inviscid flow, which is extended to the airfoil surface and the wake streamline as the location of the viscous boundary conditions for the inviscid flow.

14.2.1 Equivalent inviscid flow and viscous boundary conditions

The equivalent inviscid flow with prescribed viscous boundary conditions is computed using the transonic full potential method of Jakob [7] with an entropy correction in the shock region from Mertens et al. [8]. The full potential equations are transformed from the physical domain into the incompressible streamline and potential line plane and solved by a successive line over-relaxation (SLOR) iteration method in the given H–type grid. Because the potential flow solution is only valid for isentropic conditions, a shock operator is introduced satisfying the Rankine–Hugoniot relation. Here, a controlled mass flow increase simulates the entropy rise in the potential equation. The interaction of the inviscid flow with the boundary layer, resulting in a weakening of the shock, can be reasonably approximated by prescribing the oblique shock relation with maximum stream deflection in the shock operator.

Following the defect formulation the displacement effect of the viscous layer is achieved by normal velocities v_{iw} on the airfoil and normal velocity jumps $< v_{iw} >$ across the wake. In addition, 2nd order corrections for the curvature effects are taken into account by a jump in the tangential velocities along the wake $< u_{iw} >$ and a correction between the inviscid and viscous wall pressure ($p_{iw} - p_w$), Sketch 1.

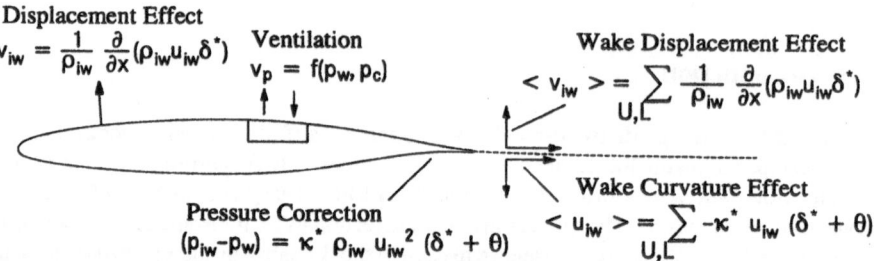

Sketch 1: Viscous boundary conditions of the equivalent inviscid flow

14.2.2 Interactive boundary layer and coupling method

Restricted to airfoil flows and to 1st order boundary layer theory, the viscous flow solution is obtained from the compressible boundary layer equation in which the turbulence is given by the algebraic eddy viscosity ε:

$$\frac{\partial}{\partial x}(\rho u) + \frac{\partial}{\partial z}(\rho v) = 0 \tag{1}$$

$$\rho u \frac{\partial u}{\partial x} + \rho v \frac{\partial u}{\partial z} = -\frac{\partial p}{\partial x} + \frac{\partial}{\partial z}\left[(\mu - \rho \varepsilon)\frac{\partial u}{\partial z}\right] . \tag{2}$$

Instead of solving the energy equation, Crocco's relationship is used for the density profile at adiabatic conditions:

$$\frac{\rho_e}{\rho} = 1 + r\frac{\kappa-1}{2}M_e^2\left[1-\left(\frac{u}{u_e}\right)^2\right]. \tag{3}$$

The boundary conditions are for an adiabatic surface with mass flow transfer

$$z = 0: \qquad u = 0, \quad v = v_w \qquad\qquad (4)$$

$$z = \delta: \qquad u = u_e(x) \qquad\qquad (5)$$

and for each half of the wake with symmetric conditions at the wake center line :

$$z = 0: \qquad \frac{\partial u}{\partial z} = 0. \qquad\qquad (6)$$

To perform calculations for flows with separation, the boundary layer equations are solved in an inverse mode, and the external velocity is computed as part of the solution. Applying Veldman's quasi–simultaneous coupling technique [9] in a difference formulation between two successive viscous inviscid iteration sweeps ($v,v-1$), the edge boundary condition, eq (5), is written as

$$u_e^v(x) = u_{iw}^v(x) + (\Delta u_{iw})^{v,v-1} \quad , \qquad\qquad (7)$$

where Δu_{iw} is computed from the Hilbert integral containing the displacement effect $\partial(u_e\delta^*)/\partial x$

$$\Delta u_{iw}(x) = \frac{1}{\pi\sqrt{1-M_\infty^2}} \int_0^\infty \frac{\partial}{\partial x'}\left(u_e\delta^*\right) d\frac{x'}{x-x'} \quad . \qquad\qquad (8)$$

The turbulence model used is expressed in terms of the Cebeci and Smith (CS) eddy viscosity formulation [10]. Boundary layer transition is assumed to occur discontinuously.

The boundary layer equations are transformed using a stream function formulation and solved by an implicit marching procedure with 1st order discretization in streamwise direction and 4th order Hermitian polynomials in normal direction,Thiele [11], along the airfoil contour. In the wake the boundary layer computation of an airfoil side is continued for each wake side separately. Due to the upstream influence of the displacement effect in the Hilbert integral, eq (8), additional iteration sweeps are necessary to solve the interactive boundary layer equations.

To speed up the viscous inviscid iteration, Carter's relaxation formula [12] for the displacement thickness is applied before the viscous boundary conditions are computed.

14.2.3 Forces

Lift and pitching moment coefficients are obtained by appropriate integration around the airfoil contour. The total drag is determined from its components: the wave drag from the entropy jump along the shock wave and the viscous drag from the momentum thickness in the far wake. This method for determining drag is found to be less sensitive to numerical errors than predicting drag from integration.

14.3 Extension of the VII Method to Passive Shock Control

The prediction of the complex airfoil flow field with passive shock control requires a clear physical understanding of the phenomena involved in the control process and realistic physical models to represent the effect of control on the flow field mean and turbulent properties.

So far the extension of the present method to passive SC could only rely on certain simplifying assumptions in the shock control and turbulence modelling since improved models developed in Task 1 of EUROSHOCK were not available due to the limited time of the project.

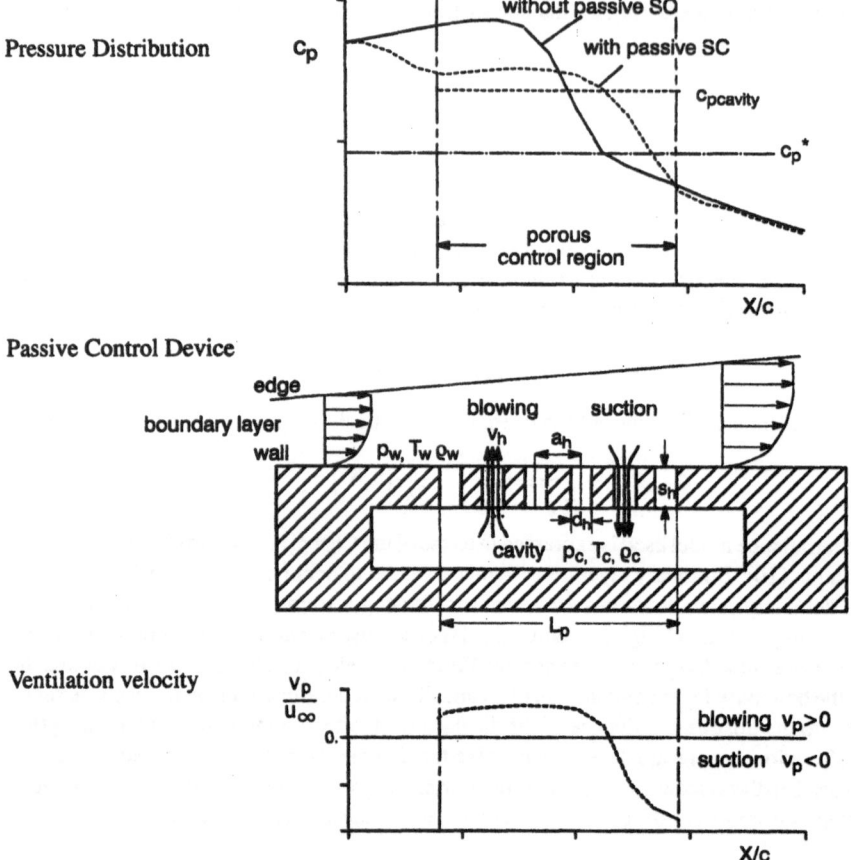

Pressure Distribution

Passive Control Device

Ventilation velocity

Sketch 2: Passive shock control region

14.3.1 SC Law procedure

The passive SC device, Sketch 2, consists of a porous surface with a cavity underneath. A typical porous surface consists of n_h holes with a specified diameter d_h and pitch a_h, giving a porosity n_p of

$$n_p = \frac{n_h A_h}{A_p} . \tag{9}$$

Due to the pressure difference between the flow side p_w and the cavity p_c, mass flow transfer occurs through the holes and changes the boundary condition in the region of the interaction between the shock and the boundary layer. The characteristic of the porous surface, i.e., the relationship between the mass flow rate through and the pressure drop across the surface affects the ventilation velocity. An average mass flow rate $\mathcal{M}_p$ can be defined by

$$\mathcal{M}_p = n_h \mathcal{M}_h = v_p \varrho_w A_p, \tag{10}$$

giving an average ventilation velocity v_p referenced to the stagnation conditions of the free stream flow:

$$\frac{v_p}{a_o} = \frac{1}{\varrho_w/\varrho_o} \, n_p \, \frac{\dot{\mathcal{M}}_h}{a_o \rho_o A_h}$$

with $\quad \dfrac{\rho_w}{\rho_o} = \dfrac{1}{\left(1 + r\frac{2}{\kappa-1}M_e^2\right)} \left(\dfrac{p_w}{p_o}\right)^{\frac{1}{\kappa}}$. \hfill (11)

The SC laws given below determine the hole velocity as function of the pressure drop with some empirical corrections, however, without taking the boundary layer flow along the control region into account.

14.3.1.1 Isentropic law

Under the assumption that the hole is treated as an isentropic diffuser or nozzle, a theoretical mass flow rate through a hole is predicted by the isentropic law as used by Breitling [4] and Bur [13] in case of
– suction $p_w > p_c$

$$\frac{\dot{\mathcal{M}}_{ht}}{a_o \, \varrho_o A_h} = \frac{p_c/p_o}{\sqrt{T_w/T_o}} M_{ht} \sqrt{\left(\frac{p_w}{p_c}\right)^{\frac{\kappa-1}{\kappa}}}$$

where M_{ht} is the theoretical hole Mach number

$$M_{ht} = \frac{2}{\kappa-1}\left[\left(\frac{p_w}{p_c}\right)^{\frac{\kappa-1}{\kappa}} - 1\right] \text{ restricted to } M_{ht} < 1.0$$

and $\quad \dfrac{T_w}{T_o} = \left(1 + r\dfrac{2}{\kappa-1}M_e^2\right) \left(\dfrac{p_w}{p_o}\right)^{\frac{\kappa-1}{\kappa}}$ \hfill (12)

– blowing $p_c > p_w$

$$\frac{\dot{\mathcal{M}}_{ht}}{a_o \, \varrho_o A_h} = \frac{p_w/p_o}{\sqrt{T_c/T_o}} M_{ht} \sqrt{\left(\frac{p_c}{p_w}\right)^{\frac{\kappa-1}{\kappa}}}$$

with $\quad M_{ht} = \dfrac{2}{\kappa-1}\left[\left(\dfrac{p_c}{p_w}\right)^{\frac{\kappa-1}{\kappa}} - 1\right]$ restricted to $M_{ht} < 1.0$

and $\quad \dfrac{T_c}{T_o} = \left(\dfrac{p_c}{p_o}\right)^{\frac{\kappa-1}{\kappa}}$. \hfill (13)

Due to the viscous and inlet/outlet effects of the hole flow and the imperfect hole drilling process, a loss coefficient η_v for the mass flow rate is introduced by Breitling [4] and Braun [14]

$$\eta_v = \frac{\dot{\mathcal{M}}_p}{\dot{\mathcal{M}}_{pt}},$$ \hfill (14)

that can be evaluated by calibration tests of the porous control device as function of the pressure drop.

14.3.1.2 Poll's calibration law

Poll investigated the aerodynamic properties of laser drilled titanium sheets employed in laminar flow studies [15]. Under the assumption of a laminar and incompressible pipe flow, he developed a flow model for an ideal hole, resulting in a quadratic relationship between the pressure drop $\Delta p = (p_w - p_c)$ and the mass flow rate $\mathcal{M}_h$ through a hole:

$$\frac{\Delta p}{p_o} \frac{\rho_h}{\rho_o} \frac{1}{\kappa} = \left[\frac{32}{R} \left(\frac{\mathcal{M}_h}{a_o \rho_o A_h} \right) + 1.203 \left(\frac{\mathcal{M}_h}{a_o \rho_o A_h} \right)^2 \right] \frac{1}{K}$$

$$\text{with} \quad R = \frac{a_o \rho_o c}{\mu_o} \frac{d_h/c}{s_h/d_h \; \mu_h/\mu_o}$$

$$\text{and} \quad K = \left[\frac{\text{effective } d_h}{\text{measured } d_h} \right]^4 . \tag{15}$$

The quadratic equation is solved for the mass flow rate $\mathcal{M}_h$ by imposing $|\Delta p|$ for suction and blowing and taking the positive value of the root. Average values are taken for ρ_h and μ_h, using the properties from the flow and cavity side. The ventilation velocity is finally obtained from eq (11) with the actual porosity. The correction factor K can be used to approximate the measured flow characteristic of the porous surface from calibration tests.

14.3.1.3 Prediction of the cavity pressure

It was observed in the channel flow experiments with passive SC that the static pressure is nearly constant in the cavity. With this assumption the cavity pressure p_c can be iteratively defined by the condition of zero net mass flow along the porous surface with the length L_p in the case of passive SC:

$$\int_{L_p} \left(\frac{v_p \rho_w}{a_o \rho_o} \right) d \left(\frac{x}{c} \right) = 0 . \tag{16}$$

Also in a second way, the SC law procedure can be applied by prescribing the cavity pressure which corresponds to the active SC application that can be achieved by connecting a suction system to the cavity.

14.3.2 Modifications of the viscous solver

The introduction of the ventilation velocity, due to the presence of the porous control region, has necessitated a slight modification of the wall boundary layer condition, eq(4): $v_w(x) = v_p(x)$.

In addition, the inner region of the turbulent boundary layer is disturbed by the mass flow transfer as observed by many investigators. In the present simulation the influence of a distributed ventilation velocity on the turbulent structure is taken into account neglecting the fact that the mass flow transfer is produced by an array of single hole flows, blowing or suction.

Since the algebraic eddy viscosity turbulence model is applied, the inner eddy viscosity ε_i

$$\epsilon_i = (0.4 \; z \; D)^2 \left| \frac{\partial u}{\partial z} \right| \tag{17}$$

is extended to mass flow transfer by modification of the van Driest wall damping function

$$D = 1 - \exp\left(-\frac{z\frac{v}{u_{\tau max}}}{A^+}\right),$$ (18)

suggested by Cebeci [10]:

$$A^+ = \frac{26}{N}\left(\frac{\rho}{\rho_w}\right)^{1/2}$$

$$N^2 = \frac{\mu}{\mu_e}\left(\frac{\rho_e}{\rho_w}\right)^2 \frac{p^+}{v_w^+}\left[1 - \exp\left(11.8\frac{\mu_w}{\mu}v_w^+\right)\right] + \exp\left(11.8\frac{\mu_w}{\mu}v_w^+\right)$$

with pressure gradient term $\quad p^+ = \frac{v_e u_e}{u_{\tau max}}\frac{du_e}{dx}$

and mass flow transfer term $\quad v_w^+ = \frac{v_w}{u_{\tau max}}.$ (19)

$u_{\tau max}$ represents the maximum friction velocity instead of the wall friction velocity in order to avoid infinite values at separated flow conditions.

An advantage of this formula is that the mixing rate is influenced by the mass flow rate together with the pressure gradient in the shock boundary layer interaction region. The modification yields skin friction values in the blowing region higher than those calculated by the basic model, because the blowing increases the mixing rate by a reduced damping parameter A^+. Similarly, a strong adverse pressure gradient reduces A^+. Opposite effects occur in the suction region and in favourable pressure gradients.

In the later stage of the EUROSHOCK project, the CS turbulence model was replaced by the non–equilibrium algebraic model of Johnson/Coackley (JC) [16] in order to obtain improved turbulence modelling in the shock boundary layer interaction region with non–equilibrium effects of the boundary layer.

14.3.3 Modifications of the solution procedure

To compute the flow over an airfoil with passive SC, the cavity parameters, as location and porosity, have to be defined in addition to the input of the standard airfoil prediction. An essential task before starting is the approximation of the measured flow characteristic of the porous surface by the applied SC law procedure, determining the empirical function η_v, eq (14), or the correction parameter K, eq (15).

The solution procedure of the VII method with SC involves four nested iterations, Sketch 3. The first iteration solves the non–linear potential equation with fixed viscous boundary conditions by SLOR method, the second one the SC law procedure with zero net mass flow rate condition for the cavity pressure, the third one the interactive boundary layer equation with the quasi–simultaneous coupling condition, and the fourth one the overall viscous inviscid coupling procedure (outer VII loop).

The inviscid calculation is carried out on a sequence of three grids. The inviscid iterative solution is interrupted periodically to determine the ventilation velocities, to solve the boundary layer flow with mass flow transfer and to update the viscous boundary conditions by the coupling procedure using underrelaxation.

With the given external pressure distribution the ventilation velocities are computed from the SC law that is incorporated in the iteration cycle for the cavity pressure. A relaxation factor of about 0.3%–0.5% is applied for the ventilation velocity per outer iteration cycle in order to avoid instabilities in the computations. At the leading and trailing edge of the control region a blending zone of 10% of the ventilation region is added in order to smooth the rapid change in the boundary condition for the inviscid and viscous flow solvers. Numerical investigations have shown that it is reasonable to start the passive SC calculation after obtaining a converged solution of the VII method without SC.

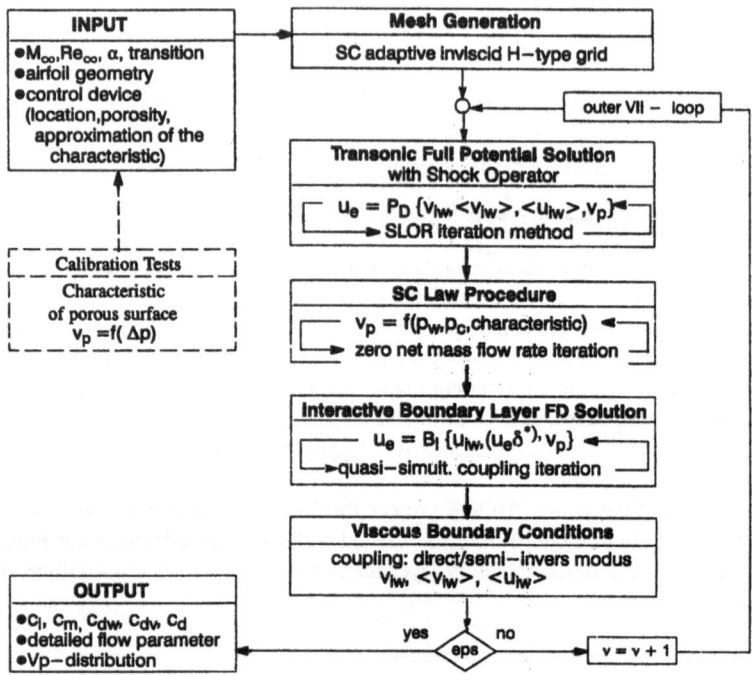

Sketch 3: Flow chart of SC VII prediction method

14.3.4 Grid adaptation

The H–type grid used for the inviscid flow solution is generated by the incompressible potential lines and streamlines of the inviscid flow past the airfoil at given incidence in a box of 5x5 chords. In the case of SC the grid can be adapted by using a grid refinement in the range of the control region, Fig. 1, or by increasing the total number of grid points, Figs. 8 and 14.

The criteria for both adaptation techniques is a mesh size of 1% chord in the control region. This mesh size is a compromise between accuracy and stability of the SC calculations. Of course, a smaller mesh size in the shock region improves the prediction of the interaction of the shock with the boundary layer, but it also originates stability and convergence problems.

For the SC law procedure an internal grid is used in the range of the control region in order to get an accurate integration of the mass flow rate.

The grid used for the viscous flow solution follows in streamwise direction the grid point distribution of the inviscid grid but with an additional subgrid technique in the control region. In normal direction of the boundary layer a number of 81 points is found to be reasonable for the shock boundary layer interaction. In order to maintain this constant number of grid points, a grid stretching factor is introduced which is locally controlled by the boundary layer growth along the surface.

14.4 Validation of the Numerical Results

In Task 2 the shock boundary layer interaction without and with passive control was studied in detail on three airfoils of different type (DRA 2303, MBB Va2, DA LVA–1A) by specifying a set of test cases. The computed results were validated with experimental data from Task 3.

14.4.1 DRA 2303 airfoil

The DRA tests have been carried out by Fulker et al. [17] with the DRA 2303 airfoil in the DRA 8ft x 8ft Subsonic–Supersonic Wind Tunnel within the EUROSHOCK project. The laminar–type airfoil has been equipped with a control device between $X/c=0.5 - 0.6$. It contains a perforated sheet of 1 mm thickness with a porosity of 8% formed by laser drilled holes with a nominal diameter of 0.076mm. From calibration tests DRA recommends to use Poll's formula, eq (15), for the approximation of the flow characteristic of the porous surface, using the nominal diameter and the correction factor of $K=1$.

14.4.1.1 Test Cases 1 and 2

For the validation Test Case 1 at $c_l=0.56$ and Test Case 2 at $c_l=0.81$ have been chosen at the design Mach number of $M_\infty=0.68$ and the flight Reynolds number of $R_\infty=19 \cdot 10^6$. Boundary layer transition is fixed at 5% chord on both surfaces. The H–type grid used for the inviscid flow solution is refined in the upper/lower box to 161x61/145x45 grid points, giving 197 grid points around the airfoil itself with a constant spacing of 1% chord in the control region, Fig. 1. The viscous flow is solved in normal direction at 81 points and at the streamwise stations of the inviscid grid with an additional subgrid in the control region.

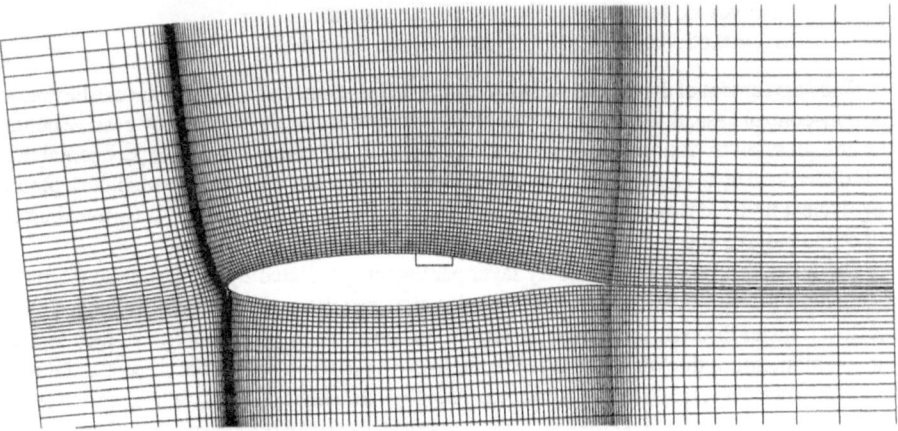

Fig. 1 :   Airfoil DRA 2303 Test Case 2
SC adapted H–type grid of the inviscid flow solution
Inviscid flow grid : $(165 \times 61)_{upper}$ / $(141 \times 45)_{lower}$
Viscous flow grid : $(165 \times 81)_{upper}$ plus SC subgrid / $(141 \times 81)_{lower}$

In Figs. 2 and 3 flow parameters as the pressure coefficient c_p, the boundary layer parameters δ^*, θ and c_f on the upper surface and the ventilation velocity v_p are plotted for the two test cases without and with passive SC, calculated with the prescribed lift coefficients of the experiments.

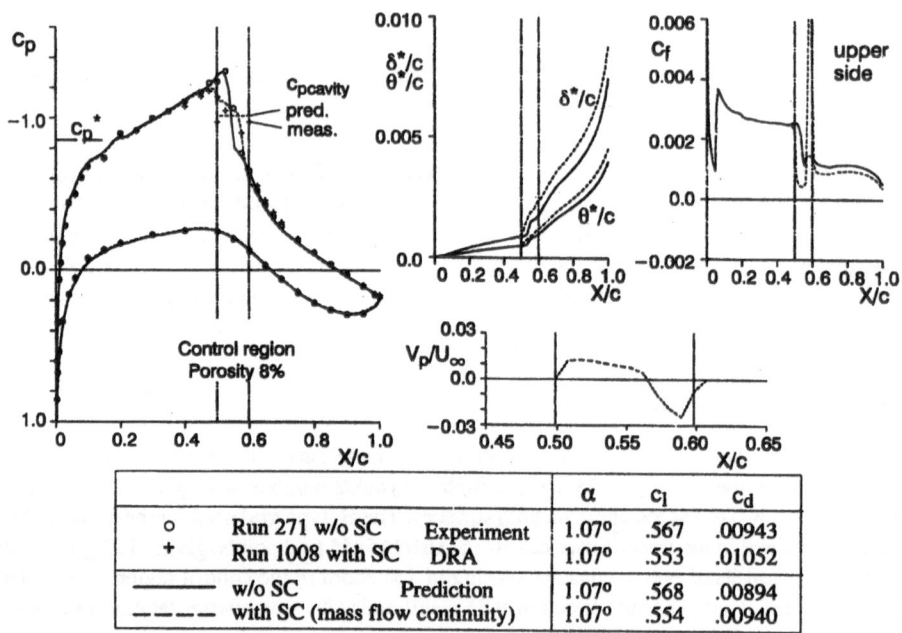

				α	c_l	c_d
o	Run 271 w/o SC	Experiment		1.07°	.567	.00943
+	Run 1008 with SC	DRA		1.07°	.553	.01052
——	w/o SC	Prediction		1.07°	.568	.00894
– – – –	with SC (mass flow continuity)			1.07°	.554	.00940

Fig. 2: Airfoil DRA 2303 SC Test Case 1
Comparison between prediction and experiment
$M_\infty = 0.68$, $Re_\infty = 19 \cdot 10^6$, fixed transition 5%/5%c

Good agreement of the shock location between experiment and computation can be noticed in the non–control case. A weakening of the shock strength is clearly obtained by passive control as the compression in front of the original shock shows in the computed and experimental data. This is underlined by the spreading of the shock in the ventilation case in the iso Mach line plots of the equivalent inviscid flow field for the Test Case 2 in Fig. 4. A reduction of the wave drag is a result of the passive shock control device application.

On the viscous side, the boundary layer parameters predicted by the algebraic CS turbulence model with modification of the damping function, eq (19), are dramatically changed by the mass flow transfer through the porous surface in the control region. An increase of the displacement and momentum thickness is caused by the blowing in the front part of the control device, combined with a decrease of the skin friction. Although the suction yields high skin friction values, the boundary layer is thicker at the trailing edge of the control region and is subject to nearly the same adverse pressure gradient up to trailing edge, consequently increasing the viscous drag.

When the code is running in the passive SC mode, the pressure in the cavity is a result of the computation. It seems that the present SC law procedure with Poll's formula is unable to predict the measured cavity pressure, see Figs. 2 and 3, causing some deviations between the measured and predicted pressure distributions in the control region. By switching to the active SC mode, i.e., prescribing the measured cavity pressure, a better agreement in the pressure distribution can be noticed between experiment and computation for Test Case 2 in Fig. 3.

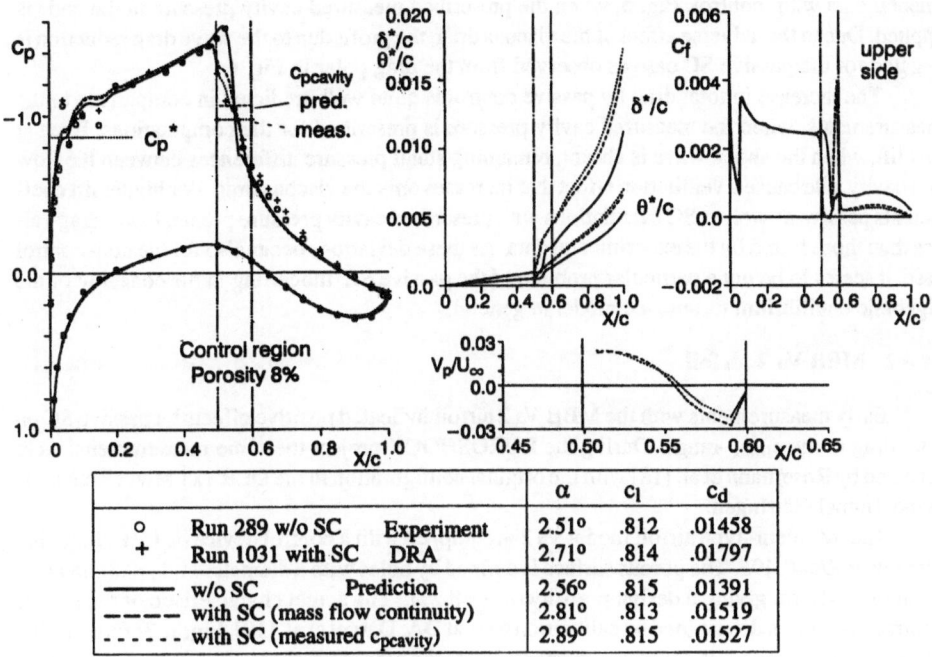

			α	c_l	c_d
o	Run 289 w/o SC	Experiment	2.51°	.812	.01458
+	Run 1031 with SC	DRA	2.71°	.814	.01797
——	w/o SC	Prediction	2.56°	.815	.01391
––––	with SC (mass flow continuity)		2.81°	.813	.01519
......	with SC (measured $c_{pcavity}$)		2.89°	.815	.01527

Fig. 3: Airfoil DRA 2303 SC Test Case 2
Comparison between prediction and experiment
$M_\infty = 0.68$, $Re_\infty = 19 \cdot 10^6$, fixed transition 5%/5%c

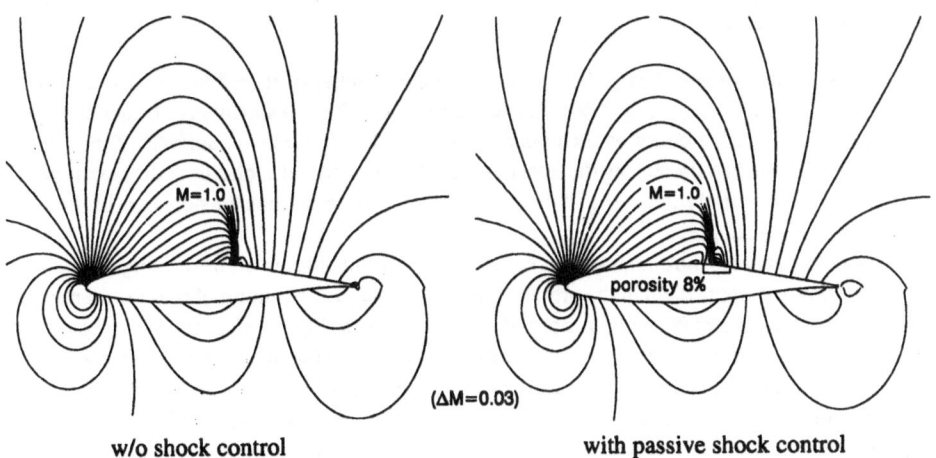

w/o shock control   with passive shock control

Fig. 4: Airfoil DRA 2303 Test Case 2
Iso Mach–lines of the equivalent inviscid flow field

14.4.1.2 Variation of pressure distribution and drag with lift

For a series of lift coefficients the predicted pressure distributions agree well with experimental data with control, Fig. 5, when the prescribed measured cavity pressure in the code is applied. Due to the adverse effect of the viscous drag, the profit due to the wave drag reduction is negated for the passive SC case as observed from the drag polar in Fig. 6.

The increase in total drag by passive control is quite well predicted in comparison to the measurements, when the measured cavity pressure is prescribed for the computations. Even at low lift, when the shock wave is absent, remaining small pressure differences between the flow and cavity side cause a ventilation effect that increases only the viscous drag. For higher lift coefficients passive as well as SC calculations with prescribed cavity pressure predict lower drag values than those found by the experimental data. As these deviations occur also for the non–control case, it seems to be not a particular problem of the passive SC modelling in the code, but of the algebraic equilibrium turbulence model in general.

14.4.2 MBB Va 2 airfoil

Early measurements with the MBB Va 2 airfoil indicated positive effects by passive SC on total drag in the c_{lmax}–range. During the EUROSHOCK project the same measurements were repeated by Rosemann et al. [18] with the original configuration in the DLR 1x1 Meter Transonic Wind Tunnel Göttingen.

The turbulent–type airfoil model Va 2 is equipped with a control device of 15% chord, beginning at $X/c=0.495$. The porous surface is defined by holes with a diameter of $d_h=0.3$mm and a pitch of $a_h=1$mm, giving a design porosity of $n_p=8.2\%$. The actual characteristic of the porous control region was determined by calibration tests at DA, Dargel et al. [19], Fig. 7. Before starting the SC computations it is necessary to approximate the measured characteristic by the SC laws. Poll's formula gives the best fit of the measured curves when applying the design hole parameters and a correction factor of $K=1$, while the isentropic law yields a stronger mass flow for the same pressure drop.

14.4.2.1 Test Cases 1 and 2

At the Mach number of $M_\infty=0.743$ and Reynolds number of $R_\infty=2.5\cdot10^6$, Test Case 1 is defined at the lift coefficient of $c_l=0.645$ and Test Case 2 at $c_l=0.80$, respectively. Boundary layer transition is fixed at 25% chord on the upper and 30% chord on the lower surface. The H–type grid used is generated by increasing the total number of grid points on each side at the condition of 1% spacing in the control region, Fig. 8. For the following computations the non–equilibrium JC turbulence model is applied.

For both test cases the computations of the turbulent transonic airfoil flow, using the passive SC mode, show a compression in the pressure distribution at the beginning of the control region, followed by a pressure plateau and a normal shock at the end, Figs. 9 and 10. At $c_l=0.645$ the predicted reduction of wave drag is completely compensated by the increase of the viscous drag. The high lift case is characterized by a λ–shock system with a remaining strong shock in the rear part of the control region as observed from the iso Mach–line plot in Fig. 11, yielding a weak wave drag reduction of about 16 cts. A strong increase of the viscous drag, as indicated by the distribution of the boundary layer thickness, is responsible for a total drag growth of about 17 cts compared to the non–control case. Even in the suction part of the control region negative skin friction values are observed in Fig. 10.

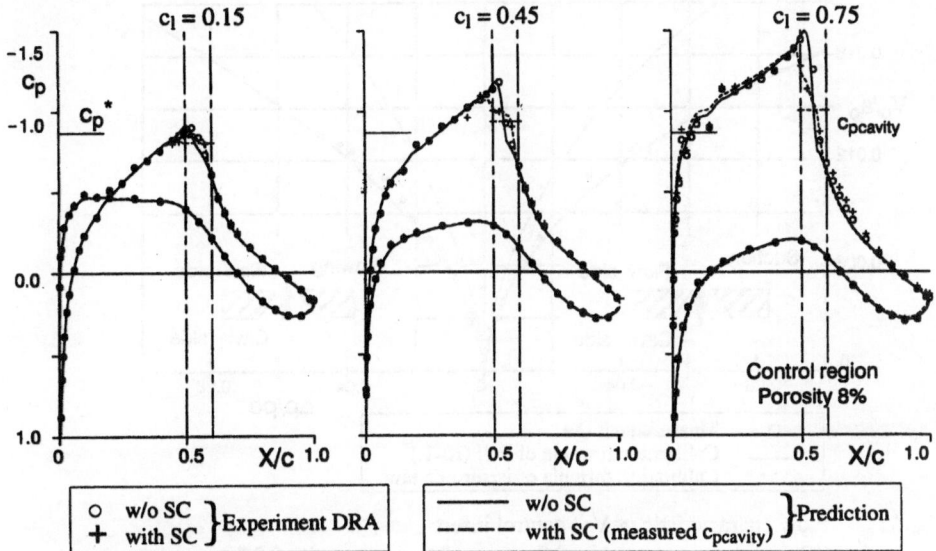

Fig. 5: Airfoil DRA 2303: Variation of pressure distributions with lift
$M_\infty = 0.68$, $Re_\infty = 19 \cdot 10^6$, fixed transition 5%/5%c

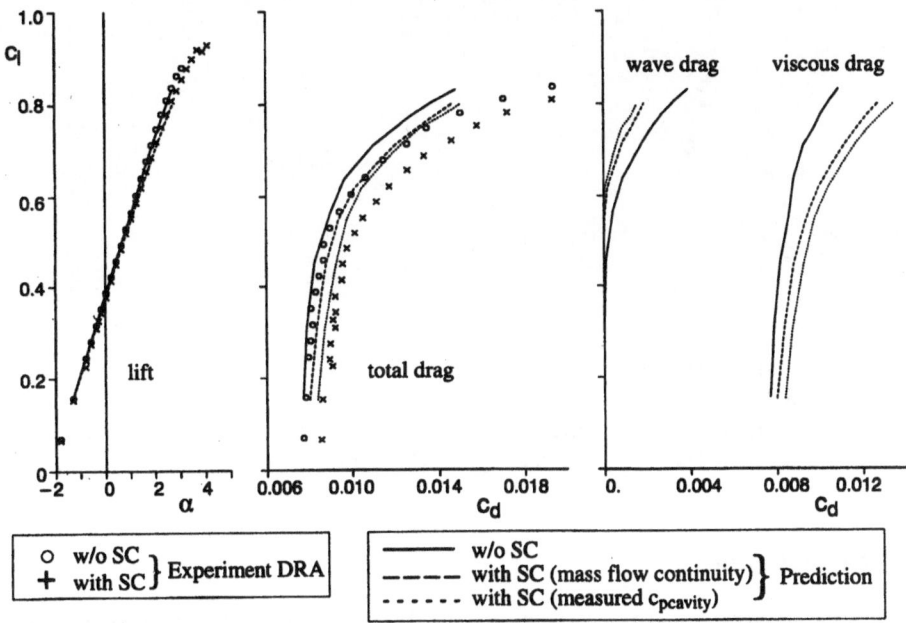

Fig. 6: Airfoil DRA 2303: Predicted and measured lift and drag polars
$M_\infty = 0.68$, $Re_\infty = 19 \cdot 10^6$, fixed transition 5%/5%c

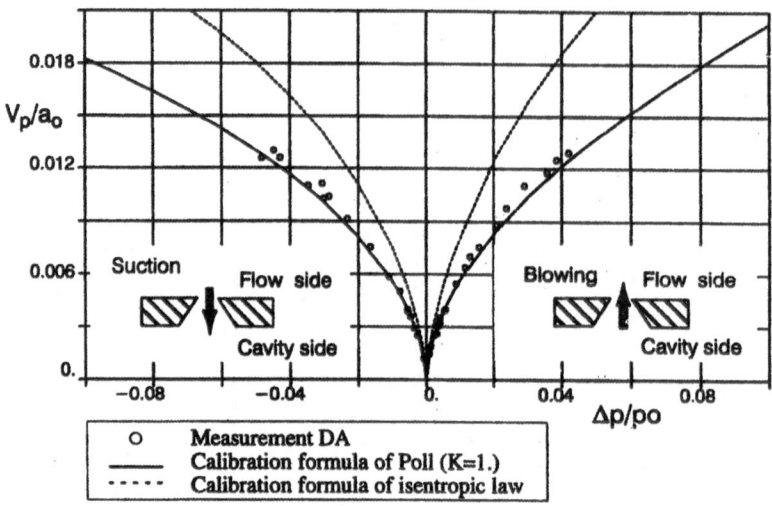

Fig. 7: Flow characteristic of Va2 control insert
SC design parameters: d_h=0.3mm, a_h=0.3mm, porosity 8.2%

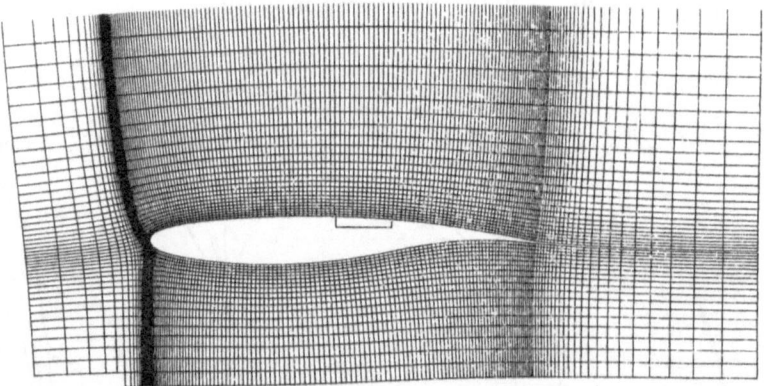

Fig. 8: Airfoil MBB Va 2 SC test case 2
SC adapted H–type grid of the inviscid flow solution
Inviscid flow grid : (120 x 45) on upper/lower airfoil side
Viscous flow grid : (120 x 81) plus SC subgrid

14.4.2.2 Drag polar

A lift sweep from c_l=0.25 up to 0.80 was carried out, using the passive SC mode of the code. Up to a lift coefficient of c_l=0.50 shock waves are absent on the datum airfoil as shown in the pressure distribution in Fig. 12. Nearly no additional drag is computed, which is in contrast to the results of the laminar–type airfoils DRA 2303 and DA LVA–1A.

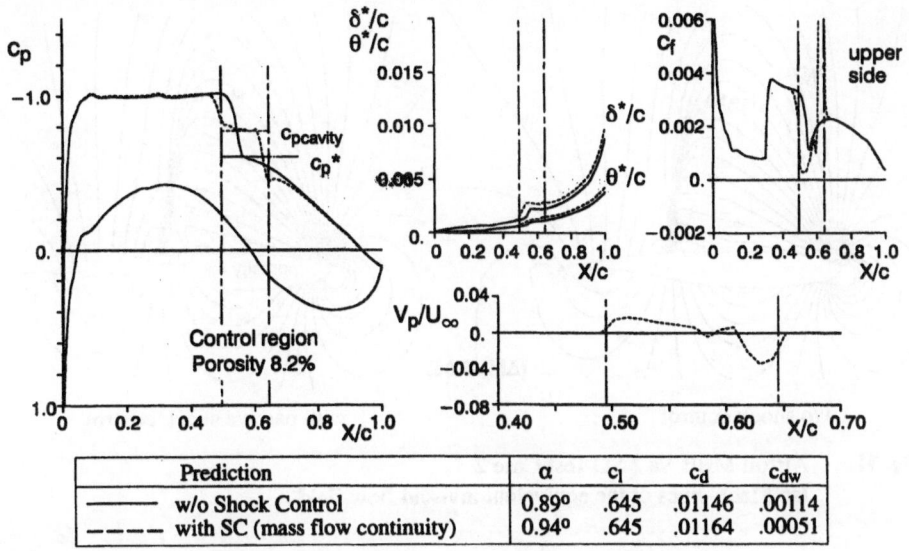

Fig. 9: Airfoil MBB Va2 SC Test Case 1: Airfoil flow parameters
$M_\infty = 0.743$, $Re_\infty = 2.5 \cdot 10^6$, fixed transition 30%/25%c

Prediction		α	c_l	c_d	c_{dw}
———	w/o Shock Control	0.89°	.645	.01146	.00114
– – – –	with SC (mass flow continuity)	0.94°	.645	.01164	.00051

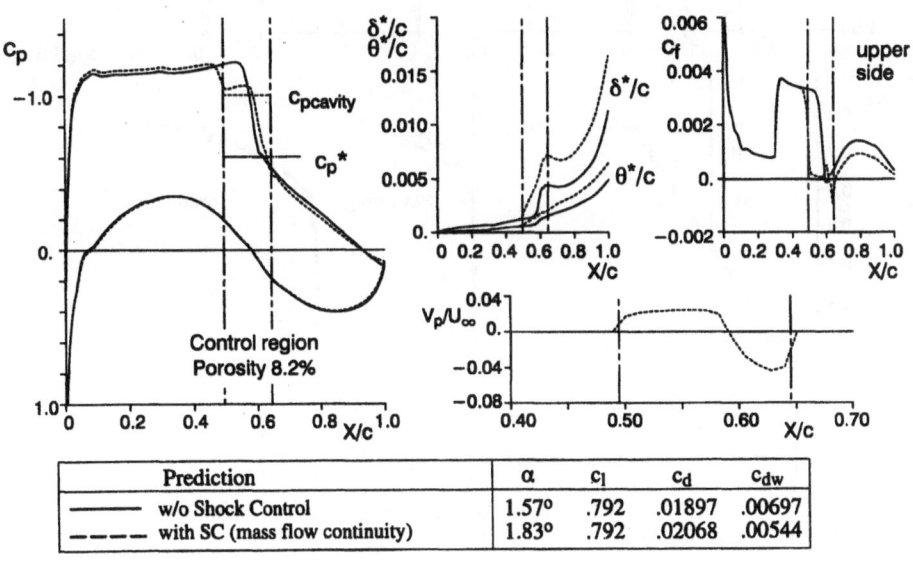

Fig. 10: Airfoil MBB Va2 SC Test Case 2: Airfoil flow parameters
$M_\infty = 0.743$, $Re_\infty = 2.5 \cdot 10^6$, fixed transition 30%/25%c

Prediction		α	c_l	c_d	c_{dw}
———	w/o Shock Control	1.57°	.792	.01897	.00697
– – – –	with SC (mass flow continuity)	1.83°	.792	.02068	.00544

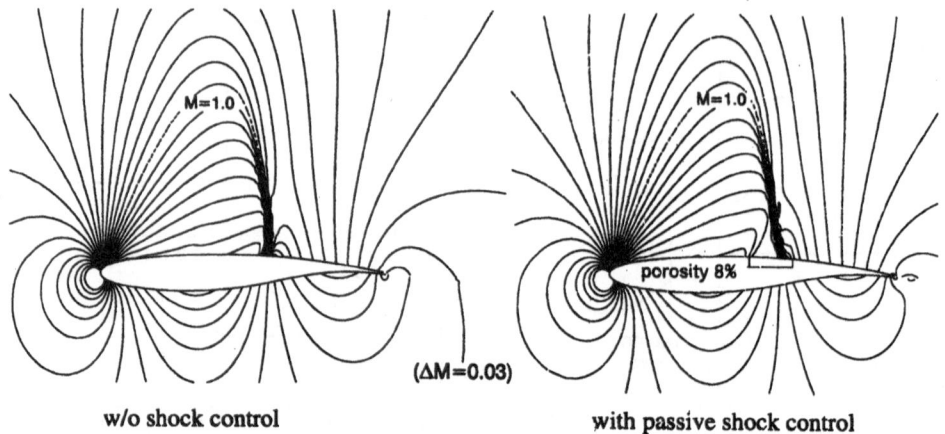

w/o shock control   with passive shock control

Fig. 11:   Airfoil MBB Va 2 SC Test Case 2
            Iso Mach–lines of the equivalent inviscid flow field

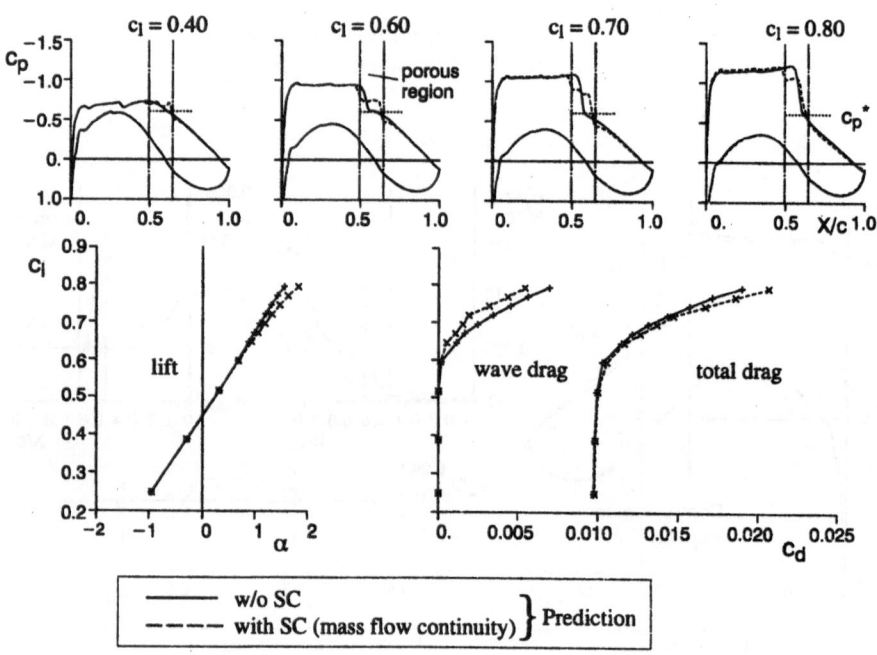

Fig. 12:   Airfoil MBB Va2: Variation of drag and pressure distributions with lift
            $M_\infty = 0.743$, $Re_\infty = 2.5\cdot10^6$, fixed transition 30%/25%c

This behaviour may be linked to the roof–top pressure type of the datum airfoil. For lift coefficients up to $c_l=0.70$ wave drag reduction occurs due to passive SC. But this gain in drag is compensated by the increase of the viscous drag, giving finally a small increase of the overall drag. By increasing lift further close to c_{lmax} an overproportional increase in viscous drag is predicted with the consequence of large growth of the total drag. No further calculation with the passive SC could be performed in the c_{lmax}–region of the drag polar due to convergence problems.

Comparisons with the experimental data of the repeated Va 2 test campaign were not performed, because too large corrections of the freestream Mach number and the angle of attack are needed to match the shock position between experiment and computation for a given lift coefficient.

14.4.3 DA LVA–1A airfoil

The DA airfoil model LVA–1A contains a perforated control region with a length of 11.5% chord and an actual porosity of 5.1%, starting at $X/c=0.58$. The electron–beam drilled holes have an average diameter of $d_h=0.1123$mm, arranged in equilateral triangles with a pitch of $a_h=0.4756$mm. A detailed description of the control device is given in Chapter 19.

Based on the experience of the other test cases, Poll's calibration formula is applied, Fig. 13, to approximate the flow characteristic of the porous surface that was measured by DA. The correction factor K is found to be different for the blowing and suction case.

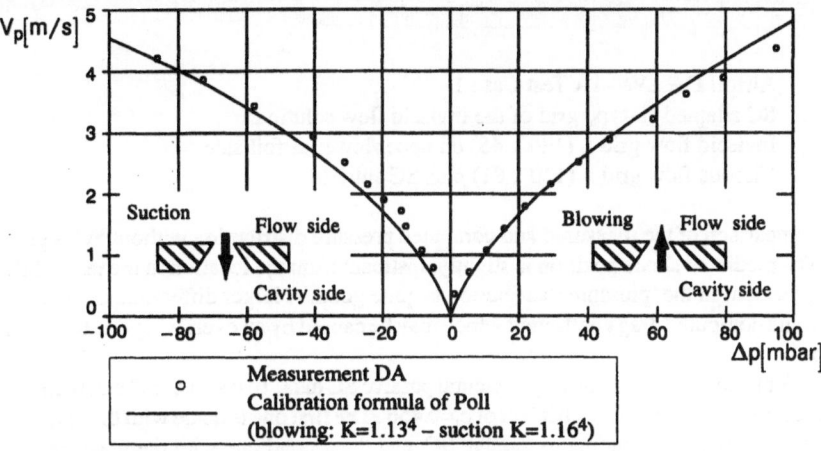

Fig. 13: Flow characteristic of the LVA–1A control insert
Measured SC parameters: $d_h=0.1123$mm, $a_h=0.4756$mm, $s_h=1.2$mm,
porosity 5.06%

The model was tested in the two cryogenic wind tunnels, ONERA T2 by Archambaud [20] and DLR KRG by Rosemann et al. [18], within the frame of EUROSHOCK. Due to the time limit of the project, only few experimental results were available for validation.

14.4.3.1 Test Case 1

Test Case 1 at $c_l=0.478$ has been chosen from ONERA T2 experimental data at the corrected freestream Mach number of $M_\infty=0.7613$ (correction due to side wall effects) and a Reynolds number of $R_\infty=4.6\cdot10^6$. Boundary layer transition for this laminar–type airfoil was fixed at 48% chord on both surfaces in order to establish a turbulent shock boundary layer interaction.

For the given lift coefficients computations with and without passive SC were carried out in the inviscid flow grid with a high resolution in the nose as well as in the interaction region of this laminar–type airfoil by increasing the total number of grid points, Fig. 14.

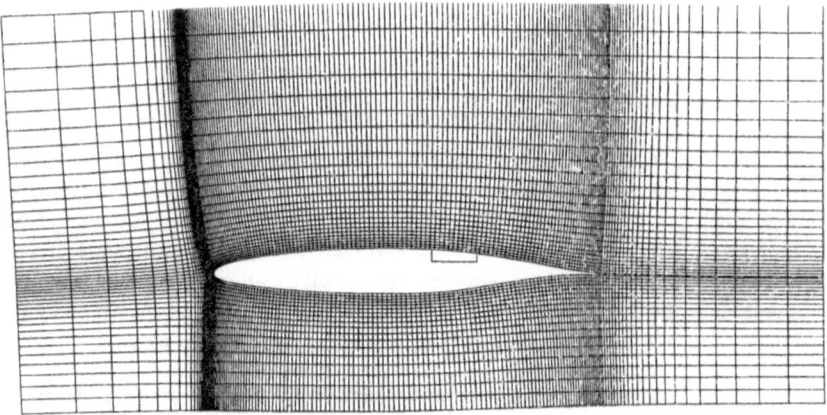

Fig. 14:   Airfoil DA LVA–1A Test Case 1
SC adapted H–type grid of the inviscid flow solution
Inviscid flow grid : (120 x 45) on upper/lower airfoil side
Viscous flow grid : (120 x 81) plus SC subgrid

A comparison of the measured and computed pressure distribution without SC is given in Fig. 15. The predicted shock position is slightly upstream from the location in the test, while the overall agreement in the pressure distribution is quite good. A larger difference is found in the measured and computed drag coefficient which may be caused by an overfixing due to the transition strip.

In order to compare with the experimental passive SC data, two sets of SC calculations are carried out at the measured lift coefficient of $c_l = 0.456$. The first one is made with the condition of zero net mass flow corresponding to passive SC, and the second one with the measured cavity pressure, Fig. 15. As in the experiment the predicted pressure distribution indicates a reduction of the shock strength by the ventilation which is clearly observed from the iso Mach–line plot in Fig. 16. However, while the experimental data show nearly a linear distribution across the control region, a pressure plateau with a remaining shock in the rear part of the control region is predicted by the code. Concerning the experimental pressure distribution, an inspection of the control region after instrumentation showed blocked holes around the pressure taps, leading to an increase of the pressure loss across the porous sheet which possibly effects the measured pressure distribution in this section, see [20] and Chapter 19.

As also found during the validation phase of the DRA experiments, the predicted cavity pressure differs from the measured one when the passive SC mode is applied in the code, yielding stronger deviations of the pressure distribution in the control region compared to the experimental data. The predicted increment in overall drag due to control effects is about 6 cts when using the SC mode with the measured cavity pressure in the computation, but its increase is lower than the experimental one.

212

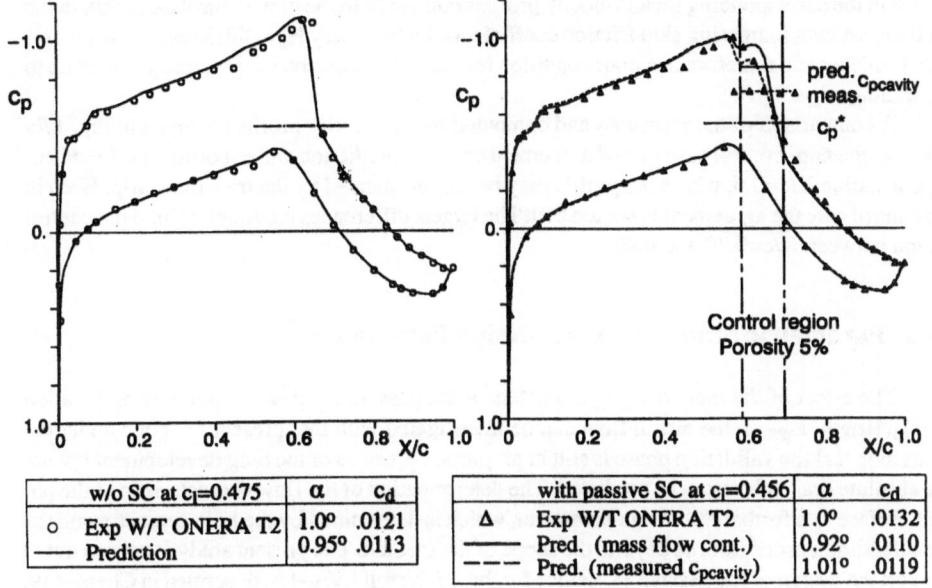

w/o SC at c_l=0.475		α	c_d
o	Exp W/T ONERA T2	1.0°	.0121
—	Prediction	0.95°	.0113

with passive SC at c_l=0.456		α	c_d
Δ	Exp W/T ONERA T2	1.0°	.0132
—	Pred. (mass flow cont.)	0.92°	.0110
---	Pred. (measured $c_{pcavity}$)	1.01°	.0119

Fig. 15: Airfoil DA LVA–1A SC Test Case 1: Pressure distribution comparison
$M_\infty = 0.7613$, $Re_\infty = 4.6 \cdot 10^6$, fixed transition 48%/48%c

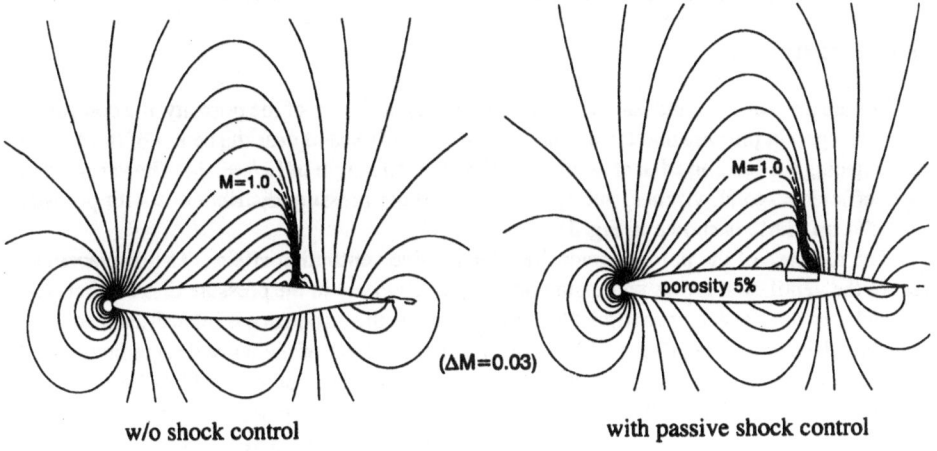

w/o shock control   with passive shock control

Fig. 16: Airfoil DA LVA–1A SC Test Case 1
Iso Mach–lines of the equivalent inviscid flow field

14.4.3.2 Boundary layer development

The predicted development of the boundary layer of Test Case 1 without and with shock control is compared in Figs. 17a and b, using the non–equilibrium JC turbulence model in the computation. The strong increase of the boundary layer in the blowing part of the control region is visible in the corresponding mean velocity profiles compared to the non–control case. Still in the suction part with increasing skin friction coefficients the boundary layer thickness of the control case is thicker and therefore in a poor condition for the following adverse pressure gradient up to the trailing edge.

A comparison of the measured and computed mean velocity profiles is given in Fig. 17b. While in the non–control case a good agreement between prediction and experiment is found except at station X/c=0.6, where the profile may be still influenced by the transition strip. Even in the control case the agreement is not too bad. The largest differences are found behind the control region between X/c=0.70 and 0.80.

14.5 Parametric Study of the SC Design Parameters

The effect of the main design parameters of the passive SC device –porosity n_p, location $(X/c)_p$, length L_p– on the airfoil flow can be investigated with the present SC code under the restriction that the validation phase is still in progress, i.e, trends of the drag development but not the absolute drag values can be predicted. The determination of the flow characteristic of the porous surface as a further SC design parameter, which is determined by the hole geometry and the manufacturing procedure, is beyond the scope of the method. The present study is carried out to support the design of the passive SC device for the DA airfoil LVA–1A, described in Chapter 19.

For the aerodynamic design of the passive SC device an airfoil flow with a lift coefficient of c_l=0.45 at M_∞=0.76 and Re_∞=6·10^6 is selected as a characteristic off–design case for the LVA–1A airfoil. The extent of the laminar flow is restricted up to 50% chord on the upper side by a transition strip. On the lower side boundary layer transition is fixed at X/c=0.05.

14.5.1 Porosity

For the prediction of the passive control effects by variation of the porosity, the control region is arranged in the way that the original shock wave is located near the center of the control region, placing the leading edge at $(X/c)_p$=0.593 at a given length of L_p/c=0.14. The variation of drag coefficients –total, viscous and wave drag– and the pressure distributions with porosity from 0.5% up to 6% is shown in Fig. 18.

The increase of the porosity intensifies the passive ventilation effects. The displacement effect of the ventilation grows with rising porosity as observed in the pressure distributions and hence reduces the wave drag coefficients. Simultaneously the turbulence is intensified leading to increasing viscous drag coefficients with porosity. The gradient of the wave drag reduction and viscous drag increase becomes more gentle at higher porosity. For a porosity of 2% the wave drag is already reduced by about 10 cts and the increase of the total drag remains small, while for a porosity of 4% the reduction in wave drag is about 15 cts with an increase of the total drag of about 4 cts. The predicted cavity pressure remains nearly unchanged with variation of the porosity.

14.5.2 Perforation location

Selecting a porosity of 2% and a length of L_p/c=0.14, the cavity pressure and drag are strongly affected by the location of the control region in relation to the shock wave position, as shown in Fig. 19.

If the original shock is located in the rear part of the control region, i.e. the most upstream cavity location, a larger negative cavity pressure is computed with a small profit in the wave drag reduction due to the weaker displacement effect, combined with a smaller viscous drag increase. A shock location in the forward part of the control region, i.e., the most downstream cavity location, yields an increase of the cavity pressure with strong ventilation effects, giving a larger wave drag reduction but with a strong increase in viscous drag.

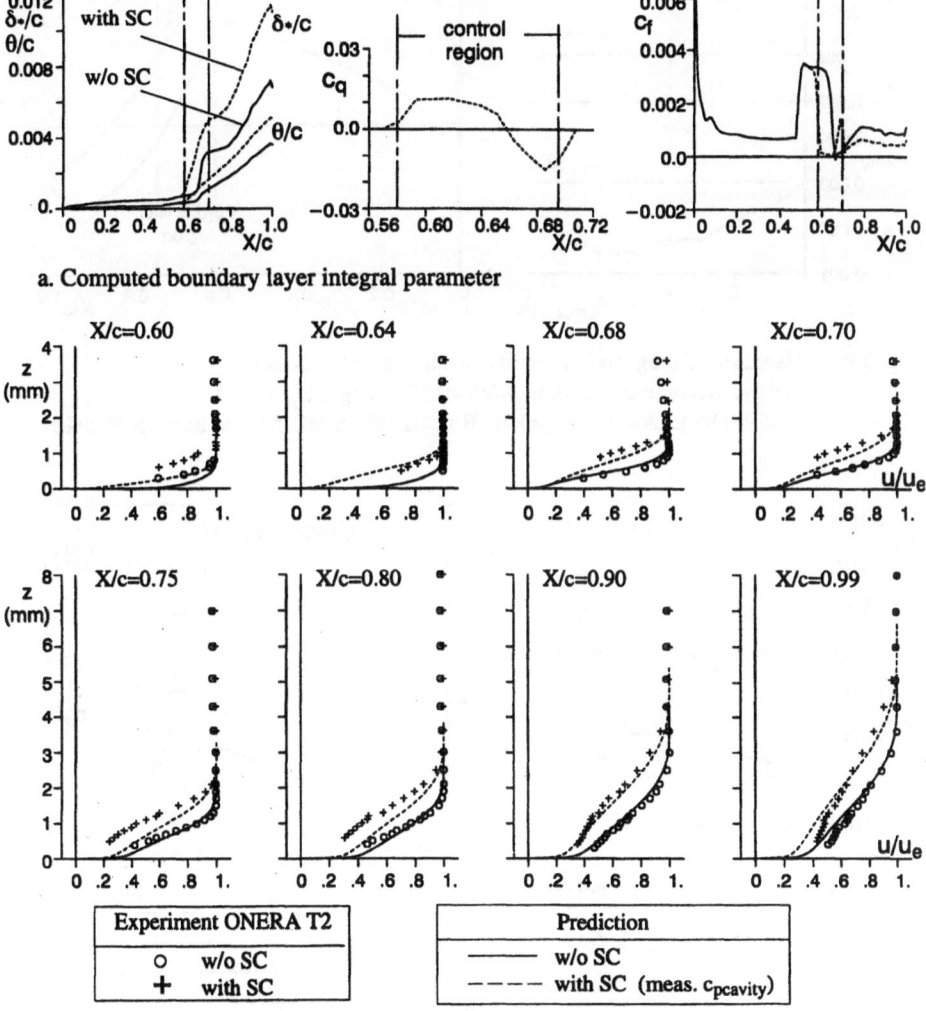

a. Computed boundary layer integral parameter

b. Measured and computed boundary layer mean velocity profiles

Fig. 17: Airfoil DA LVA–1A SC test case 1: development of boundary layer $M_\infty = 0.7613$, $Re_\infty = 4.6 \cdot 10^6$, $c_l \approx 0.46$, fixed transition 48%/48%c

215

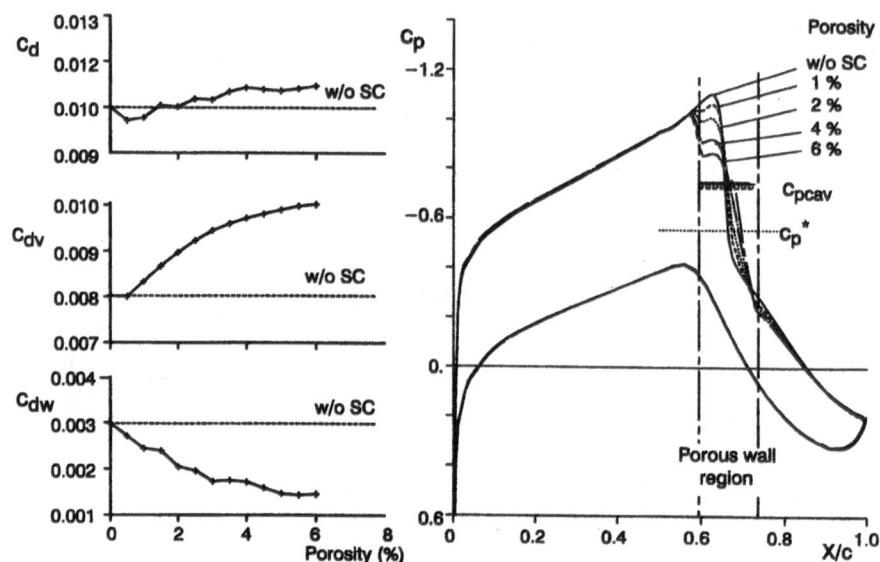

Fig. 18: Variation of drag and pressure distribution with porosity
Passive SC region: location $(X/c)_p = 0.593$, length $L_p/c = 0.14$
Airfoil DA LVA–1A: $M_\infty = 0.76$, $Re_\infty = 6 \cdot 10^6$, $c_l = 0.45$, fixed trans. 50%/5%c

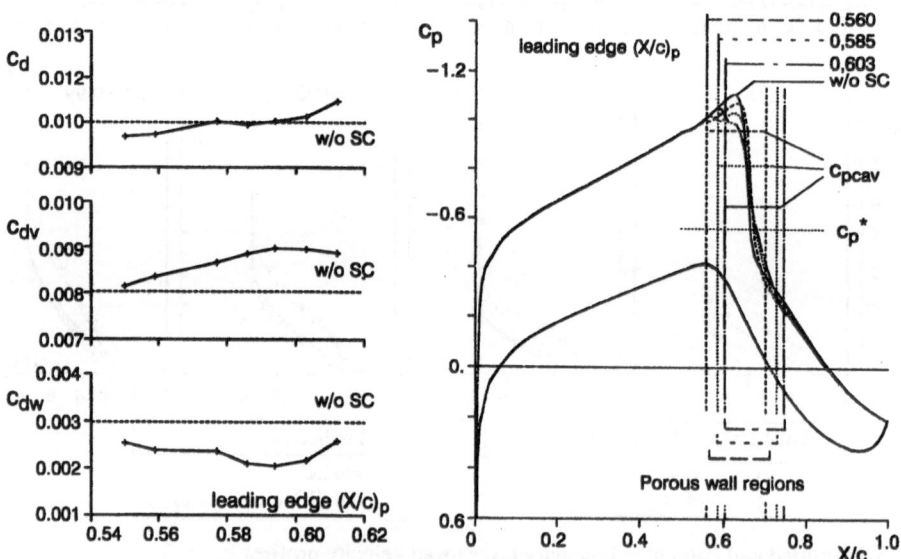

Fig. 19: Variation of drag and pressure distribution with perforation location
Passive SC region: porosity 2%, length $L_p/c = 0.14$
Airfoil DA LVA–1A: $M_\infty = 0.76$, $Re_\infty = 6 \cdot 10^6$, $c_l = 0.45$, fixed trans. 50%/5%c

14.5.3 Perforation length

The length of the control region also strongly affects the intensity of the ventilation velocity in the blowing and suction region by the variation of the cavity pressure as shown in Fig. 20 for a porosity of 2% and fixed leading edge of the control region at $(X/c)_p=0.593$. With increasing length, wave drag reduction increases, but is offset by the growing of the viscous drag.

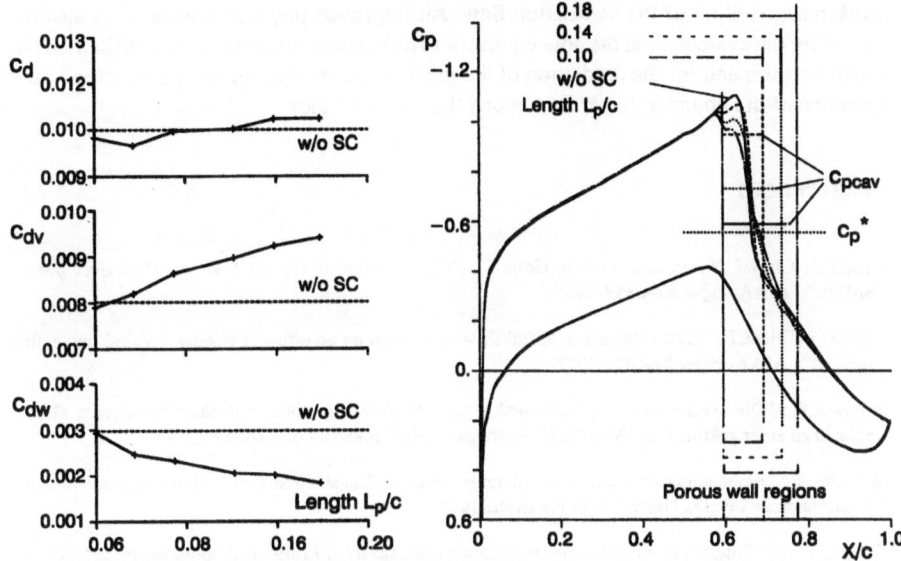

Fig. 20: Variation of drag and pressure distribution with perforation length
Passive SC region: porosity 2%, location $(X/c)_p = 0.593$
Airfoil DA LVA–1A: $M_\infty=0.76$, $Re_\infty=6\cdot10^6$, $c_l=0.45$, fixed trans. 50%/5%c

14.6 Conclusion

The numerical simulation of the passive shock control effects, contributed by Dasa Airbus to the EUROSHOCK project within Task 2.1, has generated the following conclusions:

- The computational approach to simulate passive SC effects could be substantially improved in the present version of the prediction method by introducing an interactive boundary layer finite difference method compared to the VII method with the incorporated triple deck solution, used in the beginning of the project.
- The main features of ventilation by a passive SC device, i.e., the reduction of the wave drag by the displacement effect of the ventilation in combination with an increase of the viscous drag due to the intensification of the turbulence by blowing, could be correctly predicted in agreement with the experimental data in Task 3.
- The numerical investigation of the transonic airfoil test cases shows that total drag is not reduced by passive shock control until close to maximum lift for the one turbulent–type airfoil (Va2) and the two laminar–type airfoils (DRA 2303, DA LVA–1A), although differ-

217

ent behaviour in the drag development at low lift coefficients is found between the two airfoil types.

- The present version of the SC prediction method allows to predict qualitatively the effects of the main passive SC design parameters, but the results depend on the flow characteristic of the porous surface which is only approximated in the code.
- The observed drag differences as well as the principal deviation between the predicted and measured cavity pressure when the passive SC mode is applied, indicate the need for further work on modelling of the ventilation flow. An improved physical modelling is also required for the evaluation of the non–equilibrium turbulence structure within and behind the control region and for the prediction of the pressure losses through the perforation in the presence of the boundary layer flow along the control region.

14.7 References

[1]   Chen,C.L.,Chow,C.Y.,van Dalsem,W.R.,Holst,T.L.:"Computation of viscous transonic flow over porous airfoils", AIAA–Paper 87–0359, 1987.

[2]   Dargel,G.,Thiede,P.:"Viscous transonic airfoil flow simulation by an efficient viscous inviscid interaction method", AIAA–Paper 87–0412,1987.

[3]   Bohning,R.: "Die Wechselwirkung eines senkrechten Verdichtungsstoßes mit einer turbulenten Grenz-schicht an einer gekrümmten Wand", Habilitationsschrift, University of Karlsruhe,1982.

[4]   Breitling,T.: "Berechnung transsonischer, reibungsbehaftete Kanal– und Profilströmungen mit passiver Beeinflussung", Ph.D., University of Karlsruhe,1990.

[5]   Dargel,G.:"Ein Programmsystem für die Berechnung transsonischer Profil– und konischer Flügelströmun-gen auf der Basis gekoppelter Potential– und Grenzschichtlösungen", DGLR–Bericht 92–07,1992.

[6]   Le Balleur,J.C.: "Strong matching method for computing transonic viscous flows including wakes and separations. Lifting airfoils", La Rech. Aerosp. 1981–3, 1981.

[7]   Jakob,H.:"Ein Verfahren zur Berechnung der ebenen transsonischen Strömung in Stromlinienkoordina-ten", MBB LFK 81117 (IFAS 11), 1984 (not published).

[8]   Mertens,J.,Klevenhusen,K.D.,Jakob,H.:"Accurate transonic wave drag prediction using simple physical models", AIAA Paper 86–0512 (1986).

[9]   Veldman,A.E.P.:"New, quasi–simultaneous method to calculate interacting boundary layers", AIAA Jour-nal, Vol. 154, 1981.

[10] Cebeci,T.,Smith,A.M.O.:"Analysis of turbulent boundary layers", Academic Press, New York, 1974.

[11] Thiele,F.:"Ein effektives Differenzenverfahren zur Berechnung komplexer zweidimensionaler Grenz-schichtströmungen", Institutsbericht Nr. 01/87, Hermann–Föttinger–Institut, TU Berlin, 1987.

[12] Carter,J.E.:"A new boundary–layer inviscid iteration technique for separated flow", AIAA Paper No. 79–1450,1979.

[13] Bur,R.:"Passive control of a shock wave/turbulent boundary layer interaction in a transonic flow", Rech. Aerosp. 1992–6, 1992.

[14] Braun,W.:"Experimentelle Untersuchung der turbulenten Stoß–Grenzschicht–Wechselwirkung mit pas-siver Beeinflussung", Ph.D., University of Karlsruhe,1990.

[15] Poll,D.I.A.,Danks,M.,Humphreys,B.E.: "The aerodynamic performance of the laser drilled sheets", In Proceedings 'First European Forum on Laminar Flow Technology', DGLR–Bericht 92–06,1992.

[16] Johnson,D.,Coackley,T.:"Improvements to a nonequilibrium algebraic turbulence model", AIAA Journal, Vol.28, No.11, 1990.

[17] Fulker,J.L.,Simmons,M.J.:"An experimental investigation of passive shock/boundary–layer interaction control on an aerofoil", EUROSHOCK TR AER 2–92–49/3.2, 1995.

[18] Rosemann,H.,Knauer,A.,Stanewsky,E.:"Experimental investigation of the transonic airfoils DA LVA–1A and Va2 with shock control", EUROSHOCK TR AER 2–92–49/3.5, 1996.

[19] Dargel,G.,Kühn,W.,Resagk,P.:"Pressure loss measurement on SC–insert of model MBB Va2", Technical Note DA–EF12–109/94, 1994.

[20] Archambaud,J.P.:"Qualification by laser measurements of the passive control on LVA–1Ae airfoil in the T2 wind tunnel", EUROSHOCK TR AER 2–92–49/3.4, 1996.

14.8 List of Symbols

a_o	stagnation speed of sound
a_h	pitch between the holes
A_h	area of single hole
A_p	area of perforated surface
A^+	factor in damping factor relationship, eq.(14.19)
c	chord of the airfoil
c_d	total drag coefficient
c_{dv}	viscous drag coefficient
c_{dw}	wave drag coefficient
c_f	skin friction coefficient
c_l	lift coefficient
c_p	static pressure coefficient
$c_{pcavity}$	static pressure coefficient in the cavity
c_p^*	critical static pressure coefficient
c_q	non dimesional local mass flow
D	van Driest damping factor, eq.(14.18)
d_h	diameter of hole
K	correction factor in Poll's calibration law, eq.(14.15)
L_p	length of perforation
M	Mach number
$\mathcal{M}_b$	mass flow rate
n_h	number of holes on the porous surface
n_p	porosity, eq. (14.9)
p	static pressure
r	recovery factor
Re_∞	free stream Reynolds number based on chord
s_h	thickness of sheet
u,v	mean flow velocity components
v_h	hole velocity
v_p	ventilation velocity
x,z	surface orientated coordinates
X	Cartesian coordinate
X_p	leading edge of perforation
α	angle of attack

δ	boundary layer thickness
δ*	displacement thickness
ε	eddy viscosity
$ε_i$	inner eddy viscosity
$η_v$	loss coefficient of mass flow rate, eq. (14.14)
θ	momentum thickness
ϰ	adiabatic exponent
ϰ*	curvature of displacement surface
ϱ	density
μ	dynamic viscosity
ν	kinematic viscosity

Subscripts

c	cavity
e	boundary layer edge
h	hole
ht	hole, theoretical
iw	wall, equivalent inviscid flow
p	porous surface
w	wall, viscous flow
ν	viscous–inviscid iteration number
o	stagnation
∞	free stream condition

15 DEVELOPMENT OF VISCOUS-INVISCID INTERACTION CODES FOR PREDICTION OF SHOCK BOUNDARY-LAYER INTERACTION CONTROL (SBLIC) AND BUFFET OVER AIRFOILS

J.C. Le Balleur, P. Girodroux-Lavigne, H. Gassot

ONERA Aerodynamics Department
29, avenue de la Division Leclerc, 92320 Châtillon, France

Summary: The CFD computational work performed at ONERA on Shock-Boundary Layer Interaction Control (SBLIC), within the Task 2 of the EUROSHOCK cooperation, is described.

The Viscous-Inviscid Interaction (VII) approach has been developed for passive or active shock-control over porous walls. A double computational work has been performed, with two complementary VII codes, steady/unsteady. The two codes are almost identical for the viscous methodology (hybrid field/integral FD/IBL viscous method, steady strong-coupling or time-consistent strong-coupling), for the viscous and turbulent modelling, for the grid refinement, but one contains simpler inviscid equations in unsteady flow.

Drag reduction seems possible in steady flow, but is rather exceptional. Unsteady flows, together with buffet damping by passive control, have been computed.

15.1 Introduction

A CFD computational work has been performed at ONERA for Shock-Boundary Layer Interaction Control (SBLIC) on transonic airfoils, within the EUROSHOCK Task-2 project, over the three years of the EUROSHOCK cooperation.

The Viscous-Inviscid Interaction (VII) approach has been developed for passive or active shock-control over porous walls. A double steady/unsteady analysis has been achieved with two complementary VII codes. The two codes are almost identical for the viscous methodology (hybrid field/integral FD/IBL viscous method, steady strong-coupling or time-consistent strong-coupling), for the viscous and turbulent modelling, and for the grid refinement.

The first computational work (code VIS05c, *J.C. LeBalleur*, [1] [2] [3] [4]) is based on steady viscous equations, the selected Viscous-Inviscid Solver (VIS) introducing the higher-order inviscid approximation of the full-potential equation.

The second work performed (code VIS15, *J.C. Le Balleur, P. Girodroux-Lavigne*, [5] [6]), gives access to time-accurate solutions with a VIS solver that uses a lower-order inviscid approximation of TSP-small-perturbations, but that generates with a single code the steady as well as the buffet solutions.

The numerical results of the two codes then deliver a complementary insight for analysis with, in the steady test-cases, a cross-comparison between the time-consistent solutions, that self-discriminate buffet, and the higher-order calculations in steady flow.

A wall-ventilation procedure has been developed and implemented in both the steady code VIS05c and the time-consistent code VIS15.

All mandatory "steady test-cases" have been computed, with both the steady code VIS05c and the time-accurate code VIS15, with and without control. The code VIS15 was also used to calculate all the "unsteady test-cases", with and without control.

The "passive control" option of the wall-ventilation procedure, with a Newton iteration for the prediction of the constant cavity pressure, has been used for the configurations with control.

15.2 Viscous-Inviscid Interaction (VII) Numerical Methods

The steady code VIS05c and the unsteady code VIS15, for transonic airfoils, have the common methodology and the similar coding of the Viscous-Inviscid-Solvers "VIS", initiated around 1979 (code VIS05, [3]), and including specific adaptative grids [1] [3].

The advantage of the VIS methodology, which splits the overall numerical solution into two distinct viscous and inviscid schemes, is to have a very low numerical viscosity for a given grid, and also to give access to the finer grids necessary for the physics of Shock-Bondary Layer Interaction, thus minimizing the uncontrolled numerical truncation of the equations.

The present viscous-inviscid approach is based on Le Balleur's "Defect-Formulation Theory" for the full Navier-Stokes (NS) equations [1] [2] [3], that replaces the single-field "NS" by a double "VII" field ("Defect-NS" plus interacting "Pseudo-inviscid") *with full overlay* .

A thin-layer approximation of the theory only, is solved at present by the codes VIS05c and VIS15. Nevertheless, the thin-layer approximations used are neither wall-prescribed as in boundary layer theory, nor simply grid-prescribed as they are in the classical "Thin-layer Navier-Stokes" equations. They are here better governed by the aerodynamic field itself, the thin-layer truncations being performed in the "Displacement Reference Frame" [1] [2]. This extends the validity of the thin-layer equations to massively separated flows.

15.2.1 Steady code VIS05c (VII + Full Potential)

The viscous part is a "Hybrid Field/Integral" method (FD/IBL). It is solved as a marching thin-layer 2D-numerical technique, with non-linearly implicit schemes, in direct/inverse modes [1] [2] [3]. At each viscous station, the method *discretizes* in the direction normal to the local interacting inviscid streamlines parametric turbulent velocity profiles, designed for attached or deeply separated flow. The turbulence is computed either with an algebraic model (mixing length plus velocity profiles) or with an original 2-equations $k - u'v'$ model [3] [2] [1], "forced by the velocity profiles", for the Reynolds stresses and the entrainment.

The steady code VIS05 uses the "Semi-Inverse" Le Balleur coupling algorithm [4], defined in 1978, with the extension of 1990 "Semi-inverse Massive-separation", [2] [1]. The code solves the full-potential equation for the inviscid field.

The code VIS05, in two first versions VIS05a ([3], 1981) and VIS05b ([2], [1], 1989), has been developed successfully for the computation of attached/separated flows, or even of the deep stall over airfoils, at subsonic or mildly transonic speeds.

The new version VIS05c, used for the present EUROSHOCK computations, is a slightly more recent version (still experimental), designed for strongly interacting transonic flows. The version introduces several improvements at the level of the full-potential solver and of the coupling algorithm. The new "potential" VIS05c solver can take into account the stiff metric tensors which appear when the mesh is locally highly clustered. A rotative scheme has been introduced (as well as a "quasi-conservative" option) and the convergence has been accelerated by a doubly implicit $i - k$ successive line over-relaxation (SLOR) including boundary conditions. The new VIS05c "coupling" introduces, first, a more compact discretization of the "Semi-inverse" coupling algorithm. Moreover, an additional new coupling algorithm has been introduced in the shock area, whereas the "semi-inverse" algorithm is maintained elsewhere. This advanced version VIS05c permits both to solve the potential equation at transonic speed on highly clustered grids and to calculate shock-induced separated flows as well as the deep stall range.

15.2.2 Unsteady code VIS15 (VII + TSP, time-consistent coupling)

The time-consistent code VIS15 [5] [6] uses the same viscous methodology as the steady code VIS05c.

The viscous hybrid field/integral method (FD/IBL) includes here the full unsteady viscous terms at the difference of the "quasi-steady" viscous approximation frequently used in most of unsteady VII methods. Moreover, the full viscous time-consistency (full viscous unsteady terms plus full "time-consistent coupling") is introduced. However, as compared to the steady code VIS05c, a cruder approximation is introduced for the inviscid flow, solving an unsteady transonic small perturbations equation (TSP).

The time-accurate code VIS15 uses the "Semi-Implicit" time-consistent coupling algorithm defined in 1984, Le Balleur, Girodroux [5], well adapted to the alternate direction implicit ADI-scheme of the inviscid TSP part.

The full "time-consistent coupling" is obtained by discretizing and converging the VII-coupling at each new time step, iterating the viscous and inviscid parts. This VII convergence within the time step is necessary to take into account the full viscous upstream influence at a given time-step (fully time-parabolic system, equivalent to a NS system). This permits to compute the strong viscous interaction phenomena such as shock-boundary layer interaction and separation, in unsteady flows, with the same properties as a full Navier-Stokes solver.

The time-consistent code VIS15 has been used to compute unsteady shock-induced separated flows over airfoils undergoing forced oscillations (pitch or flap oscillations) as well as buffet type flows over fixed airfoils, where the unsteady flow is self-induced by separation.

An important point is that the fully time-consistent code VIS15 is used exactly in the same way independently of the fact whether the flow is steady or unsteady. This makes it possible for the code to discriminate between a steady or an unsteady solution and so to compute the buffet boundary, as it has been shown in the case of the NACA0012 airfoil [7].

15.2.3 Self-adaptive grids of codes VIS05c and VIS15

The code VIS05 has introduced also a self-adaptive VII-grid. The self-adaptation of the grid is achieved by adaptive clusterings both in the tangential and normal directions. In the normal k-direction, within the viscous hybrid FD/IBL solver, the normal meshes are self-adapted to the thickness and to the gradient of the velocity profiles. In the tangential i-direction, i.e. the direction for which the overlaying viscous-inviscid discretizations have coincident i-stations at the wall (or at the wake-center line), i-clusterings are introduced at the strong viscous interactions.

All the steady VIS05c calculations have been performed using fine meshes in i-direction, with a clustering of the C-grid in the control region and with an aspect ratio of order unity for the inviscid meshes near the wall. The clustering has been self-adapted to the position of the shock for the cases without control and has been located in the middle of the porous region for the cases with control.

Rather very small space-steps Δx near the shock have been used in all cases, but without exactly the same clustering in the different test-cases, depending on the width of the control area and on the limits of convergence of the codes. The refinement is generally such that $\Delta x \simeq 0.3\delta$ (to $\Delta x \simeq 0.5\delta$), where δ is the boundary layer thickness. Such grids at the shock ($\Delta x = 0.3\delta$) are not devoted to improve the inviscid shock treatment, but are just reaching the minimal Δx step size required for actually discretizing at the scale of the "free-interaction" viscous process which governs the physics and the longitudinal extent of the shock-boundary layer interaction.

This minimal grid refinement is believed to be necessary, in any method, for making calculations without a dominant effect of artificial viscosity at shock-boundary layer interaction, especially in view of closely studying the effects of the control (by suction/blowing) in the neighbour-hood of the shock-boundary layer interaction. This demand is probably still higher for laminar-type airfoils, where the boundary layer thickness upstream of the shock is again reduced by a longer laminar growing rate.

The grid-adaptation technique of the code VIS05c has been used also in the time-consistent code VIS15. This self-adaptation was, nevertheless, used simply as a grid-adaptation to the steady state, not as an unsteadily adaptive grid. The same levels of spatial discretization as in the case of the code VIS05c have been used in the VIS15 computations. The self-adaptive cartesian H-grid is however simplified by the TSP approximation of the inviscid solver.

15.3 Shock-control Extension in Codes VIS05c and VIS15

The steady code VIS05c and the unsteady code VIS15 have been modified in order to com-pute transonic flows with shock control. The control device that we consider is a cavity with porous surface, located in the shock region. This control device is modelled in the computations through a continuously distributed ventilation velocity, connected with the pressure field and the cavity pressure by a semi-empirical control law.

15.3.1 Implementation of a ventilation velocity at the wall

The first extension step has been to introduce, in the hybrid field/integral (FD/IBL) viscous numerical method, a term of ventilation velocity at the wall in the mass and momentum equations at the level of the integral balances, with no change of the outer-edge momentum equation, i.e. of the entrainment model.

The ventilation velocity is assumed as a continuous distribution, which is an average of the wall mass-fluxes introduced by the porous wall. In a first approximation, the parametric velocity profile modelling of the hybrid FD/IBL method has been maintained unchanged.

Both normal and tangential components of the ventilation velocity have been taken into ac-count. The introduction of both components will permit to handle the case of porous regions with normal as well as inclined holes.

In the case of the steady code VIS05c, the ventilation velocities, defined from the pressures, the selected control law, and the active/passive control selection have been iteratively updated at each VII coupling iteration.

In the case of the time-consistent code VIS15, the update of the ventilation velocities is de-layed during a first initial phase, during which the geometry and flow-field are rapidly develop-ing. The update is then introduced when the physical solution has begun to take form. Thereafter, two updates of the ventilation velocity per time-step (but two only) have been performed.

15.3.2 Control laws

In a second extension-step, a procedure has been developed for the computation of the ven-tilation velocity distribution, following the two proposed shock-control modellings. The mass flux distribution over the porous region is computed either using the isentropic SC-law with the mass flow loss coefficient obtained from calibration experiments of porous sheets, or using Poll's formula calibrated for laser drilled holes. A key allows also to choose between active shock con-trol, where the pressure in the cavity is prescribed, and passive shock control, for which the net mass flow through the perforated surface is equal to zero.

Denoting by q_h the hole velocity, d the hole diameter, t the thickness of the porous sheet, X the hole spacing, $S_h = \pi d^2/4$, $S = X^2$, $\dot{m}$ the mass flow through the hole, the ventilation velocity q_v is defined by:

$$\dot{m} = \rho_h q_h S_h = \rho q_v S \tag{1}$$

so $q_v = \rho_h/\rho q_h \eta_P$, where the porosity η_P is equal to the ratio S_h/S. If θ is the hole inclination with respect to the airfoil surface, the tangential and normal components of the ventilation velocity are:

$$u_v = q_v \cos\theta \qquad w_v = q_v \sin\theta. \tag{2}$$

The hole velocity q_h depends on the chosen control law and is a function of the geometry of the holes, of the outer flow characteristics (pressure P, density ρ), and of the cavity pressure, P_{cavity}, assumed constant.

15.3.3 Isentropic law

At isentropic conditions and assuming a zero velocity in the cavity, we have:

$$\frac{P_{cavity}}{P} = \left(\frac{\rho_{cavity}}{\rho}\right)^\gamma \tag{3}$$

and the theoretical hole velocity is:

$$\rho q_{h_{theo}} = \left\{\frac{2\gamma}{\gamma - 1}\rho P \left[\left(\frac{P_{cavity}}{P}\right)^{\frac{\gamma}{\gamma-1}} - 1\right]\right\}^{1/2}. \tag{4}$$

Due to the pressure loss of the porous surface, the real hole velocity is:

$$q_h = \eta_V q_{h_{theo}} \tag{5}$$

where the reduction factor $\eta_V = \eta_V(P_{cavity}/P)$ can be obtained from calibration experiments of porous sheets.

15.3.4 Poll's formula

The other control law investigated is based on Poll's formula for laser drilled titanium sheets, see [12]. This law is based on the following assumptions:

- the interference between holes is small ($d \ll X$),

- the flow in each hole is a fully established "pipe" flow,

- the flow in the hole is laminar and compressibility effects are small.

The theoretical pressure loss in a hole with cylindrical walls, $\Delta P = P_{cavity} - P$, is then related to the mean value of the hole velocity q_h by the following linear relationship:

$$\Delta P = 32\mu_h \frac{t}{d^2} q_h. \tag{6}$$

This simple law has to be modified in order to take into account the fact that the real flow in the hole is not a fully developed pipe flow and that the holes are conic. Experimental results have shown that a better relationship is given by:

$$\Delta P = K(\frac{C}{2}\rho_h q_h^2 + 32\mu_h \frac{t}{d^2}q_h) \qquad (7)$$

where $C \simeq 2.42$ and K is related to the hole conicity. With $\dot{m} = \rho_h q_h \pi d^2/4$ and defining

$$Y = \frac{\Delta P}{\rho_h} \frac{d^2}{\nu_h^2} \frac{d^2}{t^2} \quad \text{and} \quad X = \frac{\dot{m}}{\mu_h t} \qquad (8)$$

we obtain the following quadratic equation:

$$Y = \frac{1}{K}(40.7X + 1.95X^2). \qquad (9)$$

15.3.5 Active or passive shock-control

In the case of active control, the ventilation distribution q_v over the perforated region is directly deduced from one of the previous laws, prescribing the cavity pressure P_{cavity}. This leads to a non-zero net mass flow through the porous surface:

$$Q(P_{cavity}) = \int_S \rho q_v dS \neq 0. \qquad (10)$$

For passive control through the cavity, the net mass flow must be zero, leading to an iterative procedure for the prediction of the constant cavity pressure. A Newton iteration has been implemented in both codes:

$$Q(\tilde{P}_{cavity}) = 0 = Q(P_{cavity}) + \frac{\partial Q}{\partial P_{cavity}}(\tilde{P}_{cavity} - P_{cavity}) \qquad (11)$$

where the derivative $\partial Q/\partial P_{cavity}$ is estimated numerically.

15.4 Computation of the Steady Test-cases

15.4.1 Pre-calculations: DA-LVA-1A airfoil without control

The steady pre-calculations without control for the DA-LVA-1A airfoil (Test-cases 1 and 2) have been performed with both the code VIS05c and the code VIS15.

The VIS05c results have been obtained prescribing the lift coefficient via two Newton iterations. The viscous computations have been performed with the 2-equation $k - u'v'$ turbulence model, with "base" effect, and with wake equilibration. Exceptionally in these pre-calculations, the VII "wake-curvature" effect has been switched on.

The time-consistent VIS15 computations use the same 2-equation $k - u'v'$ turbulence model as the code VIS05c, but do not introduce a "wake-curvature" effect.

An adaptive C-grid (301x48 inviscid + 301x37 viscous nodes) strongly clustered at the shock, Figure 1a, has been used in the VIS05c computations. The numerical thickness of the shock (see the Mach number field of Figure 1b) is then much smaller than the incoming boundary layer thickness δ at shock interaction.

The mesh size at the shock ($\Delta x = 0.3\delta$) is, however, not yet luxurious, but just reaching the minimal Δx step size required for actually discretizing at the scale of the physical "free-interaction" viscous process which governs the compression and longitudinal extent of the shock-boundary layer interaction.

226

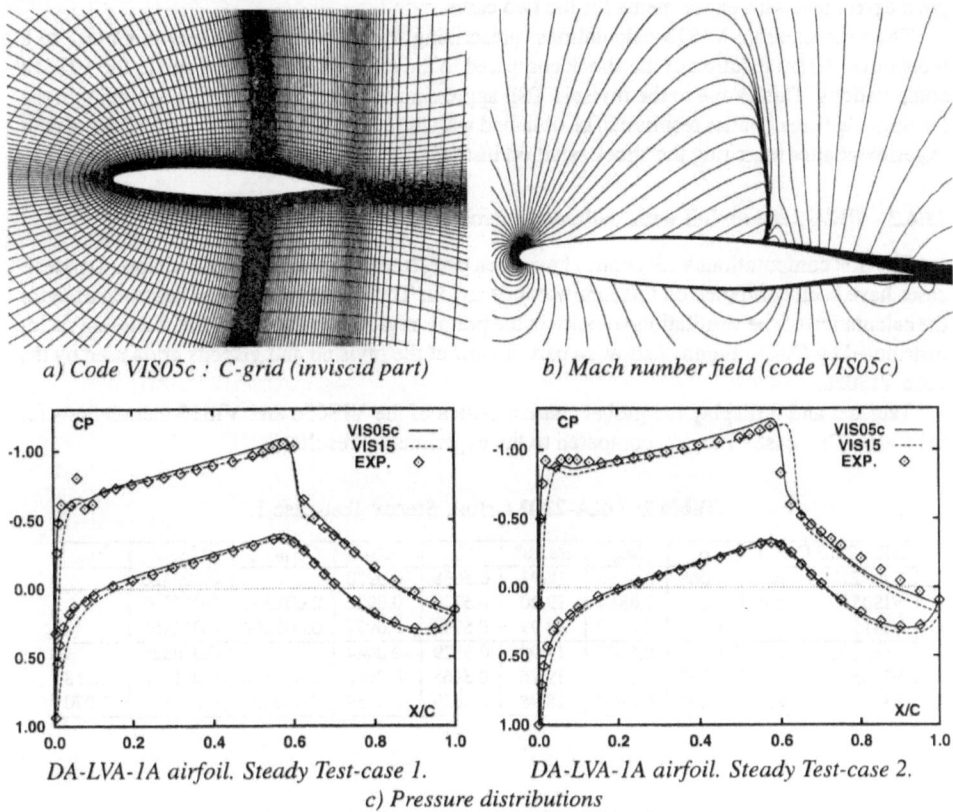

a) Code VIS05c : C-grid (inviscid part)   *b) Mach number field (code VIS05c)*

DA-LVA-1A airfoil. Steady Test-case 1.   *DA-LVA-1A airfoil. Steady Test-case 2.*

c) Pressure distributions

Figure 1: *Steady computation (code VIS05c) and unsteady computation (code VIS15).*

A 160x100 inviscid cartesian H-grid with a self-adaptive clustering at the shock was used in the VIS15 computations. However, the mesh size near the shock ($\Delta x = 0.5\delta$) is greater than in the VIS05c calculations.

The lift and moment coefficients shown in Table 1 are computed by numerical integration at the surface. The drag is here computed from the viscous wake drag (momentum thickness) and from a wave drag which is estimated from the computed Mach number field (potential) and the Rankine-Hugoniot relations.

Table 1: DA-LVA-1A airfoil: Steady Test-cases 1 and 2.

Code VIS05c	C_L	CM	C_{DWake}	C_{DWave}	C_{Dtot}
DA case 1	0.5339	-0.078	0.0114	0.0024	0.0138
DA case 2	0.6920	-0.084	0.0133	0.0071	0.0204

Figure 1c compares the pressure distributions obtained by the codes VIS05c and VIS15 for the two angles of attack. The pressure distributions predicted by the steady code VIS05c are in good agreement with experiments for the two cases.

The time-accurate VIS15 calculations, prescribing the lift coefficient, lead in Case 2 to a more downstream location of the shock compared to the experimental results and to the VIS05c computations. This is due to the inviscid TSP approximation and to the airfoil specificity in the leading-edge area. Better results can be obtained with the code VIS15, however, comparing with experiment after matching the shock position instead of the lift coefficient.

15.4.2 DRA-2303 airfoil with/without control

The first computations with control have been performed for the DRA-2303 airfoil. Two test-cases have been computed on this laminar-type airfoil. The isentropic SC-law has been used for the calculation of the ventilation velocity in the porous region, together with the reduction factor distributed by CIRA. Figure 2 shows a partial view of the inviscid and viscous grids used by the code VIS05c.

Tables 2 and 3 display the global characteristics of the VIS05c and VIS15 calculations for the steady Test-cases 1 and 2, compared to the experimental results.

Table 2: DRA-2303 airfoil: Steady Test-case 1.

DRA-2303 Case 1		α	M_∞	$Re.10^6$	C_L	C_M	C_{DWave}	C_{Dtot}	C_{Pcav}
Exp. 271	n.c.	1.068	0.6816	18.97	0.5668	-0.0958		0.009428	
VIS05c	n.c.	1.068	0.6884	19.20	0.5720	-0.0926	0.001560	0.010790	
VIS15	n.c.	1.068	0.6952	18.97	0.5759	-0.0977	0.001289	0.012468	
Exp. 1008	w.c.	1.066	0.6807	18.98	0.5529	-0.0944		0.010525	
VIS05c	w.c.	1.066	0.6875	19.20	0.5465	-0.0926	0.000520	0.011300	-1.118
VIS15	w.c.	1.066	0.6943	18.98	0.5629	-0.0959	0.000606	0.012543	-1.070

Table 3: DRA-2303 airfoil: Steady Test-case 2.

DRA-2303 Case 2		α	M_∞	$Re.10^6$	C_L	C_M	C_{DWave}	C_{Dtot}	C_{Pcav}
Exp. 289	n.c.	2.507	0.6795	18.91	0.8115	-0.1002		0.014576	
VIS05c	n.c.	2.507	0.6863	19.10	0.8069	-0.1054	0.005820	0.016220	
VIS15	n.c.	2.507	0.6863	18.91	0.8119	-0.0998	0.003578	0.015793	
Exp. 1031	w.c.	2.710	0.6806	18.98	0.8142	-0.0984		0.017972	-0.974
VIS05c	w.c.	2.710	0.6874	19.20	0.7632	-0.0954	0.003060	0.017920	-1.061
VIS15	w.c.	2.710	0.6874	18.98	0.8124	-0.0960	0.002891	0.017968	-1.041
Exp. 290	n.c.	2.712	0.6795	18.91	0.8427	-0.1014		0.016557	
VIS05c	n.c.	2.712	0.6863	19.10	0.8396	-0.1067	0.006920	0.017790	
VIS15	n.c.	2.712	0.6863	18.91	0.8501	-0.1009	0.004985	0.017378	

A reduction of the wave drag is observed with control in both cases, but also an increase of the total drag. This increase of the total drag is particularly obvious for Test-case 2, compared at the same lift, see Table 3. The cavity pressures predicted by the codes VIS05c and VIS15 for Test-case 2 with passive control are nearly identical and very close to the experimental value.

The pressure distributions, with and without control, are well predicted by both codes, see for example the VIS05c results in Figure 3a. The code VIS15 overpredicts slightly, however, the pressure levels just before the shock for the higher lift (Case 2). In agreement with experiment, the calculations with SBLIC indicate a spreading of the shock and a decrease of the intensity of the shock.

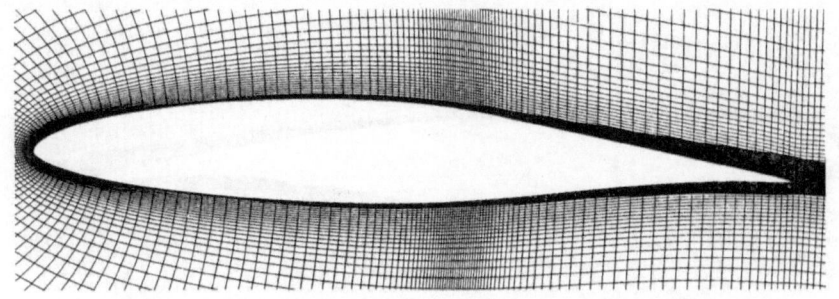

Figure 2: *DRA-2303 airfoil. Viscous and inviscid grids. VIS05c.*

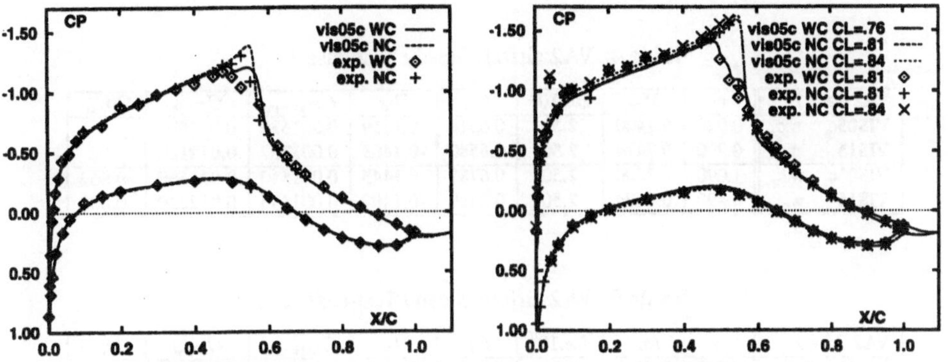

a) Pressure distributions

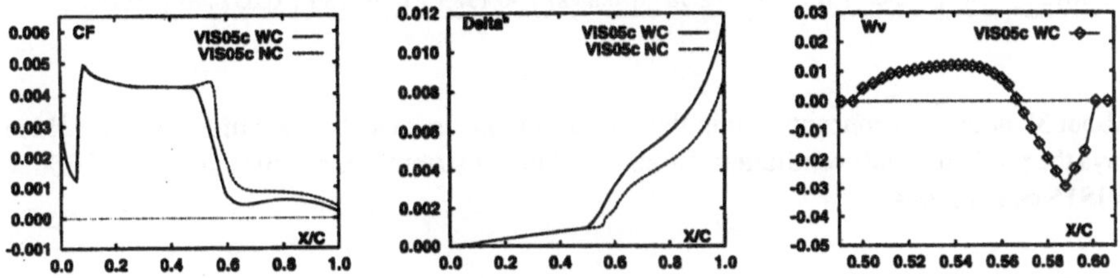

Steady Test-case 1

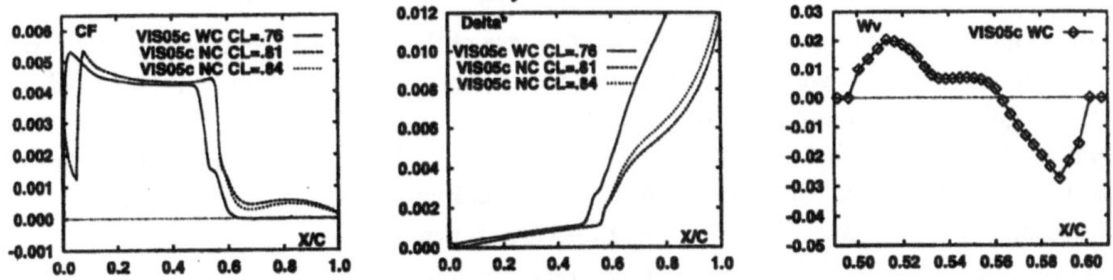

Steady Test-case 2

Steady Test-case 1

Steady Test-case 2

b) Skin-friction, displacement thickness, ventilation velocity (upper surface)

Figure 3: *DRA-2303 airfoil: Steady Test-cases with and without control. Steady calculations (code VIS05c).*

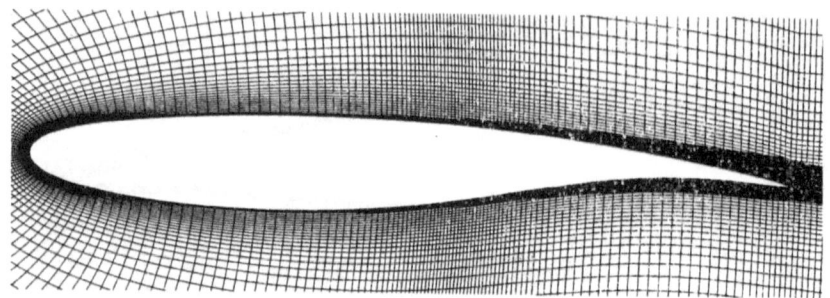

Figure 4: *VA2 airfoil: Partial view of the VIS05c mesh.*

Table 4: VA2 airfoil: Steady Test-case 1.

VA2 Case 1		α	M_∞	$Re.10^6$	C_L	C_M	C_{DWave}	C_{Dtot}	Cp_{cav}
VIS05c	n.c.	0.940	0.7400	2.50	0.6516	-0.1419	0.001630	0.013990	
VIS15	n.c.	0.890	0.7474	2.50	0.6580	-0.1403	0.001757	0.012180	
VIS05c	w.c.	1.090	0.7400	2.50	0.6787	-0.1445	0.001540	0.011650	-0.862
VIS15	w.c.	0.890	0.7474	2.50	0.6530	-0.1402	0.001455	0.012230	-0.802

Table 5: VA2 airfoil: Steady Test-case 2.

VA2 Case 2		α	M_∞	$Re.10^6$	C_L	C_M	C_{DWave}	C_{Dtot}	Cp_{cav}
VIS05c	n.c.	2.390	0.7400	2.50	0.9047	-0.1571	0.013240	0.027820	
VIS05c	w.c.	2.590	0.7400	2.50	0.8784	-0.1468	0.009930	0.025800	-1.079
VIS15	w.c.	2.500	0.7400	2.50	0.8900	-0.1383	0.012730	0.032440	-0.953

About 30 nodes are concentrated in the porous region, as it can be seen in Figure 3b, which displays the predicted wall ventilation velocities. These distributions are similar in both VIS05c and VIS15 calculations.

15.4.3 VA2 airfoil with/without control

Shock control has also been investigated for the case of the VA2 turbulent-type airfoil. Two steady test cases have first been proposed for this airfoil, corresponding to $C_L = 0.65$ and $C_L = 0.90$. Here again, the computations with control use the isentropic SC-law for the calculation of the ventilation velocity with a reduction factor distributed by CIRA. The geometry of the airfoil and the C-grid used by the VIS05c code are displayed on Figure 4.

The results of the steady VIS05c computations are shown in Figures 5a and 5b.

For this turbulent-type airfoil, a small reduction of the total drag has been found with SBLIC by the steady code VIS05c for both test-cases, see Tables 4 and 5.

With control, a double transonic pocket has been found with the code VIS05c for the lower lift coefficient, whereas a more common λ-shock is predicted for the higher lift. There may be a connection between the existence of a double transonic pocket and a possible drag reduction, because this test-case is the only one among all the test-cases for the different airfoils investigated, for which a significant drag reduction has been predicted.

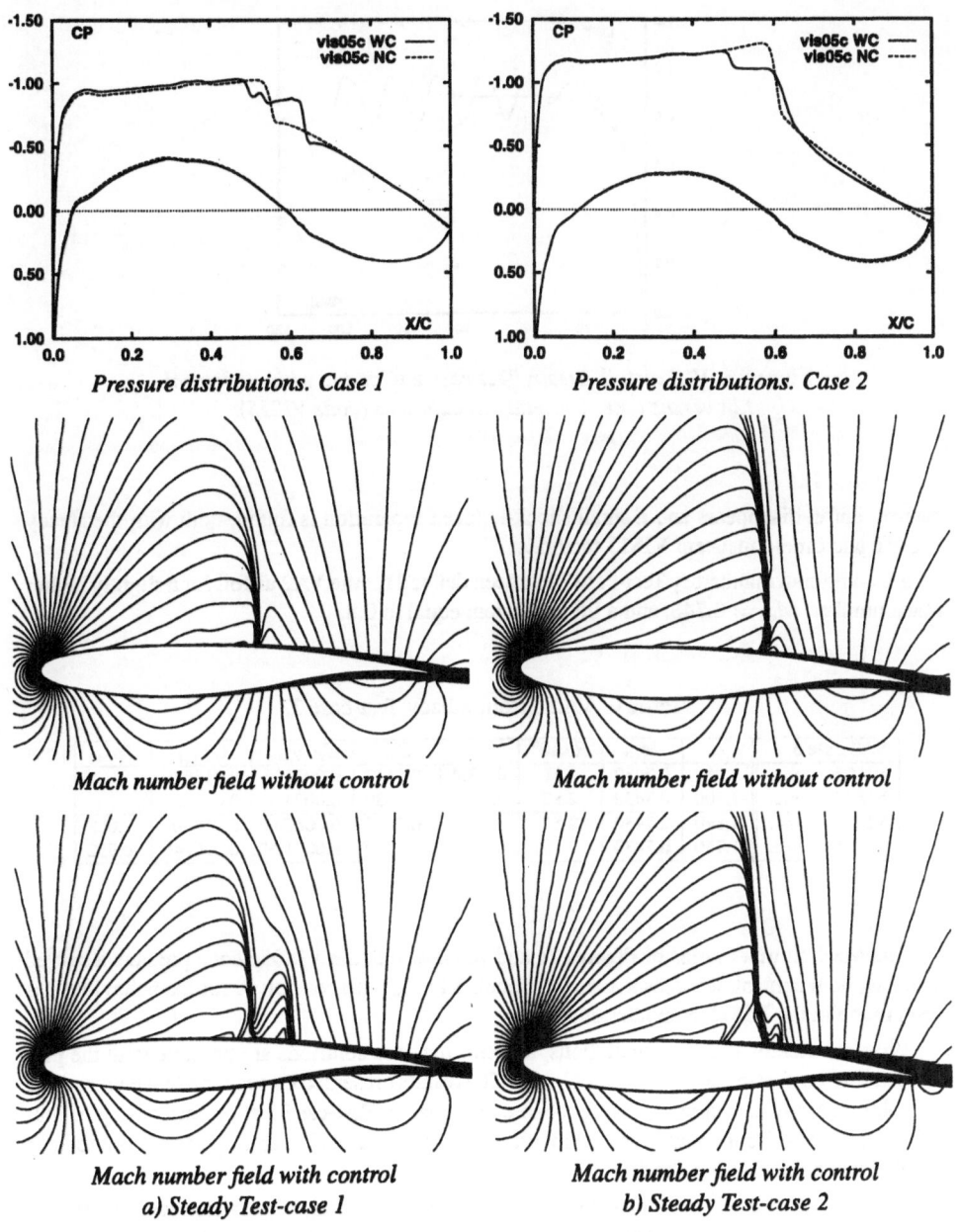

Pressure distributions. Case 1

Pressure distributions. Case 2

Mach number field without control

Mach number field without control

Mach number field with control
a) Steady Test-case 1

Mach number field with control
b) Steady Test-case 2

Figure 5: *VA2 airfoil. Steady test-cases with and without control.*
Steady calculations (code VIS05c).

With the time-consistent VIS15 calculation for the higher lift case (Test-case 2) without control, an unsteady buffet solution is predicted, as shown by the lift history of Figure 6. With

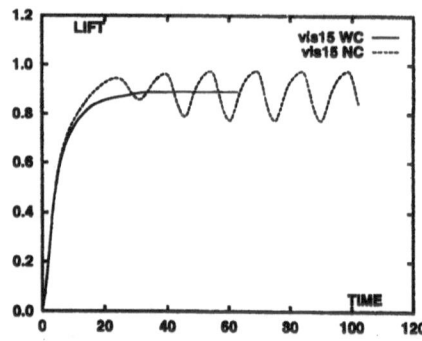

Figure 6: *VA2 airfoil. Steady Test-case 2 with and without control. Lift versus time. Unsteady calculations (code VIS15).*

control, buffet disappears and a small shock-induced separation is found, both with the steady VIS05c and time-consistent VIS15 methods.

An additional mandatory Test-case 3 has been defined for the VA2 airfoil for a slightly higher Mach number, $M_\infty = 0.743$, and a lift coefficient equal to 0.8.

Table 6: VA2 airfoil. Steady Test-case 3.

VA2 Case 3		α	M_∞	$Re.10^6$	C_L	C_M	C_{DWave}	C_{Dtot}	Cp_{cav}
VIS05c	n.c.	1.780	0.7430	2.51	0.8004	-0.1512	0.007126	0.018070	
VIS15	n.c.	1.500	0.7430	2.50	0.7967	-0.1439	0.007170	0.017260	
VIS05c	w.c.	1.840	0.7430	2.51	0.7993	-0.1499	0.006034	0.018600	-1.088
VIS15	w.c.	1.750	0.7430	2.50	0.7963	-0.1370	0.007139	0.020640	-0.952

The Mach number fields of Figures 7c and 7d show that the VIS05c code predicts a curved shock without control, whereas a λ-shaped shock is found with control. A similar behaviour has been found in the VIS15 computations.

From the steady VIS05c calculations, it seems that two solutions are possible with the prescribed lift, the Test-case 3 being near C_{Lmax}. The results presented correspond to the prescribed lift solution at lower incidence. The probable existence in Test-case 3 at prescribed-C_L of four solutions (two without control, and two with control), is presumably an important point to keep in mind when considering the drag.

With this choice of lower incidence, a decrease in wave drag is predicted with control by the steady code VIS05c, but there is a small increase of the total drag, see Table 6. The wave drag computed by the time-consistent code VIS15 is nearly identical with and without control, leading to a larger increase of total drag.

As in all previous computations, SBLIC leads here again to larger values of the displacement thickness and to smaller values of skin friction behind the shock, increasing the tendency to flow separation. The VIS05c solution is very close to separation with control, and a small shock-induced separation is predicted by the code VIS15.

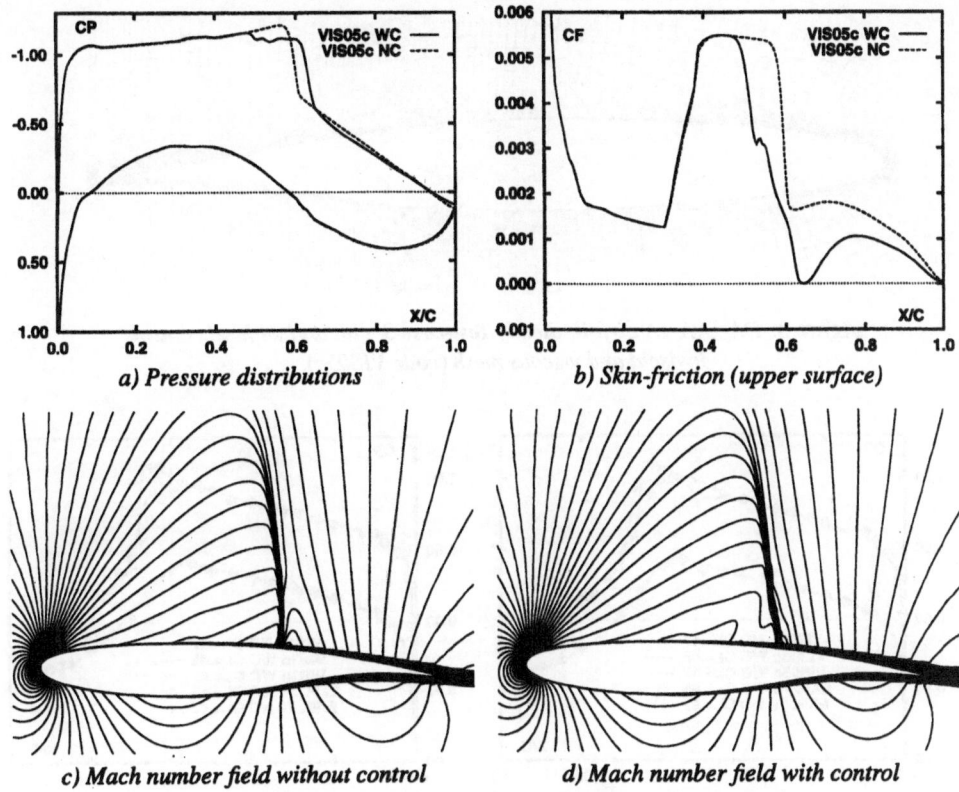

a) Pressure distributions

b) Skin-friction (upper surface)

c) Mach number field without control

d) Mach number field with control

Figure 7: *VA2 airfoil: Steady Test-case 3 with and without control.*
Steady calculations (code VIS05c).

15.4.4 DA-LVA-1A airfoil with/without control

At the end of the EUROSHOCK cooperation, a third test case, with and without control, has been defined for the DA-LVA-1A geometry. Due to the rear position of transition (x/c=48%) for this test case, the boundary layer thickness just before the shock is very small, as shown by the viscous grid of Figure 8. For this reason, special care has been given to the number of nodes and clustering of the grid in the perforated area (34 nodes in the VIS05c computations, 40 points in the VIS15 calculations, with $\Delta x \simeq 0.25\%$).

Poll's law has been used for the computation of the ventilation velocity in the perforated zone. All the computations have been performed using the corrected values of the Mach number and adjusting the angle of attack in order to match the experimental lift coefficient, see Table 7.

At the same lift, both codes predict here again a decrease of wave drag with control, but an increase of total drag. Comparing at the same incidence, VIS05c predicts, as the experiments, an increase of total drag with SBLIC, whereas a small decrease is found by VIS15.

The pressure distributions obtained by both codes (Figures 9a and 9b) agree well with the experimental results, which were not given at the time the computations were performed , with nevertheless a plateau of larger extent in the computed compression at control conditions ; the

233

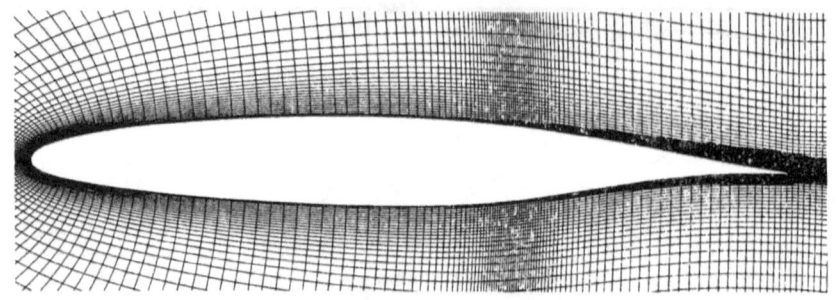

Figure 8: *DA-LVA-1A airfoil. Steady Test-case 3 with and without control.*
Inviscid and viscous mesh (code VIS05c).

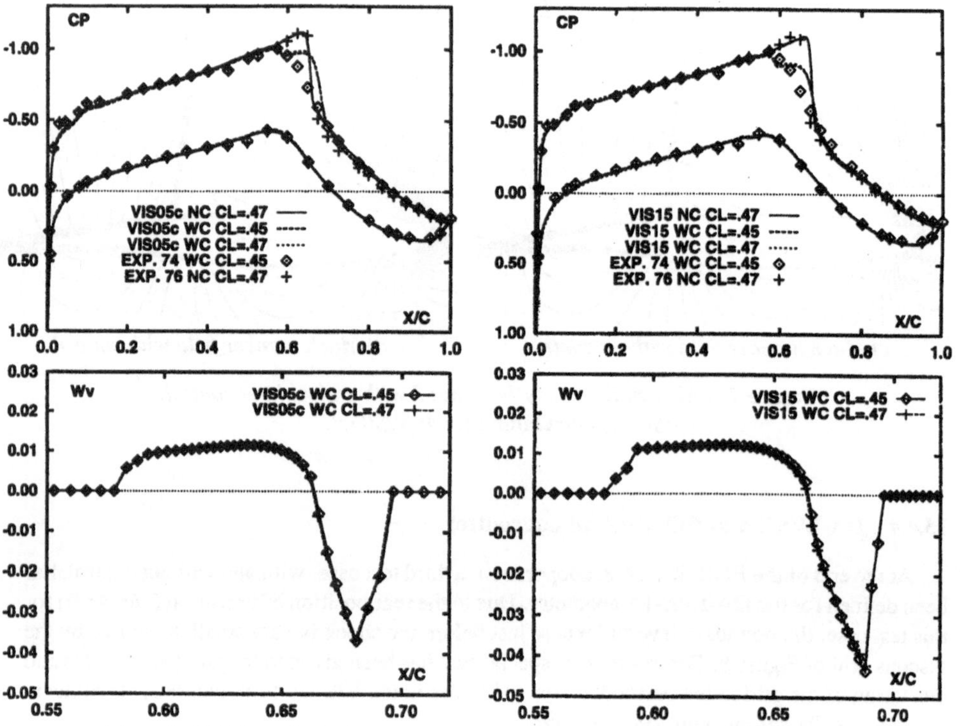

a) *Steady VIS05c calculation*

b) *Unsteady VIS15 calculation*

Figure 9: *DA-LVA-1A airfoil: Steady Test-case 3 with and without control.*

distributions of the ventilation velocity are very similar in the two calculations. The location of
the shock in the VIS15 computation is , however, slightly too much downstream.

As for the VA2 airfoil, the computations without control are characterized by a curved shock,
and the calculations with control exhibit a λ-shaped shock. Figure 10 displays the boundary layer
velocity profiles, computed by the code VIS05c, at different streamwise stations. The first three

234

Table 7: DA-LVA-1A airfoil: Steady Test-case 3.

DA-LVA-1A case 3		α	M_∞	$Re.10^6$	C_L	C_M	C_{DWave}	C_{Dtot}	Cp_{cav}
exp 76	n.c.	1.000	0.7698	4.64	0.4736	-0.0858		0.011720	
VIS05c	n.c.	0.890	0.7613	4.61	0.4732	-0.0935	0.004919	0.011170	
VIS15	n.c.	0.788	0.7613	4.64	0.4736	-0.0877	0.004712	0.011280	
exp 74	w.c.	1.000	0.7692	4.66	0.4556	-0.0833		0.012330	-0.7148
VIS05c	w.c.	0.850	0.7614	4.61	0.4538	-0.0908	0.003230	0.011540	-0.9097
VIS15	w.c.	0.840	0.7613	4.64	0.4538	-0.0828	0.002594	0.010730	-0.8416
VIS05c	w.c.	0.950	0.7614	4.61	0.4726	-0.0922	0.003599	0.012100	-0.9221
VIS15	w.c.	0.940	0.7613	4.64	0.4739	-0.0840	0.002964	0.011510	-0.8530

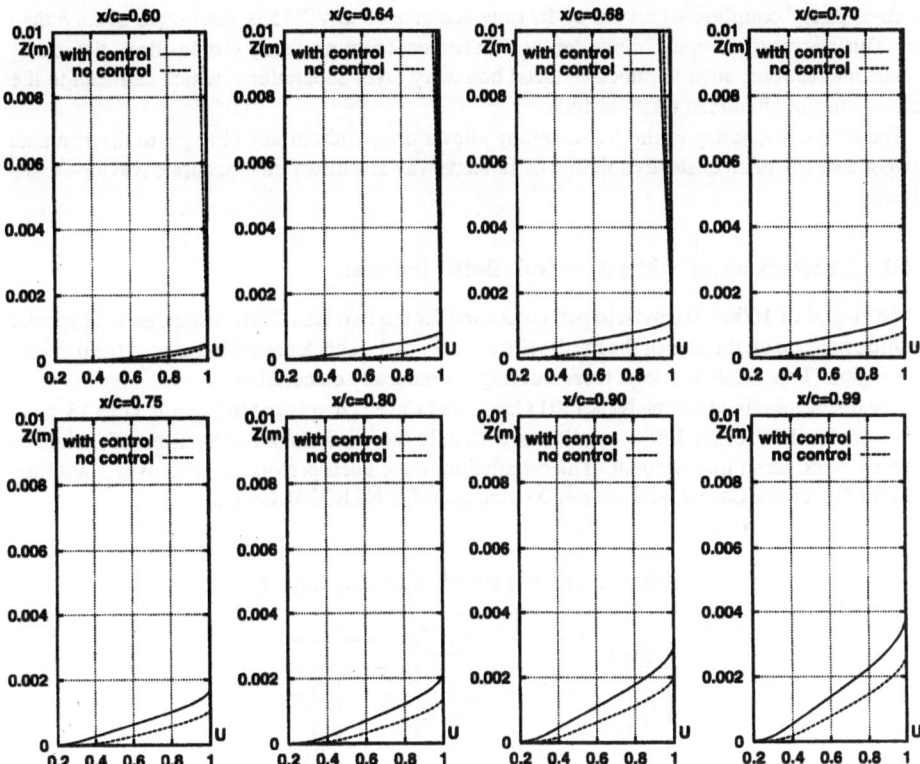

Figure 10: *Boundary layer velocity profiles with and without control.*
Steady calculations (code VIS05c).
DA-LVA-1A airfoil. Steady Test-case 3.

stations are in the perforated zone, and the other stations are between the porous region and the trailing-edge. The VIS15 results are roughly identical and are not shown here.

The comparison with and without control shows that the trends are similar in the computation and in the EUROSHOCK experiments. However, quantitative differences remain between the computed and experimental velocity profiles. It seems that the calculation underestimates the boundary layer thickness for this test case (see Chapter 3). This may be partly due to the need for introducing in the computation the thickening of the boundary layer that is induced by the transition trip and neglected in the present calculation. This may also be partly due to the intrinsic rear position of transition which is close to the shock, so cumulating the difficulties of low Reynolds number modelling and of very small scalings for the numerical methods.

15.5   Computation of the Unsteady Test-cases

The unsteady test-cases have been computed with the time-consistent VII code VIS15, using the 2-equation $k - u'v'$ turbulence model. For all the present unsteady flow calculations, the "Semi-implicit" coupling algorithm of the time-accurate code VIS15 is *converged at each time step*. This allows to compute, consistently with respect to an unsteady viscous flow, the strong viscous interactions, such as shock-induced boundary layer separations, which can lead to the transonic buffet phenomenon on airfoils.

The time-consistency of the VII-coupling allows also, without any change, to discriminate between a steady or an unsteady solution in the same way as a direct time-accurate Navier-Stokes solver.

15.5.1   NLR7301 airfoil without control: Buffet test-cases

An H-grid of 160x100 (inviscid part) was used for the two NLR7301 test-cases with special attention focused on the size of the mesh cells not only in the shock region but also in the trailing-edge region. This mesh was kept frozen during the unsteady calculation.

The results obtained for the NLR7301 Case 2 and Case 3, displayed in Figures 11 to 13, have been calculated at nominal flow conditions with only the wind-tunnel corrections of incidence given by NLR taken into account. The calculations have been performed with over 3000 time steps for NLR7301 Case 2, and over 4000 time steps for NLR7301 Case 3.

Table 8: NLR7301 airfoil: Unsteady Case 2.

	Lift			
Mean	0.671			
	Mod	Phase	Real	Imag
Harm 1	.0925	120.8	.0794	-.0474
Harm 2	.0206	-204.2	.0084	-.0188
Harm 3	.0052	68.9	.0049	.0019
Harm 4	.0055	94.3	.0055	-.0004

Table 9: NLR7301 airfoil: Unsteady Case 3.

	Lift			
Mean	0.272			
	Mod	Phase	Real	Imag
Harm 1	.0146	101.9	.0143	-.0030
Harm 2	.0006	88.1	.0006	.0000
Harm 3	.0006	110.0	.0006	-.0002
Harm 4	.0001	66.7	.0000	.0000

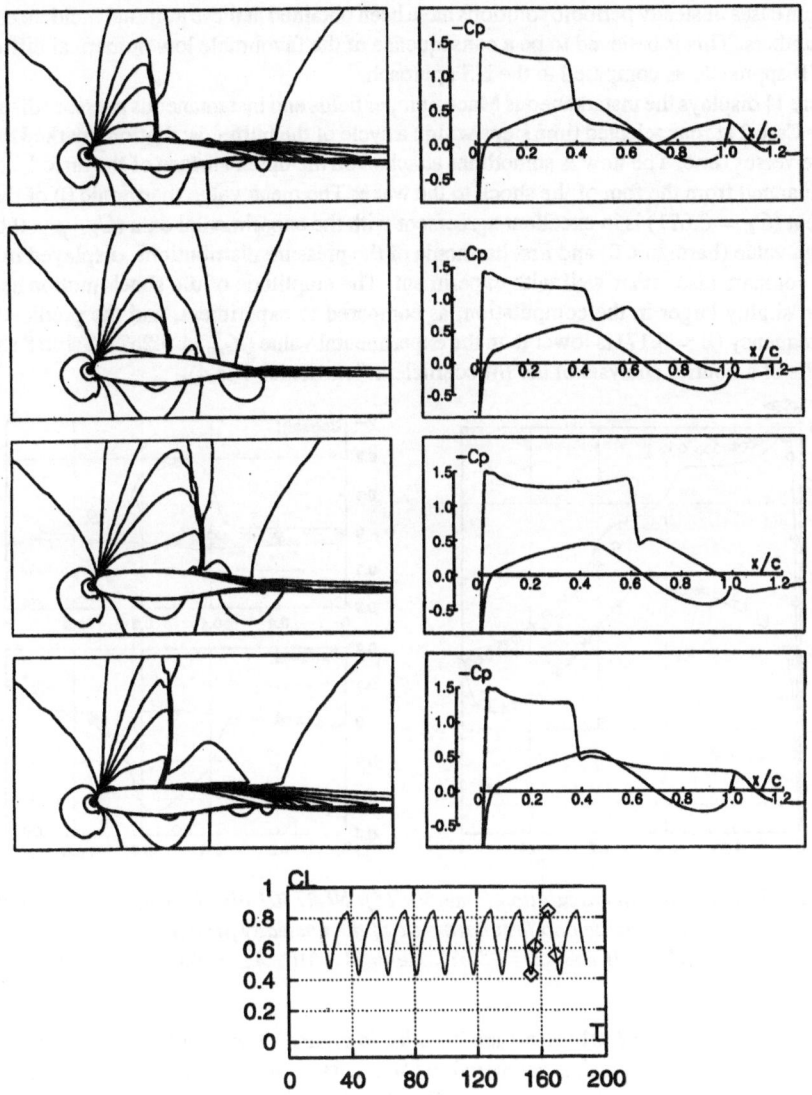

Figure 11: *Time-consistent calculation (code VIS15). NLR7301 airfoil. Unsteady Test-case 2. Lift history. Instantaneous Mach number fields and pressure distributions.*
$M_\infty = 0.738$, $\alpha = 2.257°$, $Re = 11.7\ 10^6$, $X_{tr} = 0.07$

In both cases unsteady periodic solutions have been obtained at these nominal incidences and Mach numbers. This is believed to be a consequence of the favourable low numerical diffusion of the VII approach, as compared to the NS approach.

Figure 11 displays the instantaneous Mach number fields and instantaneous pressure distributions for Case 2 at four selected time steps within a cycle of the buffet oscillation, marked on the lift curve versus time. The flow is sometimes attached on the upper surface of the airfoil, sometimes separated from the foot of the shock to the wake. The mean value (harmonic 0) of the lift coefficient ($C_L = 0.671$) is in excellent agreement with the experimental data ($C_{Lexp} = 0.678$). The mean value (harmonic 0) and first harmonic of the pressure distributions, displayed in Figure 12, compare also rather well with experiment. The amplitude of the shock motion seems, however, slighly larger in the computation, as compared to experiment, and the predicted reduced frequency ($k = 0.17$) is lower than the experimental value ($k_{exp} = 0.257$). Table 8 shows the results of a Fourier analysis of the lift coefficient (harmonics 1 to 4).

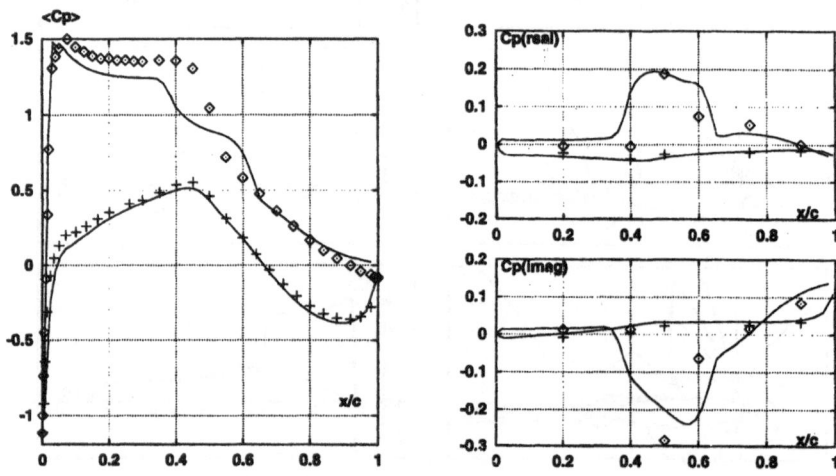

Figure 12: *Time-consistent calculation (code VIS15). NLR7301 airfoil. Unsteady Test-case 2. Mean value and first harmonic of the unsteady pressure.*
$$M_\infty = 0.738, \; \alpha = 2.257°, \; Re = 11.7 \; 10^6, \; X_{tr} = 0.07$$

The results of the NLR7301 Case 3 are included in Figure 13 and Table 9. The amplitude of the shock motion is much smaller compared to Case 2. Here again, a good agreement is found with experiment not only in the mean value of the lift coefficient ($C_L = 0.272$, $C_{Lexp} = 0.259$) but also in the unsteady pressure distributions (harmonic 0 and 1). The calculated reduced frequency, $k = 0.27$, remains smaller than the experimental frequency, $k_{exp} = 0.32$

15.5.2 NACA0012 airfoil without control: Severe buffet test-case

The unsteady test-case for the NACA0012 airfoil has been computed with the same 160x100 inviscid H-grid, previously used for the NLR7301 cases. The computation has been performed with the nominal values of incidence, Mach number, and Reynolds number, in over 900 time steps.

The lift evolution versus time, pressure distributions and Mach number fields at four time steps selected within one cycle are displayed in Figure 14. The first four harmonics of the un-

238

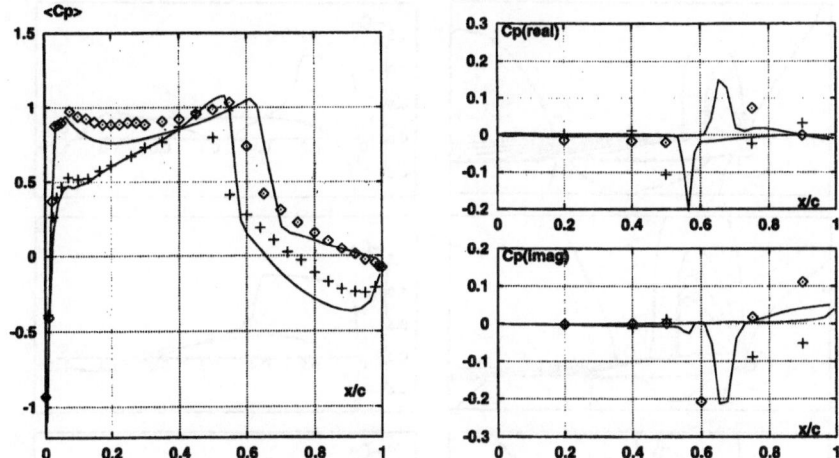

Figure 13: *Time-consistent calculation (code VIS15). NLR7301 airfoil. Unsteady test-case 3. Mean value and first harmonic of the unsteady pressure.* $M_\infty = 0.789$, $\alpha = 0.409°$, $Re = 12.6 \; 10^6$, $X_{tr} = 0.07$

steady lift coefficient are shown in Table 10.

A very large amplitude of the lift has been found for this test case. The instantaneous Mach number fields show that there is a strong variation of shock intensity with respect to time, and that a separation of large extend appears during the unsteady oscillation. It is clear that computations with a frozen mesh are not fully satisfactory when the flow exhibits such a large shock motion, and buffet calculations with a self-adaptive grid at each time step and clustering at the shock are therefore envisaged in the future.

Table 10: NACA0012 airfoil: Unsteady Case 1.

Mean		Lift		
		0.492		
	Mod	Phase	Real	Imag
Harm 1	.1378	129.8	.1058	-.0882
Harm 2	.0038	14.2	.0009	.0037
Harm 3	.0044	39.0	.0028	.0034
Harm 4	.0050	-129.4	-.0039	-.0032

15.5.3 DRA-2303 airfoil with/without control: Damping of buffet

The time-consistent code VIS15 has been used to calculate the mandatory unsteady Test-case 3, with and without control, for the DRA-2303 airfoil. 3000 time steps have been computed.

Without control, a transonic buffet is predicted by the method, as shown by the lift evolution versus time on Figure 15a. This is in good agreement with the experimental data which predicts also a buffet phenomenon without shock control.

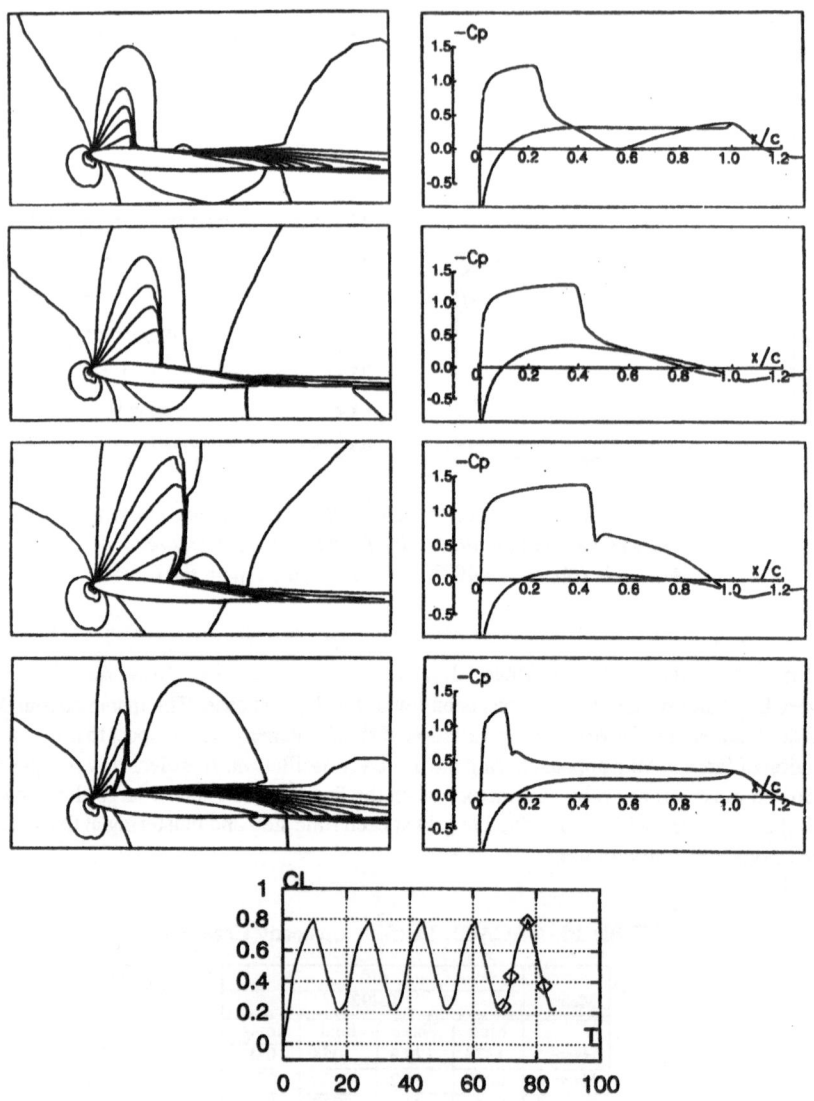

Figure 14: *Time-consistent calculation (code VIS15).*
NACA0012 airfoil. Unsteady Test-case 1.
Lift history. Instantaneous Mach number fields and pressure distributions.
$M_\infty = 0.75$, $\alpha = 5°$, $Re = 10.10^6$, *fully turbulent.*

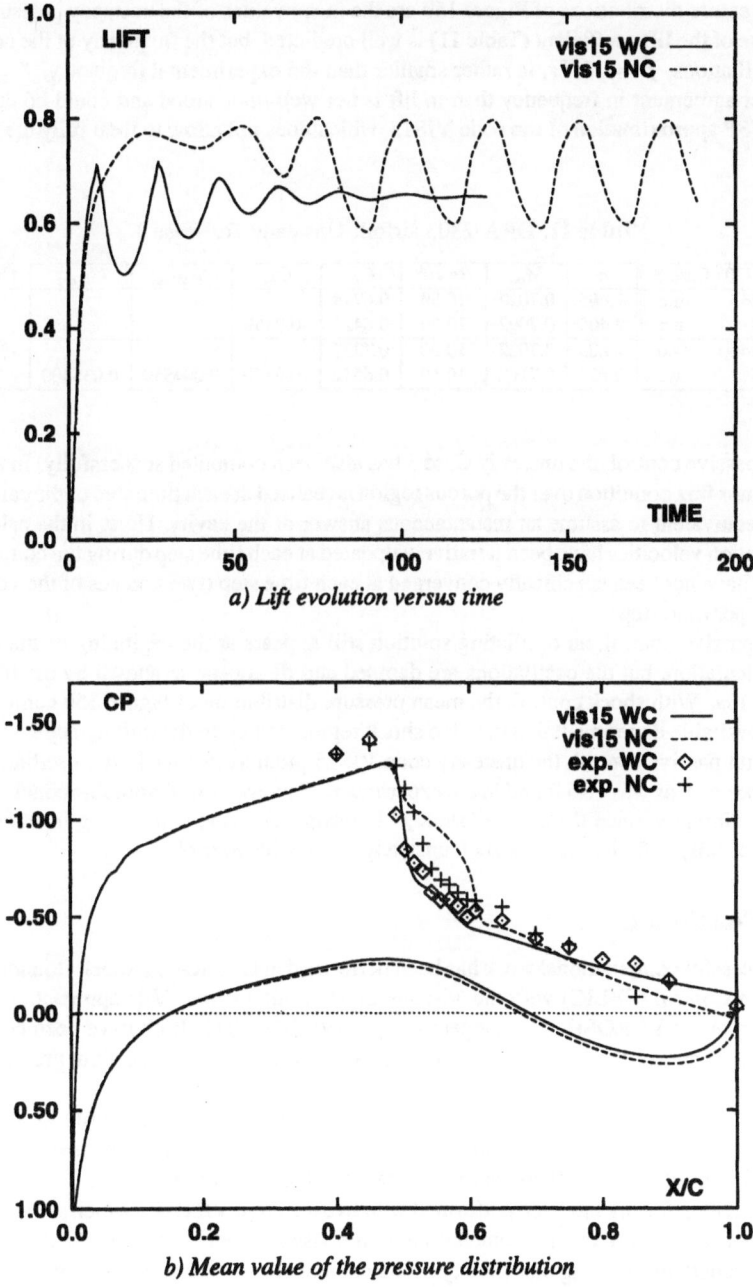

a) Lift evolution versus time

b) Mean value of the pressure distribution

Figure 15: *DRA-2303 airfoil: Unsteady calculations (code VIS15). Unsteady Test-case 3.*

The pressure distributions of Figure 15b are the mean values of the unsteady pressures. The mean value of the lift coefficient (Table 11) is well predicted, but the frequency of the computed buffet oscillations, $f = 15Hz$, is rather smaller than the experimental frequency, $f = 36Hz$. This lesser agreement in frequency than in lift is not well understood and could be due to the inviscid TSP approximation of the code VIS15, which does not allow to treat fully the leading-edge area.

Table 11: DRA-2303 airfoil: Unsteady Test-case 3.

DRA-2303 Case 3		α	M_∞	$Re.10^6$	C_L	C_M	C_{DWave}	C_{Dtot}	C_{Pcav}
Exp. 543	n.c.	2.405	0.7022	10.50	0.6714				
VIS15	n.c.	2.405	0.7092	10.50	0.6427	-0.0859			
Exp. 1469	w.c.	2.805	0.7022	10.50	0.7327				-0.613
VIS15	w.c.	2.805	0.7162	10.50	0.6512	-0.0870	0.004550	0.035500	

With passive control, the unsteady Case 3 has also been computed successfully. In this case, the zero mass flux condition over the porous region is realised at each time step of the calculation, which is equivalent to assume an instantaneous answer of the cavity. Here, in the calculation, the ventilation velocities have been iteratively updated at each time step during the coupling iteration, but have not been specifically converged at each time step (two updates of the ventilation velocities per time step).

With passive control, an oscillating solution still appears at the beginning of the unsteady VIS15 calculation, but the oscillations are damped and disappear, as shown by the lift history in Figure 15a. With shock control, the mean pressure distribution of Figure 15b compares well with the available experimental data in the shock region and up to the trailing edge.

So, with passive control, the unsteady code VIS15 predicts that the buffet oscillations have been suppressed, as was also found in the experiment. However, this favourable conclusion may still be preliminary since the simple "steady" isentropic SC-law, with no lag in the cavity, is perhaps not fully sufficient to treat such unsteady flows with control.

15.6 Conclusion

A successful computational work has been performed at ONERA on Shock Boundary Layer Interaction Control (SBLIC) with the Viscous-Inviscid Interaction (VII) approach, during the three years of the EUROSHOCK cooperation [8], [9], [10], [11]. Both have been considered, the aspects of steady flow computation and the aspects of time-accurate flow prediction with buffet.

The Viscous-Inviscid Interaction approach has been developed successfully for the simulation of passive and active shock-control over porous walls, with the advantage of the very low numerical viscosity of the VII solvers as compared to Navier-Stokes solvers.

Computational work has been performed with two different VII codes which are almost identical in the adaptive VII grid-refinements and in the viscous methodology (hybrid field/integral FD/IBL method, discretizing the velocity profiles, 2-equation $k - u'v'$ turbulence model, steady strong coupling or time-consistent strong coupling). The first code VIS05c provides an advanced steady viscous flow solver (VII + Full-Potential) for severe transonic conditions with the better inviscid approximation. The second time-consistent code VIS15 uses a cruder inviscid approximation (VII + TSP-small-perturbations + time-consistent coupling), but generates with a single code the steady as well as the buffet solutions.

The two codes deliver then complementary insights for flow analysis and drag-estimate with, in the steady test-cases, an overlap for cross comparison between the time-accurate code, that is also able to predict the buffet occurrence, and the higher-order steady code.

A shock control procedure has been developed and introduced in the two codes, following the EUROSHOCK directives for the modelling of control laws. The wall ventilation velocity distribution over a perforated zone is computed using either Poll's formula or the isentropic SC-law with a local reduction factor. Both active and passive control have been implemented and normal as well as inclined holes can be taken into account.

Computations have been performed, with both the steady and unsteady codes, using very fine grids with adaptive clusterings in the shock area, with and without control, for all the mandatory steady EUROSHOCK test-cases on the three airfoils DRA-2303, VA2, and DA-LVA-1A.

For the steady solutions, it has been found that passive shock control leads in general to a decrease of wave drag and an increase of total drag. Also, passive control seems to increase the boundary layer thickness and the tendency of flow separation behind the shock.

With control, a modification of the structure of the shock has been found, leading in general to a λ-shock, also found in the basic experimental studies of EUROSHOCK. In some cases for the turbulent-type VA2 airfoil, a double transonic-pocket structure has been predicted by the code VIS05c together with a reduction of total drag. This possible alternate structure seems also confirmed by some of the VIS15 computations.

With C_L prescribed and near C_{Lmax}, the probable existence of double solutions in the steady computations would also require further investigations before a final conclusion concerning drag reduction. The C_L prescribed solutions at lower incidence have been considered here. Unsteadiness may also interfere, as shown by VIS15 computations.

Time-consistent computations with code VIS15, which allows to discriminate without any change between steady and unsteady solutions, have also been performed for all the mandatory unsteady test-cases of EUROSHOCK.

Buffet solutions have been predicted, at the nominal conditions, for the NLR7301 and NACA-0012 airfoils. This seems to be a favourable result of the low numerical diffusion of the VII method. A good agreement is found with experiment on lift and unsteady pressure distributions, but the frequencies are underpredicted.

The unsteady Test-case 3 with and without control for the DRA-2303 airfoil has been computed with success. As in the experiment, the code VIS15 predicts an unsteady buffet solution without control which is damped with passive control.

15.7   References

[1] Le Balleur J. C. Viscous-inviscid calculation of high-lift separated compressible flows over airfoils and wings. Proceedings AGARD-CP-515, Paper 26, Symposium AGARD/FDP "High-lift aerodynamics", Banff (Canada), 5-8 Octobre 1992, (or ONERA TP 1992-184).

[2] Le Balleur J. C. New possibilities of viscous-inviscid numerical techniques for solving viscous flow equations, with massive separation. Proceed. Fourth Symp. Numerical-Physical Aspects of Aero. Flows, Long-Beach, USA, Jan. 16-19, 1989, Selected papers, chap. 4, p. 71-96, Cebeci ed., Springer-verlag, 1990. (or ONERA TP 1989-24).

[3] Le Balleur J. C. Strong matching method for computing transonic viscous flows including wakes and separations. Lifting airfoils. *La Recherche Aérospatiale 1981-3, p.21-45, English and French editions*, March 1981.

[4] Le Balleur J. C. Viscous-inviscid flow matching : Numerical method and applications to two-dimensional transonic and supersonic flows. *La Recherche Aérospatiale 1978-2, p.67-76, French, or English transl.ESA-TT-496*, March 1978.

[5] Le Balleur J. C., Girodroux-Lavigne P. A semi-implicit and unsteady numerical method of viscous-inviscid interaction for transonic separated flows. *La Recherche Aérospatiale 1984-1, p.15-37, English and French editions*, Jan. 1984.

[6] Le Balleur J. C., Girodroux-Lavigne P. Viscous-Inviscid Strategy and Computation of transonic Buffet. Proceed. Symp. IUTAM Transsonicum III, Göttingen, May 24-27, 1988, Springer-Verlag 1988, or ONERA TP 1988-111.

[7] Girodroux-Lavigne P., Le Balleur J. C. Time-consistent computation of transonic buffet over airfoils. Proceed. of the 16th ICAS, Jerusalem, Israel, August 28-September 2, 1988, or ONERA TP 1988-37.

[8] Le Balleur J. C., Girodroux-Lavigne P., Blaise D. , Gassot H. Development of Viscous-Inviscid Interaction Codes for Prediction of Shock Boundary Layer Interaction Control (SBLIC) and Buffet over Airfoils. Brite-Euram EUROSHOCK Contract AER2-92-49/2.9, ONERA TASK-2.1 and TASK-2.2, Annual Report 1993, December 1993.

[9] Le Balleur J. C., Girodroux-Lavigne P. , Gassot H. Development of Viscous-Inviscid Interaction Codes for Prediction of Shock Boundary Layer Interaction Control (SBLIC) and Buffet over Airfoils. Brite-Euram EUROSHOCK Contract AER2-92-49/2.9, ONERA TASK-2.1 and TASK-2.2, Annual Report 1994, January 1995.

[10] Le Balleur J. C., Girodroux-Lavigne P. , Gassot H. Development of Viscous-Inviscid Interaction Codes for Prediction of Shock Boundary Layer Interaction Control (SBLIC) and Buffet over Airfoils. Brite-Euram EUROSHOCK Contract AER2-92-49/2.9, ONERA TASK-2.1 and TASK-2.2, Annual Report 1995, January 1996, RT 11/3073 AY, February 1996.

[11] Le Balleur J. C., Girodroux-Lavigne P. , Gassot H. Development of Viscous-Inviscid Interaction Codes for Prediction of Shock Boundary Layer Interaction Control (SBLIC) and Buffet over Airfoils. Brite-Euram EUROSHOCK Contract AER2-92-49/2.9, ONERA TASK-2.1 and TASK-2.2, Final Task-2 Report, March 1996.

[12] Poll D., Danks M. , Humphreys B. The Aerodynamic Performance of Laser Drilled Sheets. First European Forum on Laminar Flow Technology, Hamburg, March 1992, Paper 92-02-02.

16 TRANSONIC AIRFOIL FLOW PREDICTION WITH SHOCK BOUNDARY LAYER INTERACTION CONTROL (SBLIC) BY A TIME-ACCURATE NAVIER-STOKES CODE

W. Geissler

DLR - Institut für Strömungsmechanik
Bunsenstraße 10, D - 37073 Göttingen, Germany

Summary: Shock wave / boundary layer interaction and its control has been investigated within Task 2 of EUROSHOCK by means of a 2D time-accurate Navier - Stokes code. With this code it was demonstrated that it can successfully be used below (steady flow) as well as above (unsteady flow) the buffet boundary. The code has been modified to account for shock control taking into account a porous surface close to the shock wave. Several control laws have been implemented in the code. Steady as well as unsteady calculations have been carried out and the results are compared with corresponding experimental data. In the steady cases a reduction of the wave drag could be achieved by SBLIC. However, the overall drag was increased due to a considerable increase of viscous drag in the control cases. This general trend was also found in the experiments as well as in most of the numerical investigations of EUROSHOCK.

In the unsteady cases a surprisingly good correspondence between calculation and experiment was found with respect to the buffet frequency. However, the start of shock oscillations was predicted at slightly higher Mach numbers / incidences compared to the experiment. In general the 2D Navier-Stokes code has proven to be a sufficient and versatile tool for the investigation of the complex shock wave / boundary layer interaction problem and its control.

16.1 Introduction

First numerical investigations of shock boundary layer interaction and its control have been carried out by means of the triple deck approach [1]. Within this approach the details of the flow at the foot of the shock have been investigated. This concept has later been implemented into a viscous/inviscid coupling procedure [2] to calculate the flow about a complete airfoil. First numerical calculations using a thin layer Navier-Stokes code have been performed in Ref. 3. Corresponding experimental investigations [4] have shown the positive effects of active/passive control on shock boundary layer interaction, at least for a turbulent airfoil.

The present numerical investigation has concentrated on the application of a 2D time-accurate Navier-Stokes code [5] which has specifically been developed for the dynamic stall problem of oscillating helicopter airfoils [6]. Compared to the computer-time efficient coupling procedures between boundary layer and outer inviscid flow, which have been applied in EUROSHOCK the present Navier-Stokes code works in a **time-accurate** mode. This feature has the disadvantage of considerable computing time, but the advantage of handling both steady and unsteady (buffet) cases without any code modifications. Thus, a calculation on both sides of the buffet boundary is possible and hence the buffet boundary itself can be determined by this code. To take into account the possibility of shock control, the corresponding modifications have been implemented:

- normal flow velocity variation at the porous region by means of different control laws,
- changes in turbulence modeling for the Baldwin-Lomax model.

Due to the applicability of the present N.S. -code for both steady and unsteady flows, DLR contributed to both Subtasks of Task 2:

- Subtask 2.1: Steady Flow Predictions
- Subtask 2.2: Unsteady Flow Predictions

For the different test cases defined as mandatory in EUROSHOCK as much as five airfoil sections with a total of 12 test cases have been calculated. The large amount of numerical data has been compressed into a number of representative figures, showing pressures, skin friction and transpiration velocity distributions in the steady cases. Corresponding experimental data are used for comparison. In the unsteady cases four instantaneous pressure distributions and Mach-contours for a selected period of oscillation are being displayed.

In the unsteady case of the DRA-2303 airfoil, a video movie has been developed to show the effect of **active** control on the time dependent flow around the airfoil.

16.2 Numerical Method

16.2.1 2D Time - accurate Navier-Stokes code

For the present investigations of shock boundary layer interaction a 2D time - accurate Navier-Stokes code [5] has been used. This code is based on the Beam/Warming [7] approximate factorization implicit methodology using central differences in the space coordinates with additional numerical damping terms. These terms are eigenvalue - scaled, i.e., they take into account the local eigenvalues for control of the damping coefficients. Due to the implicit character of the code, the Courant number can be considerably increased to a maximum value of about 10^2. Nevertheless, the time-steps in the present cases, specifically if separation is expected, have to be much smaller as would be necessary for stability constraints. This is even of greater importance in unsteady (buffet) situations in order to resolve physical time scales sufficiently. An elliptic mesh solver has been applied to resolve the flowfield by means of a C-mesh topology. The first grid size at the airfoil surface has to be controlled such that the first meshpoint is located inside the laminar sublayer at normalized distances

$$y^* < 5$$

and that about 25 meshpoints remain inside the boundary layer. The mesh is stretched outside to 10 chord length from the airfoil into all spatial directions.

In all present calculations a 157 x 59 C-mesh has been used. To take into account transition on the airfoil, prescribed transition points on both upper and lower surface are taken into account. To represent turbulent flows, the Baldwin/Lomax algebraic turbulence model has been used. It has already been mentioned that the code is always operated in a time-accurate mode. Local time - stepping procedures or other accelerating measures have not been applied. A change of the code - methodology from steady to unsteady flow or vice versa is not necessary.

16.2.2 Implementation of shock control laws

One of the major efforts in EUROSHOCK was the modification of the codes to account for shock boundary layer interaction **control** (SBLIC). Two main steps were necessary for this implementation:

1. Modification of the surface boundary condition along the control region to account for transpiration velocities
2. Modification of the turbulence model

For the solution of the first problem different rather simple and empirical control laws are described in the literature. Three different control laws have been implemented into the code:

a) Darcy's Law [8],
b) Poll's Formula [9] and the
c) Isentropic Law

where the user decides which law will be used for the calculations. In the present studies Poll's Formula was taken as the standard law.

a) Darcy's Law:

Following the work of Bur [8], the law of Darcy relates the transpiration velocity, v_p, along the control region to the pressure difference between the outer surface pressure, $p(x)$, and the cavity pressure, p_c. The latter is assumed to be constant,

$$v_p = a \, (p_c - p \, (x)), \tag{1}$$

with the proportionality coefficient

$$a = \frac{P \cdot d^2}{32 \, e \, \mu} \tag{2}$$

where

P = porosity of the porous plate; open area/total area,
d = diameter of the holes,
e = thickness of the porous plate and
μ = molecular viscosity.

Experience with this rather simple model was in some cases not very satisfactory.

b) Poll's Formula:

The second empirical law has been developed by Poll et. al. [9]. This law is based on experimental investigations of a large number of samples of laser drilled titanium plates in order to obtain a relation between the mass flow and the pressure drop through the porous surface. As a final result this relationship can be expressed by the quadratic expression

$$Y = \frac{1}{\kappa}[40.7 \cdot X + 1.95 \cdot X^2] \tag{3}$$

with X proportional to the mass flux $\dot{m}$ and Y proportional to the pressure difference:

$$X = \frac{\dot{m}}{\mu \cdot e}; \; \dot{m} = \rho \cdot \pi \cdot \frac{d^2}{4} \cdot \upsilon \tag{4a}$$

$$Y = \frac{(p_c - p)d^2}{\rho \cdot \upsilon^2}\left(\frac{d^2}{e}\right). \tag{4b}$$

The laser drilling of a porous sheet does not produce perfect holes. If the spacing is uniform the holes themselves are not strictly circular and they have a conical shape instead of a cylindrical one. This causes differences in suction and blowing properties due to the corresponding diffuser or jet effects of the holes.

Due to these deficiencies it is necessary to specify an effective diameter (Eq. 3)

$$\kappa = \frac{effective\ diameter}{measured\ diameter}, \tag{5}$$

with κ as an average value for all holes. In the present calculations $\kappa = 2.5$ was assumed in correspondence with Poll's recommendations [9].

c) Isentropic Law:

Assuming isentropic flow through a single hole the isentropic relations between points outside (Index b) and inside (Index a) of the cavity are:

$$\frac{p_b}{p_a} = \left(\frac{\rho_b}{\rho_a}\right)^{\gamma} \tag{6}$$

and for the corresponding velocity ($v_a \equiv 0$)

$$\frac{\upsilon_b^2}{2} = \frac{\gamma}{\gamma - 1}\frac{p_a}{\rho_a}\left[1 - \left(\frac{p_b}{p_a}\right)^{\frac{\gamma-1}{\gamma}}\right]. \tag{7}$$

From Eqs. (6) and (7) the mass flow across one hole yields

$$\rho_b \cdot \upsilon_b = \sqrt{\frac{2}{\gamma - 1}p_b \cdot \rho_b\left[\left(\frac{p_a}{p_b}\right)^{\frac{\gamma-1}{\gamma}} - 1\right]}. \tag{8}$$

So far no losses have been taken into account in this model. Therefore, the calculated velocities are larger compared to the real ones. A scaling factor has to be taken into account to model this effect. This scale factor can either be measured or be determined from one of the pure empirical laws. In the present code the scaling factor is calculated by means of Poll's formula.

The different control laws described before have to be implemented into the overall calculation procedure of the N.S. - code in such a way, that at each time step zero net mass flow through the surface of the cavity is achieved, i.e., inflow into and out of the cavity must balance (passive control). To fulfill this condition an additional iteration loop for each time step has to be carried out.

In addition to passive control where the final (constant) cavity pressure p_c is an output value, also **active** control can be initiated. This is achieved by prescribing the cavity pressure, i.e., p_c serves then as an input value. In the latter case a mass balance can, of course, not be achieved.

16.2.3 Modification of turbulence model

Along the interaction region the flow is assumed as fully turbulent. Surface mass transfer as well as the longitudinal pressure gradient within the active region make some modifications of the turbulence model necessary. In the Baldwin/Lomax [10] model used in the present numerical code, the generalised Van Driest wall damping function is defined as

$$D = 1 - \exp\left(-\frac{y^+}{A^+}\right) \tag{9}$$

where

$$A^+ = 26, \qquad y^+ = y \cdot \frac{\bar{u}_\tau}{\upsilon}, \qquad \bar{u}_\tau = \sqrt{\frac{\bar{\tau}_p}{\rho}}, \tag{10}$$

with $\bar{u}_\tau$ being the friction velocity. However, in separated boundary layers the friction velocity may be zero and terms in the turbulence model referred to $\bar{u}_\tau$ are no longer meaningful. Cebeci, [11], has proposed to take in cases of mass transfer not the friction velocity, but the corresponding velocity $\bar{u}_{\tau s}$ at the edge of the viscous sublayer at $y_s^+ = 11.8$. If both effects of mass transfer and longitudinal pressure gradient are taken into account, the coefficient A^+ in Eq. (9) can be modified after [11]:

$$A^+ = 26\left[\exp\left(11.8\upsilon_p^+\right) - \frac{p^+}{\upsilon_p^+}\left\{\exp\left(11.8\upsilon_p^+\right) - 1\right\}\right]^{-\frac{1}{2}} \tag{11}$$

where

$$\upsilon_p^+ = \frac{\upsilon_p}{\bar{u}_{\tau s}}, \quad p^+ = -\frac{d\bar{p}}{dx}\frac{\upsilon}{\rho\bar{u}_{\tau s}^3}.$$

16.3 Airfoils and Test Cases

Fig. 1 shows plots of the different airfoils, which have been investigated during the EUROSHOCK research. A total of 13 steady and unsteady mandatory cases had to be calculated. These cases are in detail:

DA LVA - 1A Airfoil (only steady flow):

Case 1	$M = 0.75$	$C_L = 0.53$	$Re = 6 \cdot 10^6$	$X_{tr} = 5\%/5\%$	no control
Case 2	$M = 0.75$	$C_L = 0.69$	$Re = 6 \cdot 10^6$	$X_{tr} = 5\%/5\%$	no control
Case 3	$M = 0.7613$	$C_L = 0.4736$	$Re = 4.64 \cdot 10^6$	$X_{tr} = 48\%/48\%$	w/wo control

DRA - 2303 Airfoil (steady)

Case 1	$M = 0.680$	$C_L = 0.5730$	$Re = 19 \cdot 10^6$	$X_{tr} = 5\%/5\%$	w/wo control
Case 2	$M = 0.680$	$C_L = 0.811$	$Re = 19 \cdot 10^6$	$X_{tr} = 5\%/5\%$	w/wo control

DRA -2303 Airfoil (unsteady)

Case 3	$M = 0.720$	$\alpha = 3.0°$	$Re = 10.5 \cdot 10^6$	$X_{tr} = 5\%/5\%$	w/wo active control

NLR 7301 Airfoil (steady)

Case 1	$M = 0.745$	$\alpha = 0.2°$	$Re = 12.5 \cdot 10^6$	$X_{tr} = 7\%/7\%$	no control

NLR 7301 Airfoil (unsteady)

Case 2	$M = 0.738$	$\alpha = 3.50°$	$Re = 12 \cdot 10^6$	$X_{tr} = 7\%/7\%$	no control
Case 3	$M = 0.789$	$\alpha = 0.41°$	$Re = 12.6 \cdot 10^6$	$X_{tr} = 7\%/7\%$	no control

VA2 Airfoil (steady)

Case 1	$M = 0.74$	$C_L = 0.65$	$Re = 2.5 \cdot 10^6$	$X_{tr} = 30\%/25\%$	w/wo control
Case 2	$M = 0.74$	$C_L = 0.90$	$Re = 2.5 \cdot 10^6$	$X_{tr} = 30\%/25\%$	w/wo control
Case 3	$M = 0.743$	$C_L = 0.8$	$Re = 2.5 \cdot 10^6$	$X_{tr} = 30\%/25\%$	w/wo control

NACA 0012 Airfoil (unsteady)

Case 1	$M = 0.75$	$\alpha = 5°$	$Re = 10^6$	turbulent	no control

16.4 Results

16.4.1 Steady flow

DA LVA - 1A airfoil, w/wo control

The mandatory DA LVA-1A airfoil Test Cases 1 and 2 were precalculations without control to compare results of the various numerical codes involved and, in addition, to compare with corresponding experimental data from the DLR TWB. These data have been discussed in the EUROSHOCK Final Technical Report [12] and briefly in Chapter 3.2.1. For Case 3 numerical as well as experimental investigations have been carried out with and without control. Fig. 2 displays pressure distributions, skin friction distributions as well as transpiration velocities along the cavity area. The measured pressure distributions are from the T2 - wind tunnel of ONERA/CERT. Finally, the table in Fig. 2 shows lift and drag coefficients for both calculation and experiment. The measured C_L - values with and without control and the corresponding computations show deviations so that due to the difference in C_L, the C_D - values are not directly comparable.

Therefore, the calculation with control was repeated with the same C_L as in the no control case. Several conclusions may be drawn from these figures:

- The pressure distributions show good correspondence between calculation and experiment.
- The strength of the shock wave has been reduced considerably due to control.
- The skin friction distribution shows no separation without control. With control the flow is locally separated behind the shock.
- The drag coefficient (overall drag) is increased in both experimental and numerical cases, for the latter, however, only when comparing at equal lift.
- A difference in drag - level can be observed for measurement and calculation.

These features can also be observed for the other airfoils and flow cases as will be discussed in the following subsections.

DRA - 2303 airfoil, w/wo control

Fig. 3 and 4 show the steady Cases 1 and 2 for the laminar DRA airfoil with and without control. Fig. 3 displays the lower lift case ($C_L \approx 0.55$). In this case the shock wave is rather weak and the influence on lift and drag is small as can be seen in the table of Fig. 3. Nevertheless, the same trend occurs as in the previously discussed DA LVA-1A Case 3: The drag is increased in the control case in both the numerical and experimental results. The skin friction distribution does not show any indication of separation in this case.

In Fig. 4 the lift is considerably higher ($C_L \approx 0.81$) and the effect of shock control, i.e., a reduction of the shock strength, is increased. Now, the skin friction touches the zero-line at the shock and shows further downstream a small separation area with reattachment towards the trailing edge. The increase of drag occurs again in the predicted as well as in the experimental data. The drag levels are in both cases the same, although the lift coefficients differ somewhat.

251

VA2-airfoil, w/wo control

Fig. 5 shows again pressures, skin friction and transpiration velocity, here for the super-critical VA2 airfoil. No comparable experimental data were available for this case. With control the strong shock wave is considerably reduced. The skin friction distributions show a small separation bubble behind the shock without control which is extended noticeable in the control case. A drag increase is also observed in the VA2 airfoil case at the present conditions.

Fig. 6 shows as an example Mach number contours for the VA2 airfoil, Case 2 (C_L = 0.9). Two common features can be observed in these plots:

1) The shock strength is reduced with control and shows a λ-structure at its foot.
2) The boundary layer thickness behind the shock is considerably increased.

Specifically the second feature seems to be responsible for the total drag increase. The viscous drag in the area behind the shock is much larger compared to the no-control case. The reduction of the wave drag is more than compensated by the viscous drag increase.

16.4.2 Unsteady flow, buffet

NLR 7301 airfoil, no control

Also within the scope of precalculations without control, the NLR 7301 supercritical airfoil has been investigated in buffet situations. Emphasis was placed on Case 2, which has been investigated experimentally by NLR.

Employing the set of parameters from the experiment, the numerical calculation did not show any unsteadiness but reached a steady - state result. After increasing the incidence from α = 2.26° (experiment) to α = 3.5° but keeping the other parameters unchanged, a buffet situation occurred. The reason for this insensitivity of the code may be due to the relatively course mesh, which does not resolve the shock wave region sufficiently, (see Chapter 15).

The results for the higher incidence are displayed in Fig. 7. Fig. 7a shows time-dependent lift, drag and pitching moment distributions. A representative period of buffet oscillation has been chosen from the C_L - curve and four pressure distributions and the corresponding Mach number contours are plotted in Fig.7b.

The shock movement is only small in this case. The frequency of oscillation is f = 42.8 Hz for a chord length of c = 0.5 m. The corresponding experimental value as measured by NLR is f = 39.9 Hz which is in remarkable agreement in this case.

Again, it is assumed that the spacing of mesh points at the shock is too coarse. But to develop a flow adaptive mesh is a very complicated task in these unsteady cases, because the clustering of mesh points must move with the shock. Such a procedure remains to be solved in the future.

NACA 0012 airfoil, no control

A further precalculation has been performed for the NACA 0012 airfoil also in a buffet condition. The corresponding results are displayed in Fig. 8. Now the shock wave shows considerable movement along the airfoil upper surface. In correspondence, the shock strength is also varying with time. At the time-instant T = 23.1 the contour plot (Fig. 8b) shows a severe separation which is extended over the whole upper surface of the airfoil and further into the wake. The shock and the supersonic region are reduced and the lift (C_L - curve in Fig. 8a) shows a minimum at a very low lift level.

DRA - 2303 airfoil, w/wo control

Case 3 of the DRA - 2303 airfoil is the only mandatory unsteady case where experimental data are available for both control and no - control cases. To get a buffet situation in the calculations, the Mach number had to be increased from M = 0.702 (experiment) to M = 0.72 (calculation). Taking into account this increase, buffet oscillations are observed in the no - control case. The corresponding unsteady force and moment coefficients are displayed in Fig. 9 (upper figure). This figure shows that it takes considerable time until the coefficients start a periodic dependence with respect to time.

To simulate numerically the unsteady case with control leads to severe difficulties. In the experimental case the oscillations could be damped by unsealing the porous area on the airfoil upper surface. The question arises what control methodology has to be applied in the numerical simulation of this case?

If zero net mass flow is established at each time instant, the numerical results do not show any reduction of the oscillation amplitude. However, if **active** control is applied, the buffet oscillations could be influenced favorably as can be seen in Fig. 9: The lower four figures show again the force and moment distributions with respect to time at conditions where, corresponding to active control, the cavity pressure has been reduced successively from c_{pc} = -1.3/-1.4/-1.5 to finally - 1.6. Because of the same scaling in the figures it can be seen that due to the reduction of the cavity pressure first the amplitude of lift is reduced and the frequency is increased. At c_{pc} = -1.5 the curve remains only partly unsteady and finally, at c_{pc} = -1.6, reaches a steady condition.

This investigation shows that the buffet boundary can indeed favorably be influenced by active control.

Fourier decomposition of unsteady data

Fig. 10 shows as a final result for both the unsteady NLR 7301 and NACA 0012 cases results of a Fourier decomposition of the unsteady pressures. From the first four harmonics only the first harmonic pressure distributions are displayed in Fig. 10. The upper figures show the mean pressures followed by the real and imaginary parts and finally by the modulus and phase angles of the pressures. Only in the NLR case are the corresponding experimental data indicated.

The smeared mean pressure distribution in the NACA 0012 case as well as the strong and extended real and imaginary parts of the pressure are indicators of large amplitude shock movements and time dependent shock-strength variations in this case.

16.5   Final Discussion of Results

16.5.1 Steady results

The numerical investigations with the time - accurate Navier Stokes code were performed on both the steady and unsteady sides of the buffet boundary. After implementation and testing of different empirical control laws, the calculations have also been extended to control cases. Two laminar airfoils, viz., the DA LVA-1A, and DRA - 2303 and the turbulent supercritical VA 2 airfoil have been investigated with and w/o passive control. In all cases it has been found that the wave drag was influenced favorably by passive control; however, a tendency to local separation and a thickening of the boundary layer in the control case have increased the **viscous** drag downstream of the shock. This increase of viscous drag has more than compensated the reduction of the wave drag and, therefore, led to an increase in total drag. These trends were found in all calculations carried out with the present N.S. - code but has also been found by most of the other contributors to EUROSHOCK, at least for the laminar - type airfoils.

The experimental data for the airfoils DRA-2303 and DA LVA-1A also show the same trends as discussed above. Only the levels of the measured drag differs from the calculated ones.

16.5.2 Unsteady results

Calculations beyond the buffet boundary were also possible with the present code. However, the initiation of buffet oscillations started at slightly higher Mach numbers and / or incidences compared to the experiments. This effect is not surprising because all calculations were carried out with a prescribed C - mesh of 157 x 59 mesh points. The mesh point distribution was unchanged during the calculations. The resolution of the shock region was not very high in the present computations but an increase of mesh points was not assumed as useful, because large amplitudes of shock wave motion make an adjustment of mesh points with time necessary which was considered beyond the scope of the present investigation. Time - dependent adaptive grids are of great interest for future investigations.

In situations where shock induced separation occurs the applicability of the rather simple algebraic turbulence model of Baldwin/Lomax is questionable. Improved models of one or two equations taking some history effects into account are recommended to be used in future applications. The importance of turbulence modeling has also been pointed out in conjunction with the basic investigations (see, e.g., Chapters 2, 9 and 10) of EUROSHOCK: the effects of the turbulence model seem to be of even greater concern as the control laws.

The case of **unsteady flow control** is the problem least investigated with the greatest question marks. No experimental data are available for passive / active flow control from **fundamental** tests. So it can only be assumed what type of control law has to be used. The present numerical investigations for the DRA - 2303 airfoil, Case 3, have shown that passive control as applied in the steady flow cases is not sufficient, if zero net mass flow has been applied at each time-step. This assumption is too restrictive: The shock wave may move very rapidly along the upper surface and partly leave the control region. Only with massive active control, i.e., with a low prescribed cavity pressure could it be demonstrated that the amplitude, strength and frequency of the shock wave could favorably be influenced.

Although a number of problems remain to be solved in the future, the numerical results give some detailed insight into the very complex unsteady, separated flow structure during buffet oscillations. This has been shown by means of video sequences of the DRA - 2303 airfoil (Case 3) where the very large amount of numerical data has been compressed to a short but informative movie.

16.6 Conclusion and Future Activities

A numerical code based on the 2D time - accurate Navier-Stokes equations has been applied to the investigation of shock boundary layer interaction and its control during the 3-year period of EUROSHOCK. The code has been successively extended to take into account different control laws with a corresponding necessary modification of the algebraic turbulence model.

The time-accurate code is able to cover flow conditions below (steady flow) as well as above (unsteady flow) the buffet boundary without any code modifications.

In all steady cases it has been found by the present calculations that with passive control a reduction of the wave drag could be achieved but in the control cases the viscous drag behind the shock was increased considerably due to a thickening of the boundary layer. This increase of viscous drag overcompensated the reduction of the wave drag and led to increased total drag. These trends are confirmed by other numerical calculations within EUROSHOCK as well as by experiments.

These experiences may lead to new control devices, as, for instance, surface - bumps or a combination of perforated surfaces or bumps with suction behind the shock, in order to reduce the boundary layer thickness or avoid local separation.

In unsteady (buffet) situations the numerical results show quite good agreement with the few experimental data available. A detailed study of the complex flow involving moving shocks and shock - induced separation is possible.

For control of the buffet boundaries it is highly recommended for future investigations to make corresponding fundamental tests in channel flows to determine control rules for moving shock waves.

16.7 References

[1]   Bohning, R., Zierep, J.: "Calculation of 2D-turbulent shock/boundary layer interaction at curved surfaces with suction and blowing", IUTAM Symposium on Turbulent Shear Layer/Shock Wave Interaction, Palaiseau, France, 1985.

[2]   Dargel, G., Thiede, P.: "Viscous transonic airfoil flow simulation by an efficient viscous/inviscid interaction method", AIAA Paper 87 - 0412, 1987.

[3]   Chen, C.L., Chow, C.Y., Van Dalsem, W.R., Holst, T.L.: "Computation of viscous transonic flow over porous airfoils", AIAA - paper 87 - 0359, 1987.

[4]   Thiede, P., Krogmann, P., Stanewsky, E.: "Active and passive shock/boundary layer interaction control on supercritical airfoils", AGARD - CP - 365, Brüssel, 1984.

[5]   Geissler, W.: "Instationäres Navier-Stokes-Verfahren für beschleunigt bewegte Profile mit Ablösung", DLR-FB 92-03, 1992.

[6]   Geissler, W., Vollmers, H.: "Unsteady Separated Flows on Rotor Airfoils, Analysis and Visualization of Numerical Data", 18th European Rotorcraft Forum, 15.-18. Sept., Avignon, France.

[7]   Beam, R., Warming, R.F.: "An Implicit Finite Difference Algorithm for Hyperbolic Systems in Conservation Law Form", J. Comp. Phys., Vol. 22 (Sept. 1976) pp. 87 - 110.

[8]   Bur, R.: "Passive Control of a Shock Wave/Turbulent Boundary Layer Interaction in a Transonic Flow", J. of Comp. Physics, Vol. 22, 1976.

[9]   Poll, D.I.A., Danks, M., Humphreys, B.E.: "The Aerodynamic Performance of Laser Drilled Sheets", First European Forum on Laminar Flow Technology, Hamburg, March 1992, paper 92-02-028.

[10]   Baldwin, R.S., Lomax, H.: "Thin Layer Aproximation and Algebraic Model for Separated Turbulent Flows", AIAA - J., Vol. 99 (1978).

[11]   Cebeci, T.: "Behaviour of turbulent flow near a porous wall with pressure gradient", AIAA - J., Vol. 8, No. 12, (1970).

[12]   Geissler, W.: "EUROSHOCK Final Technical Report, Appendix C", EUROSHOCK TR AER2 -92-94 / F1, 1996.

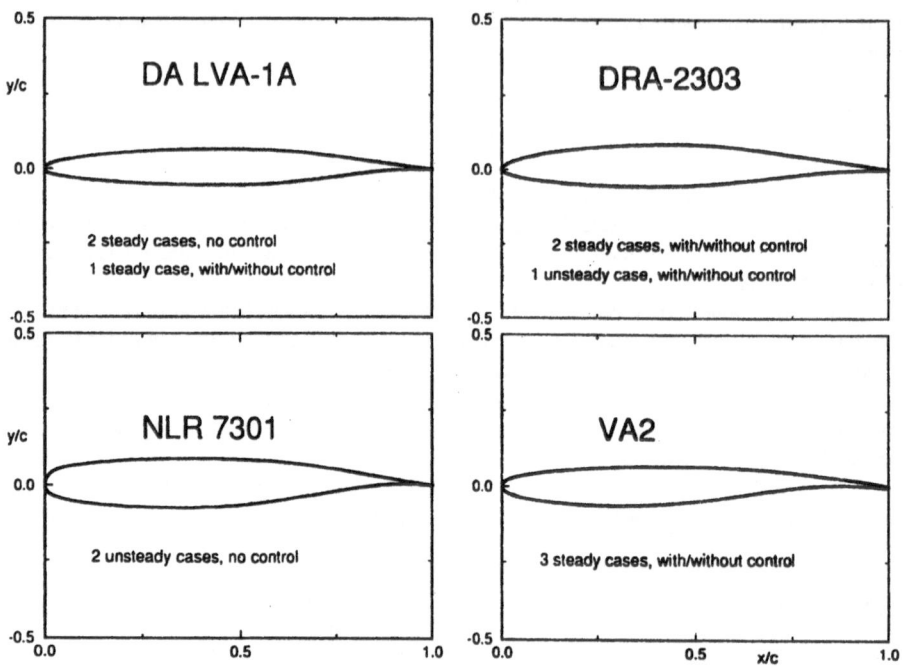

Fig.1 : Investigated airfoil sections: Mandatory cases plus one optional unsteady case

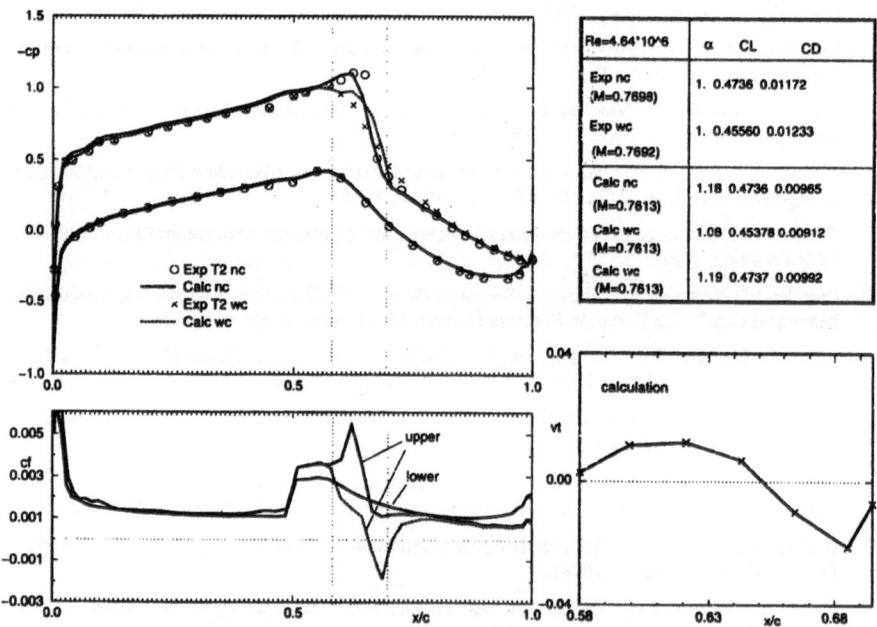

Fig. 2: DA LVA-1A airfoil: Pressure, skin friction, transpiration velocity distribution
Comparison with T2-experiment, lift and drag coefficients

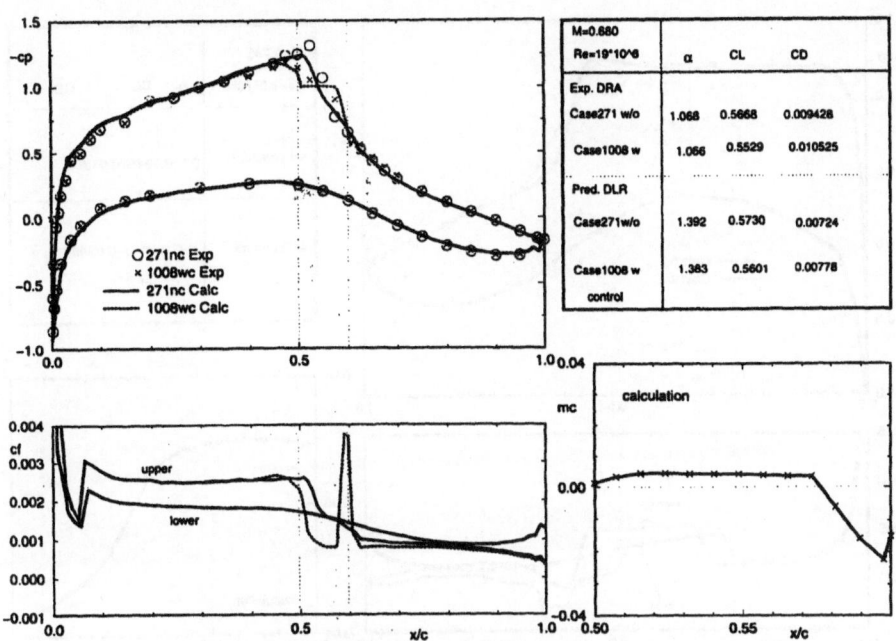

M=0.680 Re=19*10^6	α	CL	CD
Exp. DRA			
Case271 w/o	1.068	0.5668	0.009428
Case1008 w	1.066	0.5529	0.010525
Pred. DLR			
Case271w/o	1.392	0.5730	0.00724
Case1008 w control	1.383	0.5601	0.00778

Fig.3 : DRA-2303 airfoil: Pressure, skin friction and mass flow ratio distribution. Comparison with experiment, lift and drag coefficients (Case 1)

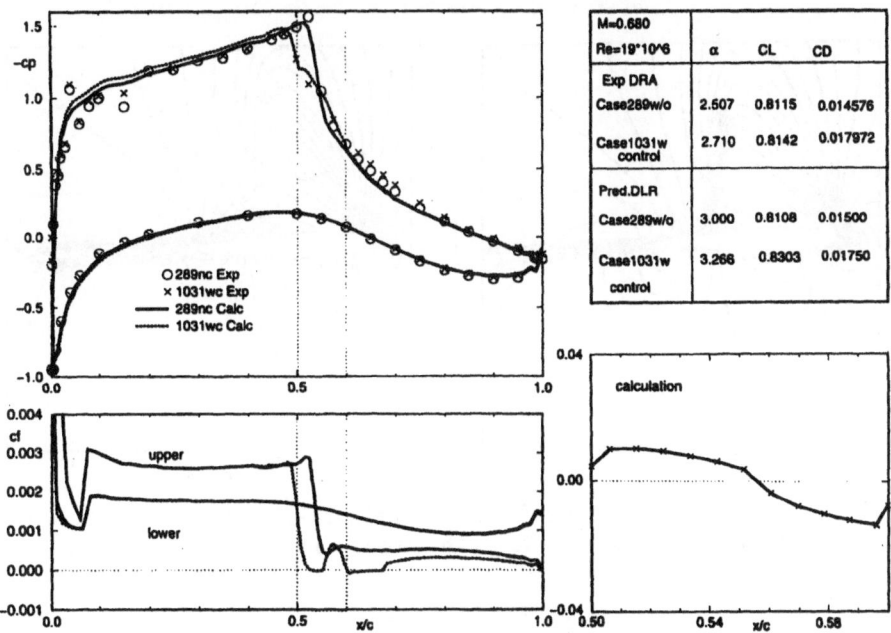

M=0.680 Re=19*10^6	α	CL	CD
Exp DRA			
Case289w/o	2.507	0.8115	0.014576
Case1031w control	2.710	0.8142	0.017972
Pred.DLR			
Case289w/o	3.000	0.8108	0.01500
Case1031w control	3.266	0.8303	0.01750

Fig. 4: DRA-2303 airfoil: Pressure, skin friction and mass flow ratio distribution. Comparison with experiment, lift and drag coefficients (Case 2)

257

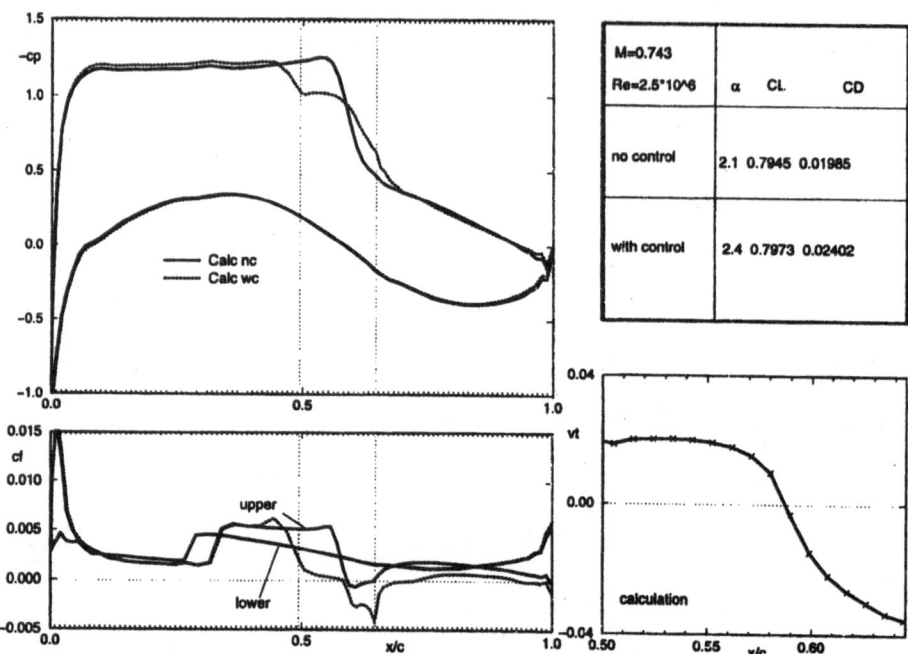

M=0.743 Re=2.5*10^6	α	CL	CD
no control	2.1	0.7945	0.01985
with control	2.4	0.7973	0.02402

Fig.5 : VA2 airfoil: Pressure, skin friction and transpiration velocity distribution.

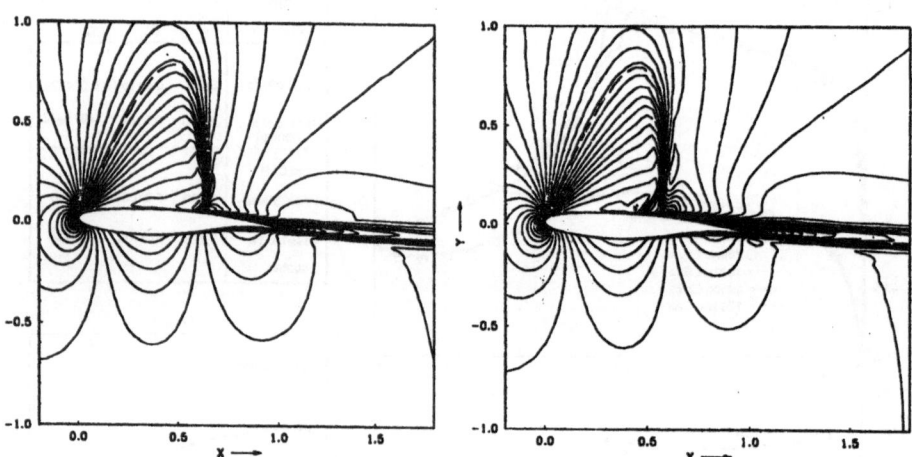

Fig. 6: Mach number contours for the VA2-airfoil, $M_\infty = 0.74$, Re = 2.5×10^6; left: no control, $C_L = 0.9$ ($\alpha = 2.75°$); right: passive control, $C_L = 0.9$ ($\alpha = 3.40°$)

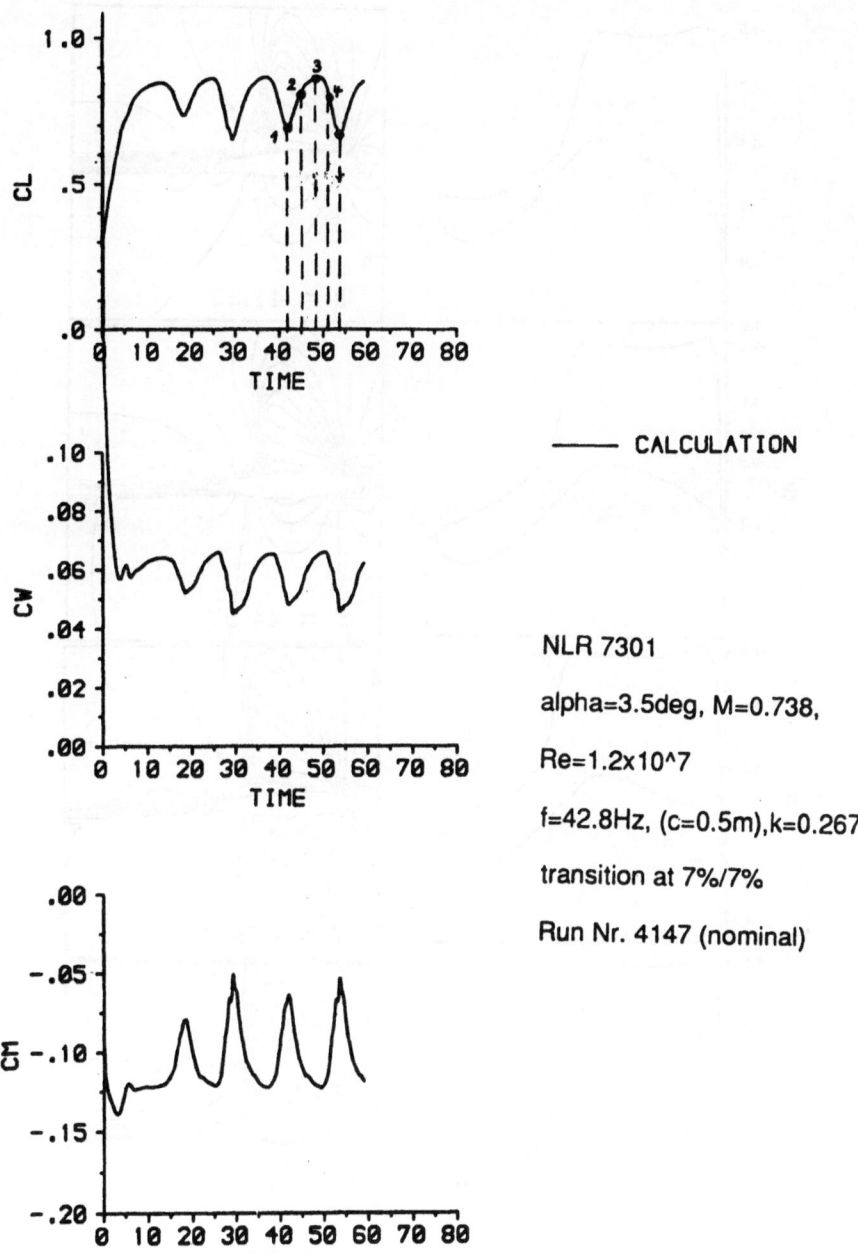

NLR 7301

alpha=3.5deg, M=0.738,

Re=1.2x10^7

f=42.8Hz, (c=0.5m),k=0.267

transition at 7%/7%

Run Nr. 4147 (nominal)

Fig.7a: Time-accurate Navier-Stokes computations for the airfoil NLR 7301
Time-dependent force and moment coefficients.

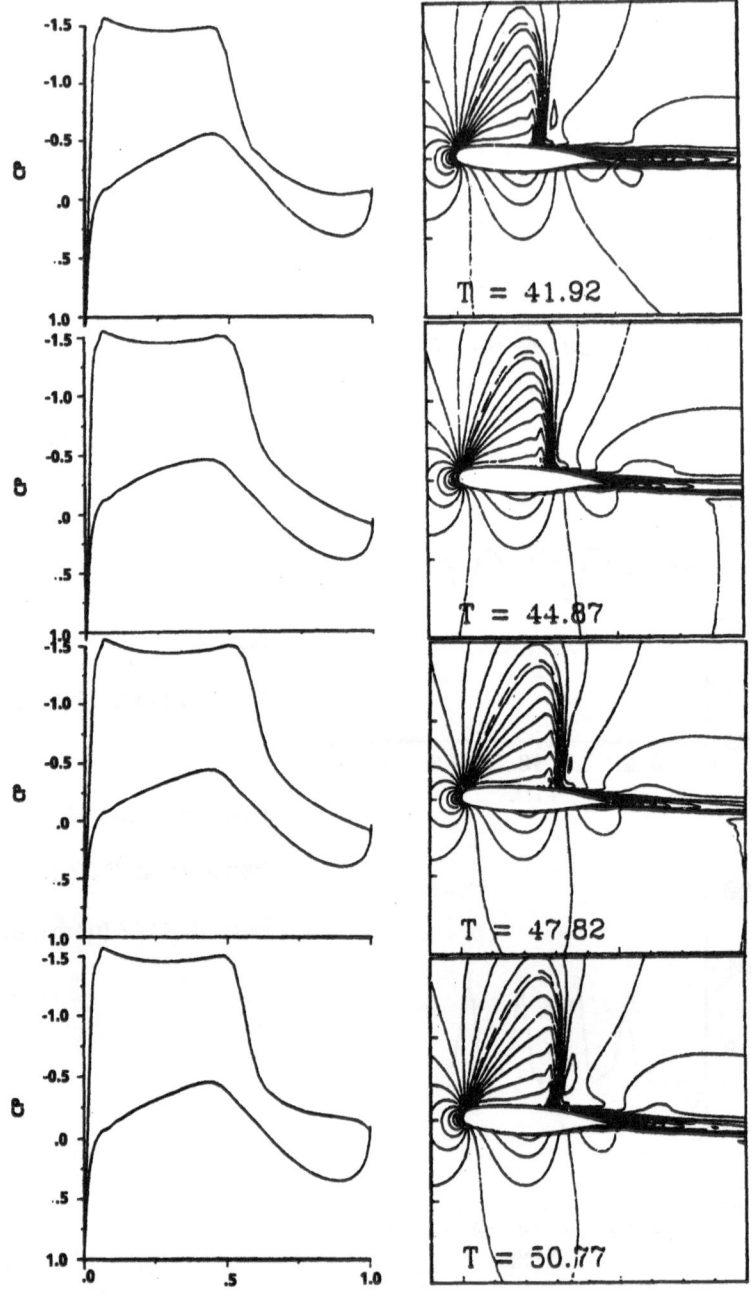

Fig.7b: Time-accurate Navier-Stokes computations for the airfoil NLR 7301.

Instantaneous pressures, Mach contours for a characteristic period.

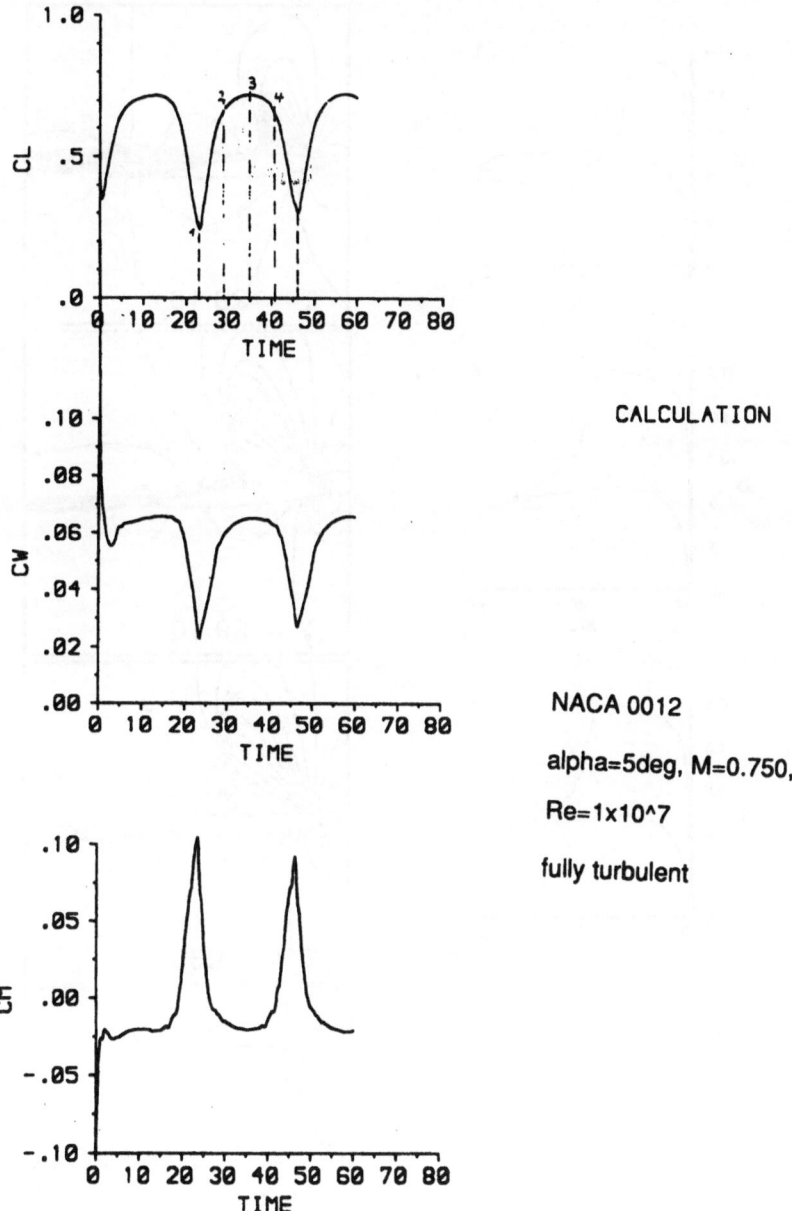

CALCULATION

NACA 0012

alpha=5deg, M=0.750,

Re=1x10^7

fully turbulent

Fig.8a: Time-accurate Navier-Stokes computations for the airfoil NACA 0012.

Time-dependent force and moment coefficients.

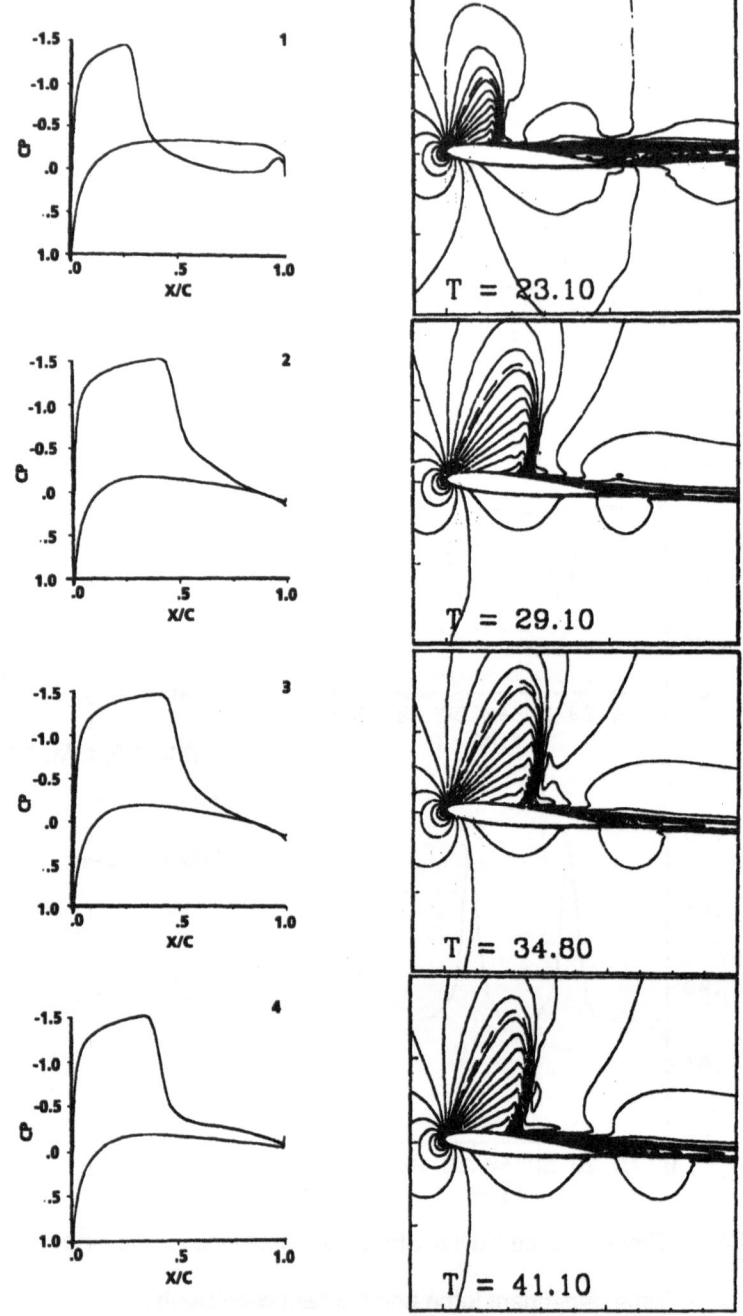

Fig.8b: Time-accurate Navier-Stokes computations for the airfoil NACA 0012. Instantaneous pressures, Mach contours for a characteristic period.

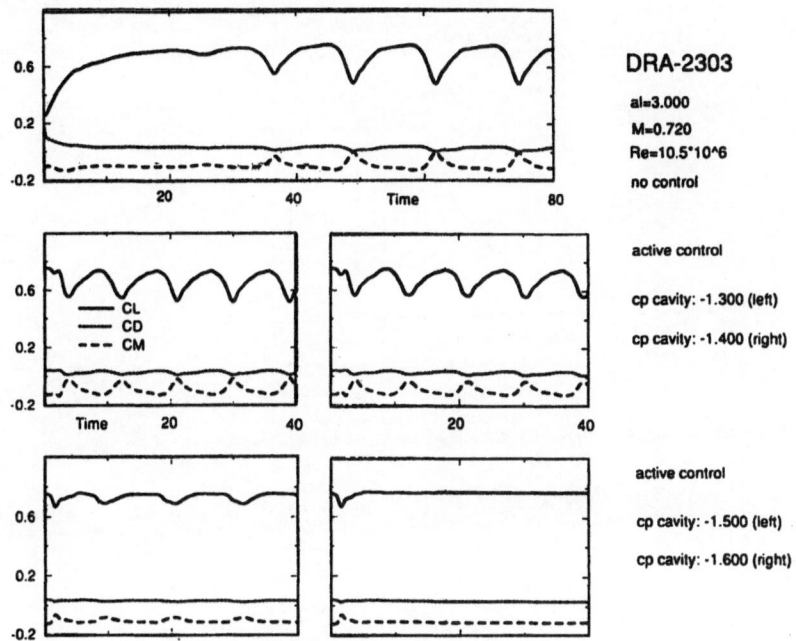

Fig.9 : Time-accurate Navier-Stokes computations for the airfoil DRA-2303
with and w/o active control

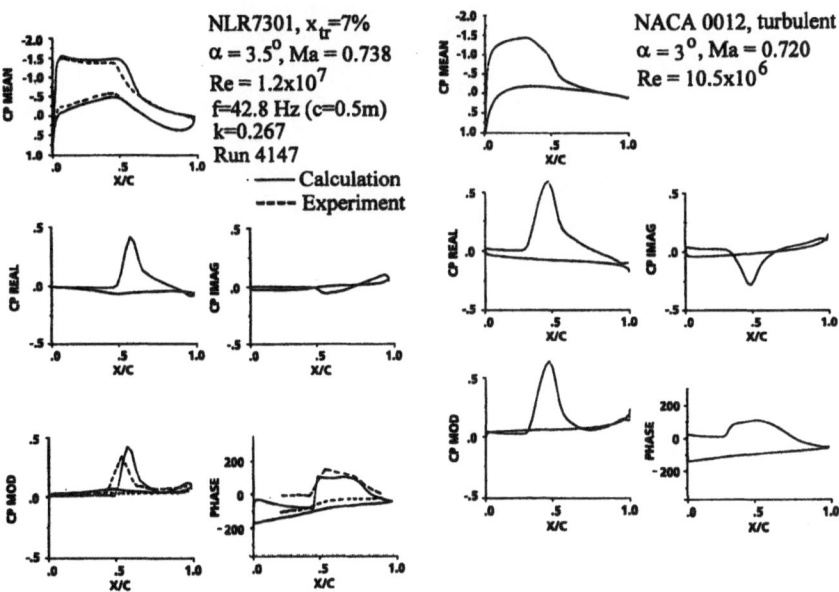

Fig. 10: Mean and first harmonic unsteady pressure distribution, modulus and phase;
left: airfoil NLR 7301, right: airfoil NACA 0012

Fig. . Laminar steady Navier-Stokes computations for the eddy ($Re = 250$) with passive/reactive control

Fig. . Mean and first harmonic unsteady pressure distribution and influence on...

17 SHOCK BOUNDARY LAYER INTERACTION CONTROL PREDICTIONS USING A VISCOUS-INVISCID INTERACTION PROCEDURE AND A NAVIER-STOKES SOLVER

G. Simandirakis, B. Bouras and K.D. Papailiou

NATIONAL TECHNICAL UNIVERSITY OF ATHENS
Laboratory of Thermal Turbomachines
P.O. Box 64069, 157 10 Athens, Greece

Summary: The present contribution describes two prediction methods for flows around transonic airfoils, including shock control devices. The whole work was carried out in the frame of the European Shock Control Investigation Project EUROSHOCK, and the global objective was the improvement of the flight performance, at transonic speed, in terms of cruise speed, fuel consumption and exhaust emissions for both laminar and turbulent wings. More specifically, the "passive" control of shock boundary layer interaction, where part of the solid surface of the airfoil is replaced by a porous surface over a shallow cavity, has been investigated with regard to improving the aerodynamic characteristics of supercritical airfoils.

17.1 Introduction

Two methods are presented here for the calculation of transonic flows around airfoils including shock control devices. The use of blowing and suction through a perforated plate underneath the shock boundary layer interaction zone is present in order to minimize the wave drag due to the presence of the shock, to prevent shock-induced separation and in order to alleviate the occurrence of the annoying buffet phenomenon. The aim of this paper is to investigate ways of predicting the complicated phenomena present in the shock boundary layer interaction region, with and without control of the shock wave, as well as to provide numerical results and comparisons with existing experiments.

First, an integral semi-inverse boundary layer calculation method is combined with a time marching Euler solver, resulting in a viscous-inviscid interaction computational tool, which can handle the flow around airfoils when shock waves are present. A substantial amount of work was performed subsequently in order to modify the relevant equations and procedure in order to account for shock control. The resulting calculation method is an iterative one, utilizing successfully the inviscid and the viscous computational components until convergence is reached. The integral momentum and kinetic energy equations, written for compressible flow, are used for the viscous computation. All normal fluctuation terms, important for the calculation of separated flows, are retained. The calculation method is operated in the direct mode until separation is approached or strong adverse pressure gradients are encountered. Then, the equations are solved in the "inverse" mode in order to avoid the well known singularity of the boundary layer equations at separation or the break down of the boundary layer calculation procedure because of unrealistic values of the streamwise pressure gradient imposed by the inviscid calculation. This calculation method copes successfully with separated flows, using a small number of iterations.

A second computational method is presented, subsequently, which solves the Navier-Stokes equations. The corresponding solver [1],[2],[3] is an explicit, time-marching, fractional-step one, which utilises multigrid acceleration.

This solver has been tested extensively and found both fast and accurate. Turbulence is modelled through either the Baldwin-Lomax or the two-equation k-e model (Jones and Launder). Both low-Reynolds number and wall function techniques can be employed. This solver was appropriately modified in order to incorporate a passive control mechanism. Both suction and blowing are possible in the shock boundary layer interaction zone, where the flow is fully turbulent.

The validity of several passive shock-control laws was tested. Firstly, as in most of the methods, the normal velocity component v_W was expressed as a function of the pressure difference across the perforated surface above the cavity in the shock boundary layer interaction region employing Darcey's law. Secondly, the calibration formula proposed by Poll[4] was used. Finally, the isentropic relations proposed by Abrahamson and Brower were utilised.

The main influence of the mass injections through the porous surface in the passive control region is restricted to the portion of the boundary layer close to the wall. Since it is preferable to treat this zone with a robust turbulence model, it was decided to use the algebraic Baldwin-Lomax one near the wall and leave the choice for the outer region, where either the same model or the more elaborate k-e model may be used, open.

Transpiration effects were taken into account by modifying van Driest's wall damping function. From the numerous modifications proposed in the literature, that of Cebeci[5] was selected and incorporated.

Numerous numerical tests were performed in order to compare the two methods with one another and both to available experimental results, first without and then with shock control.

17.2 Numerical Procedures

17.2.1 Viscous-inviscid interaction procedure

An integral semi-inverse boundary layer code was combined with a time marching Euler solver for the calculation of shock shear layer interaction. The calculation procedure is an iterative one, utilizing successively the inviscid and the viscous computations until convergence is reached.

Governing equations

In this section a brief description of the development of the integral equations will be presented. Details of the development of the method may be found in Ref.[6],[7],[8].

The basic equations are considered in a rotating frame of reference since it is easy to deduce them those valid for a stationary frame.

An axially symmetric orthogonal curvilinear system of coordinates is used. The continuity, the two momentum and the energy equations are expressed in this system. In addition, the turbulent kinetic energy equation is used for turbulent flow.

The above mentioned equations are simplified in the following way: (a) The stress terms, containing the coordinate system curvatures, as well as the terms containing the derivatives of stresses in the s-direction (parabolization) are neglected. However, all normal fluctuation terms are conserved. (b) Some simplifications are applied concerning the inertia terms of the normal momentum equation. However, the main effect of these terms which contribute to the variation of the static pressure normal to the flow direction is retained. (c) Following Lock and Firmin[9], a representative wall curvature is taken into account for each position (s). It is partly due to this curvature that the variation of the static pressure along a normal to the flow direction is accounted for. (d) The pressure term in the turbulent kinetic energy equation is neglected. At this level, the production term in the streamwise momentum equation for turbulent flow is substituted by the corresponding terms appearing in the turbulent kinetic energy equation.

With the above mentioned simplifications, the equations are written for the external inviscid and the real flow. Then, they are subtracted from each other, forming the corresponding deficit equations. These equations are then integrated along the normal to the streamwise direction resulting in the corresponding integral equation of momentum and kinetic energy.

The working equations of momentum and energy are further manipulated by introducing the non dimensional quantities L_k and X. L_k is a form factor, while X is the logarithm of a Reynolds number based on energy thickness.

We have to note that the two new variables L_k and X allow the representation of the state of the boundary layer at every step of the streamwise calculation in an image plane with L_k and X as ordinate and abscissa. This proves to be very helpful, since the basic characteristics of the boundary layer (whether laminar or turbulent, attached or separated, losses etc.) can be easily assessed.

In order to close the problem, a semi-empirical frame is necessary in order to relate all variables to the working variables L_k and X. Thus, for laminar flows the profile family is that of Falkner and Skan[10]. In the case of turbulent flows, the two-parameter velocity profile of Kuhn & Nielsen[11] was used. The above profiles work equally well for both attached and separated flows. Compressibility effects are taken into account through the Stewardson transformation[12] for the laminar profile, the Van Driest[13] transformation for the turbulent profile and the use of explicit formulae and/or calibration curves to modify the coefficients of the working equations.

In this way, the working equations take the final form

$$2F_1 \frac{dL_k}{d\Phi} + F_2 \frac{dq}{d\Phi} + F_3 \frac{dRe_{te}}{d\Phi} + F_4 \frac{Re_{me}}{d\Phi} = F_5 \qquad (1)$$

$$\frac{dX}{d\Phi} - 2 \frac{dL_k}{d\Phi} + F_6 \frac{dq}{d\Phi} + F_7 \frac{dRe_{te}}{d\Phi} = F_8 . \qquad (2)$$

In the above equations all the coefficients F_i are functions of the external flow Mach number (M_{ew}) and the working variables L_k and X. Dimensionless quantities q and Φ are defined through the relations

$$dq = \ln \frac{W_{sew}}{V_{ref}} \quad and \quad d\Phi = \frac{W_{sew}}{v_{ew}}$$

replacing in the working equations of momentum and energy the external velocity and the arc length along the airfoil surface. Re_{me} and Re_{te} are Reynolds numbers based on integral thicknesses containing normal fluctuation terms of turbulence[7].

According to the direct mode of calculation, the external velocity distribution $W_{sew}(s)$ is considered known, so Equations (1) and (2) are solved in terms of the working variables L_k and X.

Furthermore, from the distributions $L_k(s)$ and $X(s)$, all boundary layer quantities along the airfoil surface can be obtained. However, when separation is approached or strong adverse pressure gradients exist (such as in the shock region), it is preferable to solve equations in the "inverse" mode for two reasons: a) in order to avoid the well known singularity of boundary layer equations at separation, and b) to avoid failure of the boundary layer calculation because of unrealistic streamwise pressure gradients computed by the inviscid calculation. Following the inverse technique in the present method, the working equations (1) and (2) are solved simultaneously with a third equation which is the relation giving the transpiration velocity W_{new}, that is

$$\frac{W_{new}}{W_{sew}} - \frac{1}{\rho_{ew} W_{sew}} \frac{d}{ds} (\rho_{ew} W_{sew} \delta_1) = 0 . \tag{3}$$

After some algebraic manipulations, equation (3) takes the form

$$\frac{W_{new}}{W_{sew}} - \frac{d Re_1}{d\Phi} + \omega(\gamma - 1) M_{ew}^2 Re_1 \frac{dq}{d\Phi} = 0 \tag{4}$$

where Re_1 is a Reynolds number based on displacement thickness δ_1

$$Re_1 = \frac{W_{sew} \delta_1}{\nu_{ew}} .$$

During the inverse calculation, W_{sew} is considered known, while Equations (1), (2) and (4) are solved in terms of L_k, X and W_{sew}. In this way, the external velocity distribution $W_{sew}(s)$ is not any more an input to the shear layer calculation, but it is recalculated.

Before closing this section, it has to be noted that following the parabolic character of the system (1), (2) and (4), the working equations are solved at each point of the airfoil surface starting from the leading edge until the trailing edge is reached. The equations are discretized using a second order accurate numerical scheme and are solved using a Newton-Raphson technique[8].

The complete calculation procedure

As has already been pointed out, the whole procedure is an iterative one, utilizing successively the inviscid and the viscous computations until convergence is reached. When the direct calculation is performed, convergence is assessed by considering the changes after each iteration in the velocity W_{sew} and the boundary layer thickness. However, this convergence criterion cannot be applied inside the shock region where strong adverse gradients appear and the calculation is performed using the inverse technique. There, using the same distribution of the equivalent transpiration velocity W_{new}, two alternative estimations of the streamwise external velocity distribution result: a) W_{sew}^i from the direct inviscid calculation in the usual way, and b) W_{sew}^v from the inverse shear layer calculation. Convergence can be judged by examining the difference $|W_{sew}^i - W_{sew}^v|$, while a "correction" relation is needed to provide new values for W_{new} at each time step. Following Le Balleur's [14] stability analysis, we can get the relations for updating the non-dimensional transpiration velocity S defined as

$$S = \frac{W_{new}}{W_{sew}}$$

between each interaction cycle. Thus for $M_{ew} < 1$

$$S^{n+1} - S^n = \frac{\beta B \,\delta_3}{\nu B \,\delta_3 - \beta} \left\{ \frac{1}{W_{sew}^{v\,n}} \frac{d\,W_{sew}^{v\,n}}{ds} - \frac{1}{W_{sew}^{i\,n}} \frac{d\,W_{sew}^{i\,n}}{ds} \right\} , \tag{5}$$

while for $M_{ew} > 1$

$$S^{n+1} - S^n = \frac{\beta B^2 \delta_3^2}{\nu^2 B^2 \delta_3^2 + \beta^2} \left\{ \frac{1}{W_{sew}} \left[\frac{d^2 W_{sew}^{v\,n}}{ds^2} - \frac{d^2 W_{sew}^{i\,n}}{ds^2} \right] - \right.$$

$$\left. \frac{\beta}{B \,\delta_3} \frac{1}{W_{sew}} \left[\frac{d W_{sew}^{v\,n}}{ds} - \frac{d W_{sew}^{i\,n}}{ds} \right] \right\} \tag{6}$$

where n denotes the iteration number and ν n is defined as $\pi / \Delta s$ where Δs is the step size. Quantity β is defined as

$$\beta = \sqrt{|1 - M_{ew}^2|}$$

while B is a coefficient appearing in the momentum integral equation when it is written in the form[14]

$$\frac{d W_{sew}}{ds} = \alpha + \frac{W_{sew}}{\delta_3 B} S . \tag{7}$$

Using the working equations and the semi-empirical frame of the present method, B was found to be a function of the shape parameter L_k.

Summarizing, the complete algorithm reads as follows:

- An initial transpiration velocity (W_{new}^n) distribution along the airfoil surface is assumed and is given as input to the inviscid flow calculation in order to calculate the streamwise external velocity distribution around the airfoil.

- The results of the inviscid calculation are given as input to the viscous flow solver in order to calculate the shear layer development on pressure and suction side. In the shock region, where the inverse viscous calculation is performed, input to the viscous solver is the same transpiration velocity distribution given as input to the inviscid calculation.

- Distributions of the external flow streamwise velocity coming from the viscous calculation, W_{sew}^v and the velocity from the inviscid calculation, W_{sew}^i, are compared. From their difference, a new transpiration velocity is obtained using the relation (5) or (6).

- If convergence is achieved, that is $W_{sew}{}^{v} = W_{sew}{}^{i}$ in the shock region, and in the remaining region external velocity and shear layer thickness are equal to the ones of the previous iteration, the calculation procedure is completed. Otherwise, the same procedure is applied until convergence is achieved.

In order to meet the problem of boundary layer control, an adjustment of the semi-empirical background is necessary, since it is assumed that mass transpiration has no significant effect on the pressure variation normal the surface, something worth investigating in the future. The transpiration velocity on the surface for the equivalent inviscid flowfield, which is used in the interaction procedure, is obtained from the relation

$$w_{ne} = \frac{1}{\rho_{ew}} \left[\frac{\partial}{\partial s} \left(\rho_{ew} w_{sew} \delta_1 \right) \right] + \frac{\rho_w}{\rho_{ew}} w_{nw} . \tag{8}$$

The necessary modifications which must be implemented to the velocity profile family used in the shear layer calculation method are based on the following considerations: Coles[15] has shown that a simple relation

$$\frac{w_s^+}{(1/2)[(1 + w_{nw}^+ w_s^+)^{1/2} + 1]} = \frac{1}{k} \ln \left(n \frac{u_\tau}{v} \right) +$$

$$C + \frac{\Pi}{k} W \left(\frac{n}{\delta} \right) \tag{9}$$

may describe the velocity profile for flows with mass transfer with reasonable accuracy. In the present work, Equation (9) is used in combination with Kuhn & Nielsen's velocity profile family[11], taking the form

$$\frac{w_s^+}{(1/2)[(1 + w_{nw}^+ w_s^+)^{1/2} + 1]} = \frac{1}{k} \ln \left(1 + n \frac{|u_\tau|}{v} \right) -$$

$$\left(C_2 \frac{n|u_\tau|}{v} + C_1 \right) e^{an \frac{u_\tau}{v}} + C + \frac{1}{2} \frac{u\beta}{u_\tau} \left[1 - \cos \left(\pi \frac{n}{\delta} \right) \right] \tag{10}$$

with

$$C = C_0 + \frac{2K}{(1 + K w_{nw}^+)^{1/2} + 1} - K \tag{11}$$

and

$$C_0 = 5, \ K = 11, \ C_1 = C, \ k = 4, \ C_2 = 3.39 .$$

When w_{nw}, δ, u_τ are considered known, constant "a" in the exponential term in Equ.(10) may be calculated through the relation

$$\frac{1}{u_\tau} \frac{\partial w_s}{\partial n}\bigg|_w = \frac{u_\tau}{v_w} \qquad (12)$$

where the derivative $\partial w_s / \partial n|_w$ is obtained from the profile family[11]. Furthermore, the wake parameter u_β/u_τ is calculated from Equation (10) written for $n=\delta$, and the $W_s(n)$ distribution is finally obtained from the same relation.

Initial comparisons between experimental data[16] and Relation (10) for both positive and negative values of the transpiration velocity, w_{nw}, indicate that the above modified profile, family describes reasonably well the characteristics of velocity profiles on walls with mass transpiration so it can be used for the calculation of the various boundary layer quantities.

In order to calculate the coefficients of equations when the working variables L_k and X are known, an additional expression relating the quantities of the shear layer developing on the porous surface to the corresponding ones of a non-porous surface, is required. In the present work, the expression of Kays and Crawford[17] is used, that is

$$C_{fo} = C_{fo_s} \left[\frac{ln(1 + B_f)}{B_f} \right]^{1.25} (1 + B_f)^{0.25} \qquad (13)$$

where C_{fo} is the flat plate (zero pressure gradient) skin friction coefficient and C_{fos} the corresponding flat plate skin friction coefficient for a non porous surface. Quantity B_f is equal to $m_w/(C_{fo}/2)$, where

$$m_w = \frac{\rho_w W_{n_w}}{\rho_{e_w} W_{s_{e_w}}} . \qquad (14)$$

Once C_{fos} is considered known, C_{fo} is obtained from Equation (13) using an iterative Newton-Raphson procedure.

17.2.2 Navier-Stokes approach

In addition to the aforementioned work, a Navier-Stokes solver was properly modified to account for shock control in the shock boundary layer interaction region. The Navier-Stokes solver employed is an explicit, time-marching fractional-step solver for the calculation of two-dimensional, steady and unsteady compressible flows around airfoils. In the present method the conservative two-dimensional Navier-Stokes equations are split in a sequence of one-dimensional operators for the inviscid part, the viscous part and the source terms. Thus, instead of applying a pure two-dimensional scheme, a number of one-dimensional steps are executed for which numerical stability constraints are less strict.

The corresponding unsteady Favre-averaged Navier-Stokes equations, with a low-Reynolds k-e closure, are written in the form

$$\frac{\partial \vec{q}}{\partial t} + \frac{\partial \vec{f}}{\partial x} + \frac{\partial \vec{g}}{\partial y} = \vec{s} \tag{15}$$

where the unknown variable vector q, which consists of mean-flow and turbulence quantities, is given by

$$\vec{q} = \left[\rho, \rho u, \rho v, E_t, \rho k, \rho \varepsilon \right]^T$$

The flux vectors f and g are written in the form

$$\vec{f} = \left[\rho u, \rho u^2 + p_{eff} - \tau_{xx}, \rho u v - \tau_{xy}, u(E_t + p_{eff}) + q_x - u\tau_{xx} - v\tau_{xy}, \right.$$

$$\left. -\frac{\mu_t}{Pr_k} \frac{\partial k}{\partial x}, \rho u k - \left(\mu + \frac{\mu_t}{Pr_k} \right) \frac{\partial k}{\partial x}, \rho u \varepsilon - \left(\mu + \frac{\mu_t}{Pr_\varepsilon} \right) \frac{\partial \varepsilon}{\partial x} \right]^T$$

$$\vec{g} = \left[\rho v, \rho u v - \tau_{xy}, \rho v^2 + p_{eff} - \tau_{yy}, v(E_t + p_{eff}) + q_y - u\tau_{xy} - v\tau_{yy}, \right.$$

$$\left. -\frac{\mu_t}{Pr_k} \frac{\partial k}{\partial y}, \rho v k - \left(\mu + \frac{\mu_t}{Pr_k} \right) \frac{\partial k}{\partial y}, \rho v \varepsilon - \left(\mu + \frac{\mu_t}{Pr_\varepsilon} \right) \frac{\partial \varepsilon}{\partial y} \right]^T, \tag{16}$$

and the effective stress tensor is given by

$$\tau_{xx} = \mu_{eff} \left[2\frac{\partial u}{\partial x} - \frac{2}{3} \left(\frac{\partial (u)}{\partial x} + \frac{\partial (v)}{\partial y} \right) \right]$$

$$\tau_{xy} = \mu_{eff} \left(\frac{\partial u}{\partial y} + \frac{\partial v}{\partial x} \right)$$

$$\tau_{yy} = \mu_{eff} \left[2\frac{\partial v}{\partial y} - \frac{2}{3} \left(\frac{\partial (u)}{\partial x} + \frac{\partial (v)}{\partial y} \right) \right]$$

where p_{eff} should be replaced by p if the Baldwin-Lomax model is used. The heat flux vector is give by

$$(q_x, q_y) = \left(-Pr_{eff}^{-1} \frac{\partial e}{\partial x}, -Pr_{eff}^{-1} \frac{\partial e}{\partial y} \right)$$

where the effective Prandtl number, Pr_{eff}, is defined as $Pr_{eff}^{-1} = \gamma \left(\frac{\mu}{Pr} + \frac{\mu_t}{Pr_t} \right)$.

When the Jones and Launder low Reynolds number $k - \varepsilon$ turbulence model is used, non-zero source term entries for the k and ε equations also appear. Thus, the source term vector s takes the form

$$\vec{s} = \begin{bmatrix} 0, & 0, & 0, & 0, & s_k, & s_\varepsilon \end{bmatrix}^T$$

with

$$s_k = \left(P - \rho\varepsilon - 2\mu\left(\frac{\partial\sqrt{k}}{\partial n}\right)^2\right) \ , \ s_\varepsilon = \left(c_{\varepsilon 1}P\frac{\varepsilon}{k} - c_{\varepsilon 2}f_2\rho\frac{\varepsilon^2}{k} + 2\frac{\mu\mu_t}{\rho}\left(\frac{\partial^2 V}{\partial n^2}\right)^2\right) . \quad (17)$$

The production term P is given by

$$P = \mu_t\left(\frac{\partial u}{\partial y} + \frac{\partial v}{\partial x}\right)^2 + 2\mu_t\left[\left(\frac{\partial u}{\partial x}\right)^2 + \left(\frac{\partial v}{\partial y}\right)^2\right] - \frac{2}{3}\mu_t\left(\frac{\partial u}{\partial x} + \frac{\partial v}{\partial y}\right)^2 - \frac{2}{3}\rho k\left(\frac{\partial u}{\partial x} + \frac{\partial v}{\partial y}\right) ,$$

and the turbulent viscosity, m_t, is obtained from the Prandtl-Kolmogorov relation

$$\mu_t = c_\mu f_\mu \frac{\rho k^2}{\varepsilon}$$

For the selected model, the constants and functions used are given by

$$f_2 = 1 - 0.3\exp(-Re_t^2) \ , \quad f_\mu = \exp\left[\frac{-3.4}{\left(1 + \dfrac{Re_t}{50}\right)^2}\right] \ , \quad Re_t = \frac{\rho k^2}{\mu\varepsilon} .$$

The Baldwin-Lomax model is applied following the two-layer approach. Nevertheless, important modifications are introduced concerning (a) the calculation of the wall shear stress close to the separation point, (b) the calculation of the maximum of the function F(y) when the latter displays two peaks and (c) the prediction of transition.

All governing equations are transformed to the body-fitted coordinate system (x,h) defined by the grid lines of the grid generated. In the computational plane (ξ, η) the resulting equations may be cast in the following conservative form

$$\frac{\partial\vec{Q}}{\partial t} + \frac{\partial\vec{F}}{\partial\xi} + \frac{\partial\vec{G}}{\partial\eta} = \vec{S} \tag{18}$$

where

$$\vec{Q} = J\vec{q}$$

$$\vec{F} = J\left[\xi_x \vec{f} + \xi_y \vec{g}\right]$$

$$\vec{G} = J\left[\eta_x \vec{f} + \eta_y \vec{g}\right]$$

(19)

$$\vec{S} = J\vec{s}.$$

According to the fractional-step concept, the time evolution of the unknown vector array Q is obtained by applying the sequence of operators, which may be cast in the following symbolic form

$$\vec{Q}^{n+2} = L_\xi^H L_\eta^H L_\xi^P L_\eta^P L^{ST} L^{ST} L_\eta^P L_\xi^P L_\eta^H L_\xi^H \vec{Q}^n.$$

(20)

A double and inverse sequence of the one-dimensional operators leads to a second order accuracy in time, while the calculated quantities have a physical meaning only at the end of a $2\Delta t$ time interval (i.e., going from the n to n+2 iteration level), Abarbanel and Gottlieb[18]. The predictor-corrector MacCormack scheme is used to handle the hyperbolic and parabolic operators. With regard to the treatment of the source terms, a semi-implicit scheme is used to ensure numerical stability. Thus, for the solution of the intermediate step, corresponding to the L^{ST} operator, namely

$$\frac{\partial \vec{Q}}{\partial t} = \vec{S},$$

(21)

the right-hand-side array S splits in S_+, which contains the positive source terms, and S_- containing the negative ones. The negative part is Newton-linearized and the delta form of Equation (21) can be written as

$$\left[I - \Delta t^* \left(\frac{\partial \vec{S}_-}{\partial \vec{Q}}\right)^{n+\frac{4}{5}}\right]\Delta \vec{Q}^{n+\frac{4}{5}} = \Delta t^* \vec{S}^{n+\frac{4}{5}}.$$

(22)

Extra dissipation terms are explicitly added Simandirakis[1], Vassilopoulos - Simandirakis[19], to the solution array. Q at the end of a complete calculation period, corresponding to a time interval of $2\Delta t$, is as follows

$$\vec{Q}^{n+2} = \vec{Q}^{n+2} + \vec{D}_\xi + \vec{D}_\eta.$$

(23)

In the context of the present project, in order to incorporate in the solver a passive control mechanism and since the flow is fully turbulent in the shock boundary layer interaction zone, necessary modifications were performed in the solver (wall boundary conditions, turbulence modelling) in order to account for wall suction or blowing.

17.2.3 Shock-control laws

In most of the methods proposed, the normal velocity component, v_w, is expressed as a given function of the pressure difference across the perforated surface above the cavity. Thus, the ventilation velocity is given by

$$v_w = f(\Delta p) \tag{24}$$

where

$$\Delta p = p_c - p_w(x)$$

and $p_w(x)$ is the local pressure of the external flow and $f(\ )$ is a given function.

Darcy's law

The ventilation velocity is given by an expression developed from Darcy's law (linear equation)

$$v_w = \alpha \Delta p \ . \tag{25}$$

Poll's calibration formula

A first calibration technique that was used was the one proposed by Poll [4]. The ventilation velocity according to Poll's formula is given as the solution of the quadratic equation

$$Y = \frac{1}{K}\left[40.7x + 1.95x^2\right] \quad where \quad Y = \frac{\Delta p \rho_h d_h}{\mu_h^2 \left(\dfrac{d_h}{s_h}\right)^2} \quad and \quad x = \frac{\dot{m}_h}{\mu_h s_h} , \tag{26}$$

with Dp: pressure drop across the surface, d_h: hole diameter, s_h: plate thickness, μ_h: hole dynamic viscosity, ρ_h: hole density and K as a factor for the effective hole diameter.

Isentropic law

Abrahamson and Brower proposed to use isentropic relations in order to determine the ventilation velocity v_w. The ventilation velocity v_w may be defined as

$$\frac{v_w}{\alpha_t} = \frac{\dot{m}_{surf}}{\rho_w \alpha_t \ A_{surf}} = \frac{\sum\limits_{i}^{n_h} \dot{m}_i}{\rho_w \alpha_t \ A_{surf}} \tag{27}$$

where n_h: number of holes, α_t: total speed of sound.

Defining also the porosity of the surface as $\quad n_p = \dfrac{\sum\limits_i^{n_h} A_i}{A_{surf}}$ (28)

and the mass flow loss coefficient n_v as

$$n_v = \frac{\dot{m}_i\ real}{\dot{m}_i\ theor}$$ (29)

the ventilation velocity v_w is finally given by

$$\frac{v_w}{\alpha_t} = \frac{1}{\rho_w/\rho_t}\ n_p n_v\ \frac{\dot{m}_{ith}}{\rho_t \alpha_t A_{h_i}} = \frac{1}{\rho_w/\rho_t}\ n_p n_v\ S_i$$ (30)

with

$$S_i = \sqrt{\frac{2}{\gamma-1}\frac{\rho_w}{\rho_t}\frac{p_c}{p_t}\left(\frac{p_w}{p_c}\right)^{-1/\gamma}\left[\left(\frac{p_{tw}}{p_c}\right)^{\frac{\gamma-1}{\gamma}} - 1\right]}\ ,\ for\ suction\quad p_w > p_c\ (31a)$$

and

$$S_i = \sqrt{\frac{2}{\gamma-1}\frac{\rho_c}{\rho_t}\frac{p_w}{p_t}\left(\frac{p_c}{p_w}\right)^{-1/\gamma}\left[\left(\frac{p_c}{p_w}\right)^{\frac{\gamma-1}{\gamma}} - 1\right]}\ ,\ for\ blowing\quad p_c > p_w\ .\ (31b)$$

The unknown in the above formulation is the cavity pressure which is derived by the total mass flow through the porous surface. Thus $Q = \int\limits_S \rho_w v_w dS = 0 \Rightarrow \int\limits_S \dfrac{\rho_w}{\rho_t}\dfrac{v_w}{\alpha_t} dS = 0$. (32)

Turbulence modelling

According to Chen [20] and Bur [21,22], the main influence of the mass injection through the porous surface in the passive control region is restricted to the portion of the boundary layer very close to the wall. Also, as many authors suggest, it is preferable to treat this zone with a robust turbulence model such as the algebraic Baldwin-Lomax model. In the case that a more elaborate turbulence model is used, such as the k - ε model (already incorporated in the code), it is proposed to maintain it for the outer part of the boundary layer and switch to an algebraic one for the inner zone.

Having analyzed the problem that transpiration effects have to be taken into account in the turbulence modelling, numerous modifications for the Van Driest wall damping function have been proposed in the literature (Cebeci , Kinney and Sparrow , Chen , Rotta) . In the present study the extension proposed by Cebeci [5] is incorporated. So the Van Driest wall damping function is generalized for the simultaneous presence of mass transfer through the wall and a longitudinal pressure gradient.

Hence, the damping function D as used in the vicinity of solid walls is now given by

$$D = 1 - \exp\left(-\frac{y^+}{A^+}\right) \tag{33}$$

where

$$A^+ = 26\left\{(11.8\,Up^+) - \frac{p^+}{U\,p^+}\left[\exp(11.8U\,p^+) - 1\right]\right\}^{-1/2} \tag{34}$$

and

$$Up^+ = \frac{Up}{u_\tau}\;,\quad p^+ = -\frac{v}{\rho u_\tau^3}\frac{dp}{ds}\;. \tag{35}$$

17.3 Results and Discussion

17.3.1 Viscous-inviscid interaction results

In order to assess the prediction ability of the VII calculation method, two test cases were examined concerning the transonic flow around the DRA-2303 airfoil. The computational grid used for the Euler solver had 256x31 nodes.

For the first test case, data point 271 w/o control, the freestream Mach number is equal to 0.6816 the angle of attack is 1.068° and the Reynolds number, based on chord and freestream conditions is 18.97x10^6. The pressure coefficient C_p over the airfoil surface is presented in Figure 1 together with the experimental results and the initial pressure distribution resulting from the Euler solver without viscous effects. In Figure 2 the distributions of displacement thickness δ_1 and the momentum thickness δ_2 are presented; Figure 3 shows the distribution of the skin friction coefficient. The abrupt thickening of the boundary layer due to the shock can be seen very clearly from Figure 2 where the displacement thickness δ_1 increases considerably in the shock region.

It has to be noted that, as can be seen from the above figures, the prediction of the method is very good; the number of iterations needed for the achievement of convergence was not more than ten.

The second test case computed by the present method concerns the shock boundary layer interaction with passive control on the same airfoil. The freestream Mach number is equal to 0.68070 and the angle of attack is 1.8668° and the Reynolds number is 18.98x10^6. The cavity extended from 50% to 60% of the chord. The transpiration velocity, which was obtained using the isentropic law, was given as input to the viscous solver at every calculation step. The pressure coefficient C_p over the airfoil surface is presented in Figure 4 together with the experimental results. As can be seen, the prediction is quite reasonable, although some irregularities appear in the cavity region. These irregularities appear, because the ventilation velocity was not smooth enough, and the problem could not be overcome with the use of a relaxation factor during the iterative calculation procedure as as thought initially. However, the effect on the boundary layer is not so strong, as can be seen from Figures 5 and 6, where the displacement and momentum thickness distributions as well as the skin friction coefficient distribution is presented.

The number of iterations needed for convergence increased to 22 for this case because of the additional calculation of the ventilation velocity which did not remain constant during the viscous-inviscid interaction procedure.

17.3.2 Navier-Stokes results

DRA-2303: Data Point 271, w/o shock control

The first case concerns the transonic flow around the DRA - 2303 airfoil. It is a steady test case with a freestream Mach number of M_∞ =0.6816 and an angle of attack α =1.068°. The Reynolds number, based on chord and freestream conditions, is Re=18970000. Transition was fixed at 5% of the chord. The computational grid has 265x45 nodes.

Iso-Mach contours of the flow are presented in Figure 7. The pressure coefficient C_p over the airfoil surface is presented in Figure 8. In the same figure results obtained using our VII code and the experimental data are presented. It is clearly seen that both predictions may be termed very reasonable compared to the experiment. Figure 9 shows the skin friction coefficient along the airfoil surface using both methods. The Navier-Stokes computational grid maximum value of y^+ was kept below 10, and wall function techniques were used. This probably explains the differences between the skin friction coefficient predictions.

The computed lift coefficients equal C_l=0.535 C_l=0.544 and the drag coefficients C_d=0.00944 and C_d=0.0112, and the moment coefficients C_m=-0.0856 and C_m=-0.088 for the Navier-Stokes and the VII solution, respectively.

DRA - 2303: Data Point 289, w/o shock control

The second case concerns the transonic flow around the same airfoil. It is a steady test case with a freestream Mach of number M_∞ =0.6795 and an angle of attack α =2.5070°. The Reynolds number, based on chord and freestream conditions, is Re=18.907.000. The computational grid has 265x45 nodes. The pressure coefficient C_p over the airfoil surface is presented in Figure 10. In the same figure results obtained using our VII code and the experimental data are presented. The computed lift coefficients equal C_l=0.77 and C_l=0.76, the drag coefficients C_d=0.013 and C_d=0.014, and the moment coefficients C_m=-0.088 and C_m=-0.086 for the Navier-Stokes and the VII solution, respectively.

DRA - 2303: Data Point 1008, with shock control

The DRA test cases were first solved using Poll's calibration formula. The perforated plate was of thickness 1mm and the hole diameter was 0.076mm. The control surface extended from 50% to 60% of the chord. If it was found, from extensive experimental calibrations[23] that the factor K was 1.0 (the holes were behaving like ideal holes). Finally, the porosity of the surface was 8%.

The third case is a shock control test case and concerns the same airfoil. It is also a steady case with a freestream Mach number of M_∞ =0.6807, a Reynolds number of Re=18.984×10^6, and an angle of attack α =1.0660°. Transition was fixed at 5% of the chord. The computational grid has 265x41 nodes and a portion of it, close to the airfoil, is presented in Figure 11. In the same figure the clustering of the grid lines above the cavity is clearly seen. The cavity extended from 50% to 60% of the chord and we make use of 20 equidistant nodes in the shock-control region. Iso-Mach contours of the flow are also presented in the same figure.

Figure 12 shows the computed pressure coefficient along the airfoil surface. Also in the same figure the experimental data are presented. The computed lift coefficient equals C$_l$=0.522, the drag coefficient C$_d$=0.00956, and the moment coefficient C$_m$=-0.08375. The first important result from inspecting the pressure coefficient distribution is the decrease of the shock wave drag. The blowing just before the shock wave seems very effective in spreading the compression. On the other hand, an increase in viscous drag is observed and that results in a slight increment in total drag.

DRA - 2303: Data Point 1031, with shock control

The fourth case is also a shock control test case and concerns the same airfoil. It is also a steady flow case with a freestream Mach number of M_∞ =0.6806, a Reynolds number of Re=18.982×10^6, and an angle of attack α =2.71°. The computational grid was the same. The cavity extended again from 50% to 60% of the chord and we used 20 nodes in the shock-control region. Figure 13 shows the computed pressure coefficient along the airfoil surface. The skin friction coefficient is presented in the Figure 14. The computed lift coefficient equals C$_l$=0.732, the drag coefficient C$_d$=0.013, and the moment coefficient C$_m$=-0.074.

DRA - 2303: Unsteady test case, Data Point 543, w/o shock control

The next test case is an unsteady one. The freestream Mach number equals M_∞ =0.7022, a Reynolds number of Re=10.5×10^6, and an angle of attack α =2.405°. The computational grid has 251x47 nodes. Transition was fixed at 5% of the chord on both surfaces. The time evolution of the lift coefficient, drag coefficient, and moment coefficient is presented in Figure 15. No shock control is implemented, but the clustering of the grid lines in the shock region is preserved for better resolution of the shock boundary layer interaction. For time saving reasons, it was found much more preferable to treat this test case with a robust turbulence model such as the algebraic Baldwin-Lomax model. The computation switched to the time-consistent mode when the first unsteadiness appeared. After a transitional time interval, a periodic flow was established. Snapshots of the iso-Mach contours of the flow for several successive time phases are presented in Figure 16. Figure 17 shows the computed pressure coefficients and the skin friction coefficients along the airfoil surface. Considering the results a buffet-type flow is observed which is characterized by serious shock oscillations.

The aim of the unsteady calculation was to investigate the capability of the application of passive control to alleviate or to postpone the buffet onset, and, of course, to exercise our solver in unsteady buffeting situations.

A first run was carried out by using a rather coarse grid and utilizing wall function techniques. It resulted in a frequency of 36.23 Hz which was very close to the experimental value of 36 Hz. Looking at the 0th harmonic of the pressure coefficient distribution, quite a large discrepancy was observed with regard to the experimental values. Another run utilizing a fine grid gave much better results (Figure 18), while the resulting frequency was unchanged (35.9 Hz). Figures 19 and 20 show the real and the imaginary part of the 1st Harmonic of the pressure coefficient distribution.

VA - 2: polar test case with and without shock control

The final test case concerns the VA-2 airfoil. The cavity extended from 49.5 to 64.5% of the chord. We make use of 40 equidistant nodes in the shock-control region. The perforated plate thickness was 1.2mm, the nominal hole diameter was 0.5mm. The porosity of the plate equals 8%.

The flow was steady with a freestream Mach number of M_∞=0.76 (0.74 corrected), and the Reynolds number equals 2.5×10^6. Transition was fixed at 30% of the chord on the suction side, and at 25% of the chord on the pressure side. The computational grid has 265x41 nodes.

A first run was carried out without shock control. Figure 21 shows the computed pressure coefficients along the airfoil surface for several angles of attack:

$$\alpha = 1.35 \ CL = 0.72954 \ CD = 0.01502$$
$$\alpha = 1.50 \ CL = 0.76154 \ CD = 0.01663$$
$$\alpha = 2.00 \ CL = 0.86019 \ CD = 0.02355$$
$$\alpha = 2.30 \ CL = 0.91013 \ CD = 0.02742$$
$$\alpha = 2.50 \ CL = 0.93980 \ CD = 0.03258$$
$$\alpha = 2.75 \ CL = 0.97951 \ CD = 0.03681$$
$$\alpha = 3.50 \ CL = 0.91994 \ CD = 0.04244$$

Figure 22 shows the computed pressure coefficients along the airfoil surface for several angles of attack when shock control is implemented above the cavity zone. It was observed that numerically the boundary layer was much more sensitive to separation in the presence of the perforated plate, and unsteadiness occurred when the shock moved upstream of the cavity region.

$$\alpha = 0.95 \ CL = 0.6299 \ CD = 0.01336$$
$$\alpha = 1.00 \ CL = 0.6409 \ CD = 0.01360$$
$$\alpha = 1.35 \ CL = 0.7167 \ CD = 0.01473$$
$$\alpha = 1.50 \ CL = 0.7345 \ CD = 0.01619$$
$$\alpha = 1.65 \ CL = 0.7721 \ CD = 0.01718$$
$$\alpha = 2.00 \ CL = 0.8208 \ CD = 0.02158$$
$$\alpha = 2.30 \ CL = 0.8557 \ CD = 0.02461$$
$$\alpha = 2.50 \ CL = 0.8650 \ CD = 0.02854$$
$$\alpha = 3.40, \ unsteady$$

Drag coefficients are presented in Figure 23. A small decrease of the drag at high angles of attack is observed when shock control is implemented in the cavity region; however, lift is also reduced by control so that at equal lift drag is generally still slightly increased.

17.4 Conclusion

In the first part of the paper a steady viscous-inviscid interaction procedure was presented based on an already existing semi-empirical boundary layer code and an Euler solver. The method gave very satisfactory results for the no-control test cases. For the test cases with shock control, irregularities seem to exist in the distribution of the basic calculation quantities which could not quite be overcome by the use of a relaxation factor applied to the transpiration velocity.

In the second part, a Navier-Stokes solver, properly modified, was employed to compute a number of selected test cases. All the results obtained were very reasonable, compared to the experimental data, for both steady and unsteady situations. Additionally, comparisons between the Navier-Stokes and the viscous-inviscid interaction results were provided.

Summarizing, it can be said that the use of a porous surface below the shock wave region in order to implement passive shock control results in a significant increase in total drag. This arises from the large increase of the viscous drag due to the thickening of the boundary layer, mainly due to blowing through the porous surface in front of the shock wave. The benefits of applying passive shock control seem to be marginal and strongly dependent upon the airfoil geometry and also the angle of attack.

17.5 References

1. Simandirakis G., "Numerical Solutions of Navier-Stokes Equations for Transonic Flows Inside Turbine Bladings", Ph.D. Thesis, Athens, February 1992.
2. Simandirakis G., Dejean F., Vassilopoulos Ch., Giannakoglou K., Papailiou K.D., "Steady and Unsteady Two-Dimensional Flow Calculations Using an Explicit Fractional-Step Algorithm", Proc. of the ECCOMAS 94 Conference, 5-8 September, Stuttgart, Germany, pp. 701-710.
3. Giannakoglou K., Simandirakis G., Papailiou K.D., "Turbine Cascade Calculations Through a Fractional-Step Navier-Stokes Algorithm", ASME Paper 91-GT-55.
4. Paper 92-02-028 of the First European Forum on Laminar Flow Technology, Hamburg 1992.
5. Cebeci T., Smith A.M.O., "Analysis of Turbulence Boundary Layers", Academic Press, London, 1974.
6. Papailiou K.D., "Le Foll's Method and the Calculation of Attached and Separated Two-Dimensional Boundary Layers", VKI LS 1981-1, (1981).
7. Papailiou K.D., Bouras B., "Arbitrary Blade Section Design Based on Viscous Considerations", VKI LS 1990, AGARD-R-780, (1990).
8. Bouras V., "Optimization of Turbomachinery Blades Based on Shear Layer Theory", PhD Thesis, NTUA, Athens, June 1993.
9. Lock R.C., Firmin M.C.P., "Survey of Techniques for Estimating Viscous Effects in External Aerodynamics", IMA Conference, London, (1981).
10. Falkner V.H., Skan S.W., "Some Approximate Solutions of the Boundary Layer Equations", Phil. Mag. 12, 865 (1931), ARC R and M 1314, 1930.
11. Kuhn J.D., Nielsen J.N., "Prediction of Turbulent Separated Boundary Layers", AIAA Paper No 73-663, (1973).
12. Stewardson K., "Correlated Incompressible and Compressible Boundary Layer", Proc. Roy. Soc., London A 200, pp. 84-100, (1949).

13. Van Driest E.R., "Turbulent Boundary Layers in Compressible Fluids", J. Aero Sciences, Vol. 18, (1951).

14. Le Balleur J.C., "Calcus Couplés Visqueux-Non Visqueux, Incluant Decollements et Ondes de Choc en Ecoulement Bidimensionnel, AGARD LS-94, (1978).

15. Coles D., "A Survey of Data for Turbulent Boundary Layers with Mass Transfer", AGARD Conf. Proc. 93, Turbulent Shear Flows, (1971).

16. Andersen P.S., Kays W.M., Moffat R.J., "Experimental Results for the Transpired Turbulent Boundary Layer in an Adverse Pressure Gradient", J. Fluid Mech., Vol. 69, Part 2, pp. 353-375, (1975).

17. Kays W.M., Crawford M.E., "Convective Heat and mass Transfer", 2nd ed., McGraw-Hill, New York, (1980).

18. Abarbanel S., Gottlieb D., "Optimal Time Splitting for Two- and Three-Dimensional Navier-Stokes Equations with Mixed Derivatives", Journal of Computational Physics 41, 1981.

19. Vassilopoulos Ch., Simandirakis G., Giannakoglou K., Papailiou K.D., "Losses Prediction in Axial Flow Compressor Cascades using an Explicit k-e Navier-Stokes Solver", AGARD-CP-571, 1995.

20. Chung-Lung Chen, Chuen-Uen Chow et al., "Computation of Viscous Transonic Flow over Porous Airfoils", Paper AIAA-87-0359, 1987.

21. Bur R., "Etude Fondamentale sur le Controle Passif de l'Interaction onde de Choc/Couche Limite Turbulente en Ecoulement Transsonique", Thése de Doctorat de l'universitè Paris 6, 1992.

22. Bur R., "Passive Control of a Shock Wave/Turbulent Boundary Layer Interaction in a Transonic Flow", Rech Aérosp. n° 1992-6.

23. Dargel G., "Private Communication".

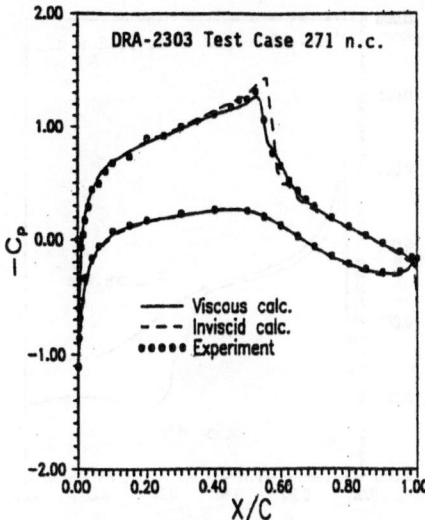

Fig. 1 Pressure coefficient C_p along DRA-2303 airfoil; Case 271

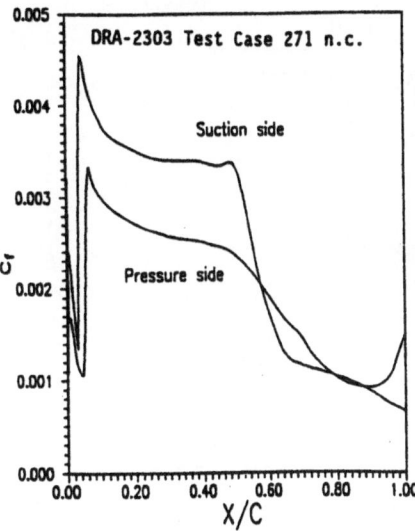

Fig. 2 Displacement and momentum thickness, δ_1 and δ_2, along DRA-2303 airfoil; Case 271

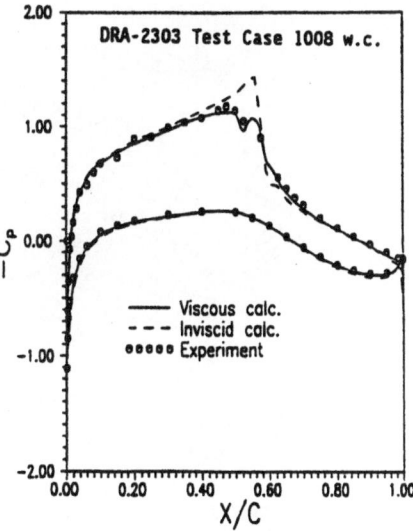

Fig. 3 Skin friction coefficient along DRA-2303 airfoil; Case 271

Fig. 4 Pressure coefficient C_p along DRA-2303 airfoil; Case 1008

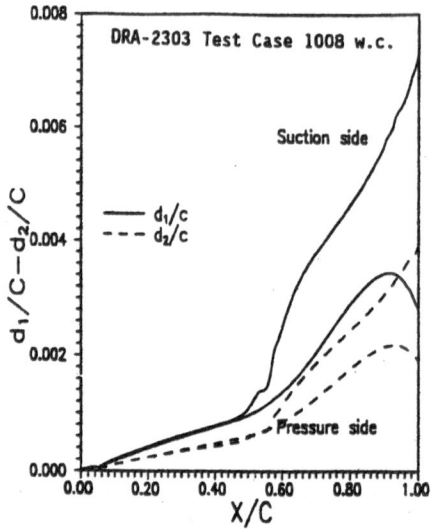

Fig. 5 Displacement and momentum thickness, δ_1 and δ_2, along DRA-2303 airfoil; Case 1008

Fig. 6 Skin friction coefficient along DRA-2303 airfoil; Case 1008

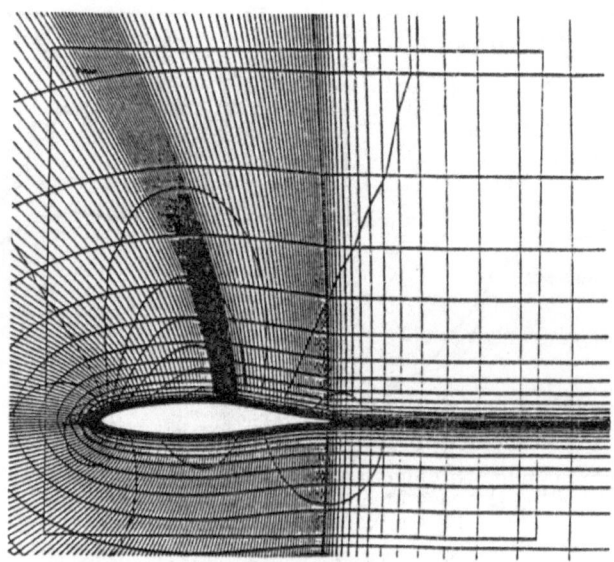

Fig.7 Part of the C-type grid used for the computation around the DRA-2303 isolated airfoil

284

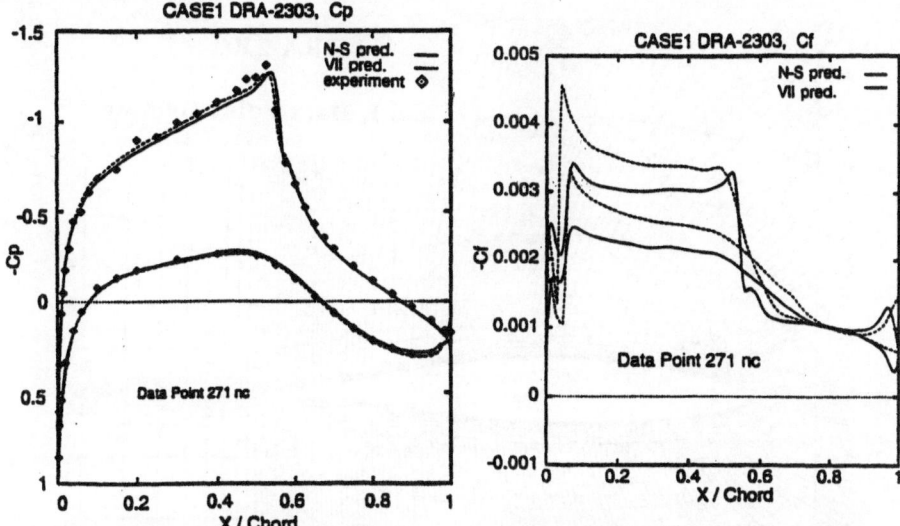

Fig. 8 Pressure coefficient C_p along DRA-2303 airfoil, Data Point 271, nc.

Fig. 9 Skin friction coefficient along DRA-2303 airfoil; Data point 271, nc.

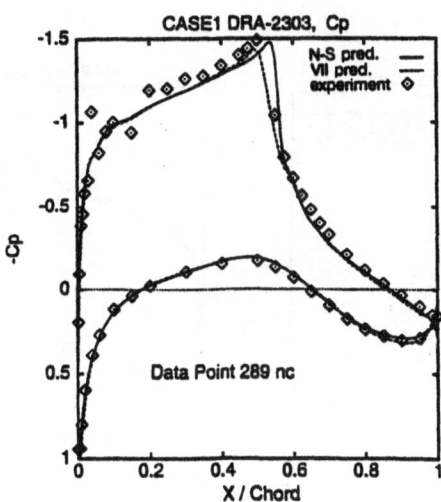

Fig.10 Pressure coefficient C_p along DRA-2303 airfoil, Data Point 289, nc.

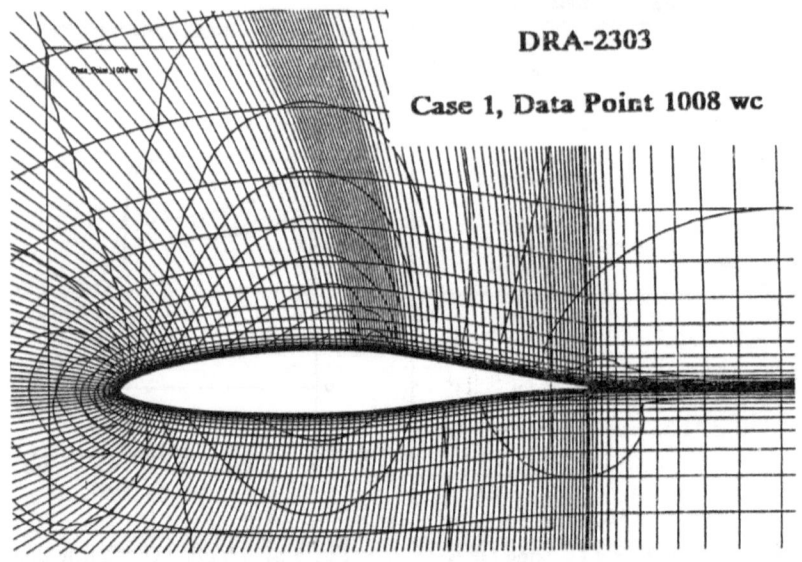

Fig. 11 Part of the C-type grid used for the computation around the DRA-2303 isolated airfoil

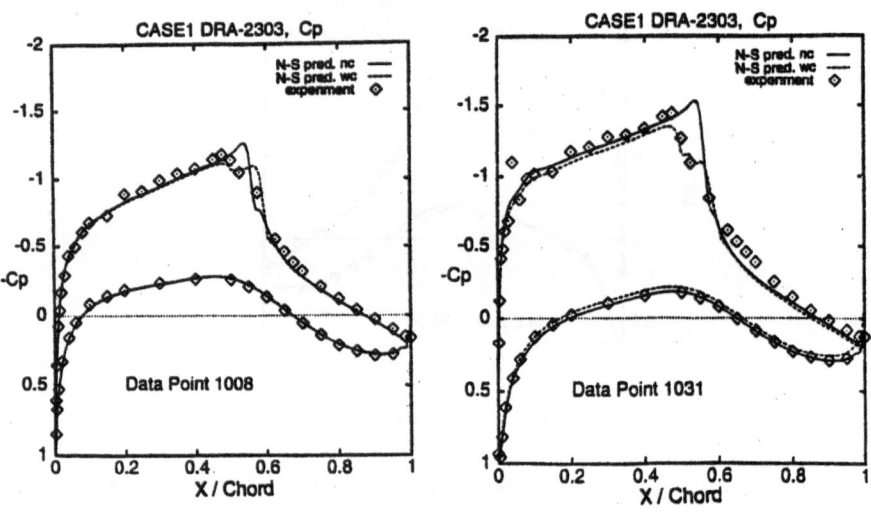

Fig. 12 Pressure coefficient C_p along DRA-2303 airfoil; Data Point 1008, wc.

Fig. 13 Pressure coefficient C_p along DRA-2303 airfoil; Data Point1031, wc.

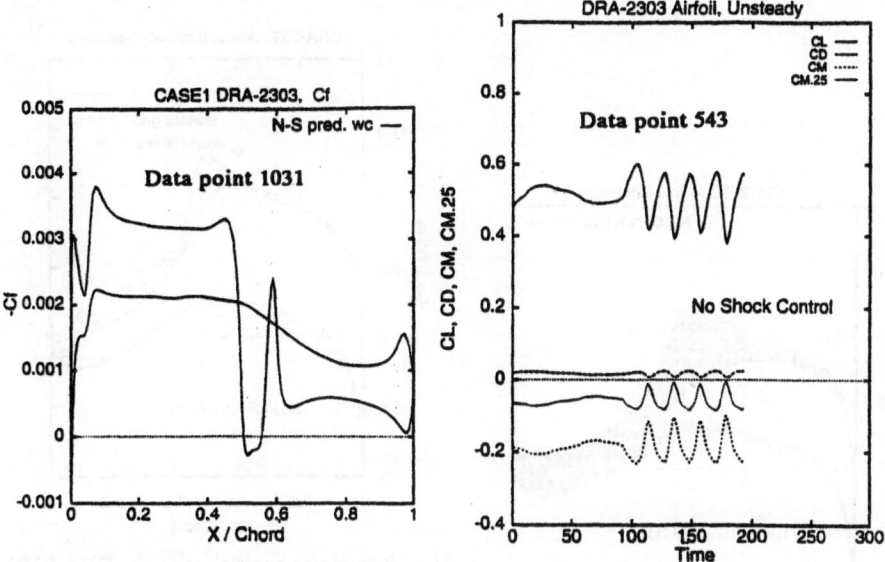

Fig. 14 Skin friction coefficient C_f along DRA-2303 airfoil; Data point 1031, wc.

Fig. 15 Time evolution of the lift, drag and moment corfficient

DRA-2303 Unsteady, Data Point 543 nc

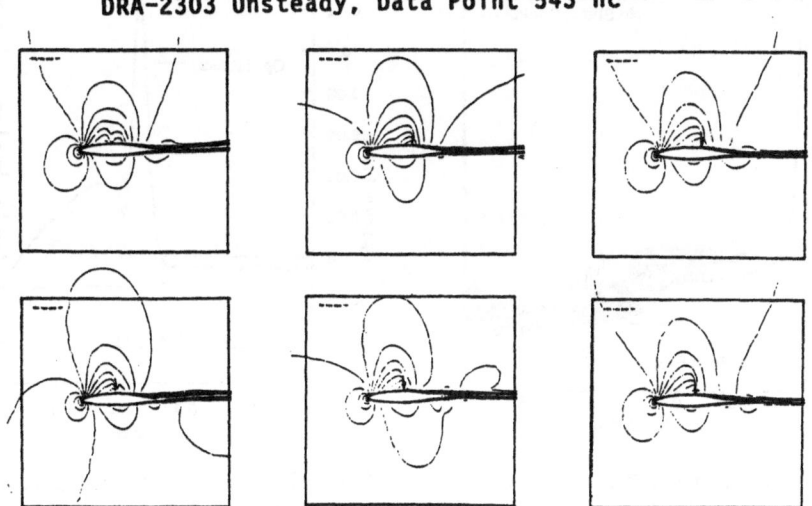

Fig.16 Iso-Mach contours around the DRA-2303 airfoil at several time phases, Data Point 543, nc.

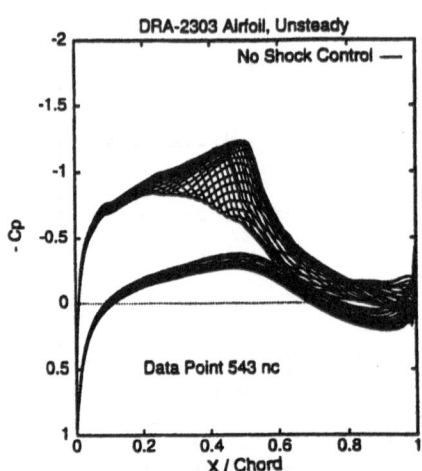

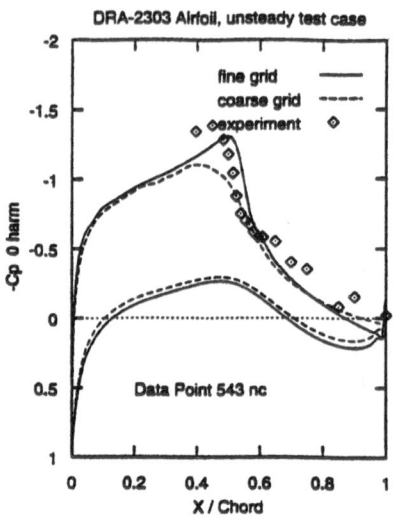

Fig. 18 Pressure coefficient C_p for DRA-2303 airfoil, unsteady test case; Data point 543, nc.

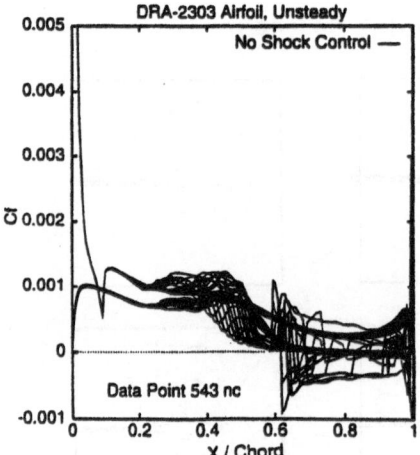

Fig. 17 Pressure coefficient C_p and skin friction coefficient C_f along DRA-2303 airfoil; Data Point 543, nc.

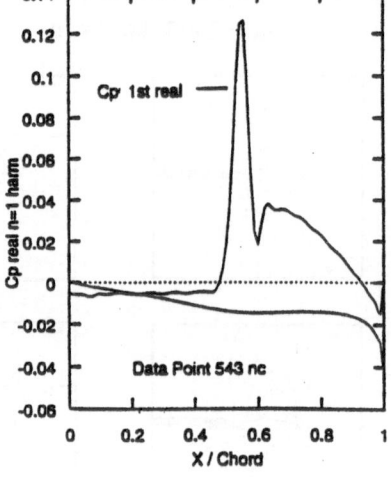

Fig. 19 Real part of the 1st harmonic of pressure coefficient C_p for DRA-2303 airfoil, unsteady test case

288

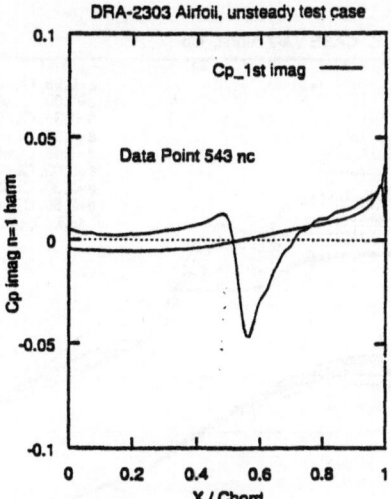

Fig. 20 Imaginary part of the 1st harmonic of the pressure coefficient C_p for the DRA-2303 airfoil; unsteady test case.

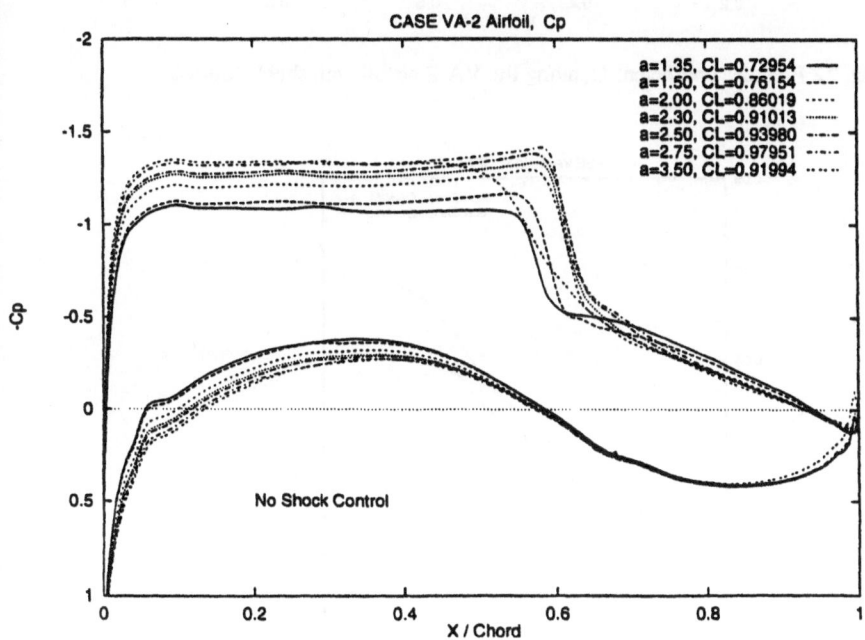

Fig.21 Pressure coefficient C_p along the VA-2 airfoil without shock control.

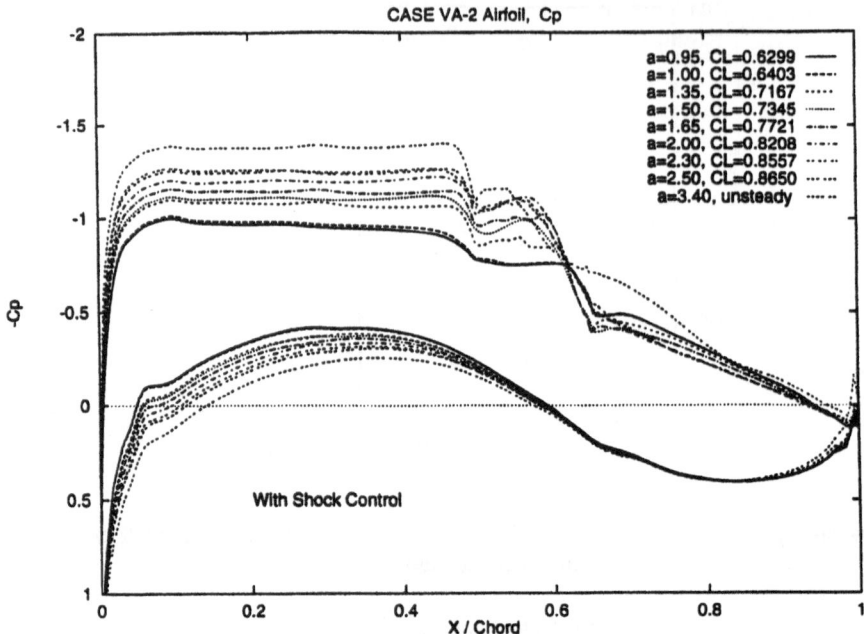

Fig. 22 Pressure coefficient C_p along the VA-2 airfoil with shock control.

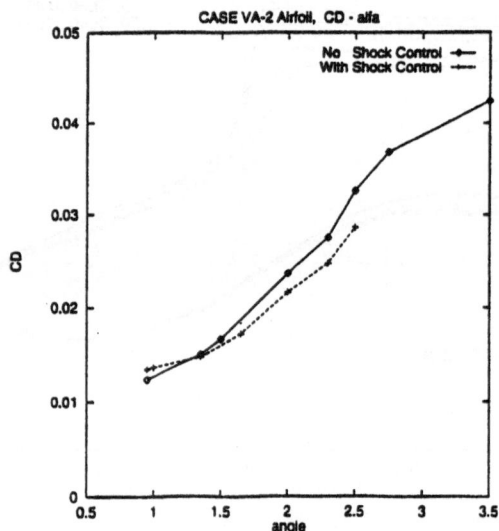

Fig.23 VA-2 airfoil: Drag coefficients for various angles of attack with and without shock control.

18 COMPUTATION OF TRANSONIC FLOWS APPLYING SHOCK BOUNDARY LAYER INTERACTION CONTROL

B.A. Wolles

National Aerospace Laboratory NLR,
Department of Unsteady Aerodynamics and Aeroelasticity
Anthony Fokkerweg 2, 1059 CM Amsterdam, The Netherlands

Summary : This article presents the results of the NLR contribution to the computational part (Task 2) of the Brite/Euram project EUROSHOCK. The aim of this work is the investigation of passive control at transonic flow conditions. For this purpose the NLR has extended its two-dimensional unsteady viscous transonic flow code ULTRAN-V with several control models in order to simulate passive control. In this article the implementation of the different control models is described. Results of the calculations performed for several test cases will be presented and discussed to evaluate the potential of passive control to improve the aerodynamic characteristics of transonic airfoil sections.

18.1 Introduction

The position and strength of a shock wave on an aircraft wing is of major concern when considering transonic flows. Wave drag and viscous drag are strongly related to shock strength and position. In general, a rearward shift of the shock is related to an increase in shock strength and wave drag, but possibly a decrease in viscous drag. Forward movement of the shock wave decreases shock strength and as a consequence decreases the wave drag and the chances for a possible shock-induced separation. With the application of shock boundary layer interaction control it should be possible to fix the shock position at an aft location, while at the same time minimizing the shock strength. As a consequence this technique has the potential to minimize the rise in drag, to prevent shock-induced separation and at the same time, due to a stable shock position, postpone buffet onset.

EUROSHOCK is a European research project, sponsored by the EC, which focuses on the application of *passive control* as one of the techniques of *shock boundary layer interaction control* (SBLIC). The primary objective of this research is the improvement of the performance of transonic transport aircraft.

The present article describes the results of the NLR effort within this project. As a member of the computational group, NLR performed several steady and unsteady flow calculations for different types of transonic airfoil sections (NLR7301, DRA-2303, DA-LVA-1A, VA2). In order to achieve this the NLR implemented several SBLIC models to simulate the effects of passive control in its two-dimensional unsteady viscous transonic flow code ULTRAN-V, a procedure based on a TSP/integral boundary layer method coupling. The ULTRAN-V code certainly is the simplest but fastest running code within the spectrum of flow models used. Such a pragmatic approach is of interest to the aircraft industry. Involvement of such a model enabled an assessment of the limits of applicability.

18.2 Basic Principles of Passive Control

The interaction of a shock wave, as explained by Bohning and Zierep [1], [2] , with a boundary layer on a curved wall is characterized by local phenomena having dramatic effects on the global fluid flow. Focusing on the region where the boundary layer interacts with the shock, one will notice a local rapid thickening of the boundary layer starting at a point near the front of the shock. Such a thickening is related to the positive pressure gradient in that region which can cause a so-called shock-induced separation. However, curvature of the profile in combination with the thickening of the boundary layer induces a local expansion behind the shock, called the post-shock expansion, which reduces the chance of the occurrence of a shock-induced separation.

Considering the thickening of the boundary layer, one should notice that this happens already in front of the shock due to the parabolic character of the boundary layer flow. Correspondingly, the compression of the fluid in the region of the boundary layer already takes place in front of the shock. As a result the strength of the shock and hence the wave drag are reduced when compared to the situation in an inviscid flow. When applying SBLIC one tries to control the

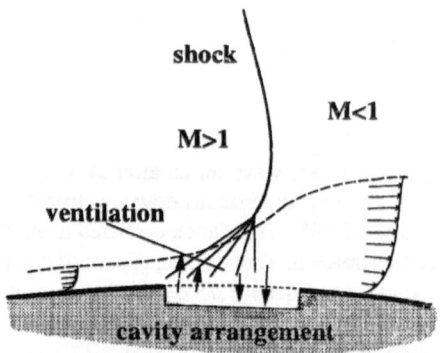

Fig. 1 The passive SBLIC mechanism

interaction effects just described in such a way that it leads to a reduction in total drag. Passive SBLIC can be achieved through the use of a cavity arrangement. For a schematic illustration of the passive SBLIC mechanism, see Figure 1. A porous plate is located in the region around the shock on the upper surface of the profile. Below the plate a cavity is present and as a result of pressure differences upstream and downstream of the shock foot, mass flows into the cavity at the downstream part of the perforation and flows out of the cavity at the upstream part of the perforation. Due to the resulting blowing the boundary layer will thicken more in front of the shock. This implies that the compression generated by the shock is spread over a larger region thereby reducing the shock wave drag. The inflow of mass into the boundary layer might, however, lead to a local increase in viscous drag. The inflow of mass, i.e., suction, behind the shock establishes a local expansion possibly compensating the increment in viscous drag generated in front of the shock.

18.3 Computational Method

18.3.1 The ULTRAN-V method

The computational method ULTRAN-V has been developed at NLR by Houwink for calculating the 2-D unsteady viscous flow about airfoils in steady or unsteady motion. The method is based on the unsteady Transonic Small Perturbation (TSP) potential equation for the inviscid flow, and an integral method for the boundary layer. Due to the strong interaction coupling between the boundary layer and the inviscid flow, the applicability of the method in practice covers a wide range of subsonic and transonic, attached and separated, steady and unsteady flow conditions (see [3] and [4]).

The TSP equation used is the fully unsteady equation. The scaling of the quantities was chosen such that it would fit applications in the subsonic and the transonic flow regime. As a basis for the Alternate Direction Implicit (ADI) finite difference solution method, the equation is linearized for the time differencing. At each time step first an intermediate solution is calculated for all lines Z=constant ("X-sweep", Z direction is perpendicular to the incoming flow, X direction is in flow direction). Each line is solved implicitly. The final solution at each time step is obtained by solving for the lines X=constant ("Z-sweep"). A downstream marching procedure is used.

For the boundary layer the integral method is used. The method of Thwaites (see Rott [5]) is used for the laminar-flow region, the method of Green [6] for the turbulent flow. The lag-entrainment equation of Green has been modified for unsteady flow. The method of Granville [7] is used to determine transition. A second-order Runge-Kutta method is used for integrating the equations. The boundary layer is coupled with the inviscid flow by two means. In the direct method the velocity of the inviscid flow at the previous time step is used for the boundary layer calculation. This is only possible if the boundary layer remains attached. In the simultaneous method the boundary layer equations are solved simultaneously with the inviscid flow during the Z-sweep, which allows the computation of separated flow. The simultaneous mode is only implemented for the turbulent boundary layer. In order to obtain second-order accuracy in space, an iterative procedure is needed at each time step in a time-accurate calculation.

18.3.2 Extension of ULTRAN-V to control

The turbulent boundary layer model implemented in the ULTRAN-V computer program is based on an unsteady version of Green's Lag Entrainment method as introduced by Houwink [8],[3]. The original ULTRAN-V employed the turbulent boundary layer equations in a form without transpiration, i.e., with zero ventilation velocity. However, representation of a cavity for SBLIC requires the modeling of ventilation at the place of the shock. In the present version these boundary layer equations were extended to a form including transpiration effects.

The entrainment equation as derived by Houwink [8] is rewritten as :

$$\frac{1}{\rho_e u_e} \frac{\partial}{\partial t} \left(\rho_e \left(H_1 + H_b \right) \Theta \right) + \frac{1}{\rho_e u_e} \frac{\partial}{\partial x} \left(\rho_e u_e H_1 \Theta \right) = C_E + \frac{v_w}{u_e} . \tag{1}$$

The transpiration form of the momentum integral equation takes the following form :

$$\frac{1}{\rho_e u_e} \frac{\partial}{\partial t} \left(\rho_e H_b \Theta \right) + \frac{\delta^*}{u_e^2} \frac{\partial u_e}{\partial t} + \frac{1}{\rho_e u_e^2} \frac{\partial}{\partial x} \left(\rho_e u_e \Theta \right) + \frac{\delta^*}{u_e} \frac{\partial u_e}{\partial x} = \frac{C_f}{2} + \frac{v_w}{u_e} . \tag{2}$$

For both equations the v_w/u_e terms represent the extension of the original form.

293

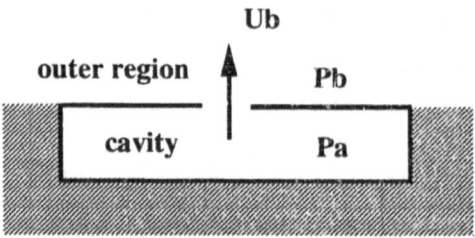

Fig. 2 Assumed cavity configuration for isentropic control law

18.3.3 Control laws

SBLIC is introduced by means of a ventilation in the shock region. Applying passive control (see Figure 1) the ventilation is generated as a result of the induced difference in pressure outside and inside the cavity. In order to calculate the ventilation terms several models have been implemented relating the ventilation velocity to the pressure jump over the porous sheet.

Darcy's law The simplest expression relating a pressure jump to a ventilation velocity is the so-called Darcy's Law (see [9] and [10]). In this law the ventilation velocity is proportional to the pressure drop across the porous sheet. In mathematical language:

$$\frac{v_w(s)}{u_e(s)} = \sigma \frac{P_b(s) - P_a}{\rho_e(s) u_e^2(s)} .$$ (3)

In this formulation σ is the permeability factor, P_a is the pressure inside the cavity, which is assumed to be a constant, and P_b is the pressure outside the cavity. The cavity pressure P_a is usually taken as the mean value of the pressure outside the cavity along the porous sheet.

Isentropic control law In case the cavity is relatively large it is justified to assume zero velocity of the fluid and a constant pressure inside the cavity. When further neglecting the influence of the outer flow, one can predict the outer flow in a rather simple way assuming isentropic flow through the porous surface (see Dargel [11]). Assuming the configuration as sketched in Figure 2 , the theoretical inviscid mass flow through the cavity holes can be predicted by means of the formulation:

$$\frac{\dot{m}_h}{A_h \rho_0 a_0} = \left\{ \frac{2}{\gamma - 1} \frac{\rho_a}{\rho_0} \frac{P_b}{P_0} \left(\frac{P_a}{P_b} \right)^{-\frac{1}{\gamma}} \left\{ \left(\frac{P_a}{P_b} \right)^{\frac{\gamma - 1}{\gamma}} - 1 \right\} \right\}^{0.5} ,$$ (4)

with:

$$\dot{m}_h = \rho_b u_b A_h .$$ (5)

In this formulation the different expressions have the following meaning:

$\dot{m}_h$: mass flux through the hole
A_h : area of the hole
ρ_0 : total density
a_0 : total speed of sound
P_a : pressure inside the cavity
ρ_a : density inside the cavity
P_b : pressure in the region just outside the cavity
ρ_b : density in the region just outside the cavity
u_b : velocity in the hole
γ : isentropic exponent $(= 1.4)$.

As noticed by Dargel [11], the pressure loss across the porous surface can be taken into account through the use of the factor η_v:

$$\frac{\dot{m}'_h}{A_h \rho_0 a_0} = \eta_v \frac{\dot{m}_h}{A_h \rho_0 a_0} . \tag{6}$$

To translate the velocity u_b in each hole into a continues distributed ventilation velocity v_w along the porous sheet, another factor called the porosity factor $\eta_p = \frac{A_h}{A}$ is introduced. It indicates the ratio between the area of the holes and the total area of the porous sheet. Hence the relation between the ventilation velocity v_w and the isentropic mass flow through the holes becomes:

$$\frac{v_w}{a_0} = \eta_p \eta_v \frac{\dot{m}_h}{A_h \rho_0 a_0} \frac{\rho_0}{\rho_b} . \tag{7}$$

Poll's law   Another formula giving a relation between the pressure drop across a porous sheet and the ventilation velocity is given by Poll and others [12] (also see Dargel [11]). Poll's

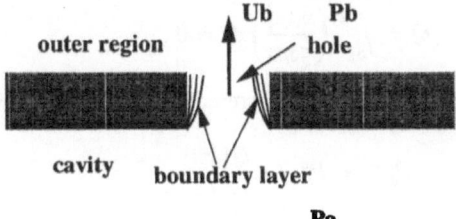

Fig. 3 Assumed cavity arrangement for Poll's law

model is the result of treating the flow through the holes as a viscous pipe flow (see Figure 3) with the following set of equations:

$$\frac{\dot{m}_h}{\rho_0 a_0 A_h} = \frac{4X}{\pi R} \tag{8}$$

$$\frac{(P_b - P_a)\rho_h d_h^2}{\mu_h^2} \left(\frac{d_h}{s_h}\right)^2 = \frac{1}{K}\left(40.7X + 1.95X^2\right) , \tag{9}$$

with:

$$\dot{m}_h = \rho_h u_b A_h \tag{10}$$

$$R = \frac{a_0 \rho_0 d_h}{\mu_h} \frac{d_h}{s_h} \tag{11}$$

The new parameters introduced in these equations have the following meaning:

d_h : hole diameter
s_h : sheet thickness
ρ_h : density in the hole
μ_h : dynamic viscosity of the fluid in the hole
A_h : area of the holes
K : factor for effective hole diameter .

Analogous to the isentropic case the factor η_p is introduced to translate the separate hole velocities u_b into a continuously distributed ventilation velocity v_w :

$$\frac{v_w}{a_0} = \eta_p \frac{1}{\dfrac{\rho_b}{\rho_0}} \frac{\dot{m}_h}{\rho_0 a_0 A_h} . \tag{12}$$

18.3.4 Implementation of the control models

Given the outer flow conditions and the specifications of the porous sheet, the ventilation v_w and the cavity pressure P_a are the unknowns for both the isentropic and Poli's formulation. To close the system one should further require zero mass flow through the porous surface:

$$Q = \int_{s_b}^{s_e} \left(\frac{v_w \rho_b}{a_0 \rho_0} \right) ds = 0 . \tag{13}$$

Starting with an estimation for P_a the actual cavity pressure P_a is predicted with a Newton-like iteration procedure:

$$P_a^{n+1} = P_a^n - \omega \frac{Q(P_a^n)}{\dfrac{\partial Q}{\partial P_a}\big|_n} \tag{14}$$

In this formulation $Q(P_a^n)$ represents the residual at iteration step n of the total mass flow through the porous sheet. ω is the relaxation factor.

In the implementation of Darcy's law the cavity pressure P_a is simply taken as the mean pressure along the porous sheet:

$$P_a = \frac{1}{s_e - s_b} \int_{s_b}^{s_e} P_b ds . \tag{15}$$

18.4 Numerical Results

18.4.1 Steady flow predictions

The main goal of the steady calculations performed within EUROSHOCK was to investigate the effectiveness of passive control to improve the aerodynamic characteristics by minimizing drag rise and preventing shock-induced separation. During the EUROSHOCK project several steady calculations have been performed for several transonic airfoils: NLR7301, DRA-2303, DA-LVA-1 and VA2. At the initial phase of the EUROSHOCK project preliminary calculations were made using a prescribed ventilation velocity for the control cases. These preliminary calculations were performed with the goal to validate the codes for the considered type of flow and to obtain a better initial insight into the physical phenomena related to control. The test cases reported in the present section are the final test cases calculated after the initial phase of the project.

DRA-2303 airfoil For this airfoil some SBLIC experiments have been carried out by the DRA in Bedford [13]. Some of the characteristics of this airfoil are:

Details of the DRA-2303 airfoil			
chord length	$c = 0.635m$	begin cavity	$x/c = 0.5$
porosity	$\eta_p = 0.08$	end cavity	$x/c = 0.6$

Two test cases with and without control were performed for this airfoil section. The flow conditions can be found in Table 1. All calculations were made for a prescribed lift coefficient C_L. The isentropic model was used to calculate the ventilation velocities for the control cases. The ventilation velocities resulting from the isentropic model were reduced by a factor which was calculated by CIRA and distributed to the partners. The reduction factor was introduced to find a better correspondence with ventilation velocities which would have been predicted with Poll's procedure. At the initial stage of the project a prescribed ventilation distribution, also provided by CIRA, was used for these test cases. Also Darcy's law was applied. The results using these latter two models will not be reported here since they were only used for demonstration purposes.

To avoid numerical problems blending regions of 10% cavity length were defined at the edges of the cavity: $x/c = 0.50$ to $x/c = 0.51$ at the beginning of the cavity and $x/c = 0.59$ to $x/c = 0.60$ at the end of the cavity. Transition was forced at 5% chord length on both surfaces of the airfoil. A 146×59 grid was used for which 15 points were present in the cavity region.

DA-VA2 airfoil For the relevant characteristics of this airfoil we note:

Details of the DA-VA2 airfoil			
hole diameter	$d_h = 0.5mm$	chord length	$c = 0.200m$
sheet thickness	$s_h = 1.2mm$	begin cavity	$x/c = 0.495$
porosity	$\eta_p = 0.126$	end cavity	$x/c = 0.645$

For this type of airfoil three different test cases with and without control were prescribed for which the flow conditions can be found in Table 2. Similar to the DRA-2303 airfoil test cases the calculations were performed for a prescribed lift coefficient C_L. Also here the isentropic model was used to calculate the ventilation velocities for the cases with control. The calculated ventilation velocities have been reduced by a reduction factor analogous to the DRA-2303 test cases. Blending regions of 10% cavity length were defined at the edges of the cavity: $x/c = 0.495$ to $x/c = 0.510$ at the beginning of the cavity and $x/c = 0.630$ to $x/c = 0.645$ at the end of the cavity for the first two cases and $x/c = 0.488$ to $x/c = 0.503$ at the beginning of the cavity and $x/c = 0.638$ to $x/c = 0.653$ at the end of the cavity for the last test case. The position of the blending regions was changed for the last case according to normalization agreements made within EUROSHOCK. Transition was forced at 30% chord length on the upper and at 25% chord length on the lower surface of the airfoil profile. For the first two cases a 146 × 59 grid and a 152 × 59 grid were used. The first grid led to 25 points in the cavity region and the last grid to 32 points in the cavity region. Several calculations showed that the 146 × 59 grid led to instabilities in the cavity region so that the 152 × 59 grid was used for the final calculations. However, both grids were unacceptably clustered in the cavity region. This problem was solved by the development of a grid generation program for the ULTRAN-V code. By using this new ULTRAN-V tool a new 158 × 81 grid with 17 points in the cavity region was generated as illustrated in Figure 4. This grid is also the one that is used for the calculation of the last VA2 test case.

After the calculations made by the different partners it was concluded that the second VA2 case was related to a maximum lift case, showing some unsteady behavior. Hence the case was rejected from any further consideration. As an optional case, a C_L sweep was computed. This sweep was computed for a prescribed C_L ranging from $C_L = 0.75$ to $C_L = 0.85$ for a step of $\Delta C_L = 0.0125$. Other flow conditions were identical to those of the mandatory VA2 cases of computation.

DA-LVA-1A airfoil   The results of the calculations performed for this test case were related to experiments which were carried out in the T2 wind tunnel at ONERA/CERT in Toulouse. The relevant details of the airfoil with its cavity are summarized below :

Details of the LVA-1A airfoil			
hole diameter	$d_h = 0.1123mm$	chord length	$c = 0.15m$
sheet thickness	$s_h = 1.2mm$	begin cavity	$x/c = 0.580$
porosity	$\eta_p = 0.0506$	end cavity	$x/c = 0.695$

Flow conditions and results of the two experimental (reference) cases are illustrated in Table 3 ; again C_L was the prescribed value for the different computations.

Calculations for the two control cases have been performed using Poll's formula. For the K-factor which was provided and calibrated by DASA, the following values were used: $K = 1.3^4$ for blowing and $K = 1.6^4$ in the case of suction. As for the previous airfoils blending regions of 10% cavity length were defined at the edges of the cavity: $x/c = 0.574$ to $x/c = 0.586$ at the beginning of the cavity and $x/c = 0.689$ to $x/c = 0.701$ at the end of the cavity.

In all cases the 158 × 81 grid has been used. The grid was refined in the cavity region where 17 points are situated.

Transition has been forced at 48% of the chord on both surfaces of the airfoil.

Discussion of results

The results of the calculated steady cases are included in the Tables 1 , 2 and 3. The most important results are also illustrated in Figures 5 to 21. More results are presented in Wolles [14]. For a comparison of the NLR results with the results of the other EUROSHOCK partners the reader is referred to Wolles [15], [16], and [17].

From the different tables shown some symbols need further explanation:

C_D : The total drag as calculated by ULTRAN-V without taking into account the excrescence drag caused by the presence of the perforated plate.

$C_{D_{excr}}$: The excrescence drag caused by the perforated plate. This drag is calculated as the clean plate drag multiplied by a factor of 2.5 which was determined experimentally.

$C_{D_{tot}}$: The drag calculated when adding C_D and $C_{D_{excr}}$.

C_{DW} : The wave drag coefficient.

C_{pcav} : The coefficient of the pressure in the cavity.

The excrescence drag was only taken into account in the DA-LVA-1A test cases, after agreements made within EUROSHOCK.

The first important result from inspecting the aerodynamic coefficients in the tables is the decrease in shock wave drag. The blowing just before the shock wave looks very effective in spreading the compression and this seems not to depend on the type of airfoil considered. Looking at the pressure distribution plots, comparing the cases with control and the cases without control, it is observed that the pressure distribution in the region of control is affected considerably (plots are given in the Figures 5, 10, 11, 16 and 20). The compression that takes place in the no-control case, represented by a one-step increase in C_p, is generally replaced in the control cases by a two-step increase in C_p. The first step starts at the begin of the cavity, the second step at the end of the cavity. In the region of the cavity itself the pressure increase is usually small or even zero as illustrated most clearly in Figure 11. This shock wave structure with two separate compressions suggests the presence of a *lambda-shock wave*. The region where the pressure remains more or less constant is often called the *pressure plateau* . With the lambda shock in the controlled cases the original shock strength is spread over two separate steps thereby reducing the wave drag.

Focusing now on the effects of control on the boundary layer, one observes that the boundary layer quantities δ^* , C_f and θ are affected in the control region, see Figure 6, 8 and 9. As a consequence the boundary layer characteristics just before the end of the cavity are, for the general control case, different compared to the ones in the no-control case. Hence, the initial conditions for the boundary layer behind the cavity are changed, which results in a permanent quantitative difference in the entire region downstream of the cavity. For this reason it is important to keep δ^* and θ low at the end of the cavity, so that the overall viscous drag does not overshadow the gain in wave drag. In principle this can be obtained with suction at the end of the cavity. However, for most cases considered this effect is, in the case of passive control, not strong enough, see Figures 8, 9 and 21. For a single case like the first VA2 test case a beneficial effect on the boundary layer was observed, see Figures 14 and 15 . It is also this case where the application of control leads to a reduction in total drag (see Table 2). The reason for the apparent minimal influence of the suction on the characteristics of the boundary layer is related to the strong positive pressure gradients between the cavity and the trailing edge. This results in a fast increase in δ^* and θ downstream of the cavity which cannot be compensated by the limited amount of suction available. As one can notice from the C_p curves, the least unfavorable pressure gradient behind the cavity is obtained with the VA2 airfoil. In the C_p curve of the first VA2 airfoil case with control, Figure 11, a local expansion occurs just behind the cavity as a result of the suction.

Considering the C_f curves an early drop right after the begin of the cavity is observed for all cases. In a single case this leads to a separation region due to control which is not present in the no-control case, Figure 12. For the first VA2 case, Figure 12, the drop of C_f is restored right after the cavity to a value higher than the one for the no-control case. In this specific case the occurrence of separation is postponed to higher incidence. In all other cases the opposite trend is observed which is illustrated in Figures 6 and 17.

The results of the C_L sweep calculation are illustrated in the Figures 18 and 19. From Figure 18, showing the C_p plots for the cases with control, it is observed that the incidence influences the ratio between the first compression at the beginning of the cavity, and the second compression at the end of the cavity. The higher the incidence the stronger the first compression compared to the second one. This trend also indicates that the suction downstream of the shock becomes less effective and the blowing upstream of the shock more effective for the higher incidence. The C_D and C_L values of the C_L sweep together with those of the first VA2 case are illustrated in Figure 19. This polar visualizes the reduction in total drag for the first VA2 case. Increasing the incidence gradually leads to an overcompensation of the reduction in shock-wave drag by an increment in viscous drag.

18.4.2 Unsteady flow predictions

The aim of the unsteady calculations performed within EUROSHOCK was the investigation of the application of passive control to alleviate or to postpone buffet onset. In the initial phase of the EUROSHOCK project two unsteady no-control test cases for the NLR7301 and one for the NACA0012 airfoil were defined for pre-calculations. Later, a final test case for the DRA-2303 airfoil was defined applying control. In the following part attention will be given to the DRA-2303 case and one NLR7301 case.

NLR7301 airfoil This case has been calculated for a Mach number of $M_a = 0.738$, an angle of attack of $\alpha = 2.26$, and a Reynolds number of $Re = 11.7 * 10^6$. Transition was fixed at 7% chord length on both surfaces of the airfoil. A relatively coarse grid of 79×59 with 64 points on the airfoil was used.

DRA2303 airfoil The two unsteady cases calculated for this type of airfoil were distinguished in the experiments by DRA [13] as data point 1469 for the case applying control and data point 543 for the case without control. A Mach number of $M_a = 0.71$ and Reynolds numbers of $Re = 10.4 * 10^6$ and $Re = 10.5 * 10^6$ were used for the cases with and without control, respectively. The Mach number was larger than the experimental value of $M_a = 0.702$ in order to obtain a buffet-type of flow. The calculations were performed for a prescribed angle of attack of $\alpha = 2.405°$ for the case without control and $\alpha = 2.804°$ for the case with control. Ventilation velocities were predicted using Darcy's law. A value of $\sigma = 0.12$ was used for the permeability factor. This value was determined by trial and error aiming at a beneficial effect by the application of control. Furthermore a grid of 98×59 mesh points was used. Transition was forced at 5% of the chord length on both surfaces of the airfoil.

Discussion of results

Visualizations of the unsteady results are presented in Figures 22 to 27. The results for the NLR7301 airfoil in Figures 22 and 23, following a computation with a steady boundary layer and unsteady outer flow method, show an instability which can be interpreted as buffet. A second calculation has been performed, modeling both unsteady inner and outer flow, which showed a more damped behavior. The main reason of the damped behavior in the second situation is probably due to the use of a relatively large time step, which was chosen to obtain a convergent result. A different solution can probably be found in performing the calculations with a finer grid. Comparing the steady mean and the first harmonic lift coefficient to the experimental values, one will notice reasonable agreement (see Geissler [18]). Concerning the reduced frequencies it is found that the second calculation led to a frequency of $k = 0.1$ (k is the reduced frequency based on the semi-chord length) which is much smaller than the experimental value of $k = 0.257$. The first calculation led to a frequency of $k = 1.07$ which is much higher than the experimental value. Comparing the C_p plots from the second computation with the C_p plots from the first computation, Figure 22, it is observed that the second case shows very small movements of the shock whereas the first one shows severe movements, illustrating a buffet-type of flow.

The results of the unsteady DRA2303 test cases, illustrated in Figures 24 to 27, were obtained by modeling unsteady inner and outer flow. Considering the results for the no-control case a buffet-type of flow is observed which is distinguished by serious shock oscillations (see Figure 24). These shock oscillations result in quite large oscillations in the values of the aerodynamic coefficients (for C_L see Figure 26). The movement of the shock wave is strongly reduced when control is applied, which is shown in Figure 25. However, it becomes clear from the C_P plots and the time history of the aerodynamic characteristics, Figure 27, that another instability occurs. This instability is probably related to the fact that the shock is located at the beginning of the cavity. Due to small oscillations of the shock around this point, large variations in the ventilation velocities occur in combination with oscillations in shock strength and so in the aerodynamic characteristics.

The unsteady results clearly demonstrate buffet. They also show a possible benefit when applying control. However, the results are very unstable and sensitive to the parameter settings of the program. Therefore their reliability cannot be fully guaranteed.

18.5   Conclusions

The present research was focused on the application of passive *shock boundary layer interaction control* (SBLIC) in transonic viscous flow about an airfoil section. The investigations demonstrated the capabilities of the simple ULTRAN-V code to produce quite reasonable results within short run times on a workstation. Also the limitations were demonstrated when using ULTRAN-V for unsteady buffet-type flow.

For all steady test cases the application of passive SBLIC led to a reduction in wave drag. This reduction was in most cases, however, absorbed by an increase in viscous drag, thereby leading to an increase in total drag. Only in Test Case 1 of the VA2 airfoil a reduction in total drag is possible by preventing a too large increment in viscous drag. For this airfoil the positive pressure gradient is low enough to let the suction behind the shock compensate the increment in viscous drag due to blowing ahead of the shock. The results produced for the steady cases are at least qualitatively correct. Quantitatively correct results were obtained for the pressure distribution outside the control region. The unsteady calculations demonstrated a buffet type of flow for the no-control cases. For the case with control a damping effect of the buffet oscillations was observed, although instabilities were still present. However, the unsteady results cannot be considered accurate due to the sensitivity of the results to the input settings.

Looking at the EUROSHOCK project in a general context, the present research led to new insights into the physics of the shock boundary layer interaction problem. The experiences obtained are undoubtedly useful for future investigations related to shock control.

18.6 Acknowledgement

Special thanks are due to ir. R. Houwink of Fokker Aircraft b.v. for his assistance during the project.

18.7 References

[1] R. Bohning and J. Zierep. Normal shock-turbulent boundary layer interaction at a curved wall. In *Computations of Viscid-Inviscid Interactions*, CP-291, pages 17–1 to 17–8. AGARD, 1980.

[2] R. Bohning and J. Zierep. Calculation of 2d turbulent shock/boundary-layer interaction at curved surfaces with suction and blowing. In J. Délery, editor, *Turbulent Shear Layer/Shock Wave Interactions*. IUTAM Symposium Palaiseau 1985, Springer-Verlag, Berlin, Heidelberg, 1986.

[3] R. Houwink. Computation of unsteady turbulent boundary layer effects on unsteady flow about airfoils. Technical Report TP 89003, NLR, 1989.

[4] R. Houwink and A.E.P. Veldman. Steady and unsteady separated flow computations for transonic airfoil. AIAA Paper 84-1618, 1984.

[5] N. Rott and L.F. Crabtree. Simplified laminar boundary-layer calculations for bodies of revolution and for yawed wings. *Journal of Aeronautical Sciences*, August 1952.

[6] J. Green, D.J. Weeks, and J.W.F. Brooman. Prediction of turbulent boundary layers and wakes in compressible flow by a lag-entrainment method. *ARC R and M*, (3791), 1973.

[7] P.S. Granville. The calculation of the viscous drag of bodies of revolution. Technical Report Rep. 849, David Taylor Model Basin, 1953.

[8] R. Houwink. Unsteady strong viscous/inviscid interaction modelling in the ULTRAN-V code. Technical Report NLR CR 91205 L, National Aerospace Laboratory NLR, The Netherlands, 1991.

[9] C.R. Olling and G.S. Dulikravich. Viscous-inviscid computations of transonic separated flows over solid and porous cascades. In *Transactions of the ASME Journal of Turbomachinery*, volume 109. 31st International Gas Turbine Conference and Exhibit, Düsseldorf, Germany, The American Society of Mechanical Engineers, April 1987.

[10] John Locke Gray Bsc. *The Passive Control of Shock Wave Boundary Layer Interaction in Transonic Flow*. PhD thesis, Faculty of Engineering Queen's University of Belfast, July 1989.

[11] G. Dargel. Private communication.

[12] D.I.A. Poll, M. Danks, and B.E. Humphreys. The aerodynamic performance of laser drilled sheets. In *First European Forum on Laminar Flow Technology*, Hamburg, March 1992.

[13] M.J. Simmons and J.L. Fulker. Notes from tests on a passive control aerofoil. Technical report, DRA, Bedford, 1993. Contract Paper AP4-CP(93)-4, EUROSHOCK TR AER2-92-49/3.1 (also see Chapter 22).

[14] B.A. Wolles. Computation of transonic flows applying shock boundary layer interaction control. Technical Report NLR CR 96093 L / EUROSHOCK TR AER-92-49/2.8, National Aerospace Laboratory NLR, The Netherlands, 1996.

[15] B. Wolles. Results of the DRA-2303 Steady Test Cases with/without Control, January 1996. In W. Geissler, *Transonic Airfoil/Wing Flow Prediction with Shock-Boundary Layer Interaction and Control (SBLIC)*, EUROSHOCK Final Technical Report TR AER2-92-94/F1, Appendix C, January 1996.

[16] B. Wolles. Results of the DA LVA-1A Steady Test Cases with/without Control, January 1996. In W. Geissler, *Transonic Airfoil/Wing Flow Prediction with Shock-Boundary Layer Interaction and Control (SBLIC)*, EUROSHOCK Final Technical Report TR AER2-92-94/F1, Appendix C, January 1996.

[17] B. Wolles. Results of the VA2 Steady Test Cases with/without Control, January 1996. In W. Geissler, *Transonic Airfoil/Wing Flow Prediction with Shock-Boundary Layer Interaction and Control (SBLIC)*, EUROSHOCK Final Technical Report TR AER2-92-94/F1, Appendix C, January 1996.

[18] W. Geissler. Transonic Airfoil/Wing Flow Prediction with Shock Boundary Layer Interaction and Control (SBLIC). EUROSHOCK Final Technical Report TR AER2-92-94/F1, Appendix C.

Table 1 Results of the DRA-2303 airfoil test cases.

Calculated/Measured Results DRA-2303 airfoil								
Case	M_a	$R_e * 10^6$	α	C_L	C_M	C_D	C_{DW}	C_{pcav}
calc. 271 n.c.	0.6816	18.97	1.061	0.5669	-0.0884	0.00860	0.00059	-
calc. 1008 w.c.	0.6807	18.98	1.072	0.5530	-0.0866	0.00900	0.00019	-0.940
exp. 271 n.c.	0.6816	18.97	1.068	0.5668	-0.0958	0.00943	-	-
exp. 1008 w.c.	0.6807	18.98	1.066	0.5529	-0.0944	0.01053	-	-
calc. 289 n.c.	0.6795	18.91	2.548	0.8114	-0.0900	0.01557	0.00533	-
calc.290 n.c.	0.6795	18.91	2.777	0.8425	-0.0906	0.01788	0.00654	-
calc.1031 w.c.	0.6806	18.98	2.845	0.8142	-0.0863	0.01765	0.00420	-0.918
exp. 289 n.c.	0.6795	18.91	2.507	0.8115	-0.1002	0.01458	-	-
exp. 290 n.c.	0.6795	18.91	2.712	0.8427	-0.1014	0.01656	-	-
exp. 1031 w.c.	0.6806	18.98	2.710	0.8142	-0.0984	0.01797	-	-

Table 2 Results of the VA2 airfoil test cases.

Calculated Results VA2 airfoil								
Case	M_a	$R_e * 10^6$	α	C_L	C_M	C_D	C_{DW}	C_{pcav}
1 n.c.	0.7400	2.50	1.128	0.6499	-0.1296	0.00979	0.00084	-
1 w.c.	0.7400	2.50	1.064	0.6499	-0.1315	0.00902	0.00028	-0.751
2 n.c.	0.7400	2.50	2.446	0.9000	-0.1472	0.02969	0.01382	-
2 w.c.	0.7400	2.50	3.475	0.8995	-0.1310	0.03886	0.01139	-0.829
3 n.c.	0.7430	2.50	1.794	0.8002	-0.1452	0.01978	0.00836	-
3 w.c.	0.7430	2.50	2.195	0.8000	-0.1339	0.02277	0.00533	-0.927

Table 3 Results of the DA-LVA-1A airfoil test cases.

Calculated/Measured Results LVA-1A airfoil					
Case	M_a	$R_e * 10^6$	α	C_L	C_M
	C_D	C_{Dexcr}	$C_{D_{tot}}$	C_{DW}	C_{pcav}
calc. 76 n.c.	0.7613	4.64	0.884	0.4738	-0.0823
	0.01227	-	-	0.00508	-
calc. 76 w.c.	0.7613	4.64	1.261	0.4737	-0.0785
	0.01175	0.00088	0.01263	0.00201	-0.765
exp. 76 n.c.	0.7698	4.64	1.000	0.4736	-0.0858
	0.01172	-	-	-	-
calc. 73 w.c.	0.7614	4.64	0.915	0.4514	-0.0766
	0.01121	0.00088	0.01209	0.00179	-0.765
exp. 74 w.c.	0.7692	4.66	1.000	0.4556	-0.0833
	-	-	0.01233	-	-

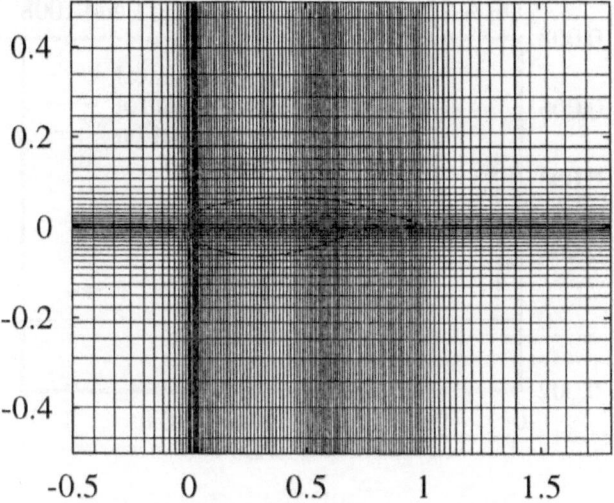

Fig. 4 VA2 airfoil with 158 × 81 grid.

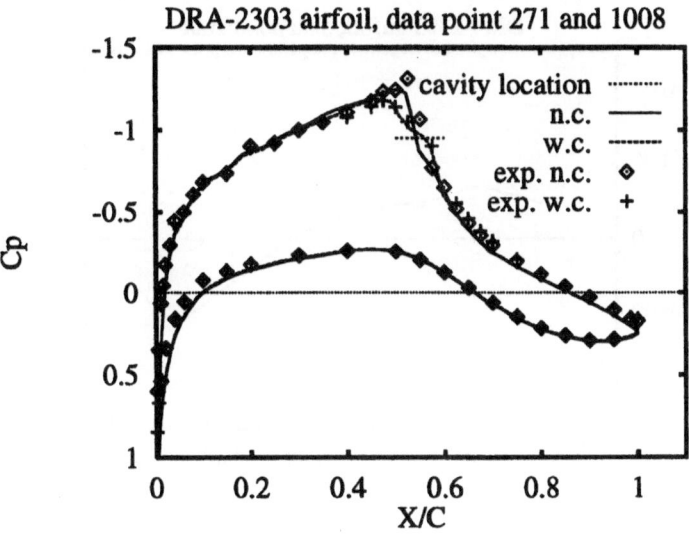

Fig. 5 Pressure distributions. DRA-2303 airfoil test case (with/without control).

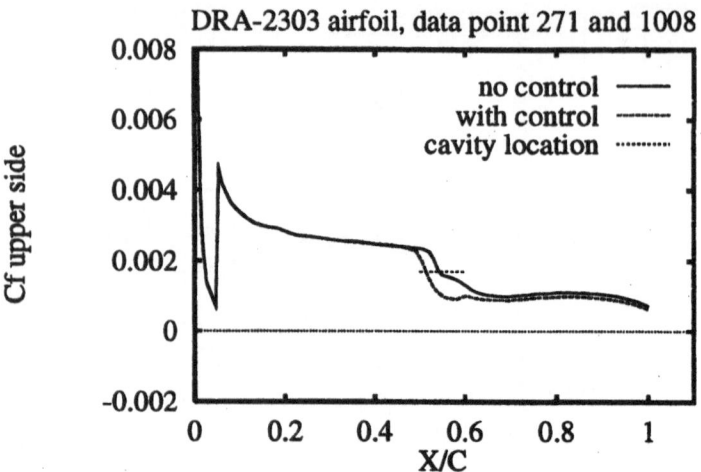

Fig. 6 Skin-friction distributions. DRA-2303 airfoil test case (with/without control).

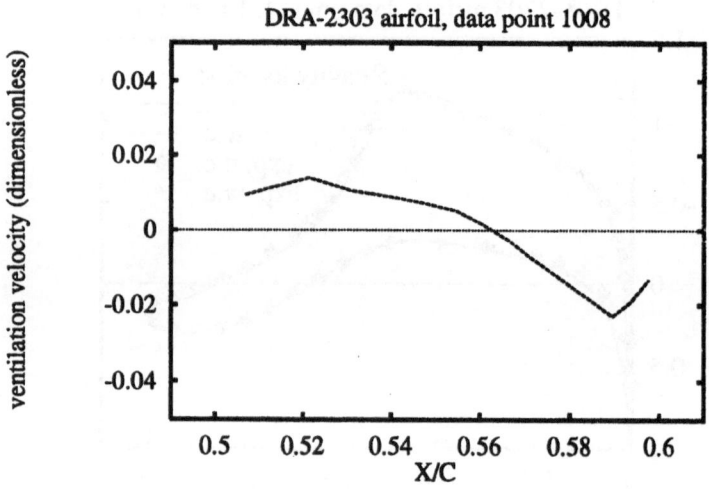

Fig. 7 Ventilation velocity distribution. DRA-2303 airfoil test case (with control).

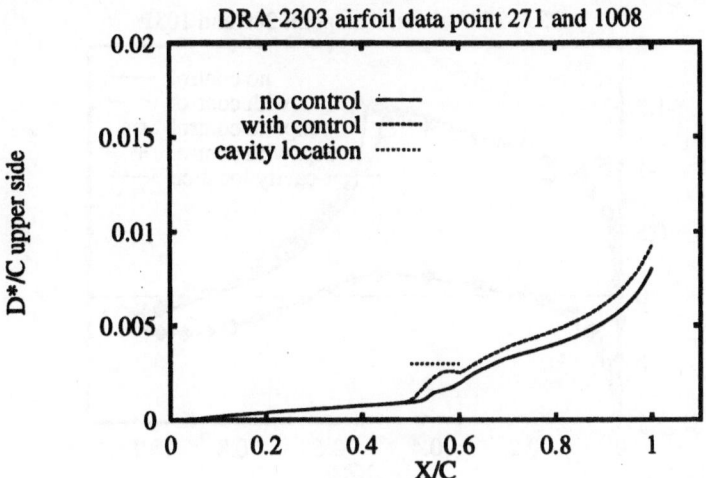

Fig. 8 Displacement thickness distributions. DRA-2303 airfoil test case (with/without control).

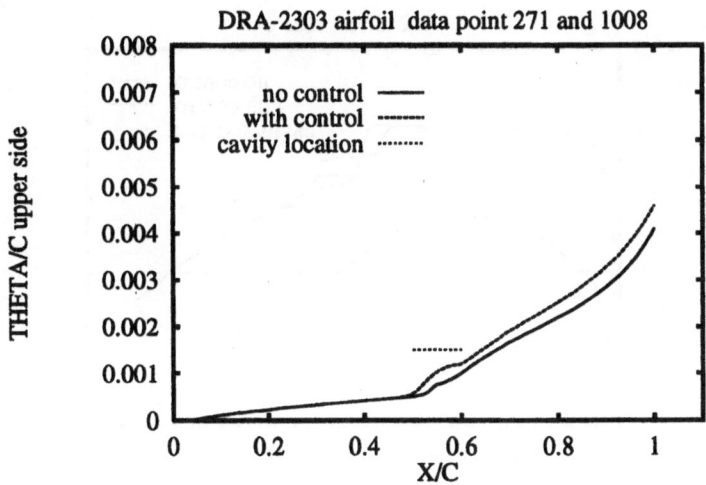

Fig. 9 Momentum thickness distributions. DRA-2303 airfoil test case (with/without control).

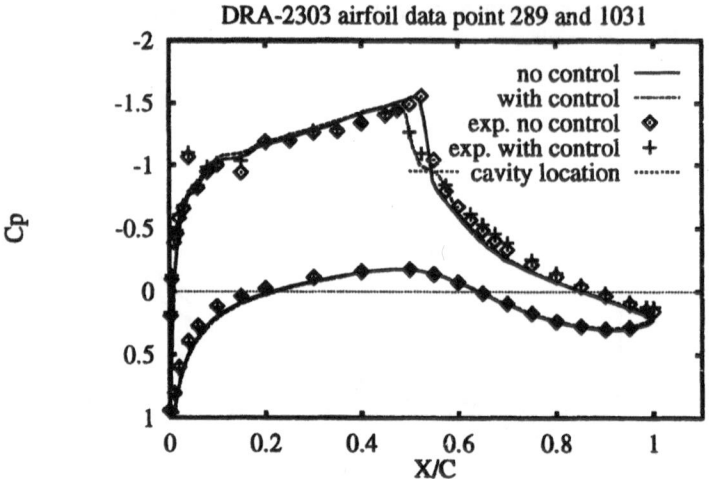

Fig. 10 Pressure distributions. DRA-2303 airfoil test case (with/without control).

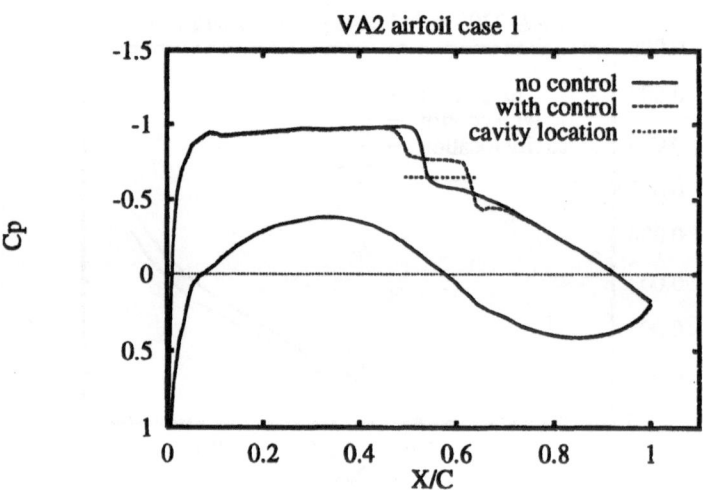

Fig. 11 Pressure distributions. VA2 airfoil test case 1 (with/without control).

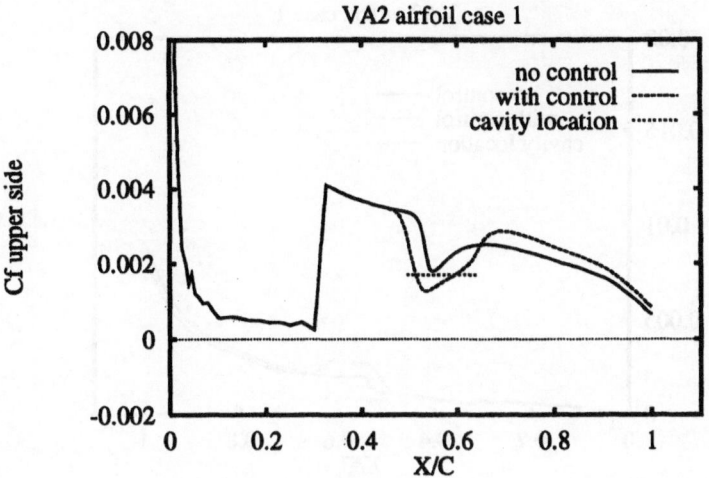

Fig. 12 Skin-friction distributions. VA2 airfoil test case 1 (with/without control).

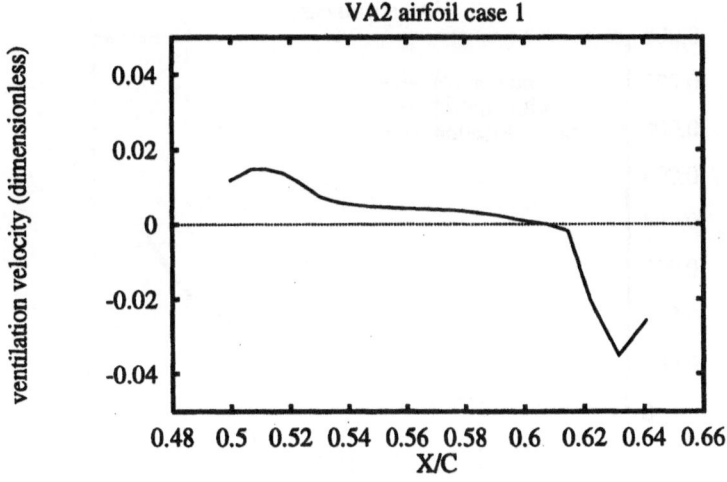

Fig. 13 Ventilation velocity distribution. VA2 airfoil test case 1 (with control).

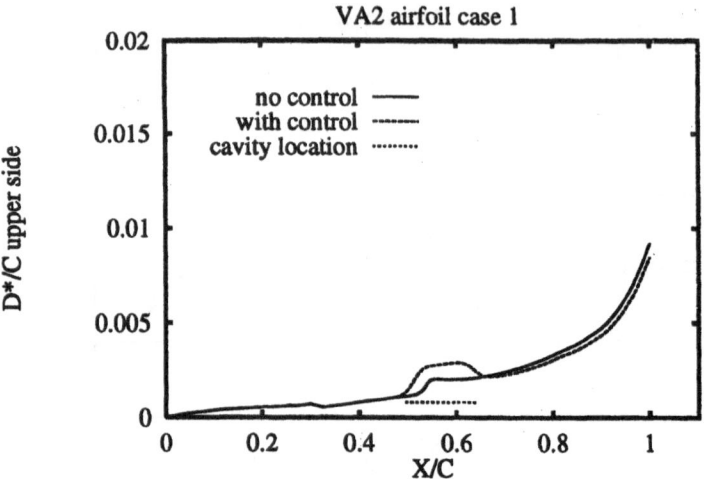

Fig. 14 Displacement thickness distributions. VA2 airfoil test case 1 (with/without control).

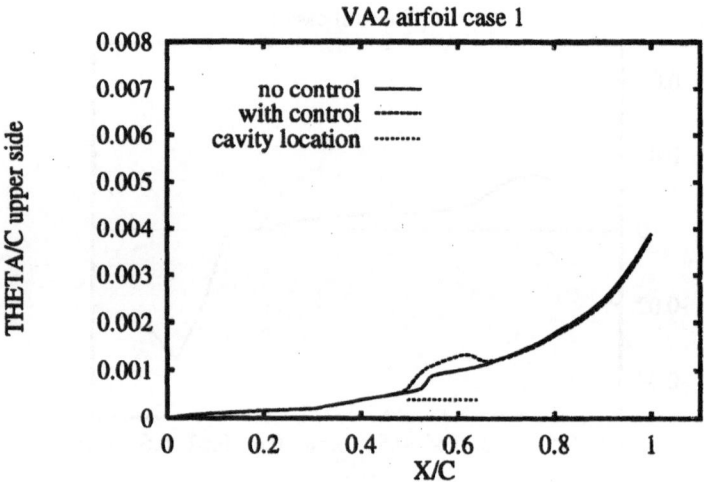

Fig. 15 Momentum thickness distributions. VA2 airfoil test case 1 (with/without control).

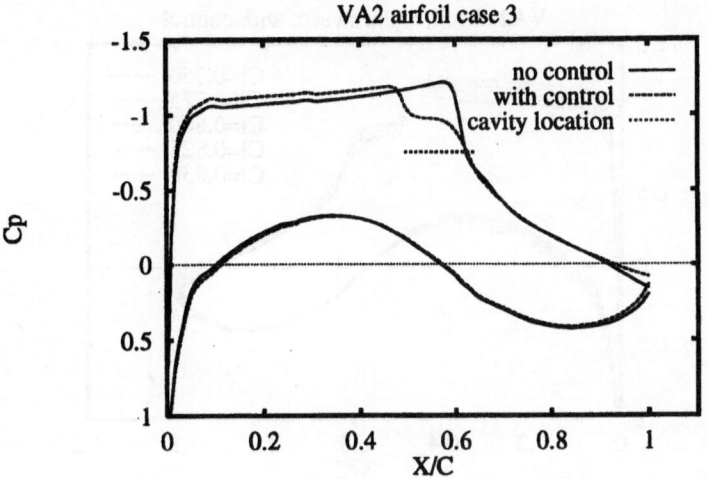

Fig. 16 Pressure distributions. VA2 airfoil test case 3 (with/without control).

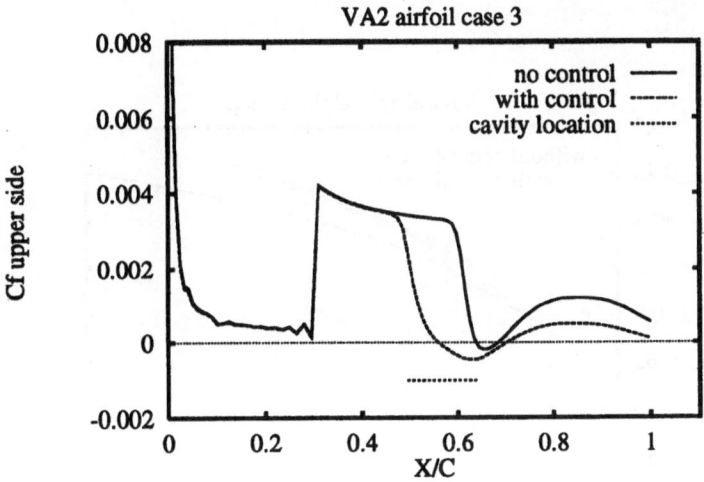

Fig. 17 Skin-friction distributions. VA2 airfoil test case 3 (with/without control).

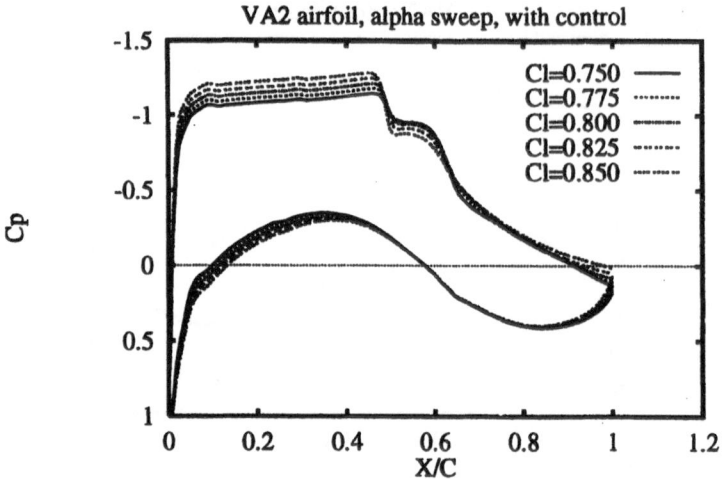

Fig. 18 Pressure distributions. VA2 airfoil polar calculations (with control).

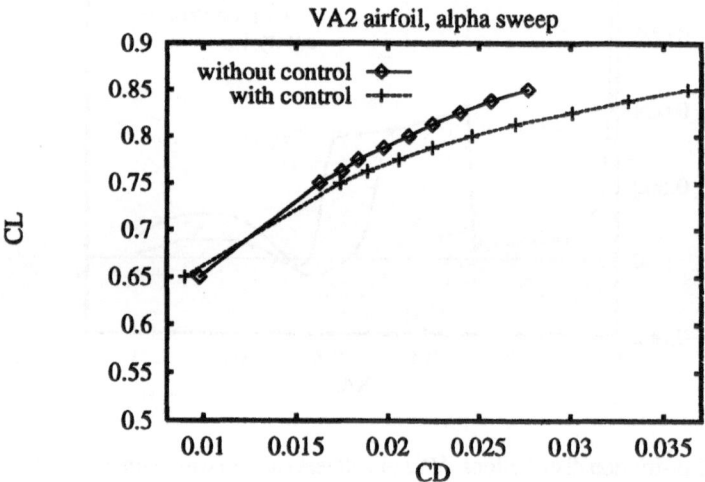

Fig. 19 Polar VA2 airfoil (with/without control).

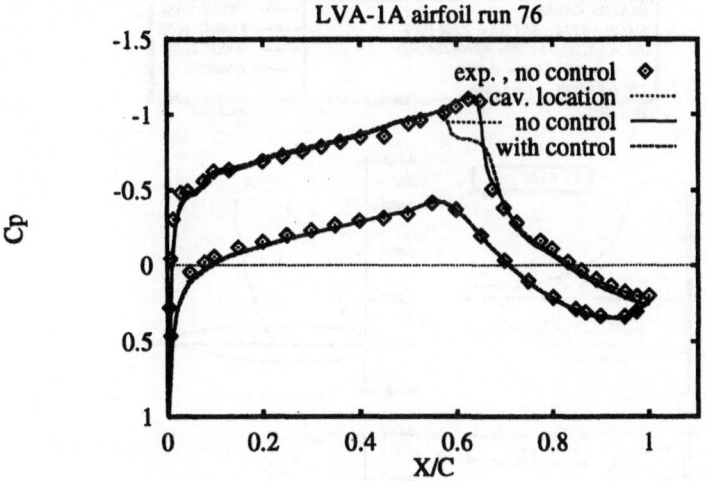

Fig. 20 Pressure distributions. LVA-1A airfoil test case (with/without control).

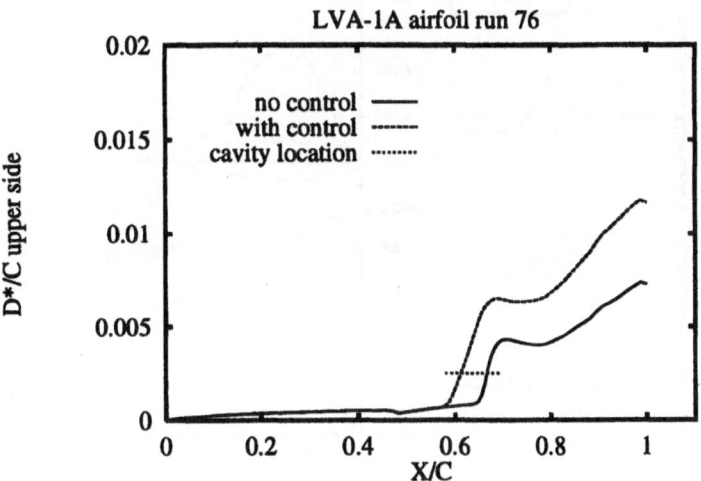

Fig. 21 Displacement thickness distributions. LVA-1A airfoil test case (with/without control).

313

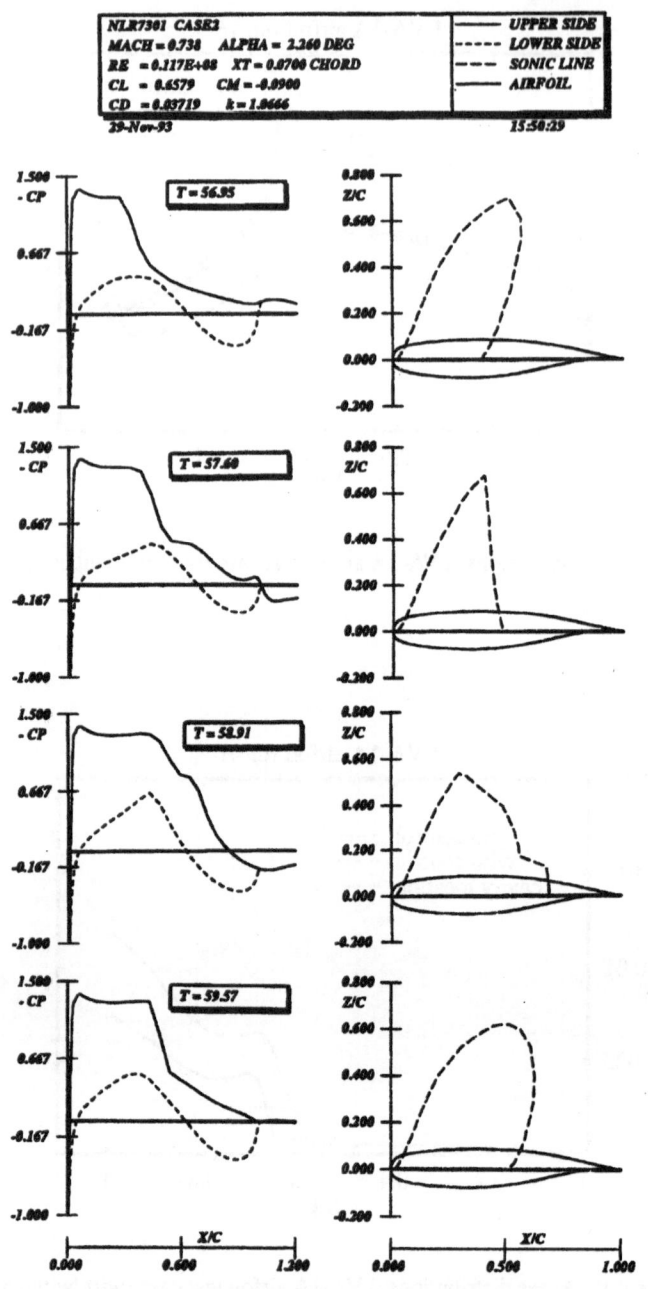

Fig. 22 Pressure distributions and sonic line for one cycle of the unsteady NLR7301 test case.

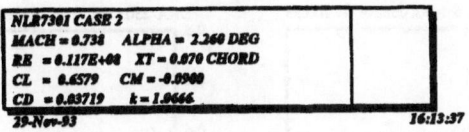

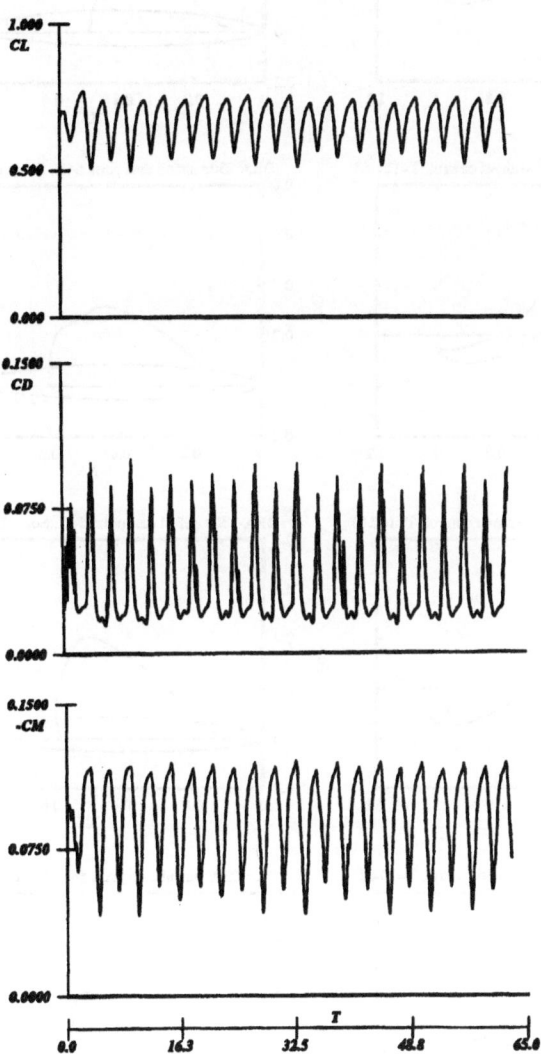

Fig. 23 Time histories of C_L, C_D and C_M for one cycle of the unsteady NLR7301 test case.

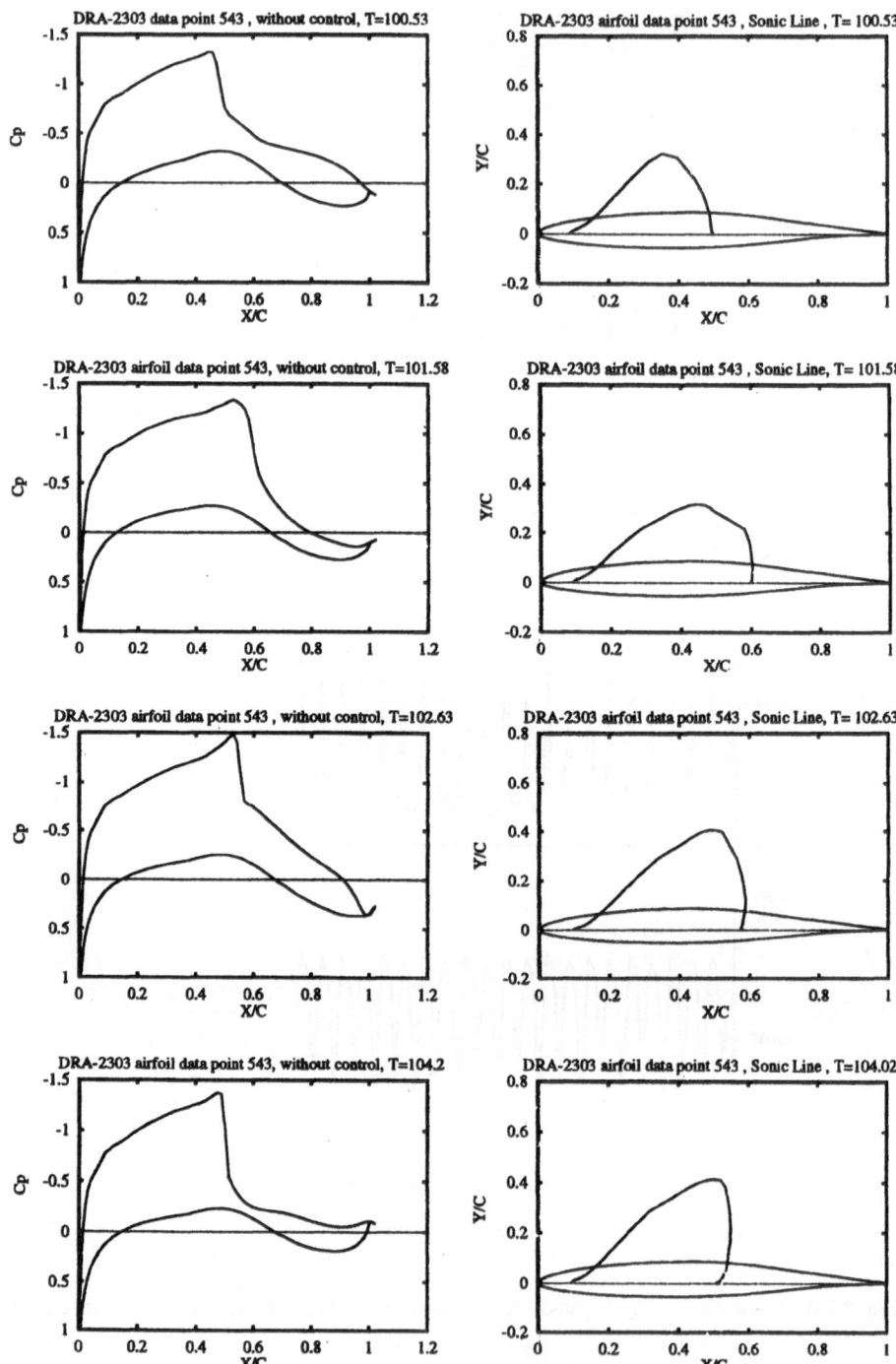

Fig. 24 Results of unsteady DRA-2303 test case without control.

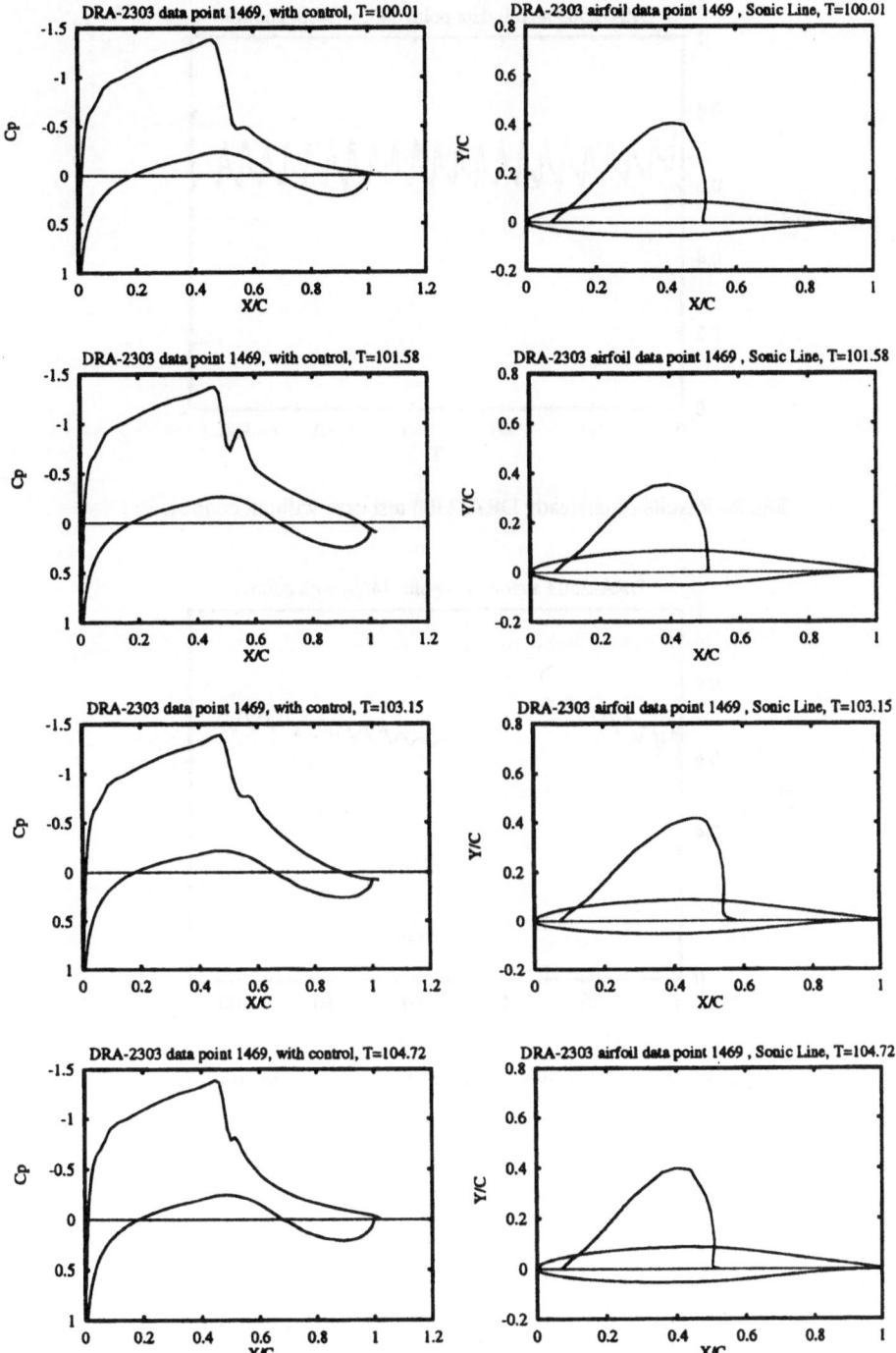

Fig. 25 Results of unsteady DRA-2303 test case with control.

317

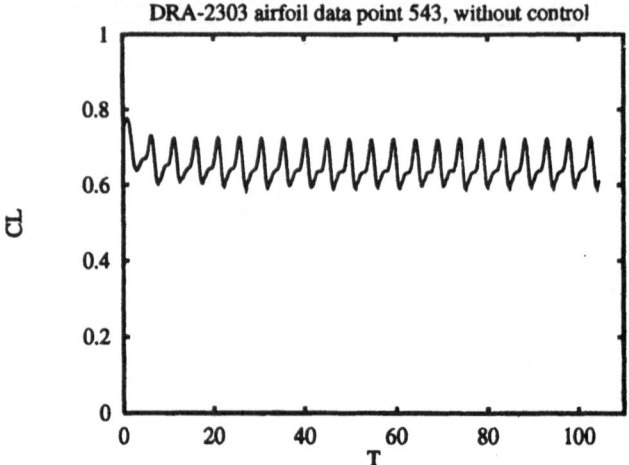

Fig. 26 Results of unsteady DRA-2303 test case without control.

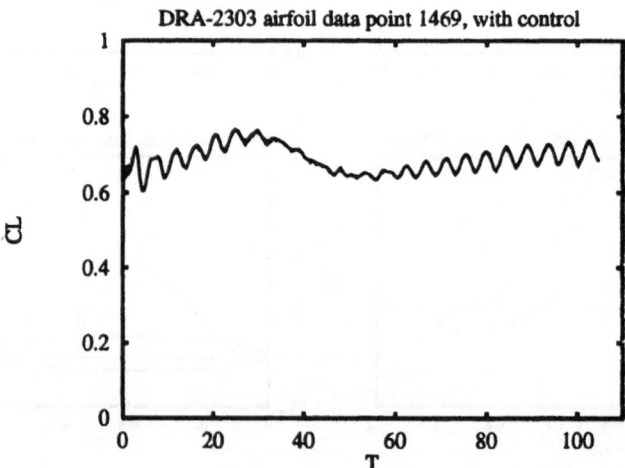

Fig. 27 Results of unsteady DRA-2303 test case with control.

19 DESIGN AND MANUFACTURE OF THE CRYOGENIC LAMINAR–TYPE AIRFOIL MODEL DA LVA–1A WITH PASSIVE SHOCK CONTROL

P. Thiede, G. Dargel
Daimler–Benz Aerospace Airbus GmbH
Hünefeldstr. 1–5, 28183 Bremen

Summary: The contribution of Dasa Airbus within EUROSHOCK Task 3.1 was to design, manufacture and equip a cryogenic laminar–type transonic airfoil model with passive shock control (SC) inserts for shock control experiments in two adaptive–wall cryogenic wind tunnels, ONERA–T2 in Toulouse and DLR–KRG in Göttingen, with the aim to generate a data–base for validation of the SC prediction methods developed in Task 2 and to assess the passive shock control concept.

The model construction had to take into account the different operational characteristics of the two cryogenic wind tunnels used and the desired laminar flow and high Reynolds number test conditions.

19.1 Introduction

In the past, various methods have been experimentally investigated to control shock wave/ boundary layer interaction by active or passive means. In principle, the effectiveness of passive shock control (SC) has been demonstrated in early 2D experiments [1–4] using SC devices with perforated/slotted surfaces and a cavity underneath. Thiede et al. [3] and Krogmann et al. [4] performed SC experiments on a turbulent–type transonic airfoil demonstrating performance improvements at off–design conditions, mainly by delaying the onset of buffet. Thibert et al. [5] published results of recent SC experiments on a laminar–type airfoil, in which the benefits of passive SC in terms of reduced shock strength and wave drag are largely offset by increased viscous drag.

In order to investigate the drag reduction potential of passive shock control on a laminar airfoil at high Reynolds numbers, the contribution of Dasa Airbus in Task 3 is devoted to the design and manufacture of a cryogenic airfoil model with exchangeable passive SC devices based on the laminar–type airfoil design DA LVA–1A.

19.2 Basic Airfoil DA LVA–1A Characteristics

19.2.1 Airfoil design

The basic airfoil LVA–1A is a transonic laminar–type airfoil with a thickness of 12% chord and a leading edge radius of 1.4% chord, Fig. 1. It has been designed by DA to establish natural laminar flow on the upper and lower surface up to 50% chord for a lift coefficient of $c_l=0.4$ at a Reynolds number of $Re_\infty = 20$ million and a Mach number of $M_\infty = 0.73$.

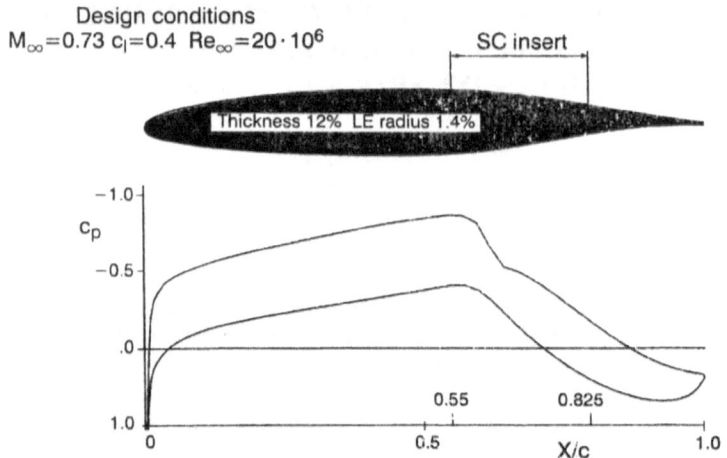

Fig. 1: Design pressure distribution of the laminar–type airfoil DA LVA–1A

The pressure distribution is characterized by a weak shock wave occurring already at design conditions. To investigate passive shock control, an exchangeable insert reaching from 55% to 82.5% chord, is foreseen. Fig. 2 shows predicted off–design pressure distributions at Reynolds numbers of $Re_\infty = 6$ and 20 million. It is obvious that with increasing lift coefficient c_l also the strength of the shocks will be increased but its location remains rather unchanged. This airfoil characteristic has shown to be a big advantage for a practicable shock control application.

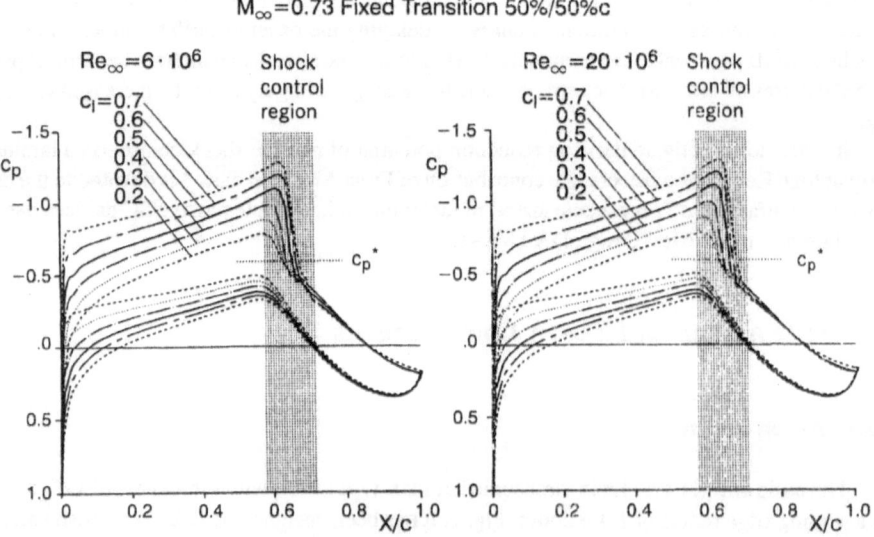

Fig. 2: Predicted off–design pressure distributions

19.2.2 Reference measurements without shock control

In 1991, a conventional LVA–1A airfoil model with a chord of c = 200 mm has been tested without shock control in the DLR Transonic Wind Tunnel Brunswick (TWB) [6]. The measured drag polars with fixed and free transition, Fig. 3, demonstrate the existence of a laminar drag buckle. Moreover, the measured pressure distributions show a rather small shock shifting with lift coefficient, this being a good condition for shock control. Two pressure distributions of these experimental data have been distributed by DA to the respective EUROSHOCK partners for CFD validation purposes at the beginning of the EUROSHOCK project.

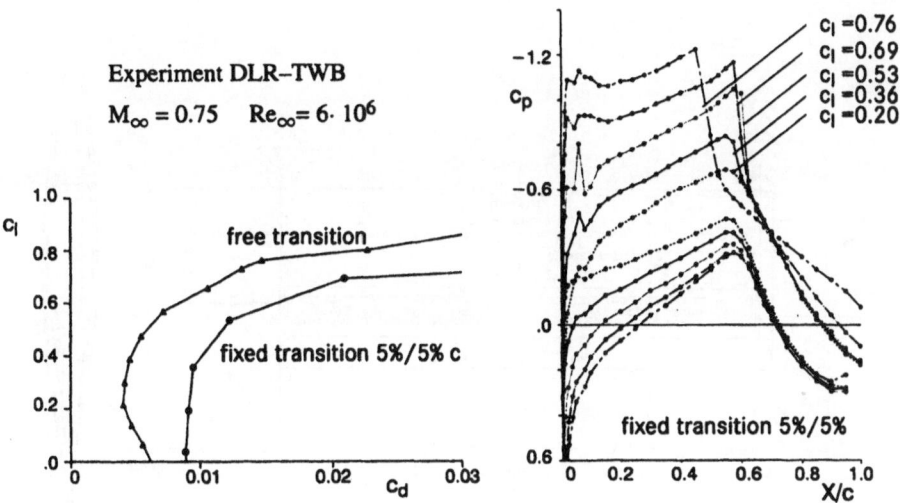

Fig. 3: Reference measurements of the LVA–1A airfoil in the DLR–TWB

19.3 Cryogenic Model Design/Manufacture

In order to carry out shock control experiments with the airfoil DA LVA–1A in the ONERA– T2 and DLR–KRG cryogenic wind tunnels, a cryogenic model has been designed by DA with a chord of 150 mm and a span of 400 mm. Exchangeable shock control inserts have been foreseen to meet different test conditions at low and high Reynolds numbers.

19.3.1 ONERA–T2/ DLR–KRG requirements

The model construction had to take into account the different characteristics of the two wind tunnels, especially the differences in the run time, the total pressure and operational aspects:
- ONERA–T2: run time 30 to 130 s, total pressure 1.5 to 5 bar, model cool–down during the initial phase of a run
- DLR–KRG: run time 0.5 to 1.0 s, total pressure 2.0 to 10 bar, model/test cool–down prior to a run

19.3.2 Model construction/manufacture

In spite of these differences it was decided to build a common model, Fig. 4, as the section dimensions and the model support of both wind tunnels are quite similar. In order to meet the ONERA–T2 requirements, the model had to be designed as a hollow body with a wall thickness of 3 mm. The model is composed of three parts: the main body with the upper and lower half-shells soldered together and the exchangeable insert. For the DLR–KRG testing, the inside of the model had to be ventilated in addition. A new technique in the manufacturing process has been applied to ensure a high surface quality with regard to accuracy ($\Delta z=0.01$mm) and waviness required for high Reynolds number testing.

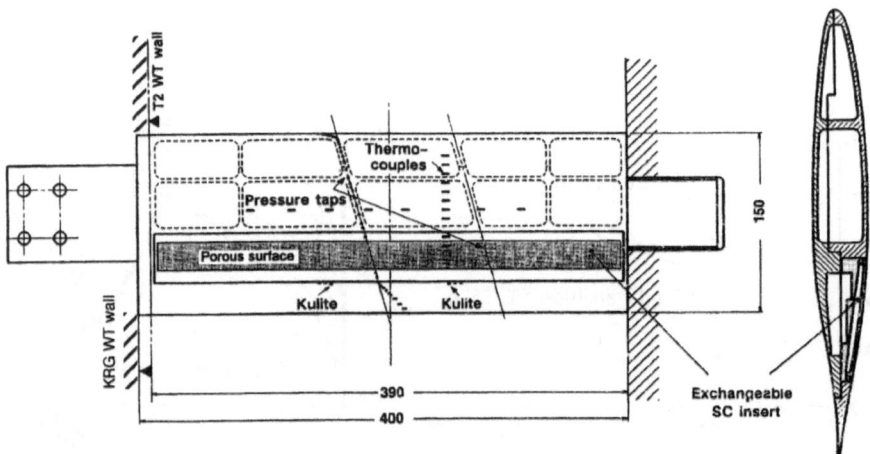

Fig. 4: EUROSHOCK cryogenic airfoil model DA LVA–1A

19.3.3 Model equipment

The model has been equipped with two chordwise sections of pressure orifices with 64 pressure taps in the center and 9 parallel to it. They are arranged under an angle of $15°$ and sized not to affect the laminar flow, i.e, they have a hole diameter of 0.1 mm between 0 and 22.5% and of 0.2mm downstream of 22.5% chord. Two dynamic pressure transducers (Kulites) have been installed to sense the pressure fluctuations downstream of the shock control insert for studying the buffet behaviour. Copper–constantane thermo–couples have been located in two lines on the inner side of the model wall for transition detection in the ONERA–T2. In addition three thermo–couples had been arranged in pockets just under the skin at $x/c=0.125, 0.325$ and 0.475 in order to investigate transition detection in the DLR–KRG.

19.4 Design/Manufacture of the Passive Shock Control Inserts

19.4.1 Aerodynamic insert design parameters

The main design parameters of the passive control inserts are
 - the location of the porous surface
 - the length of the porous surface
 - the wall porosity and its chordwise distribution
 - the hole geometry (diameter, shape and inclination).

The shock control design requirements are given by the transonic flow conditions of the laminar flow–type airfoil DA LVA–1A for a Reynolds number range from 6 to 20 million with transition fixed at 50% chord.

The location and length of the control region mainly depend on the shock position and its shifting at off–design conditions. The effect of varying wall porosity has been investigated by using the DA SC airfoil code, see Chapter 14. Fig. 18 of Chapter 14 shows increasing ventilation effects with increasing porosity. A porosity of 4% was chosen for a reasonable wave drag reduction, although an increase in viscous drag seems to occur.

Due to construction requirements the beginning of the porous control region had to be fixed at 57.5% chord. As shown in Fig. 20 of the parametric study of Chapter 14, the viscous drag increase can be affected by the length of the perforation. In order to allow sufficient latitude for the movement of the shock, a length of 12.5% chord has been finally selected so that the shock is located between the center and 2/3 of the control region.

As a further important design parameter the hole geometry had to be specified. The influence of the hole diameter on the overall shock control effect cannot numerically be simulated, since only a distributed ventilation velocity is modeled in the SC airfoil code. But the nominal hole diameter should not exceed the boundary layer displacement thickness in order to ensure that the blowing jets will dissipate within the boundary layer. With this assumption the hole geometry has been defined dependent on the flow condition.

The selected design parameters of the porous insert with normal holes are given in Table 1:

Table 1: Selected design parameters of LVA–1A passive SC insert

leading edge $(x/c)_p$	length L_p/c	porosity %	d_h mm	a_h mm	s_h mm
0.575	0.125	3.94	0.1	0.48	1.0

19.4.2 Insert construction/manufacture/equipment

Fig. 5 shows the design of the exchangeable SC inserts for the cryogenic airfoil model. It is divided into an outer and an inner part, combined by special pins. The outer part carries the porous surface with a wall thickness of 1mm over nearly the whole span of the model which can be electron–beam drilled from inside. The inner part presents the bottom of the cavity.

The cavity has a depth of about 2mm guaranteeing a constant pressure inside the cavity. The cavity is divided into three spanwise compartments to avoid a spanwise pressure gradient due to wind tunnel side wall effects. For the measurement of the static pressure inside the cavity, 16 pressure taps are drilled in the bottom of the cavity: one row with 6 taps in the center in chordwise direction, two rows with 3 taps parallel to it and one row with 4 taps in spanwise direction.

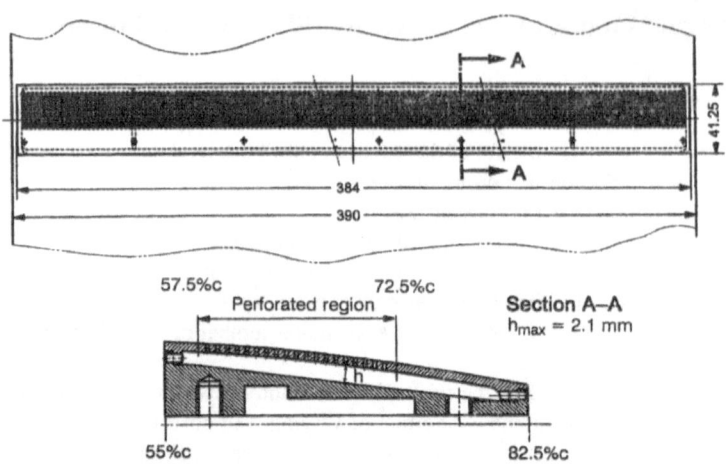

57.5%c   72.5%c

Perforated region

Section A–A

h_{max} = 2.1 mm

55%c   82.5%c

384

390

41.25

Fig. 5: Exchangeable passive shock control insert for DA LVA–1A airfoil model

Fig. 6: Photo of cryogenic airfoil model DA LVA–1A with passive SC inserts

Three inserts have been manufactured:
- insert 1: solid surface
- insert 2: porous surface with the design parameters of Table 1
- insert 3: to be defined, for later tests

Fig. 6 shows a photo of the airfoil model LVA–1A with two exchangeable SC inserts.

19.4.3 Perforation technique/check out procedure

The holes of the control region manufactured by electron–beam drilling are of conical shape but with nearly uniform spacing, as a cut through a test panel with a thickness of 1mm and a porosity of 2% in Fig. 7 shows. As a result of the burning process the hole is bigger on the side of the beam entry and the diameter varies in irregular fashion with depth. The smallest diameter appears at the outlet of the beam, which should correspond to the specified diameter.

With the electron–beam drilling technique it is hardly possible to obtain the specified diameter of an individual hole accurately. There is only a nominal diameter which can be determined by a statistical survey of a number of holes. Due to the problem in reproducing a hole of a specific diameter, the porous sheet is specified by its flow characteristic, i.e., in terms of the mass flow rate versus the pressure drop. This approach avoids the problem of knowing the hole diameter accurately.

For the manufacture of the porous surface it is necessary to specify the hole geometry that will provide the required flow characteristic. Therefore, test panels of the same material as the planned insert and with a shape approximately that of the contour in the control region will be produced, containing different hole geometries. Results of the calibration tests of two such test panels with different holes are compared in Fig. 8 with the design target (solid line) obtained by the SC law procedure of the SC prediction method assuming the design characteristic of 2% porosity.

19.4.4 Actual insert perforation/ calibration tests

Before testing in the ONERA–T2 and DLR–KRG wind tunnels the perforated SC insert used had to be checked by determining the geometrical properties of the holes and measuring their characteristics in calibration tests.

The photo in Fig. 9 shows the electron–beam drilled holes from flow and cavity side. The average hole parameters have been measured with the result of being slightly larger than the specified ones, giving an actual porosity of 5.1%. At the beginning of the control region three rows of holes could not fully be drilled due to some deviations in the wall thickness, resulting in an effective beginning of the control at x/c=.58 and in a reduced length of about 11.5% chord. The actual geometrical values are collected in Table 2.

Table 2: Actual parameters of the perforated SC insert 2.

leading edge $(x/c)_p$	length L_p/c	porosity %	d_h mm	a_h mm	s_h mm
0.58	0.115	5.1	0.112	0.476	1.2

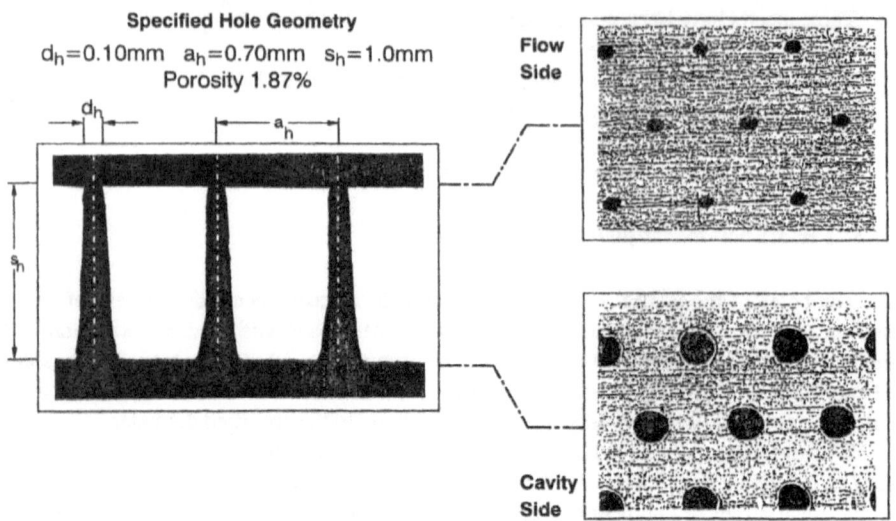

Fig. 7: Perforated test panel with electron–beam drilled holes

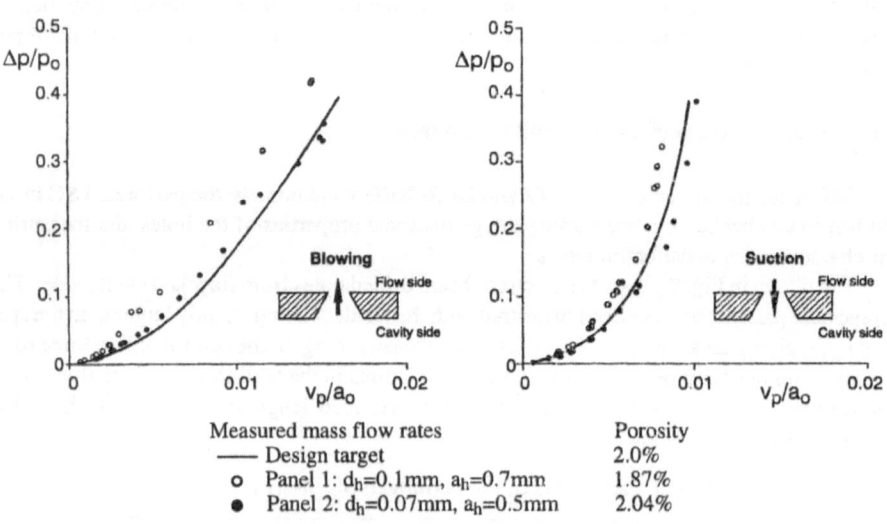

Fig. 8: Calibration of perforated test panels

326

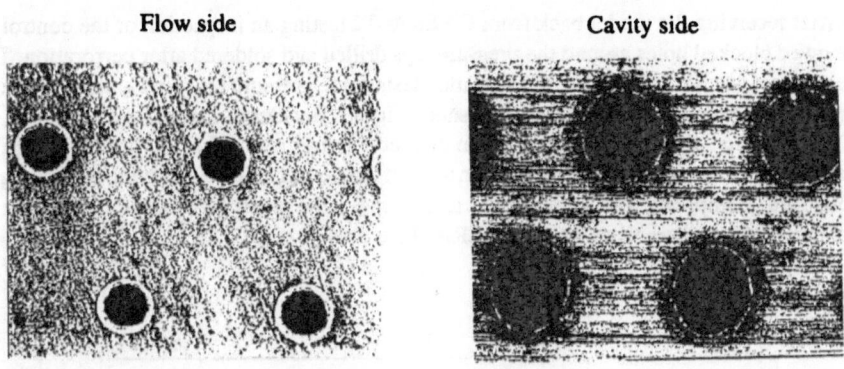

Fig. 9: Electron–beam drilled holes of SC insert (scale 100:1)

Due to the larger porosity the measured flow characteristic of the perforated SC insert shows some deviations from the specified one used for the design, Fig. 10. The actual mass flow rate versus pressure drop measured in the calibration tests has been used as input for the validation of the SC codes by the SC experiments.

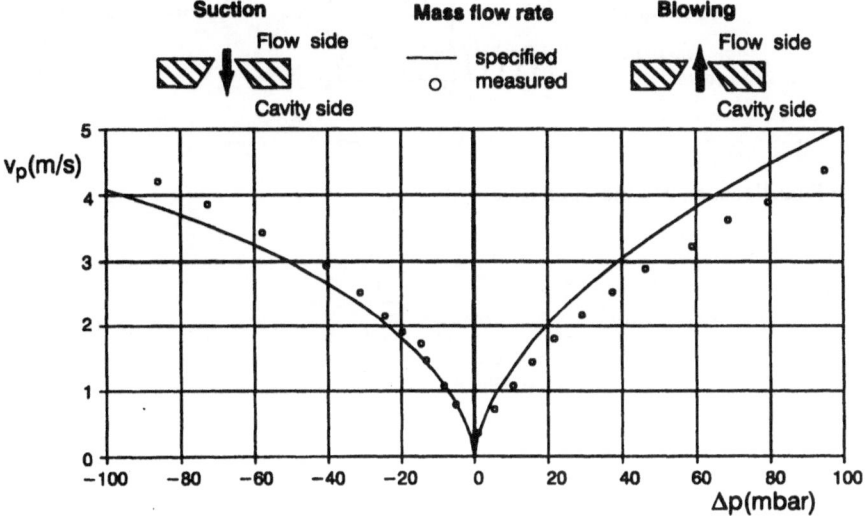

Fig. 10: Calibration of perforated SC insert

The calibration tests performed at different spanwise stations of the SC insert showed a homogenous distribution of the perforation after drilling and before assembly and instrumentation, Figs. 11 and 12. During the T2 tests a non–constant drag distribution in spanwise direction for the open insert was measured by a wake rake, containing 15 spanwise wake probes. A breakdown of the drag distribution was observed at the location of the pressure tap rows (P_1 and P_2). It was found that a similar drag distribution appears when blocking the ventilation with a locally sealed tape on the porous surface.

After receiving the model back from ONERA–T2 testing an inspection of the control region showed blocked holes around the pressure taps drilled and soldered after perforation. This blockage effect was also measured by calibration tests at the pressure tap row P_1 showing an increase of the pressure loss across the porous sheet, Fig .12. Besides its influence on the drag, the modified characteristic of the porous sheet in this section may also effect the pressure distribution. In contrast to the pressure tap position, an unchanged flow characteristic of the porous region was found after equipment with pressure taps and testing at the spanwise wake probe position 11, chosen for LDA measurements in the ONERA–T2 tunnel.

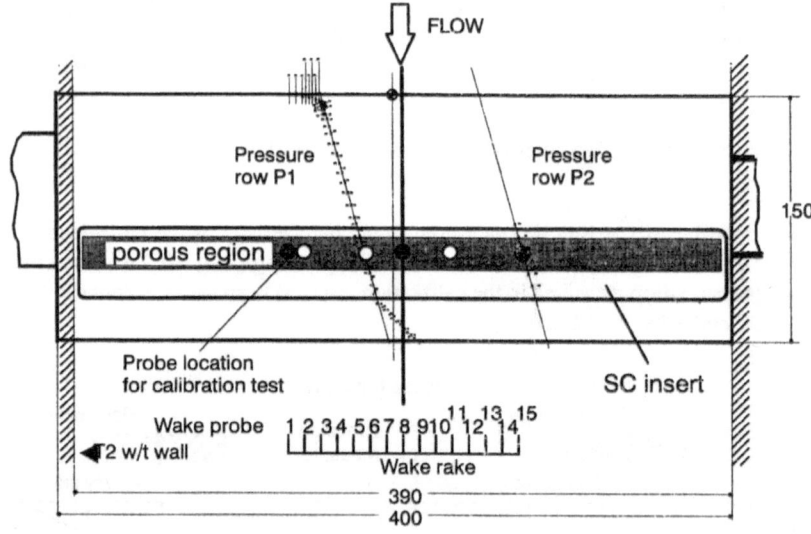

Fig. 11: Probe location of calibration tests

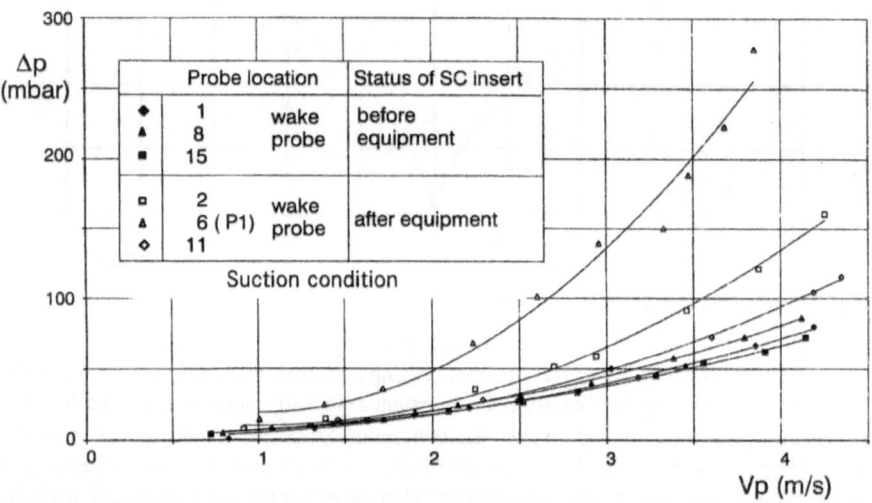

Fig. 12: Calibration tests of SC insert before and after equipment with pressure taps

19.5 Conclusions

Design, manufacture and equipment of the cryogenic laminar–type airfoil model DA LVA–1A with passive shock control inserts for SC experiments in the ONERA–T2 and DLR–KRG wind tunnels, contributed by Dasa Airbus to EUROSHOCK Task 3, have generated the following conclusions:

- To meet the different operational requirements of both adaptive–wall cryogenic wind tunnels, ONERA–T2 and DLR–KRG, a very complex cryogenic airfoil model with exchangeable passive SC devices had to be built by Dasa Airbus. Especially to meet the ONERA–T2 condition – model cool down during the initial phase of a run – the model had to be designed as a hollow body with a constant wall thickness. A new manufacturing technique has been applied to ensure a high surface quality in terms of accuracy and waviness required for high Reynolds number testing at cryogenic conditions.
- The weakest point in the manufacturing process has been proven to be the present wall perforation technique of the passive SC inserts by electron–beam drilling. In order to obtain a perforated surface with a specified flow characteristic, special calibration and check–out procedures were needed.
- After solving all specific model manufacturing problems, a working cryogenic laminar–type airfoil model is now avialable, which has been used within EUROSHOCK to generate a valuable data–base for validation of the shock control prediction methods and to investigate the efficiency of passive shock control, see Archambaud et al. [7] and Rosemann et al. [8]. Furthermore, this model will be used in EUROSHOCK II to investigate shock control by contour bumps.

19.6 References

[1]   Bahi,L.,Ross,J.M.,Nagamatsu,H.T.:"Passive shock wave/boundary layer control for transonic airfoil drag reduction", AIAA–Paper 83–0137, 1983.

[2]   Raghunathan,S.,Mabey,D.G.:"Passive shock wave/boundary layer control on a wall–mounted model", AIAA–Paper 87–0438, 1987.

[3]   Thiede,P.,Krogmann,P.,Stanewsky,E.:"Active and passive shock wave/boundary layer interaction control on supercritical airfoils", AGARD–CP–365, 1984.

[4]   Krogmann,P.,Stanewsky,E.,Thiede,P.:"Transonic shock wave/boundary layer interaction control",ICAS–Paper 84–2.3.3, 1984.

[5]   Thibert,J.J.,Reneaux,J.,Schmitt,V.:"ONERA activities on drag reduction", ICAS–Paper 90–3.6.1,1990.

[6]   Kuczmann,W.: Analyse und Vergleich mit Rechnungen der am TWB durchgeführten Messungen des DA Laminarprofils LVA–1A. DA–Bericht EF–1890, 1991, not published.

[7]   Archambaud,J.P: Qualification by laser measurements of the passive control on LVA–1Ae airfoil in the T2 wind tunnel. EUROSHOCK TR AER 2–92–49/3.4, 1996.

[8]   Rosemann,H.,Knauer,A.,Stanewsky,E.: Experimental investigation of the transonic airfoils DA LVA–1A and Va2 with shock control. EUROSHOCK TR AER 2–92–49/3.5, 1996.

20 QUALIFICATION BY LASER MEASUREMENTS OF THE PASSIVE CONTROL ON THE LVA-1A AIRFOIL IN THE T2 WIND TUNNEL

J.P. Archambaud and A.M. Rodde
ONERA/CERT, Experimental Aerodynamics Branch
2, avenue Edouard Belin, 31055 Toulouse, France

Summary: The present study concerns an experimental investigation of a passive shock/boundary layer interaction control system used on an airfoil. The passive control system is installed in the shock region and consists of a local porous part of the airfoil surface covering a small internal cavity. So the pressure jump through the shock wave is weakened by a natural fluid circulation through the porous plate, reducing the wave drag. At the same time, the viscous drag is increased, and wake measurements show that the total drag increases overall.

The first part of the study concerns pressure measurements on the model and in the wake for a varying angle of incidence at M=0.77 with and without control. The main part consists of velocity laser measurements around the passive control system and up to the trailing edge, with and without control to obtain boundary layer profiles and a far field quantification. The data constitutes a data base available for code validation.

20.1 Introduction

This report describes the work carried out by CERT/ONERA in the framework of Euroshock project, task 3 [1]. The purpose of the study is a detailed investigation of the passive control of the shock wave-boundary layer interaction.

At high subsonic or transonic speeds, a part of aircraft wings is covered by supersonic flow which is limited downstream by a shock wave. This shock wave causes wave drag and the shock wave boundary layer interaction degrades the flow until buffet occurs. These drawbacks lead to a certain acceptable limit for the aircraft flight envelope. So it seems interesting to reduce the effect of shock wave boundary layer interaction and the wave drag and to raise the limit of aircraft capability.

A natural laminar airfoil could be the place of severe shock wave-boundary layer interaction. The reason is that a favourable gradient prevents the transition to occur in the first half-chord of the airfoil, but this gradient leads to a rather strong shock wave.

The study conducted at CERT/ONERA concerns the use of a passive control system on a natural laminar airfoil. The airfoil, called LVA-1Ae, has been designed by DASA company. The passive control system consists of an insert covered by a porous plate and placed under the shock wave. The pressure jump through the shock wave produces a secondary flow circulation : viscous fluid of the boundary layer is sucked into the insert cavity downstream of the shock, and is blown out upstream of it. So the pressure jump is spread, weakening the shock wave and reducing its wave drag. In addition, the sensitivity of the flow in terms of shock-induced separation is also reduced. But this passive control system introduces perturbations : surface roughness due to the holes of the porous surface and degradation of the boundary layer by the secondary flow.

The characteristics of the passive control system used at CERT/ONERA have been defined and optimized by DASA with the experience of previous experimental investigations [2, 3]. After a rapid determination of the best configuration (α and M) for the use of passive control, testing is mainly concerned with laser measurements around the porous insert and downstream without and with control. Laser measurement results are assigned to constitute a data base which will be used to check and to improve computational methods predicting the effect of the control of the shock wave boundary layer interaction.

The experimental set-up and measurement features are presented in Sections 2 and 3, respectively. Test conditions are mentioned in Section 4. Experimental data and discussions are presented in Section 5. Conclusions are given in Section 6.

20.2 Experimental Set-up and Measurement Characteristics

20.2.1 The T2 Wind Tunnel

Tests are carried out in the T2 wind tunnel of CERT/ONERA [4]. This facility operates by runs of one or two minutes.

The T2 wind tunnel is a closed circuit tunnel, in which the flow is driven by injection of pressurized dry air. Sprayed liquid nitrogen will cool down the flow and the control of the temperature of the mixture air-gaseous nitrogen allows to adjust the Reynolds number of the test in a wide range.

The test section (Figure 1) is 0.37 m high, 0.39 m wide and 1.4 m long. It is equipped with top and bottom flexible walls. An adaptation strategy (2D or 3D according to the model geometry [5]) provides in real time the top and bottom wall shapes which cancel (2D flows) or minimize along the model axis (3D flows) the wall interferences. Wall adaptation is performed iteratively during the first part of the run before the data acquisition.

Figure 1 - T2 wind tunnel test section

The freestream Mach number is controlled by a sonic throat, downstream of the test section.

The turbulence level in the T2 wind tunnel is about $\sqrt{u'^2}/U=0.15\%$.

During the run the test parameters (stagnation pressure, stagnation temperature and Mach number) are kept constant by a regulation system controlled by a computer.

20.2.2 LVA-1Ae model

The LVA-1Ae model is a natural laminar airfoil which has been designed by DASA [6]. Figures 2 and 3 show the LVA-1Ae airfoil mounted in the T2 test section. The model has a chord of 150 mm and is mounted at mid-height of the test section on two side-wall turntables.

Figure 2 - The wake rake and the laser beam configuration

Figure 3 - The LVA-1Ae airfoil with sealed insert and transition strip

The model is hollow and is composed of three parts : the upper and lower covers, welded together and comprising the main body, and the interchangeable passive control insert.

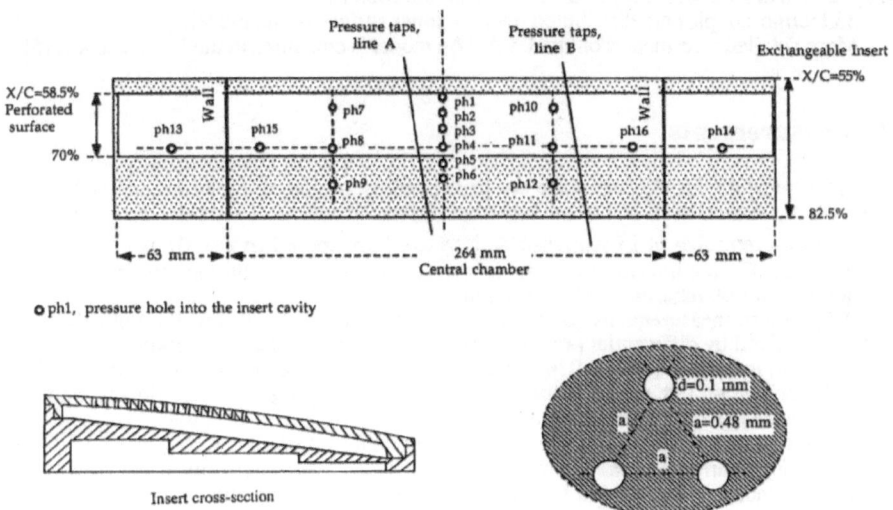

Figure 4 - Layout of the porous insert

The passive control insert (Figures 4, 5) is located between 55% and 82.5% of the chord. Perforations extend from 58.5% to 70% of the chord. Holes are normal to the surface. Nominal diameter and pitch of the holes are respectively 0.1 mm and 0.48 mm, providing a po-

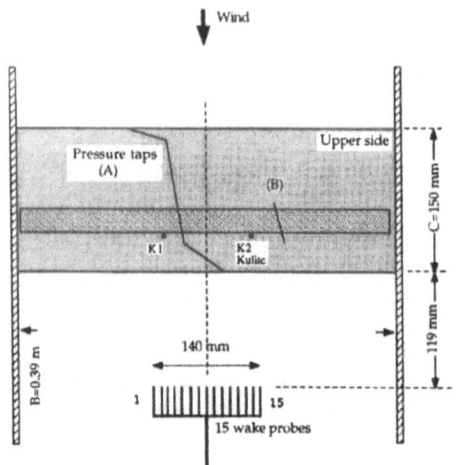

Wind

Pressure taps (A)

Upper side

(B)

K1   K2 Kulite

C =150 mm

119 mm

B=0.39 m

140 mm

1 |||||||||||||||| 15

15 wake probes

Figure 5 - Layout of the LVA-1Ae model in T2

-rosity coefficient of about 3.94%. The cavity located under the insert porous surface is divided into three separate chambers in the spanwise direction. Two small chambers at each end of the insert (63 mm wide) and the main chamber in the center of the model ($-68\% \leq Y/(B/2) \leq 68\%$; $-88\% \leq Y/C \leq 88\%$). This geometry prevents spanwise flow circulation due to side-wall effects.

The LVA-1Ae model (Figure 5) is equipped with two lines of static pressure taps. The main line A consists of 64 holes (37 holes on the upper surface, 27 holes on the lower surface).

The second line B contains 9 holes on the upper surface only. The hole diameter is 0.1 mm from the leading edge to $X/C=16\%$, 0.2 mm for $16\% \leq X/C \leq 55\%$, and 0.3 mm beyond 55%.

16 pressure taps are located inside the insert in order to check the pressure distribution in the cavity. Their arrangement is shown in Figure 4.

2 unsteady pressure transducers of Kulite-type (K1 and K2) are measuring pressure fluctuations at $X/C=83.67\%$, $Y=\pm48$ mm from the center line.

33 thermo-couples are distributed over the inner surface of the model.

More detailed information on the LVA-1Ae model is contained in the DASA report [6].

20.3 Measurements

Measurements through the wake are performed by means of a rake located at 119 mm (77% of chord) downstream of the trailing edge (figures 2 and 5). This rake is composed of stagnation pressure probes at 15 spanwise stations regularly spaced ($\Delta Y=0.01$ m), and 3 static pressure probes. During the run, this rake is moved continuously from the top down with an acquisition rate of 400 measurements / 100 mm.

All pressure measurements (on the model, along top and bottom walls, with the wake rake) are carried out by differential pressure modules modules of different working ranges, each being adapted to its pressure range. The overall accuracy of the pressure coefficients is ±0.002.

Data acquisition rate of the Kulite transducer is 2000 points/second. So the buffeting onset can be qualified in the frequency range 0-1000 Hz.

Thermo-couple temperatures were measured but not evaluated.

The second part of the testing consists of local velocity measurements in the flow, using the laser anemometer of the T2 wind tunnel. This laser is a three-colour Argon laser of 15 W power. During the tests, it is operating at 5 W power which is an optimized configuration. The assumption is made that the flow is two-dimensional in the central part of the test section where laser measurements are performed; the longitudinal and vertical components of the velocity vector are measured using only two colours (blue and green). The measuring volume (crossing of the four light beams, two green beams and two blue beams) is about 0.12 mm wide (diameter of the cross-section in a vertical plane) and 1.3 mm long (spanwise).

Oil seeding is introduced far upstream of the settling chamber with oil droplets of about 1 µm to 2 µm in diameter. The luminous signal sent by each seeding particle traversing the fringe system of the measuring volume is observed by two photomultipliers (green/blue) and then analysed by two B.S.A. (Burst Spectrum Analyser).

A first selection during acquisition consists of rejecting bursts having low signal/noise ratios, that means particles traversing only a part of the measuring volume. After this selection, 1000 or 2000 particles are acquired, determining the velocity vector at one point of the flow field. In fact 30 velocity measurements (at 30 points of the flow field) can be carried out during the same run.

Finally data consists of 2000 (or 1000 depending of the measuring time) signals (frequencies) for each colour at each point of the flow field. A second selection rejects couples of signals (blue, green) which are not synchronized in time. This selection is necessary to determine the correct cross fluctuating velocity component (u'w'). After this selection the number of valid frequencies (or particles) is usually about 600, but can be reduced near the model surface or around the shock ($\leq$ 100). In fact when the measuring volume approches the model, light reflected by the metallic surface is seen by the receiving optical system, increasing the light noise level and perturbing the quality of the measurement.

In the T2 wind tunnel, tests are performed at R_c=4.6x10^6, P_t=2.3 bar. Under the lift force (at M=0.77 and α=1°, the nominal conditions for the laser measurements), the LVA-1Ae model is slightly bent to the top. Nevertheless, all measurement coordinates are determined in the model system axis under aerodynamic forces, so *they are corrected for the model deformation* (0.03 mm accuracy/model).

The angle of attack of the model is measured before the test with an accurate air-level device (accuracy ±0.01°).

20.4 Test Conditions

All the tests were performed at a stagnation pressure of 2.3 bar, at ambient temperature, giving the expected Reynolds number R_c=4.6x10^6 (C=chord=0.15 m) at M=0.77.

Boundary layer transition was fixed at 48% on upper and lower surfaces of the model. The transition strip is a band of carborundum grains glued to the surface. This band of 0.060 mm height and 1.5 mm width has a regular geometry because of the fabrication process. The laminar nature of the boundary layer between the leading edge and X/C=48% and the effeciency of the transition strip were checked by oil visualization at M=0.77 and α=1° before the tests.

The 2D adaptation strategy [5] was used in order to cancel top and bottom wall interferences. The method converges in 3 or 4 iterations in about 15 seconds. The final wall shape takes into account the boundary layer displacement thickness of the four walls in order to have a zero Mach number gradient flow (in the empty test section).

It can be noted that the adapted wall shapes relative to the same configuration with and without passive control are quasi identical.

An estimation of sidewall boundary layer effects computed for the central test condition (M=0.77, α=1°) has given the following corrections : *ΔM=-0.008 and $\Delta\alpha$=-0.2°*. That means that the tests made at M=0.77, α=1° must be compared to computations performed at M=0.77 + ΔM=0.762 and α=1° + $\Delta\alpha$ = 0.8°.

The freestream Mach number is well adjusted by the tunnel control process. The freestream Mach number accuracy is estimated to be ±0.001.

The action of the passive control is underlined by comparison with a reference test performed at the same flow conditions with a solid-surface insert. By common consent with DASA, reference tests at the present campaign have been carried out by covering the porous insert with a 0.02 mm thick self-adhesive tape. In the past [4] the comparison between a real solid insert and the use of the 0.02 mm adhesive tape showed that the drag difference does not exceed 3 counts (ΔC_D = 0.0003), which can be neglected in a first approach.

335

20.5 Experimental Data

The experimental investigation is composed of three parts :

1/ Determination of the Mach number value providing the best configuration for the passive control. The determination is made at α=1° with sealed insert, and the result (M=0.77) will be considered as the nominal Mach number for the campaign (nominal angle of attack, α=1°)

2/ Incidence variation at nominal Mach number.

3/ Velocity measurement in the flow field with the laser anemometer at the nominal test condition (M=0.77, α=1°) without and with passive control.

20.5.1 Determination of the nominal Mach number

The nominal configuration for laser measurement must correspond

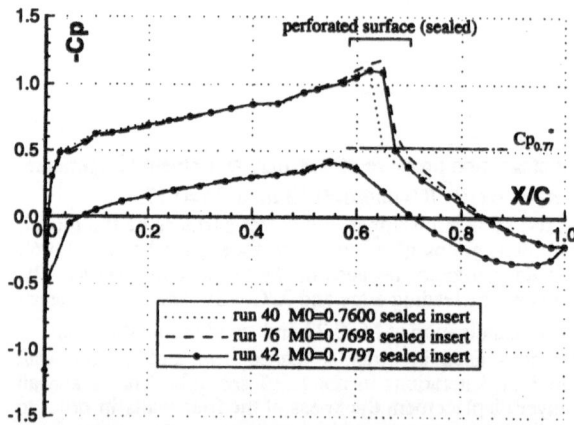

to a shock position suitable for the most significant action of passive control, that means the most downstream position on the sealed insert,
* to a case prior to the onset of flow separation.

Figure 6 shows pressure distributions for M=0.76, 0.77 and 0.78 on the reference airfoil (sealed insert). It is obvious that the increase in Mach number from 0.76 to 0.77 improves the shock location and that above 0.77 flow separation occurs.

Figure 6 - Test Mach number determination

Consequently the value **0.77 is considered as the** *nominal Mach number, and the nominal condition for laser measurements is M=0.77 and α=1°.*

20.5.2 Incidence variation at M=0.77

Tests are performed at M=0.77 and α=-1°, 0°, 0.5°, 1°, 1.5°, 2°, 2.5°. For each angle of incidence a comparison is made between sealed and open insert cases. Test parameters are compiled in the tables of the Chapter 20.7. Figures 7 to 9 present pressure distributions on the model (tap line A) and wake pressure distributions at two rake sections (8 and 11 corresponding to different flow behaviour in the case of passive control, see Figure 10.

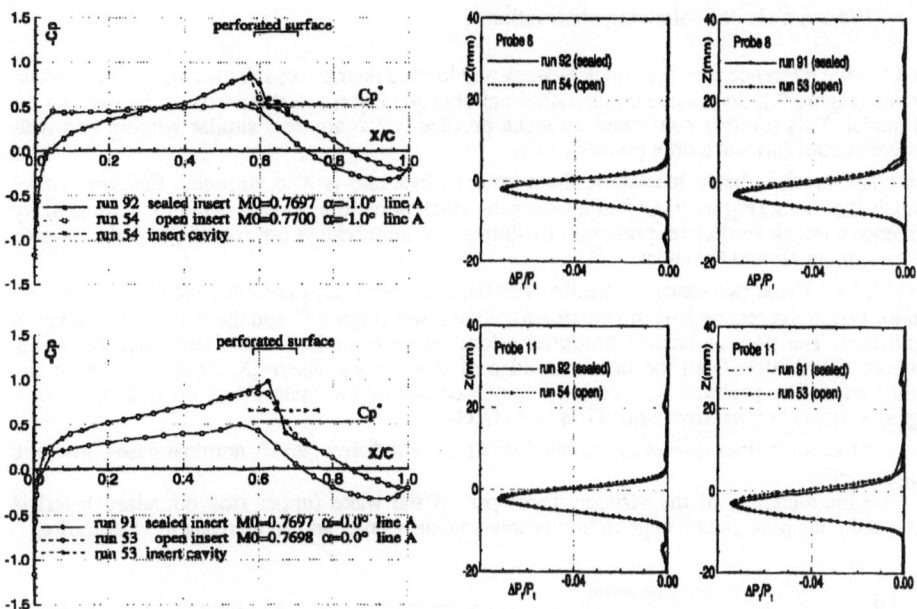

Figure 7 - Pressure distributions on the airfoil and in the wake at M=0.77, α=-1° and 0°

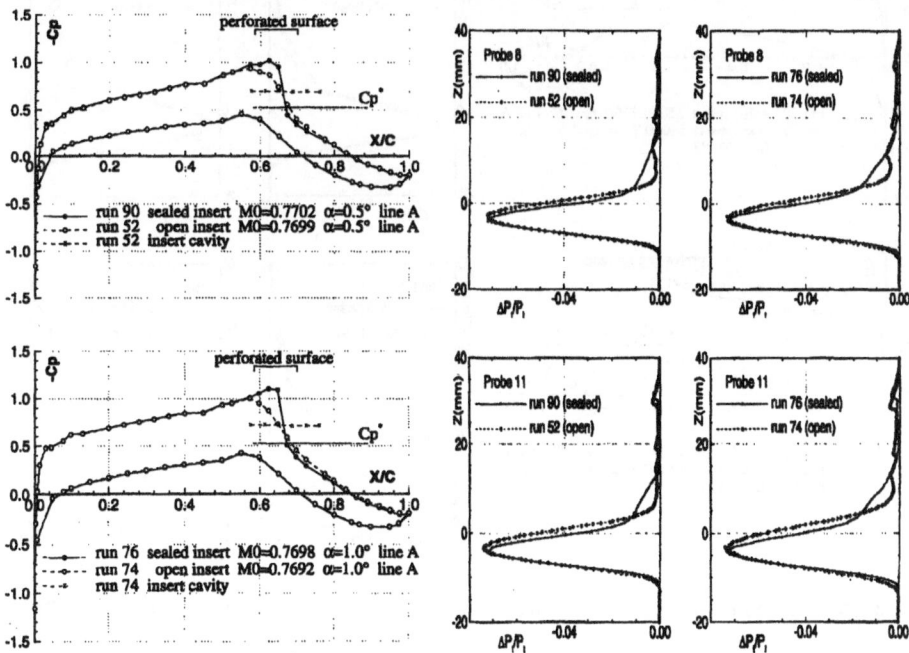

Figure 8 - Pressure distributions on the airfoil and in the wake at M=0.77, α=0.5° and 1°

337

We can make the following observations :

α=-1° In the reference case the shock is weak and located near the upstream edge of the porous surface (Figure 7). So passive control does not change very much the pressure distribution on the model. This result is confirmed by wake profiles which are very similar without and with passive control (no wave drag pocket).

α=0° The shock is better located on the porous surface and is also stronger. Passive control spreads the shock (Figure 7) and suppresses the small wave drag pocket which is exhibited by reference case. Nevertheless, pressure distribution on the model is not modified by the passive control except around the shock.

α=0.5°, 1.0° These two cases are similar. For the reference configurations, the shock is strong and located in the second half (downstream) of the insert (Figure 8) and the wave drag pocket is significant. The passive control produces a kind of more or less continuous change of the pressure distribution from the upstream edge of the porous insert (X/C=58.5%) up to the trailing edge. The pressure distribution is modified behind the open insert up to X/C=100%. Figure 8 shows that passive control has two effects :

• the wave drag pocket is removed or strongly reduced (α=1°, nominal case), giving a drag reduction.

• the thickness of the viscous upper part of the wake (upper side boundary layer) is increased ; this points out a degradation of this boundary layer and an increase in viscous drag.

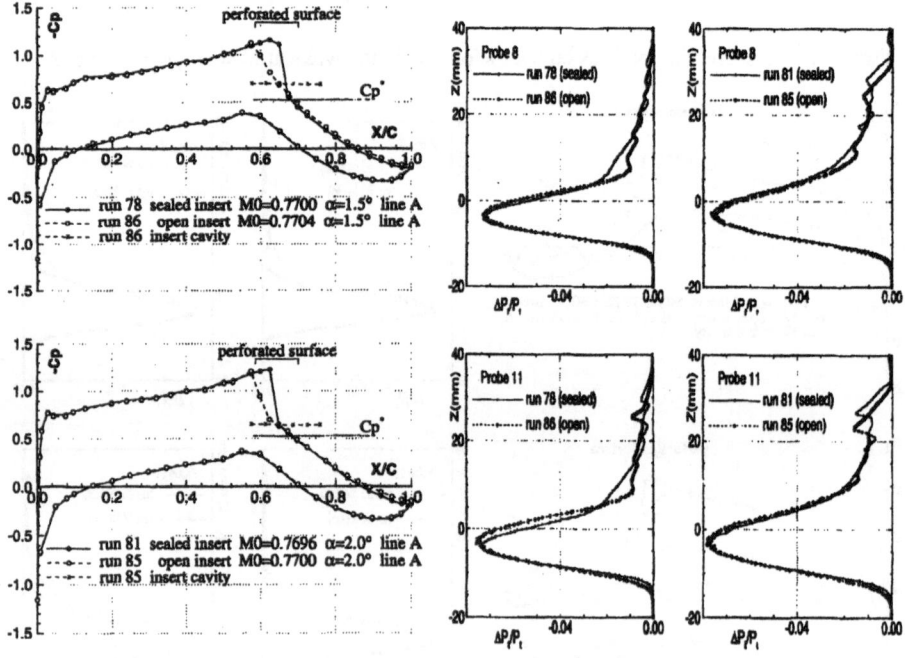

Figure 9 - Pressure distributions on the airfoil and in the wake at M=0.77, α=1.5° and 2°

338

α=1.5° This case is similar to the previous case, with an accentuation of tendencies : the change in model pressure gradient is fixed at the upstream edge of the porous insert (Figure 9) and the wave drag pocket is partially removed with passive control.

α=2° In the reference case, the shock is moving forward, compared to the α=1.5° test. Sidewall effects are rather strong and cause a partial separation of the boundary layer behind the shock on the upper surface of the model. This complex flow looks like generated by two opposite vorticies, as pointed out by visualization. At these conditions, passive control does not change very much the pressure distribution through the wake (Figure 9 indicates a very small benefit in wave drag and a small increase in viscous drag). We can mention here the beginning of the dissymmetry of the wake (increase in viscous drag for probe 11, decrease on the centerline for probe 8).

α=2.5° The forward movement of the shock continues. The reference shock position is at the upstream edge of the insert. So passive control has no effect on the pressure distribution on the model and therefore on the wake pressure profile. It is interesting to note that the asymmetry of the wake, Figure 10, is more marked for the reference case, pointing out that the open insert reduces three-dimensional behaviour of the flow on the model surface by allowing a transverse equilibrium of pressure.

Complementary information from pressure taps of the line B and in the cavity leads to the following remarks :
• Line A and B distributions are close together (2D flow), except for α=2.5°and sealed insert (3D flow on the model, related to sidewall effects), but in this case the open insert reduces the 3D flow intensity as already seen from the wake profiles. The pressure seems to be constant in the entire central chamber of the insert, as indicated by the agreement of the distributions ph1-6, ph7-9 and ph10-12 (see Figure 4).
• The pressure is constant in the central chamber for all configurations. On the contrary, the two extreme pressure taps indicate higher pressure in the small chambers located at the sides of the insert.

Figure 10 shows the variation of drag coefficient with spanwise position for all the tests mentioned above.

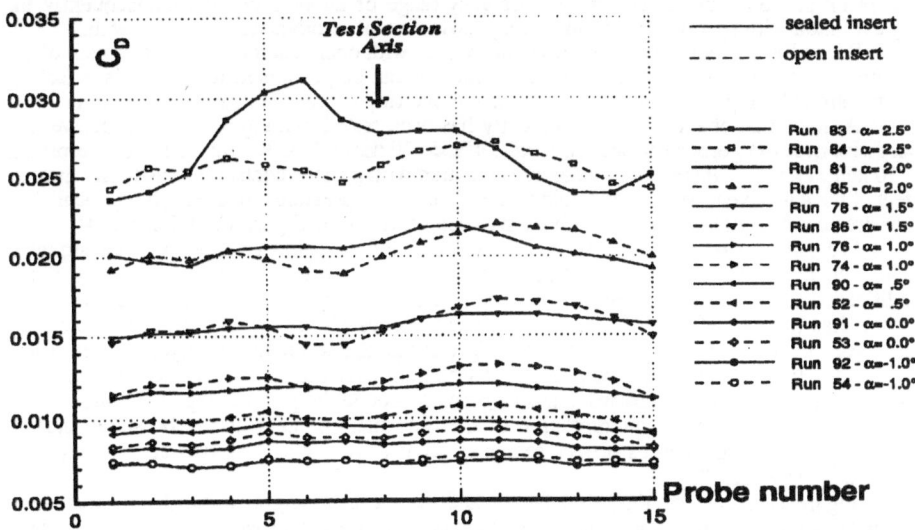

Figure 10 - Variation of drag coefficient with spanwise position for different angles of attack

The first observation is that for each type of test (sealed/open insert) curves have globally the same shape, and that the distorsion of this shape is increasing with angle of attack.

For the reference cases (sealed insert) curves are rather flat up to $\alpha=1.5°$. For $\alpha=2°$ and $2.5°$, the two-dimensional aspect of the wake is perturbed by three-dimensional effects. In addition, for $\alpha=2.5°$, the peak indicated by probes 4, 5, 6, 7 is due to a turbulent wedge produced by pressure taps located near the leading edge.

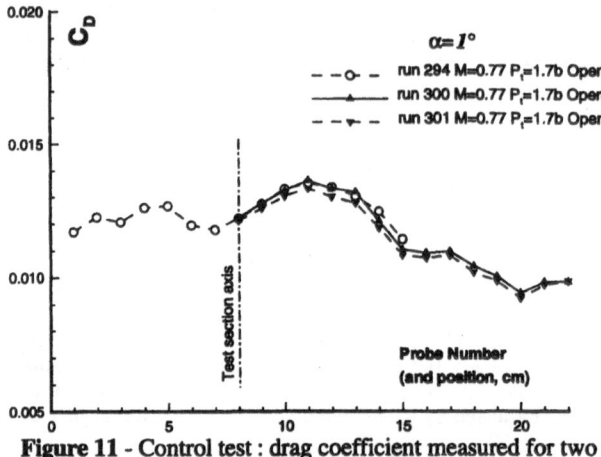

Figure 11 - Control test : drag coefficient measured for two different transverse positions of the wake rake

For passive control cases, the curves of the drag variation takes a wavy shape, the amplitude of which increases with α. The lowest level corresponds to the probes N° 6, 7 and 8 and N° 14, 15. Except at these positions, drag with passive control is higher than without for moderate angles of incidence ($\alpha \leq 1.5°$).

In order to better understand the variation of the drag with spanwise position, two investigations have been performed.

Firstly, the wake rake has been shifted by half its length in the spanwise direction. Figure 11 shows a comparison of the drag variation measured with the two wake rake configurations for the same conditions (open insert). It is obvious that the three tests give the same result in the common part with an accuracy range of about 5 counts. Consequently the drag evolution with spanwise position corresponds to a real aerodynamic phenomenom.

Secondly, we can remark that low drag values with open insert occur downstream of the pressure taps located on the porous insert surface. It can be postulated that there is a lack of porosity around the pressure taps due to the manufacturing process of this instrumentation.

A simulation of a local lack of porosity has been tested, putting a narrow adhesive tape chordwise over the porous surface (covered surface : 20 mm $\leq Y \leq$ 30 mm, ahead of the probes N° 10 and 11), where there is a large increase in drag with passive control. It was found that the effect of the adhesive tape is comparable to the effect of the pressure tap areas, giving locally the same drag coefficient without and with passive control (also see Figure 41 of Chapter 4).

We can, furthermore, remark that the presence of the adhesive tape does not suppress locally the reduction of the wave drag (probes N° 10 and 11), probably due to a minor modification of the shock geometry by the tape ; on the contrary, there is a local reduction of the viscous drag (probes N° 8, 9, 10 and symmetrically probes N° 11, 12, 13), probably connected with three-dimensional effects arising at the edges of the tape where there is a jump in porosity.

After the T2 campaign, the porosity distribution over the insert has been measured by DASA. This investigation confirms a lack of porosity around the pressure tap installation.

In conclusion to the interpretation of the drag measurement, it can be stated that the natural behaviour of the porous surface seems to occur in two locations, between wake probes N° 3 and 5, N° 9 and 13.

Figure 12 shows the time signal (left) and the spectrum (right) of the Kulite transducer K1 for some test configurations with and without control. We can observe that the amplitude of the signal is increasing with the angle of attack, and that control increases the amplitude of the signal and the spectrum level at constant conditions up to $\alpha=2°$. On the contrary, at $\alpha=2.5°$, the open insert is reducing the signal amplitude and the spectrum level at low frequency.

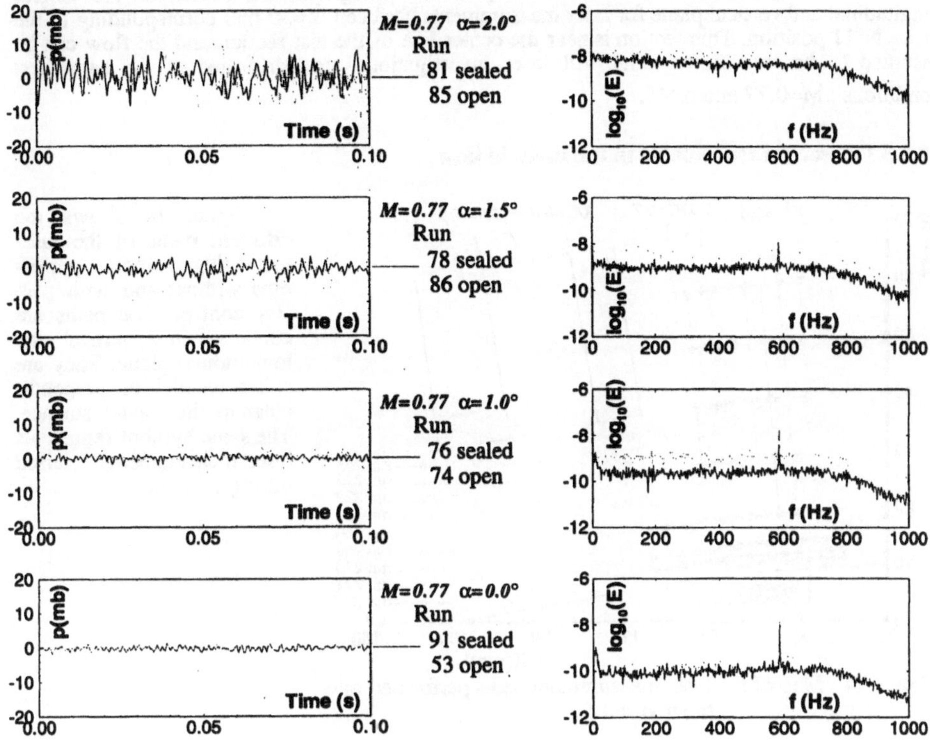

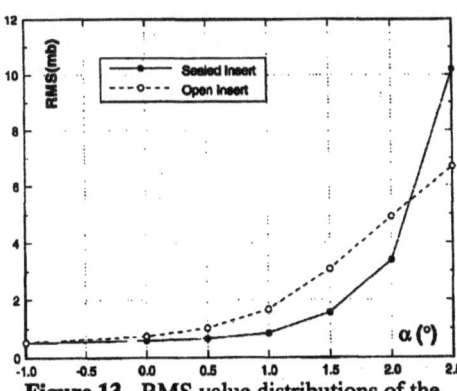

Figure 12 - Pressure fluctuations (left) measured by Kulite transducer and corresponding spectra (right) at M=0.77, $\alpha=0°$ to $2°$

The Figure 13 presents the RMS (Root Mean Square) value of the signal with the angle of attack. It points out the two opposite actions of the porous insert :
• increase of moderate pressure fluctuations for $\alpha \leq 2.0°$,
• damping of large pressure fluctuations for $\alpha=2.5°$.

Figure 13 - RMS value distributions of the Kulite transducer signal versus angle of attack

341

20.5.3 Laser measurements

According to the conclusion of the previous paragraph, the choice of the correct longitudinal and vertical plane for laser measurements has been Y=30 mm corresponding to the probe N°11 position. This section is near the center line of the test section and the flow can be assumed to be two-dimensional. All laser investigations are performed at the same test conditions : M=0.77 and α=1°.

20.5.3.1 Laser measurements in the inviscid flow

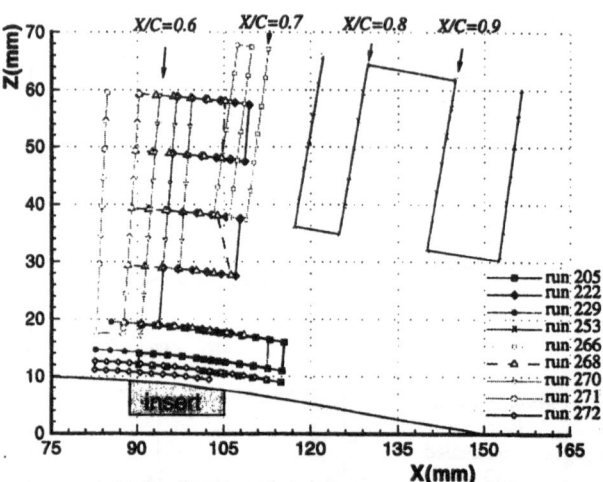

Figure 14 shows the different paths of the laser measuring volume, performed without and with passive control. The paths are contained in a vertical and longitudinal plane. They are either parallel or perpendicular to the model surface. The same symbol represents laser measurements realized during one run.

Figure 14 - Map of the laser measurement paths performed into the inviscid flow

The first set of paths is concentrated around the shock boundary layer interaction (at $\widetilde{Z} \leq$ 10 mm distance from the model surface).

Figure 15 presents the variation of the longitudinal velocity component (left) with X. Without control we observe a certain variation of the velocity level with $\widetilde{Z}$ upstream and downstream of the shock, while the shock is slightly moving downstream (the shock position for $\widetilde{Z}$ =3 mm can be estimated although one missing point prevents the drawn full line to represent the shock). Upstream of the shock, passive control equalizes the velocity levels and produces a small adverse gradient, giving a weaker shock intensity. The shock is placed more upstream. Downstream, the arrangement of the different U distributions is similar to the no-control case, however with higher velocity values.

Figure 15 (right) also shows the variation of the vertical velocity component. The main change due to passive control occurs before the shock, where blowing through the insert is deflecting the flow upward. It is interesting to note that without control the vertical velocity is increasing with the distance from the model on two sides of the shock, while with passive control this tendency is inverted upstream of the shock under the action of the blowing which is most important at the wall.

342

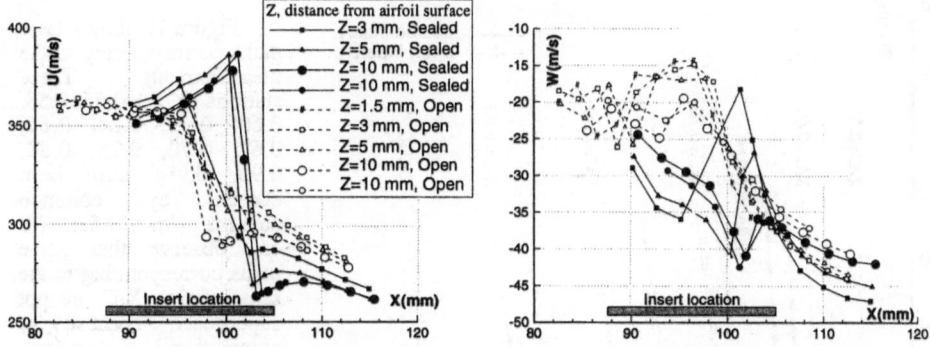

Figure 15 - Comparison of distributions of the longitudinal (left) and vertical (right) velocity components along the chord for Z≤10mm above the airfoil surface

Figure 16 (left) presents longitudinal U components versus X for measuring paths above $\widetilde{Z}$ =10 mm, pointing out the movement of the shock and the U level in front of it with the distance $\widetilde{Z}$ .The same figure also presents the variation of the vertical velocity component.

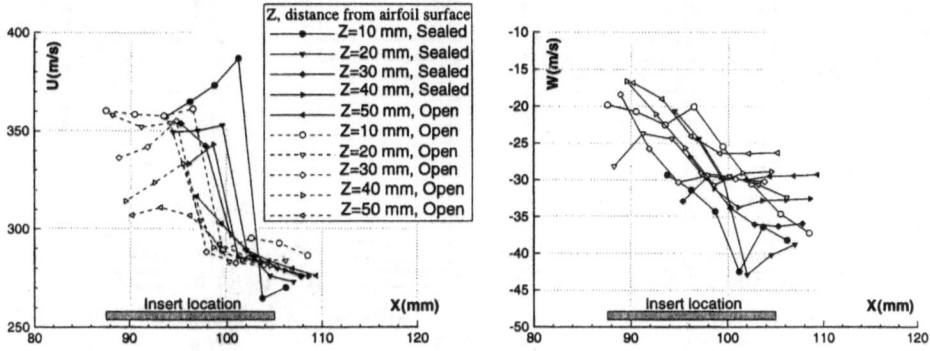

Figure 16 - Comparison of distributions of the longitudinal (left) and vertical (right) velocity components along the chord for Z≥10mm above the airfoil surface

20.5.3.2 Laser measurements of boundary layer profiles

The main objective of this section is to measure the velocity components in the viscous layer at different stations along the chord. In fact, one laser measurement is performed perpendicular to the airfoil surface up to a distance of 30 mm, with, however, the majority of points being concentrated near the wall.

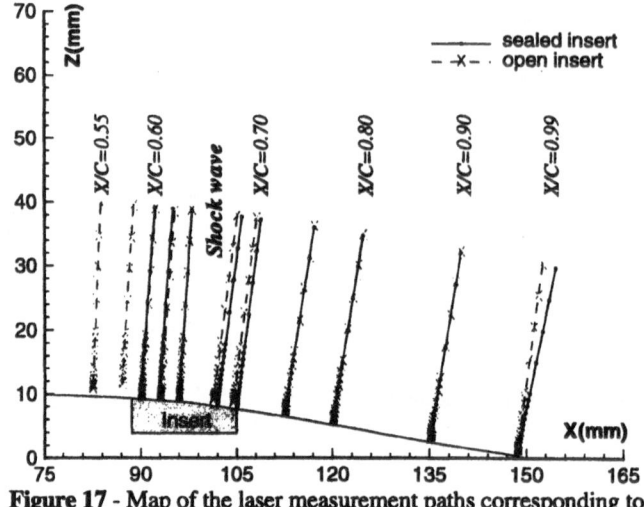

Figure 17 shows laser paths corresponding to the measurements. These stations (X/C= 0.55, 0.58, 0.60, 0.62, 0.64, 0.66, 0.68, 0.70, 0.75, 0.80, 0.90, 0.99) have been chosen by common consent with DASA. We can observe that some paths corresponding to the same station X/C are not superposed because of unintentional errors. However, the consequence of these small differences in position can be neglected.

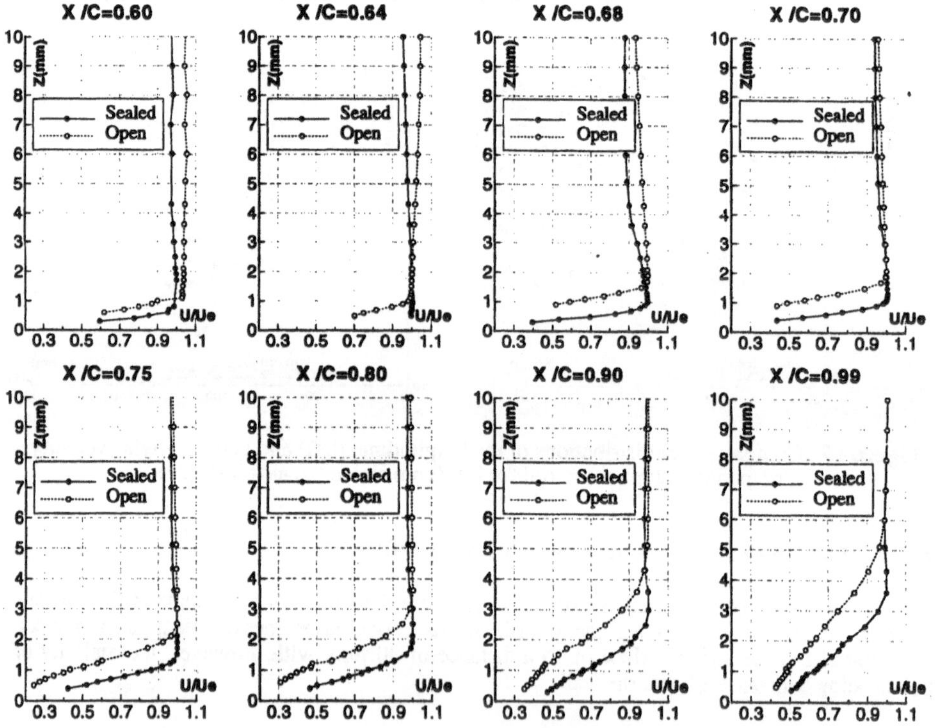

Figure 17 - Map of the laser measurement paths corresponding to the boundary layer probings

Figure 18 - Compilation of boundary layer longitudinal velocity profiles

The most interesting longitudinal velocity profiles (limited to Z=10 mm) are gathered in the Figure 18.

The first observation is the thickening of the boundary layer all along the investigated area due to the control action.

Secondly, the slight change of the profile shape at low velocity can be noted downstream of X/C=0.75 : the passive control seems to impose a curvature to the velocity profile, corresponding to the tendency to separation. Consequently, these profiles cannot be easely estimated by a simple wall law, while this treatment is possible with profiles associated with the sealed insert.

Finally, the typical overshoot of the velocity value without control, outside of the boundary layer and beyond the shock, seems to disappear with the passive control.

The Figures 19 to 30 show alternatively (except for X/C=0.60 and 0.64) the mean velocity profiles (longitudinal U/Ue and vertical W/Ue) and the turbulent stresses (u'^2/Ue^2, w'^2/Ue^2, $u'w'/Ue^2$). We can notice the local increase of the positive vertical velocity component near surface blowing (Figures 19, 20 for X/C=0.60, 0.64) and the opposite tendency in the suction part (Figure 21, X/C=0.68). The increase in turbulent stresses downstream of the insert is also pointed out in the Figures 24, 26, 28, 30.

All these results constitute a useful data base for comparison with computational results, in order to check or to adjust influencing parameters of passive control.

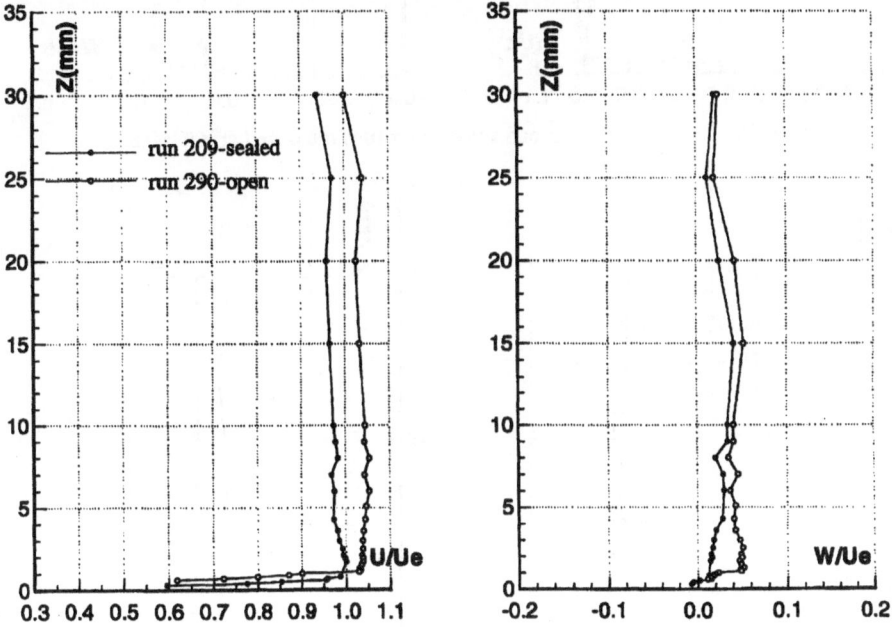

Figure 19 - Boundary layer mean velocity profiles measured at X/C=0.60

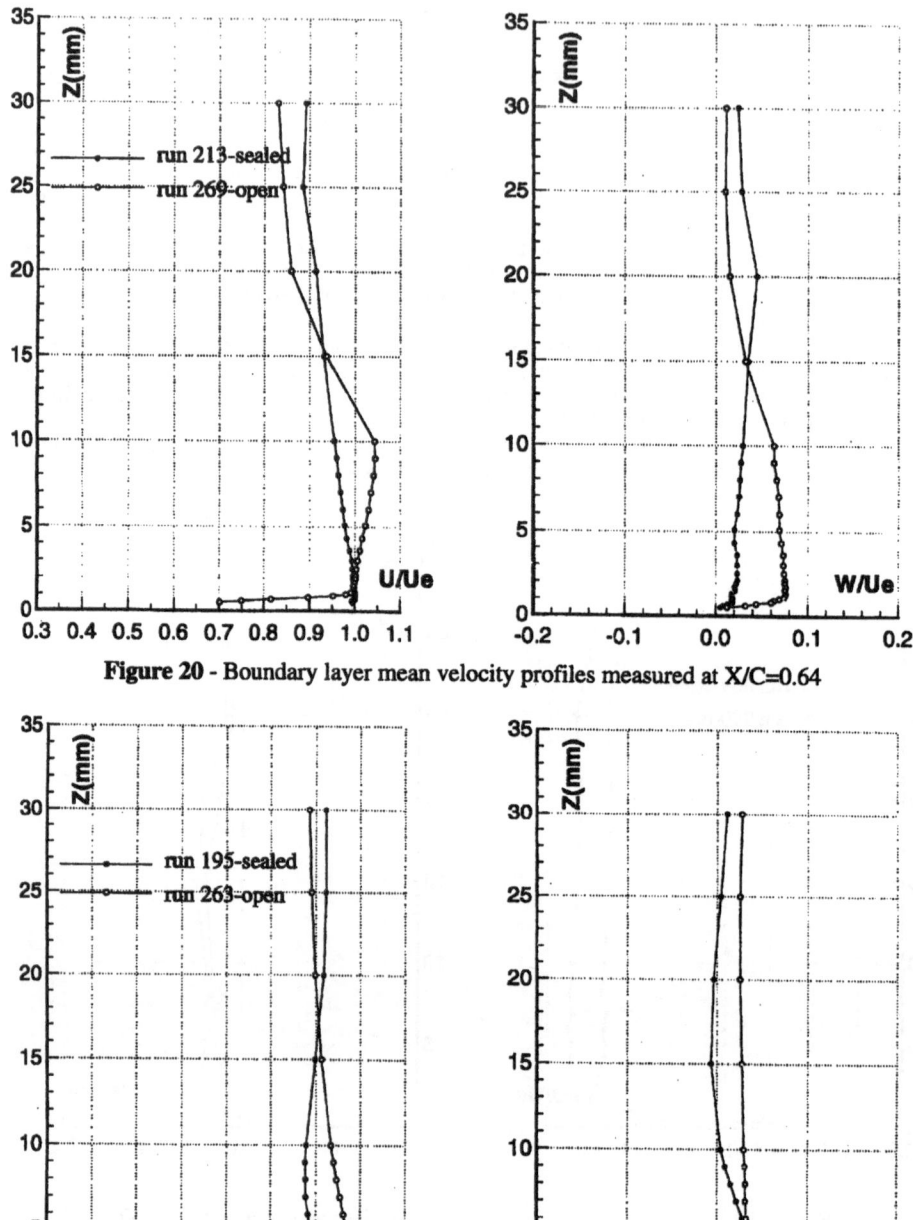

Figure 20 - Boundary layer mean velocity profiles measured at X/C=0.64

Figure 21 - Boundary layer mean velocity profiles measured at X/C=0.68

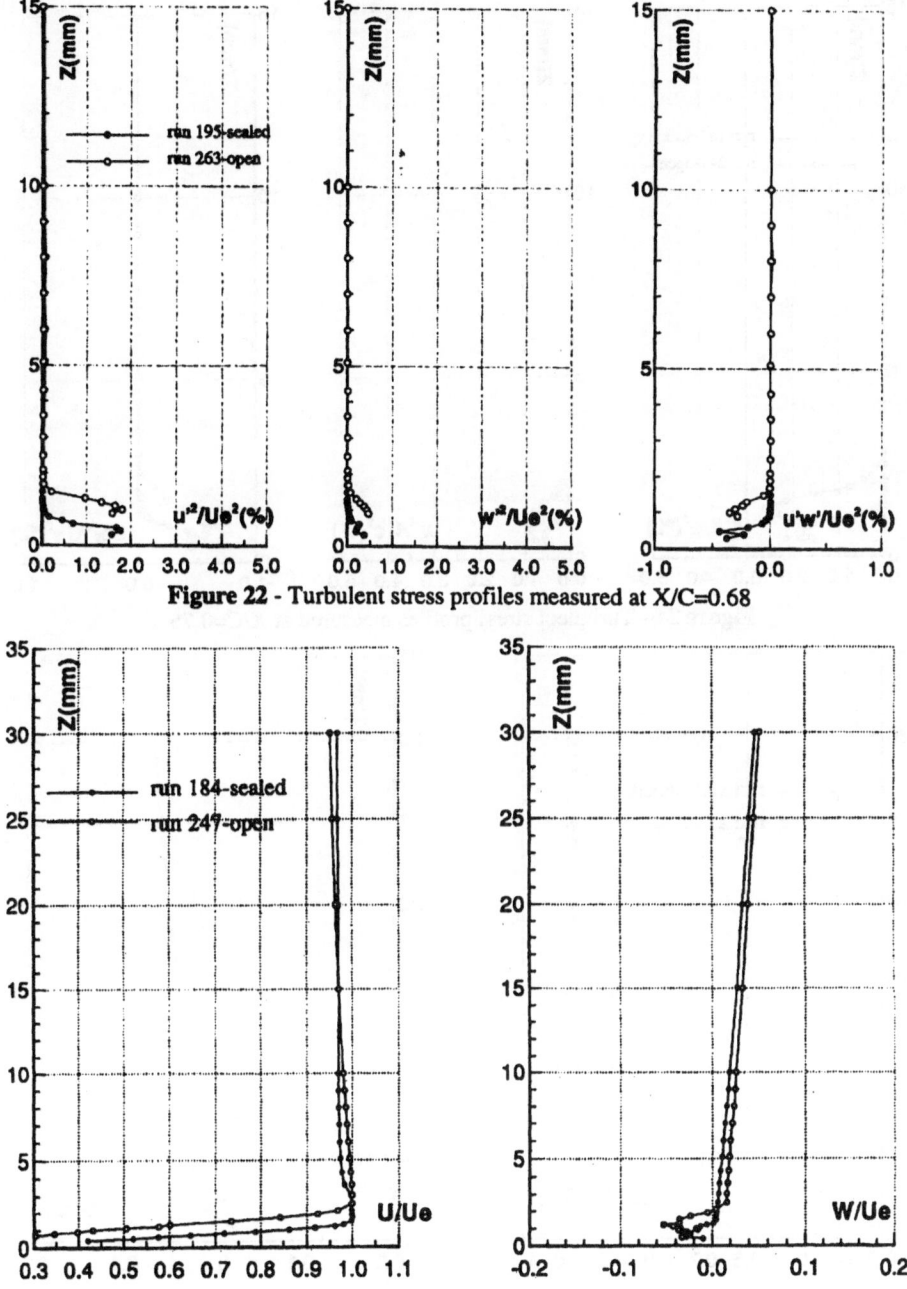

Figure 22 - Turbulent stress profiles measured at X/C=0.68

Figure 23 - Boundary layer mean velocity profiles measured at X/C=0.75

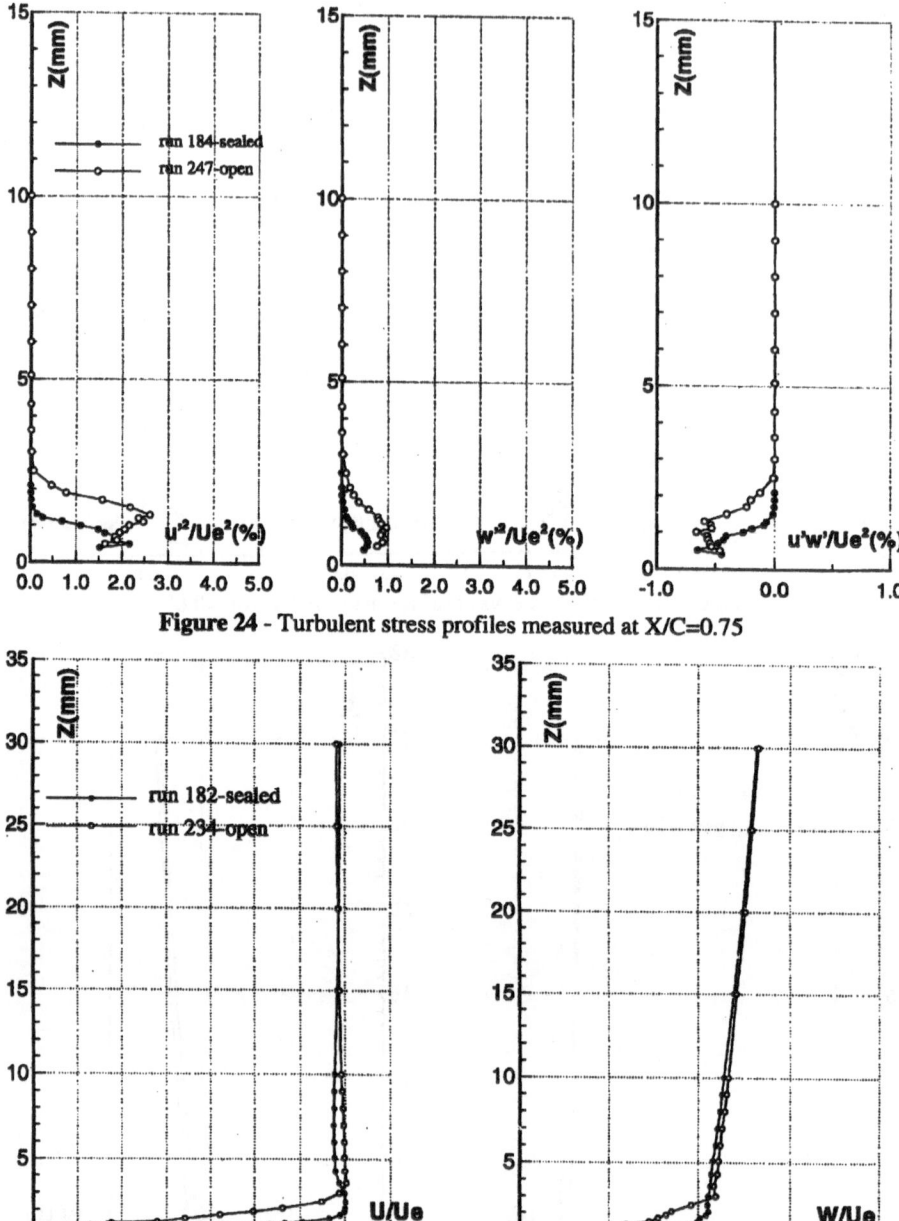

Figure 24 - Turbulent stress profiles measured at X/C=0.75

Figure 25 - Boundary layer mean velocity profiles measured at X/C=0.80

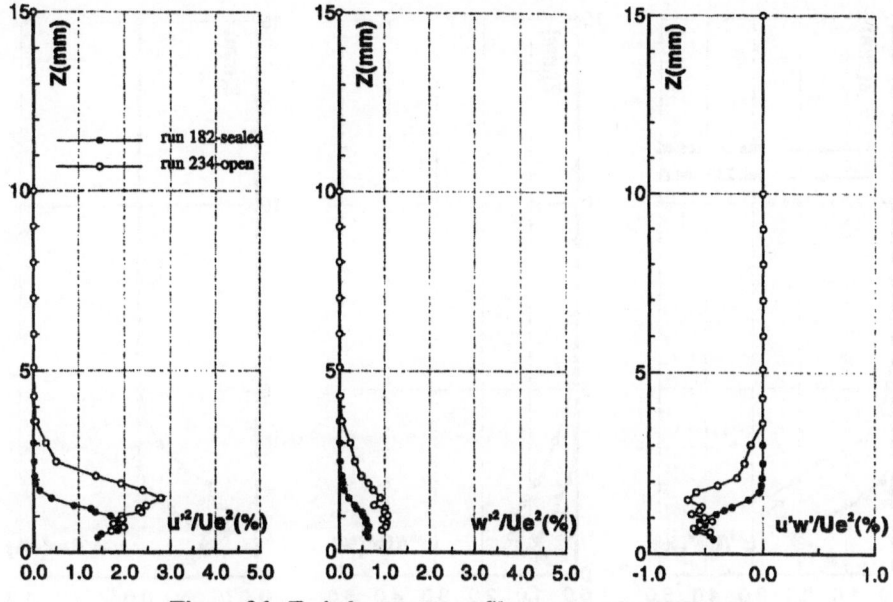

Figure 26 - Turbulent stress profiles measured at X/C=0.80

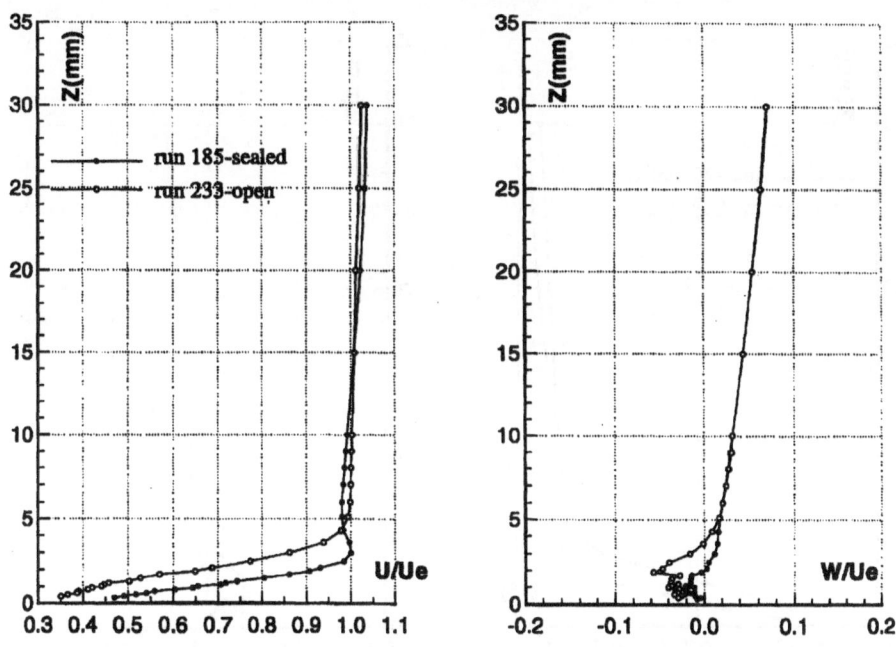

Figure 27 - Boundary layer mean velocity profiles measured at X/C=0.90

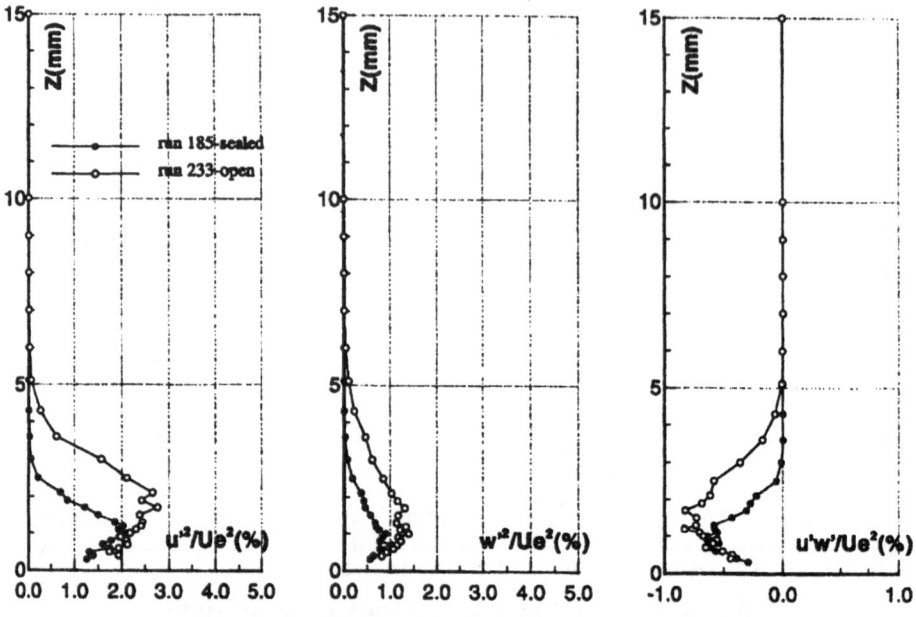

Figure 28 - Turbulent stress profiles measured at X/C=0.90

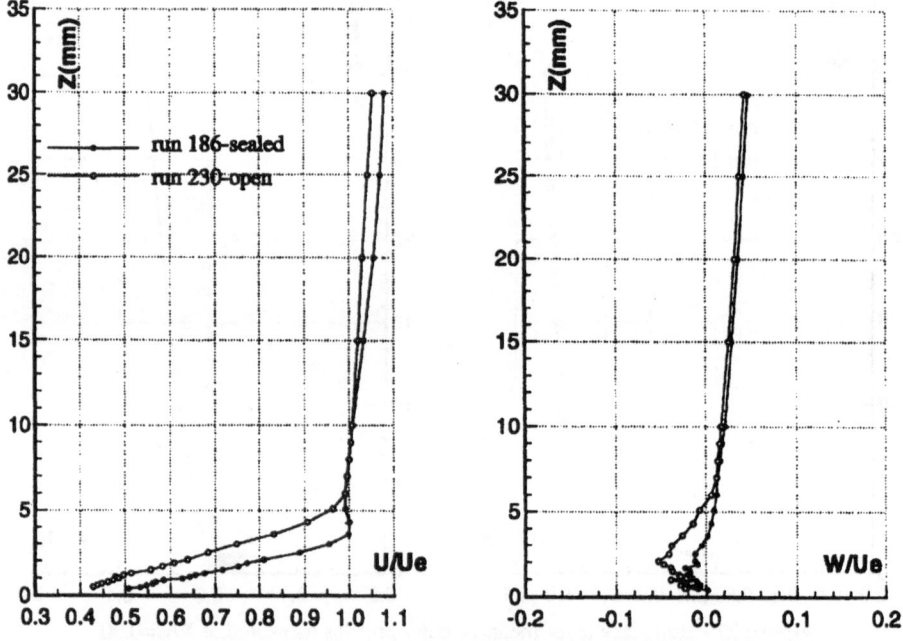

Figure 29 - Boundary layer mean velocity profiles measured at X/C=0.99

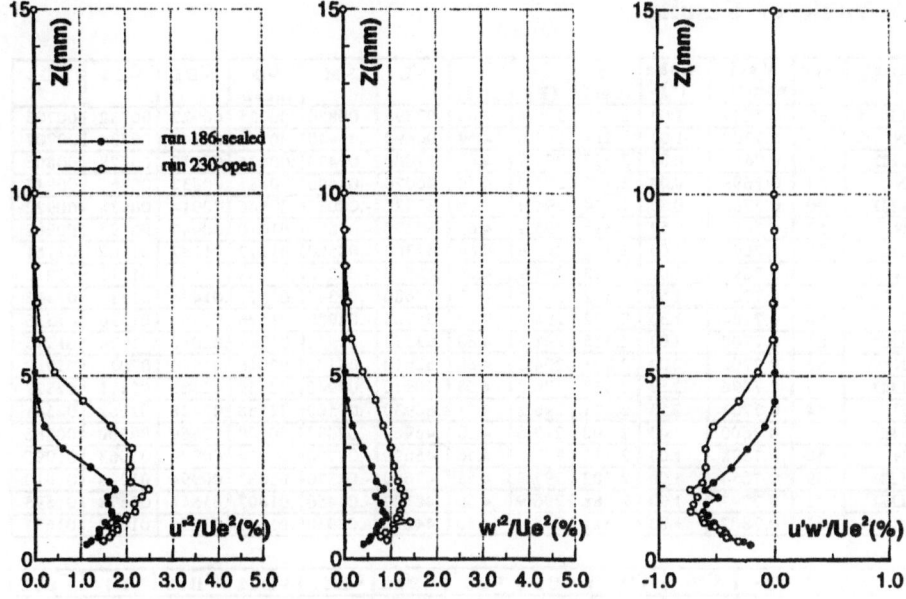

Figure 30 - Turbulent stress profiles measured at X/C=0.99

20.6 Conclusions

The present passive control experimental study leads to the following conclusions:

The passive control system used, based on a porous insert, does not give a drag reduction for the conditions tested. Having no effect at low angle of incidence, it produces an increase in drag growing with the incidence.

The passive control system reduces significantly the wave drag for moderate angles of incidence by spreading and weakening the shock.

But the passive control system increases the viscous drag (skin friction drag + pressure drag) by a large amount which exceeds the wave drag reduction. The increase in viscous drag can be associated with the manipulation of the boundary layer by the flow circulation through the insert and the porous surface.

Laser measurements have produced a data base for a large part of the model without and with passive control. This data base is composed of velocity data in the inviscid flow and the boundary layer (mean velocity components and fluctuating terms) and is available for numerical code development.

20.7 Table of Results

Insert	Run	Mach number	Alpha (°)	Pt (bar)	Tt (K)	Rc Mil.	CL	CM	CD mean	CD1	CD2	CD3
SEALED C	92	0.7697	-1.00	2.319	295.4	4.65	.07390	-.06820	.00725	.00727	.00732	.00703
OPEN C	54	0.7700	-1.00	2.311	297.6	4.59	.06890	-.06940	.00741	.00741	.00734	.00707
SEALED	91	0.7697	0.00	2.323	294.9	4.67	.27000	-.07470	.00836	.00810	.00828	.00805
OPEN	53	0.7698	0.00	2.312	297.0	4.60	.26070	-.07400	.00881	.00832	.00863	.00843
SEALED	90	0.7702	0.50	2.324	294.4	4.68	.37330	-.08030	.00946	.00914	.00938	.00922
OPEN	52	0.7699	0.50	2.315	295.9	4.63	.35690	-.07870	.01010	.00949	.00995	.00982
SEALED	76	0.7698	1.00	2.320	295.6	4.64	.47360	-.08580	.01172	.01127	.01166	.01161
OPEN	74	0.7692	1.00	2.323	294.9	4.66	.45560	-.08330	.01233	.01149	.01211	.01210
SEALED C	78	0.7700	1.50	2.322	294.5	4.67	.57680	-.09340	.01571	.01494	.01519	.01523
OPEN	86	0.7704	1.50	2.322	294.7	4.67	.54760	-.08920	.01576	.01464	.01536	.01533
SEALED	81	0.7700	2.00	2.315	297.5	4.60	.65250	-.09950	.02049	.02010	.01974	.01945
OPEN	85	0.7700	2.00	2.321	294.3	4.68	.62810	-.09400	.02040	.01920	.02011	.01980
SEALED	83	0.7696	2.50	2.313	296.9	4.60	.70910	-.09960	.02669	.02361	.02411	.02529
OPEN	84	0.7701	2.50	2.321	296.2	4.63	.68930	-.09560	.02568	.02428	.02562	.02541
SEALED	40	0.7600	1.00	2.300	299.5	4.50	.46960	-.08090	.00900	.00869	.00900	.00888
OPEN	50	0.7598	1.00	2.311	297.3	4.56	.45620	-.07790	.00974	.00909	.01063	.01095
SEALED	39	0.7654	1.00	2.302	297.9	4.55	.47600	-.08370	.01033	.00994	.01028	.01019
SEALED	42	0.7797	1.00	2.288	296.9	4.59	.48100	-.09450	.01607	.01544	.01574	.01595
OPEN	51	0.7802	1.00	2.311	298.4	4.60	.44610	-.08930	.01641	.01555	.01642	.01631

Run	CD4	CD5	CD6	CD7	CD8	CD9	CD10	CD11	CD12	CD13	CD14	CD15
92	.00712	.00745	.00737	.00743	.00718	.00725	.00735	.00744	.00738	.00705	.00709	.00700
54	.00715	.00757	.00742	.00743	.00724	.00745	.00770	.00774	.00764	.00732	.00737	.00725
91	.00822	.00865	.00850	.00861	.00838	.00850	.00863	.00861	.00851	.00814	.00805	.00808
53	.00871	.00918	.00883	.00883	.00876	.00906	.00927	.00929	.00907	.00884	.00862	.00820
90	.00937	.00970	.00970	.00957	.00947	.00963	.00980	.00975	.00953	.00939	.00915	.00897
52	.01011	.01044	.01001	.00994	.01010	.01050	.01074	.01077	.01050	.01019	.00981	.00907
76	.01173	.01189	.01196	.01171	.01179	.01189	.01207	.01207	.01180	.01164	.01143	.01116
74	.01248	.01251	.01182	.01177	.01224	.01270	.01314	.01324	.01307	.01274	.01221	.01118
78	.01550	.01558	.01559	.01533	.01552	.01608	.01634	.01634	.01634	.01609	.01588	.01571
86	.01595	.01556	.01450	.01447	.01529	.01607	.01679	.01730	.01712	.01683	.01614	.01495
81	.02040	.02062	.02063	.02052	.02093	.02182	.02197	.02136	.02054	.02013	.01978	.01924
85	.02035	.01984	.01914	.01893	.01998	.02084	.02145	.02207	.02180	.02164	.02087	.01995
83	.02866	.03035	.03114	.02866	.02776	.02792	.02794	.02678	.02494	.02398	.02396	.02515
84	.02622	.02581	.02542	.02466	.02574	.02664	.02695	.02723	.02645	.02578	.02456	.02429
40	.00902	.00911	.00925	.00909	.00898	.00919	.00928	.00914	.00900	.00878	.00865	.00898
50	.01011	.00981	.00930	.00912	.00935	.00978	.01017	.01013	.01000	.00968	.00932	.00857
39	.01034	.01048	.01059	.01036	.01034	.01052	.01061	.01052	.01030	.01012	.01000	.01030
42	.01629	.01629	.01624	.01568	.01602	.01646	.01676	.01677	.01631	.01594	.01555	.01555
51	.01696	.01654	.01538	.01524	.01620	.01686	.01746	.01771	.01741	.01696	.01611	.01497

20.8 List of Symbols

B=0.39 m : width of the T2 test section

C : model chord=0.15 m

C_D : total drag coefficient (C_{Dmean}, mean value of all probes ; C_{Di}, value of probe i)

C_L : lift coefficient

C_M : pitching moment coefficient

C_p : pressure coefficient

f : frequency (Hz)

M : freestream Mach number of the test

P_t : total pressure (bar)

R_c : Reynolds number based on model chord length

RMS : Root Mean Square

T_t : total temperature (K)

U_e : velocity (external flow to the boundary layer)

U : longitudinal velocity component

u' : longitudinal velocity fluctuation

W : vertical velocity component

w' : vertical velocity fluctuation

X : longitudinal coordinate

Y : transversal coordinate

Z : vertical coordinate

α : model angle of attack (°)

20.9 References

[1]   J.P. Archambaud, A.M. Rodde - "Qualification by laser measurements of the passive control on the LVA-1Ae airfoil in the T2 wind tunnel", EUROSHOCK TR AER2-92-49/3.4, 1996.

[2]   J.P. Archambaud, J.B. Dor, J.F. Breil, P. Barricau - "Passive shock control investigation with the LVA-1A airfoil in the T2 wind tunnel",P.V. d'Essais N° 18/5017.21 (OA N° 96/1685 AY 210 D), Août 1993.

[3]   J.P. Archambaud, J.B. Dor, J.F. Breil, P. Barricau - "Passive shock control investigation with the LVA-1A airfoil in the T2 wind tunnel-Second part : optimized inserts D and E",Rapport Final DERAT N° 19/5017.22 (OA N° 104/1685 AN 211 D), Janvier 1995.

[4]   R. Michel, C. Quemard, A. Mignosi - "The induction driven tunnel T2 of ONERA/CERT", Journal of Aircraft, Vol.16, N°3, March 1979.

[5]   J.P. Archambaud, A. Mignosi - "Two-dimensional and three-dimensional adaptation at T2 transonic wind tunnel of ONERA/CERT",AIAA, 15th Aerodynamic Testing Conference, San Diego, California (USA), May 18-20 1988.

[6]   G. Dargel, P. Thiede - "Passive shock control investigation on airfoils. Part I: numerical flow simulation. Part II: model design and manufacture", DASA Final Technical Report, DA-Report No. EF 06/96 ; EUROSHOCK TR AER2-92-49/2.7, February 1996.

21 EXPERIMENTAL INVESTIGATION OF THE TRANSONIC AIRFOILS DA LVA-1Ae AND VA-2 WITH SHOCK CONTROL

H. Rosemann, A. Knauer, and E. Stanewsky

DLR, Institute of Fluid Mechanics,
Bunsenstraße 10, 37073 Göttingen, Germany

Summary: Experiments were carried out in the DLR Cryogenic Ludwieg-Tube on the airfoil DA LVA-1Ae in order to investigate the influence of the Reynolds number on the effectiveness of passive shock control. The test results, based on pressure distributions and wake measurements, show in general an increase in total drag due to control which is, however, much smaller at the higher Reynolds numbers; at certain conditions, even a drag reduction could be achieved. The present measurements were supplemented by flow field measurements carried out with the same model at ONERA / CERT (Chapter 20) at a Reynolds number of 4.6×10^6.

During the course of the EUROSHOCK research program it was found that for the laminar-type airfoils investigated passive control generally resulted in an increase in total drag, whereas earlier measurements [1] with the turbulent VA-2 airfoil had indicated that total drag could be reduced considerably by passive control.

In order to ensure that the earlier results from the turbulent VA-2 airfoil were not erroneous due to possible faults in test procedures, these measurements were repeated. The analysis of the results of these tests confirmed the earlier measurements in trend; however, the magnitude of the gains observed earlier could not be realized which is possibly due to differences in the initial boundary layer development. The tests with active control revealed a substantial gain in performance compared to passive control.

21.1 Introduction

Early measurements with the (turbulent) airfoil VA-2 have shown that by passive shock control *via* a perforated surface with a cavity underneath, located in the shock region, total drag can be reduced over a wide range of free-stream conditions and the drag-rise and buffet boundaries can be shifted to higher Mach numbers and/or lift coefficients, Fig. 1 [1].

Based on these findings and similar results in the literature, an extensive program was launched to investigate the effect of passive shock control on the flow development over two laminar-type airfoils, *viz.*, the DRA 2303 and the DA LVA-1Ae airfoils, respectively [2].

A large-scale model of the DRA-2303 airfoil was tested by DRA in the 8 ft × 8 ft wind tunnel with one objective being to relate details in the control region to the overall flow development. A 150 mm-chord model of the DA LVA-1Ae was tested in the DLR Cryogenic Ludwieg-Tube (KRG) to investigate the effect of the boundary layer state and, more generally, Reynolds number on the effectiveness of passive control at Reynolds numbers between 4.6×10^6 and 12×10^6 including experiments with tripped boundary layers as well as with natural transition. The same model was also tested in the transonic wind tunnel T2 of ONERA / CERT with emphasis on flow field surveys.

In the present contribution we will give, after a short description of the experimental set-up in the KRG in Chapter 21.2, first a comparison of the T2 and KRG results and discuss thereafter

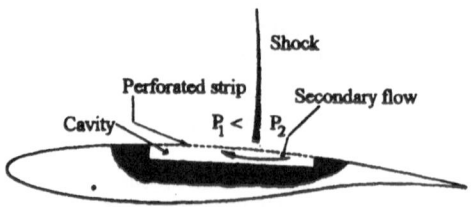

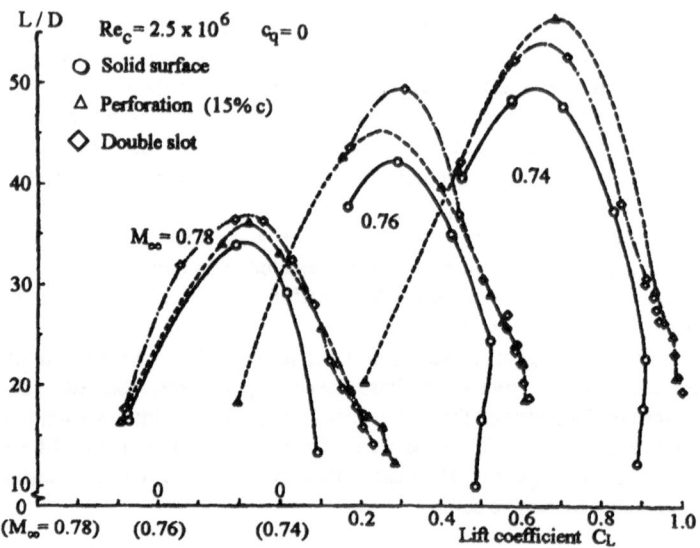

Figure 1: Influence of passive control on L/D for the VA-2 airfoil, measurements 1984 [1]

the influence of Reynolds number on the development of drag.

Initial numerical and experimental results of the investigation have shown that for the laminar airfoils investigated

- the shock oscillations were strongly damped by control, *i.e.*, the buffet boundary was shifted to higher Mach numbers and/or lift coefficients,

- the shock strength, hence wave drag, was reduced by control, but

- the total drag, contrary to the results for the turbulent airfoil VA-2, generally increased when passive control was applied, see, *e.g.* Chapter 4.3.3, Fig. 52.

Due to the opposing results on drag behavior obtained for the VA-2 airfoil and the two laminar-type airfoils, the early measurements with the VA-2 airfoil were repeated in 1995 in order to ensure that the positive results obtained in 1984 were not spurious and caused by a faulty experimental set-up but were a true representation of the flow development envisaged in conjunction with passive shock control.

A brief description of the present experiments and the results obtained including a comparison with the earlier data and a discussion of the results of the active-control tests carried out at the same time is given in Section 21.3.

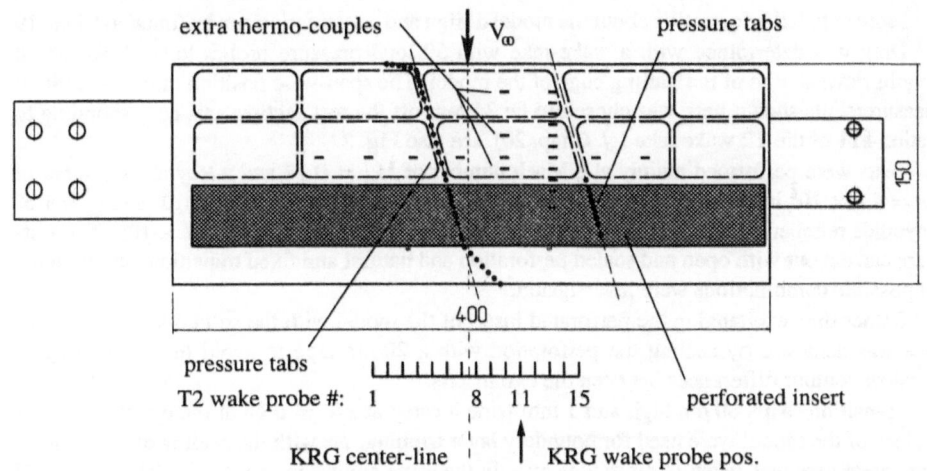

Figure 2: The DA LVA-1Ae model with inserts for passive shock control

21.2 Investigation of the Laminar-Type Airfoil DA LVA-1Ae with Passive Control at Reynolds Numbers of $4.6 \times 10^6 \leq \mathrm{Re} \leq 12 \times 10^6$

21.2.1 The Cryogenic Ludwieg-Tube

The experiments were performed in the Cryogenic Ludwieg-Tube of DLR Göttingen (KRG), which is a blow-down wind tunnel with a test time between 0.6 s and 1.0 s built according to the Ludwieg-Tube concept. High Reynolds numbers are achieved by applying stagnation pressures up to 1 MPa together with temperatures down to 100 K utilizing cryogenic technology.

The tests were carried out in the 2-D adaptive wall test section with a cross section of $0.4 \times 0.35 \ \mathrm{m}^2$ and a length of 2 m.

21.2.2 Model and test conditions

The LVA-1Ae model, shown in Fig. 2, was designed and built by DASA-Airbus. The span of the model is 400 mm and the chord-length 150 mm. The airfoil was designed for extended laminar boundary layers by maintaining accelerated flow up to about 60% and 55% chord on the upper and lower surface, respectively.

The model is equipped with a passive shock control insert with a porous surface between $x/c = 0.575$ and $x/c = 0.70$. The porosity is realized with holes of 0.1 mm diameter yielding an open area of about 5%.

Pressure distributions were measured along two lines of pressure tabs shown in Fig. 2. The pressure inside the cavity underneath the perforated surface was measured at mainly two chord-wise oriented sections and some span-wise stations.

For transition detection, the model was equipped with a number of Chromel-Alumel thermo-couples mounted on the inner side of the model wall optimized for the T2 wind tunnel test conditions. However, due to the short run time of the KRG, these transducers could not be used for the present experiments. To check the feasibility of transition detection with thermo-couples in the KRG, three supplementary transducers were installed on the upper side of the model with a distance of only 0.5 mm to the outer surface of the wall.

More detailed information about the model design and construction can be found in Chap. 19.

Drag was determined with a wake-rake with 59 total pressure probes located two chord lengths downstream of the trailing edge of the model. The span-wise position of the rake for all measurements shown here was chosen to be 35 mm off the test section axis corresponding to station #11 of the T2 wake rake (cf. Chap. 20), see also Fig. 2.

Tests were performed mainly at a Mach number of $M_\infty = 0.77$ and a Reynolds number of $Re = 4.6 \times 10^6$ in the angle of attack range of $-1.0° \leq \alpha \leq 2.5°$. At $\alpha = 1.0°$ and $\alpha = 1.5°$ Reynolds number sweeps were carried out from $Re = 4.6 \times 10^6$ to $Re = 12 \times 10^6$. The tests were carried out with open and sealed perforation and natural and fixed transition, although not all possible combinations were investigated.

Rather than exchanging the perforated insert of the model with the solid one, a closed surface was achieved by sealing the perforation with a 20 μm tape to avoid the uncertainty of possible contour differences between the two inserts.

Transition strips 60 μm high and 1 mm wide located at $x/c = 0.48$ at the upper and lower surface of the model were used for boundary layer tripping. As with the sealing of the perforation, great care was taken to reproduce exactly the same conditions on the model as in the T2 tests by using the same tape for sealing and the same material (carborundum) and procedures for the manufacturing of the transition strip.

21.2.3 Experimental results

21.2.3.1 Comparison between KRG and T2 results

Only a comparison between the data obtained in the KRG and T2 wind tunnels will be presented in this section; the effects of passive control on the results will be discussed in the next chapter together with the Reynolds number influence.

The lift and drag polars for $M_\infty = 0.77$ and $Re = 4.6 \times 10^6$ are shown in Fig. 3. Two drag polars are plotted for the T2 results, representing the span-wise average and the values measured at the span-wise station #11, respectively. For reasons, not yet determined completely, the quality of the drag measurements in the KRG was not very satisfactory for larger angles of incidence. Therefore, results for angles higher than $\alpha = 1.5°$, will not be presented here. Further investigations will hopefully make more results available.

The lift polar in Fig. 3 shows a slightly reduced slope for the KRG results compared to the T2, perhaps indicating differences in the residual wall interferences. Drag is significantly higher at low angles of attack in the KRG, but approaches the T2 polars at $\alpha = 1.0°$ and $\alpha = 1.5°$ which is the range of interest for the present investigation.

Corresponding pressure distributions for $\alpha = 1.0°$ are shown in Fig. 4. The shock is located about 2% further upstream in the KRG, which is the reason for the lower lift coefficient. Also, there is a small deviation in the pressure of the first pressure tab behind the transition strip on the lower side, indicating perhaps slight differences in the geometry of the strip. In the rear part of the airfoil, starting a little earlier on the lower side, the pressure in the KRG is generally somewhat lower than in T2, being in accordance with the higher drag measurement.

Similar differences are found comparing the results for the $\alpha = 1.5°$ case (not shown here).

The lift and drag polars obtained with the open perforation are plotted in Fig. 5. The slope of the lift polar of the KRG measurements is similarly reduced as in Fig. 3 without passive control.

The drag polars measured in T2 (average and pos. #11) exhibit a larger spread compared to Fig. 3, indicating a larger span-wise variation of drag with the perforation open. There is

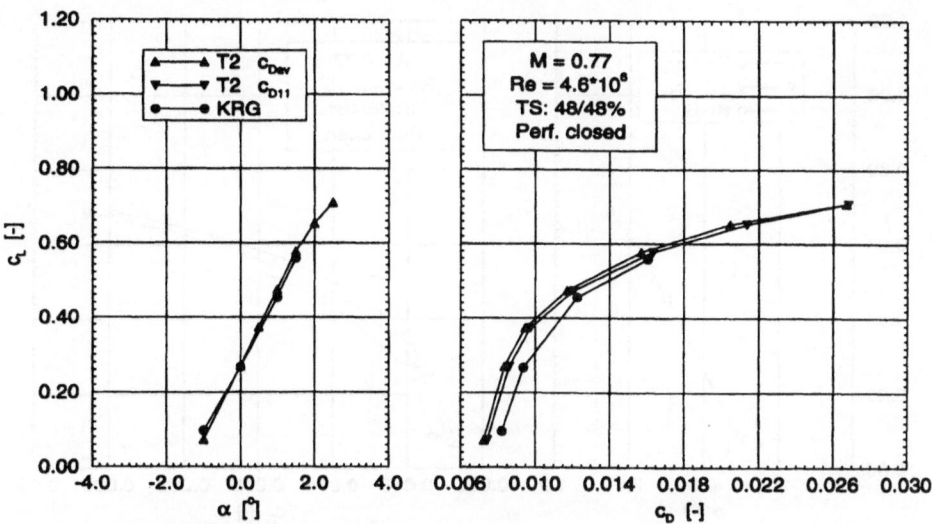

Figure 3: Comparison of lift and drag polars measured in T2 and KRG for the LVA-1Ae with closed perforation

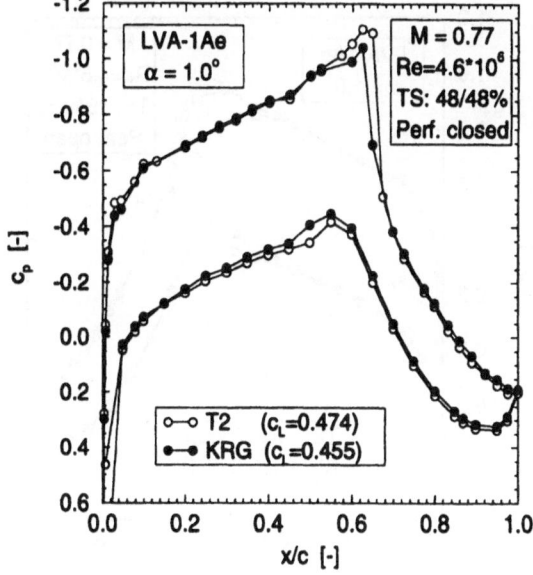

Figure 4: Comparison of pressure distribution with closed perforation and fixed transition

a nearly constant shift between the drag polars of KRG and T2 with the larger values again measured in the KRG.

The comparison of the pressure distributions at $\alpha = 1.0°$ in Fig. 6 reveal a similar, but slightly smaller, chord-wise displacement of the pressure distribution in the shock region as with the perforation closed, Fig. 4. Except for the locations affected by the transition strip,

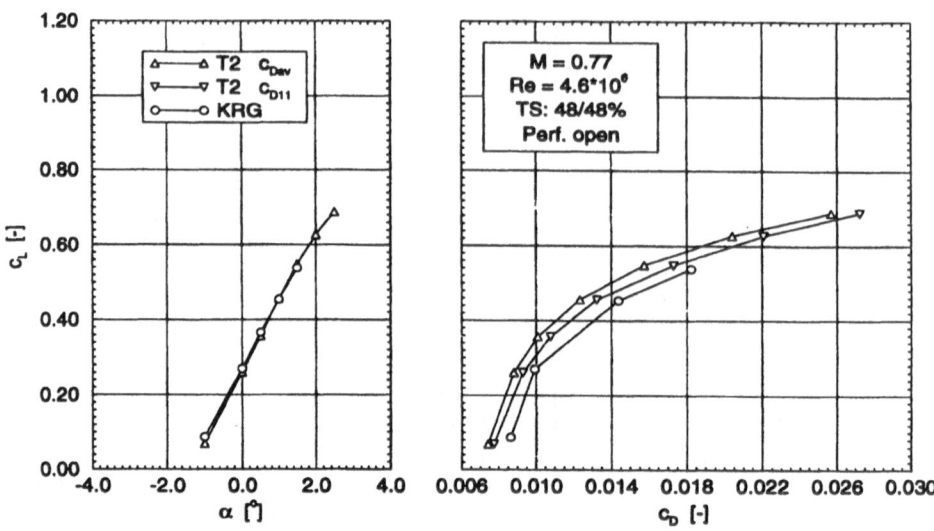

Figure 5: Comparison of lift and drag polars measured in T2 and KRG for the LVA-1Ae with open perforation

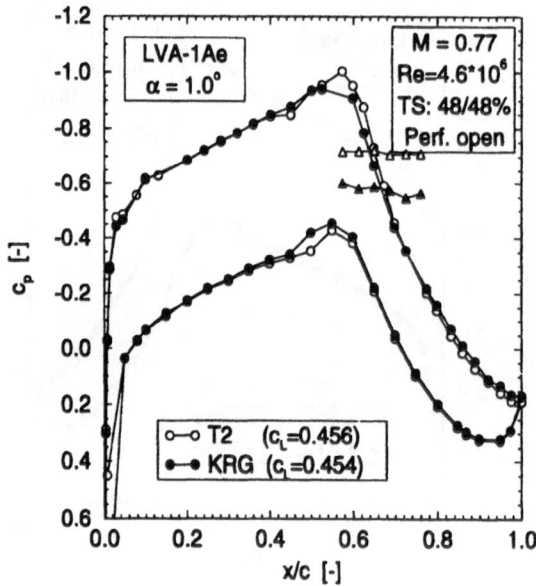

Figure 6: Comparison of pressure distribution with open perforation and fixed transition

agreement on the lower side and on the upstream part of the upper side is nearly perfect.

The pressure level in the cavity underneath the perforation is higher in the KRG results than it is in the T2 tests, probably a direct consequence of the different shock positions resulting in a higher average pressure above the perforated region of the surface in the KRG case.

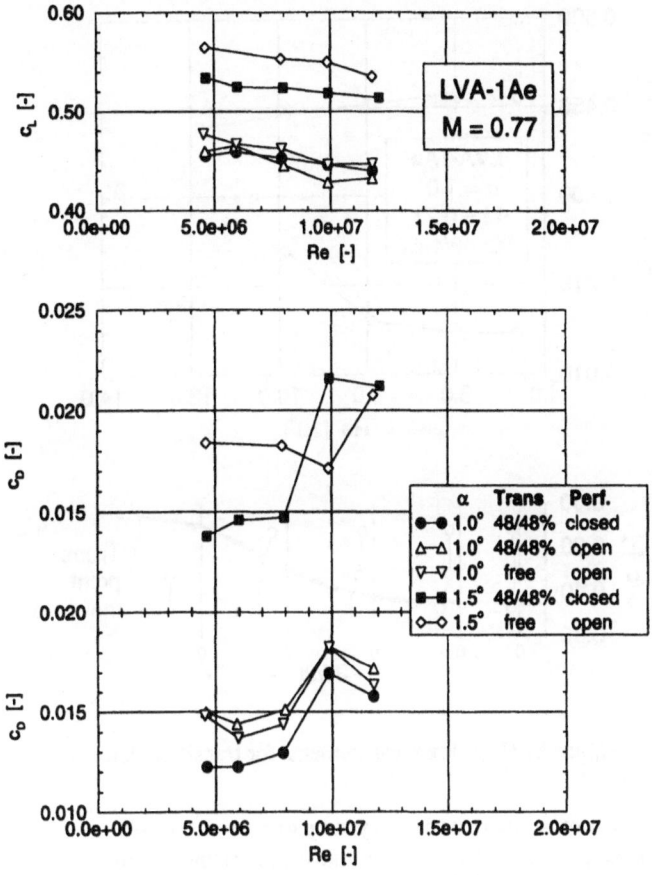

Figure 7: Reynolds number dependency of lift and drag for the various test cases

21.2.3.2 Reynolds number effects on passive control

Figure 7 shows the development of lift and drag with Reynolds number for the different test cases. The curve for $\alpha = 1.0°$ without control exhibits a general loss of lift with increasing Reynolds number. This tendency is stronger for the tests with control, but only up to $Re = 10 \times 10^6$; thereafter a recovery of lift can be observed. Due to the smaller boundary layer thickness, lift is highest for the case with free transition, and, at least for higher Reynolds numbers, lowest for the passive control case with fixed transition.

The trends for the $\alpha = 1.5°$ cases are similar, although they are lacking the lift recovery beyond $Re = 10 \times 10^6$.

Drag is nearly constant or increases slowly between $Re = 4.6 \times 10^6$ and $Re \approx 8 \times 10^6$ except for the "open perforation" cases at $\alpha = 1.0°$, where a slight minimum is observed at $Re = 6 \times 10^6$. Drag is generally lowest for the closed surface, increasing significantly by applying passive control, but decreasing somewhat when removing the transition strip. Between $Re = 8 \times 10^6$ and $Re = 10 \times 10^6$ drag shows a strong increase for all but the $\alpha = 1.5°$, "open

361

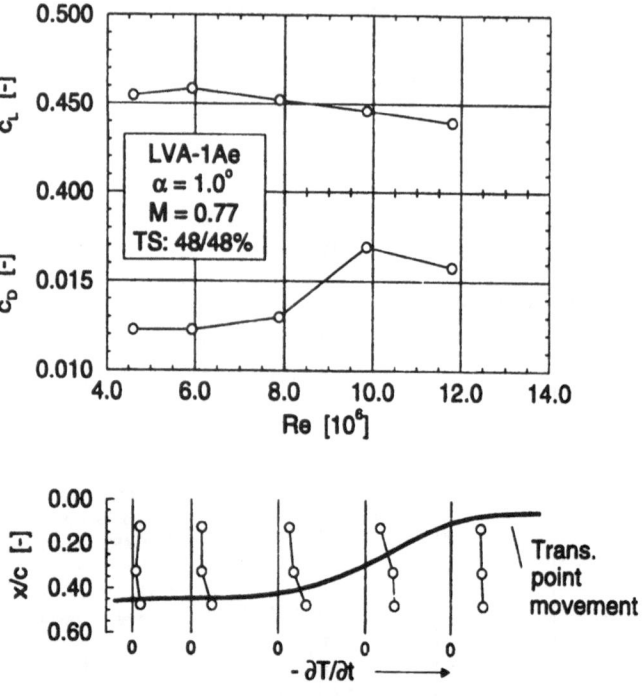

Figure 8: Heat flux measurements for transition detection

perforation" case, where the increase is delayed up to $Re = 12 \times 10^6$.

This result is explained by the upstream movement of the transition point, which mostly takes place in the range of $8 \times 10^6 \leq Re \leq 10 \times 10^6$. The thickening of the boundary layer and the corresponding deterioration of the flow conditions at the trailing edge cause a loss of circulation, hence a stronger decrease in lift and an increase in friction and viscous drag. Since the transition strip was located at 48% chord, this also holds for the case of fixed transition when the transition point moves further upstream. Beyond $Re = 10 \times 10^6$ (12×10^6 for $\alpha = 1.5°$ and open perforation) the flow is fully turbulent and the above trend is reversed by the now favorable development of the boundary layer thickness.

This explanation is supported by the results of the heat flux measurements for transition detection shown in Fig. 8 for the $\alpha = 1.0°$, "fixed transition" and "closed perforation" case.

Although the spatial resolution is very low with only three chord-wise thermo-couples, the movement of the transition point, denoted by the higher temperature gradient due to the higher heat transfer in the turbulent region, is clearly visible. At the present flow conditions, upstream movement of transition starts already at $Re = 6 \times 10^6$ and has passed the most upstream sensor only at $Re = 12 \times 10^6$.

Figure 7 shows that the increase in drag due to passive control becomes much smaller at $Re \geq 10 \times 10^6$ for $\alpha = 1.0°$. At $\alpha = 1.5°$, even a small drag reduction is observed but this was measured only for the case of free transition. Judging from the difference between fixed and free transition at $\alpha = 1.0°$ and $Re \geq 10 \times 10^6$ one might speculate, however, that also for the case with boundary layer tripping a small decrease in drag would have been found at $\alpha = 1.5°$,

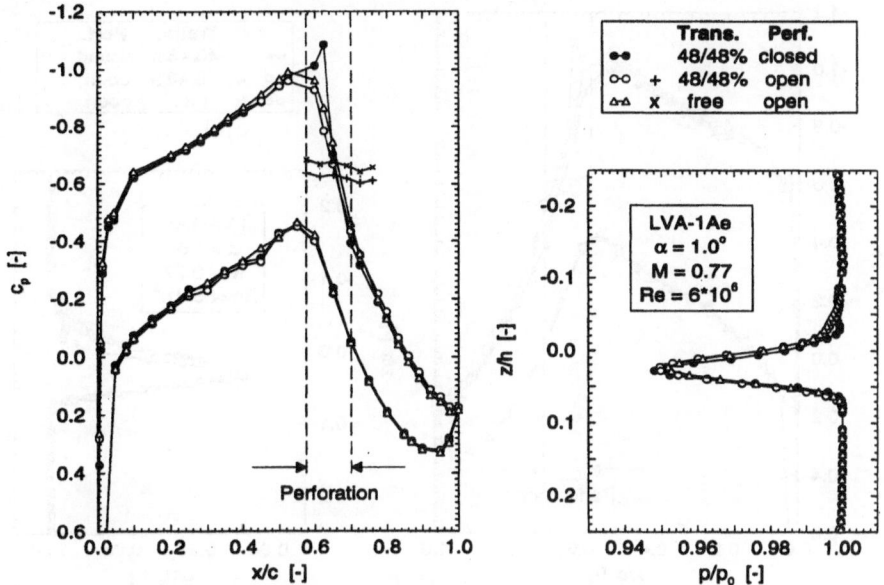

Figure 9: Comparison of airfoil and wake pressure distribution with and without ventilation for α = 1.0° at low Reynolds numbers

moreover so as for low Reynolds numbers drag for free transition and passive control has been always higher than for fixed transition without control, *cf.* Fig. 7.

For Reynolds numbers less than 10×10^6, the effects of passive control on the pressure distribution of the airfoil and in the wake is presented in Fig. 9 and Fig. 10 for α = 1.0° and α = 1.5°, respectively.

Clearly visible in Fig. 9 is the effect of the ventilation reducing the shock strength and also the pressure recovery downstream of the shock. Removing the tripping device improves pressure recovery again beyond the values for the closed airfoil and, at the same time, the pressure on the upper side upstream of the shock is reduced and the shock is shifted downstream without changing its strength resulting in a slightly higher lift coefficient.

The region of the wake corresponding to the upper side of the airfoil grows thicker by opening the perforation due to the increased boundary layer thickness associated with the perforated region. It also shows additional losses in the region where the signature of the wave drag would be expected, perhaps in contrast to the reduced shock strength. The difference between fixed and free transition is small; only a slightly reduced depth and a small change at the upper edge of the wake is observed.

The wake development is similar for α = 1.5°, *cf.* Fig. 10. Due to its greater strength, the total pressure losses related to the shock can now be seen in the wake profile more clearly. A reduction of these losses by passive control is not visible in the wake, they rather seem to be shifted upwards by an increased boundary layer displacement thickness on the upper side of the model. Without transition fixing, the pressure trace is smoother in this region.

The differences between the wake profiles of the open and closed perforation cases are much smaller at $Re = 12 \times 10^6$, see Fig. 11 for α = 1.0°, resulting in the smaller variations of the drag coefficient in Fig. 7. The pressure distributions on the airfoil exhibit a slightly larger spread,

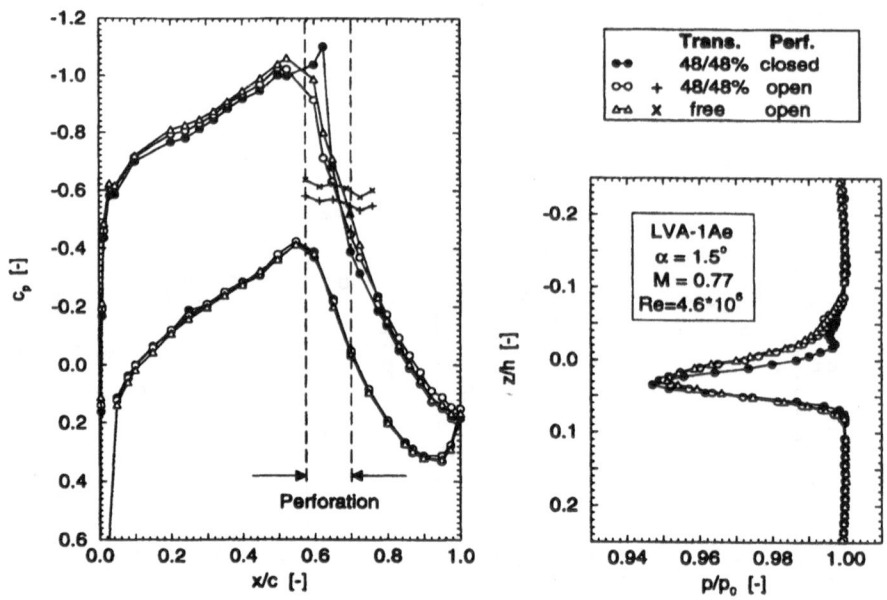

Figure 10: Comparison of airfoil and wake pressure distribution with and without ventilation for $\alpha = 1.5°$ at low Reynolds numbers

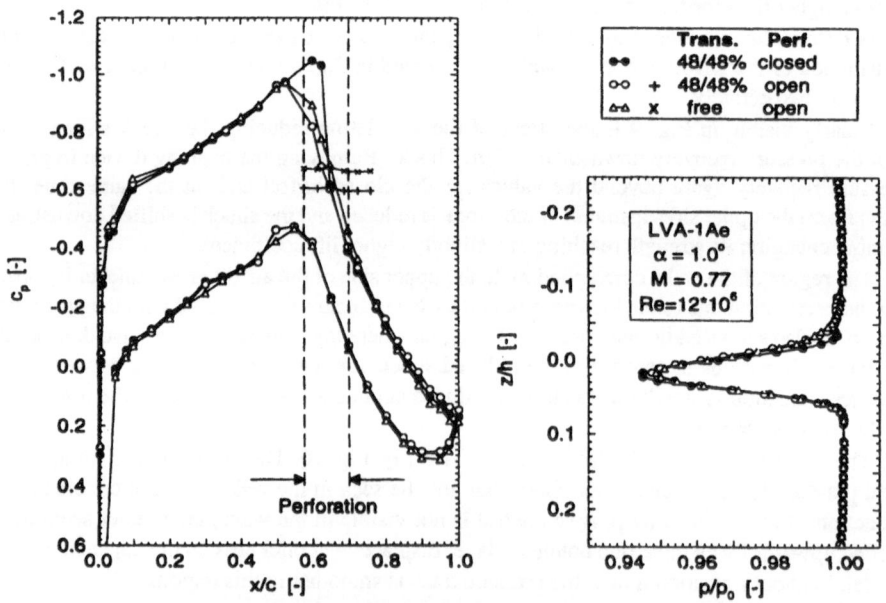

Figure 11: Comparison of airfoil and wake pressure distribution with and without ventilation for $\alpha = 1.0°$ at high Reynolds numbers

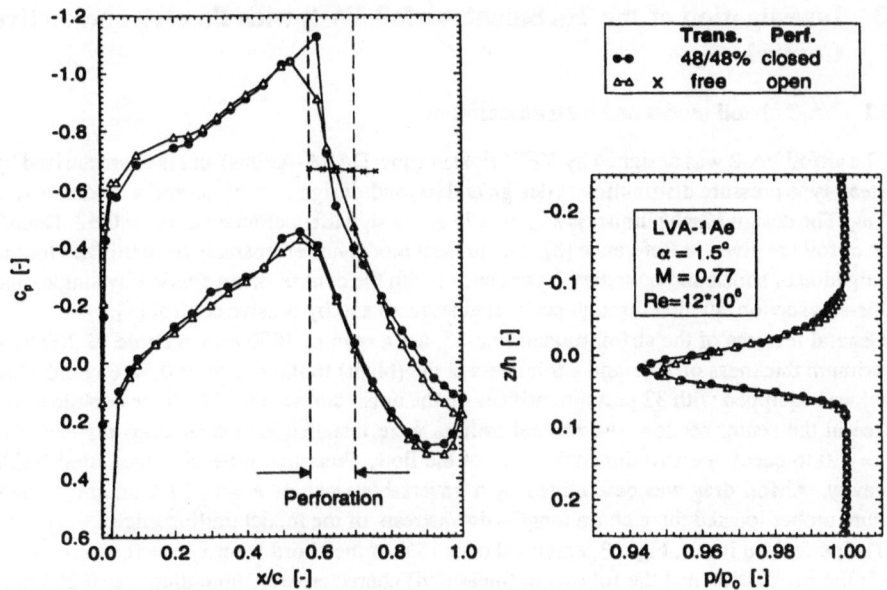

Figure 12: Comparison of airfoil and wake pressure distribution with and without ventilation for $\alpha = 1.5°$ at high Reynolds numbers

while maintaining the same relative order as in Fig. 9.

The situation is somewhat different for $\alpha = 1.5°$, see Fig. 12. At $Re = 12 \times 10^6$ the shock is located a little further upstream and probably stronger than at $Re = 4.6 \times 10^6$ (Fig. 10) leading to a trailing edge separation connected with a decrease of the trailing edge pressure. Although the shock is right at the edge of the perforated area starting at $x/c = 0.575$ (nominally), passive control reduces the shock strength enough to suppress the separation and enhance the trailing edge conditions such that a slight reduction in drag can be observed in Fig. 12 and Fig. 7.

It follows that for higher Reynolds numbers the increase of drag due to passive shock control is at least significantly reduced and under certain circumstances, as described above, even a drag reduction is possible. However, this seems to be more related to suppressing separation than to reducing wave drag.

The reduction of the drag increase was achieved only for fully turbulent boundary layers on the suction side of the airfoil as in the $Re = 12 \times 10^6$ cases; tripping the boundary layer at lower Reynolds numbers a short distance upstream of the shock did not have the same effect.

This may lead to the conclusion that the increase in viscous drag depends on the thickness of the turbulent boundary layer in the ventilated area such that a thicker turbulent boundary layer is less sensitive to additional disturbances. This would agree with the results of shock control experiments employing turbulent airfoils, where a drag reduction could be repeatedly achieved (see also Sec. 21.3).

For further insight into this mechanism, a better understanding of the interaction between the flow through the perforated surface and the boundary layer and its dependence on essential influence parameters would be necessary.

21.3 Investigation of the Turbulent Airfoil VA-2 with Passive and Active Control

21.3.1 VA-2 airfoil model and instrumentation

The airfoil VA-2 was designed by VFW-Fokker (now DASA-Airbus) and is characterized by a plateau-type pressure distribution at design and beyond-design conditions and a moderate rear loading. The design Mach number is $M_\infty = 0.73$ at a design lift coefficient of $c_L = 0.52$. Details of the airfoil are given in Reference [3]. The present model has extensively been utilized for the investigation of shock and boundary layer control with the control being applied by single- and double-slot suction, suction through perforated surfaces and by passive control [1].

General features of the airfoil model, Fig. 13, are a span of 1000 mm, a chord of 200 mm, a maximum thickness of 13% and a thickness at the (blunt) trailing edge of 0.52% chord. The model was equipped with 32 pressure orifices on the upper surface and 22 orifices on the lower surface in the center section. Additional orifices were installed in two sections off-center at $y/c = 1.0$ to check the two-dimensionality of the flow. Pressures were also measured inside the cavity. Airfoil drag was determined by a traversable wake rake with 31 total and 3 static pressure probes located three chord lengths downstream of the model trailing edge.

The perforated insert, Fig. 13, extended over 15% of the chord from $x/c = 0.495$ to $x/c = 0.645$; the perforation had the following (measured) characteristics: hole diameter 0.284 mm, distance between holes 0.753 mm, depth of the holes 1.3 mm, porosity 12.9%. The holes were electron-beam drilled. The cavity was connected to a vacuum system so that passive (pipes disconnected, cavity exits sealed) as well as active control by suction could be applied. The

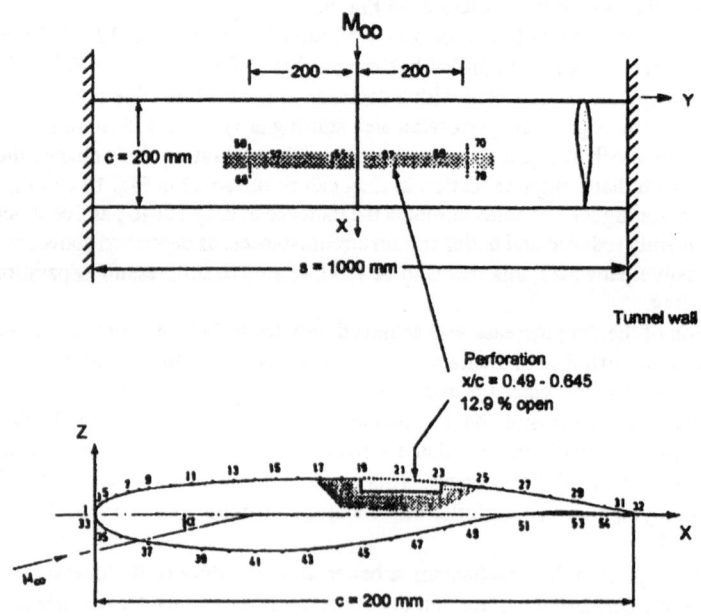

Figure 13: Airfoil Model VA-2 with perforation

Figure 14: Airfoil model VA-2 installed in the perforated test section of the 1x1m TWG

perforated section extended, as shown in Fig. 13, over 50% of the span.

21.3.2 Wind tunnel

The tests were carried out in the 1 x 1 Meter Transonic Wind Tunnel Göttingen of DLR (TWG). The TWG is a closed-circuit continuous tunnel with a cross-section area of 1×1 m². After a recent modification of the stilling chamber - test section - diffuser area, there are now three independent test sections available [4]: a perforated test section, 6% open with 60°-slanted holes, for the Mach number range 0.40 to 1.3, an adaptive-wall test section for the Mach number range 0.40 to 0.95 and a supersonic test section comprised of a Laval-nozzle and the test section proper for the 1.4 to 2.2 Mach number range.

The present investigation was performed in the perforated test section, Fig. 14, of the TWG since the previous tests were also carried out in a perforated test section (in the meantime reconstructed) of identical geometry. The disadvantage of utilizing such a test section is, of course, the occurrence of large wall interference effects since the walls essentially behave like an open-jet boundary for which corresponding corrections must be applied.

21.3.3 Test conditions

The following configurations were investigated: Airfoil model with perforation closed as reference configuration (datum), model with perforation open but cavity disconnected from the vacuum supply (passive control) and cavity connected to the vacuum system utilizing a suction coefficient of $c_q = 5 \times 10^{-4}$ (active control). Additional configurations investigated included a contour 'bump' and riblets, respectively, in the region of the shock. Results are, however, not included here.

The free-stream conditions were: nominal Mach numbers $M_{\infty N} = 0.72, 0.74, 0.76$ and 0.78; angle of attack range for all Mach numbers $\alpha = 0°$ to $\alpha = 6°$. The tests were carried out at a chord Reynolds number of $Re = 2.5 \times 10^6$ with transition fixed at 30% chord on the upper and 25% chord on the lower surface by carborundum grit. The rear transition location was chosen to simulate at the low wind tunnel Reynolds number of $Re = 2.5 \times 10^6$ higher (flight) Reynolds number conditions especially in the shock region [5].

21.3.4 Analysis and discussion of results

21.3.4.1 Comparison between present and original results

In order to judge whether wind tunnel or model inherent changes may cause differences in the flow development about the airfoil VA-2, comparisons are first carried out between the present (TWG 95) and the earlier (TWG 84) results for the datum airfoil. Corresponding data for lift and drag and representative pressure distributions are given in Fig. 15 and 16 for Mach numbers of $M_{\infty N} = 0.74$ and 0.76, respectively.

It can be observed that agreement in lift coefficients is generally good with some minor differences occurring in the maximum lift range where the earlier measurements show slightly higher lift values with a corresponding later break in the lift-curve slope. Drag coefficients are generally somewhat lower for the TWG 84-measurements which is consistent with the slightly better pressure recovery in the case of the latter, Fig. 15b and 16b. The small differences in drag and trailing edge pressure recovery are possibly a result of differences in model roughness including small differences caused by deviations in the tripping device height and roughness distribution.

A comparison of the present and earlier results with passive shock control is presented in Fig. 17, 18 and 19 for Mach numbers of $M_{\infty N} = 0.74$, 0.76 and 0.78, respectively. Consistent differences between the results are at all Mach numbers the lower total drag and the later break in the lift-curve slope in the case of the TWG 84-data. Examining corresponding pressure distributions at matching angles of attack, one observes, for instance,

- at $M_{\infty N} = 0.74$ and $\alpha = 5.5°$, Fig. 17b, a total separation for the TWG 95-measurements resulting in a more forward shock location while for the earlier measurements the flow is still attached, a development that was already noticed in the datum results, Fig. 15a,

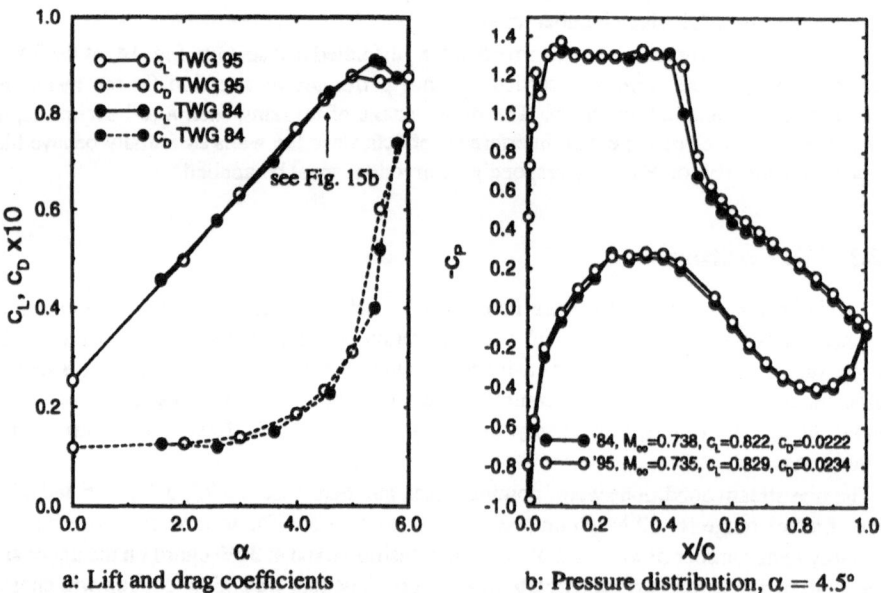

a: Lift and drag coefficients   b: Pressure distribution, $\alpha = 4.5°$

Figure 15: Comparison between 1984 and 1995 results, datum, $M_{\infty N} = 0.74$, $Re = 2.5 \times 10^6$

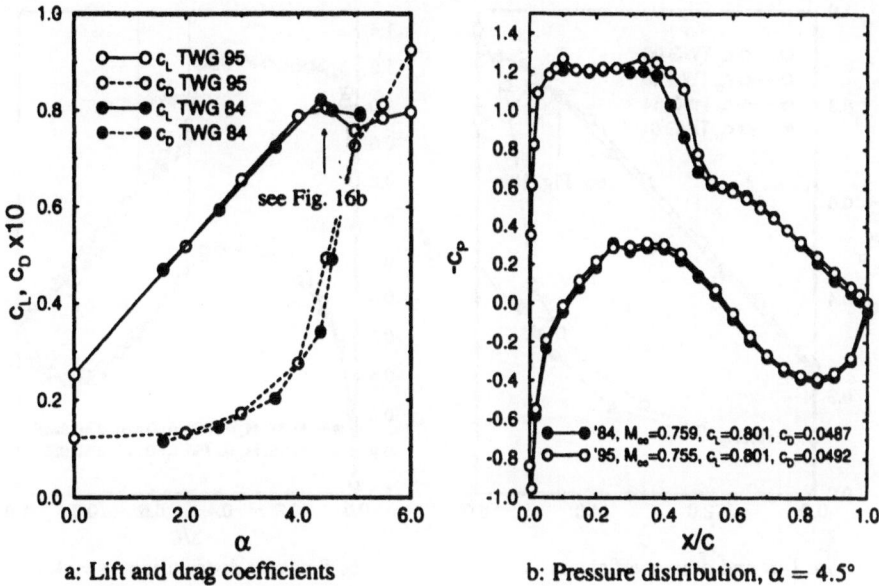

a: Lift and drag coefficients

b: Pressure distribution, $\alpha = 4.5°$

Figure 16: Comparison between 1984 and 1995 results, datum,
$M_{\infty N} = 0.76, Re = 2.5 \times 10^6$

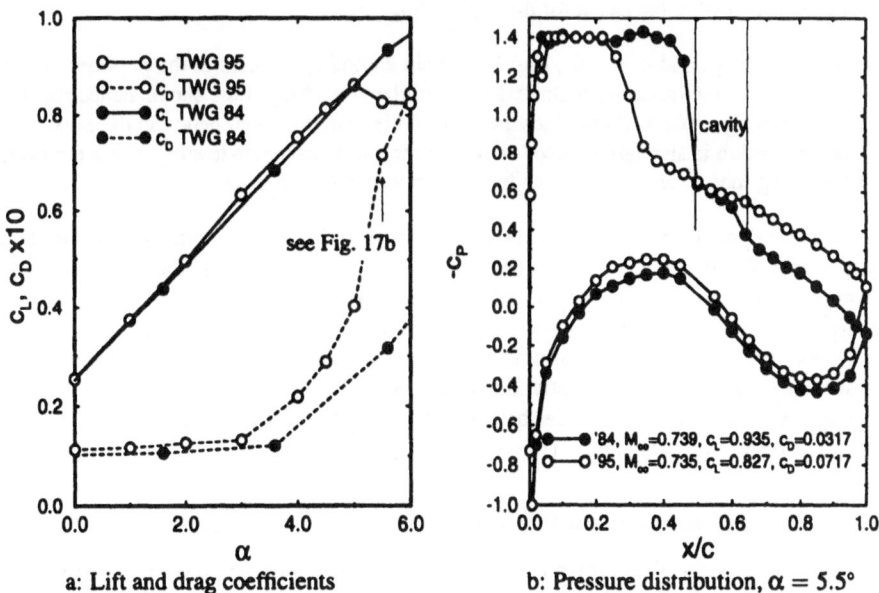

a: Lift and drag coefficients

b: Pressure distribution, $\alpha = 5.5°$

Figure 17: Comparison between 1984 and 1995 results, passive control,
$M_{\infty N} = 0.74, Re = 2.5 \times 10^6$

369

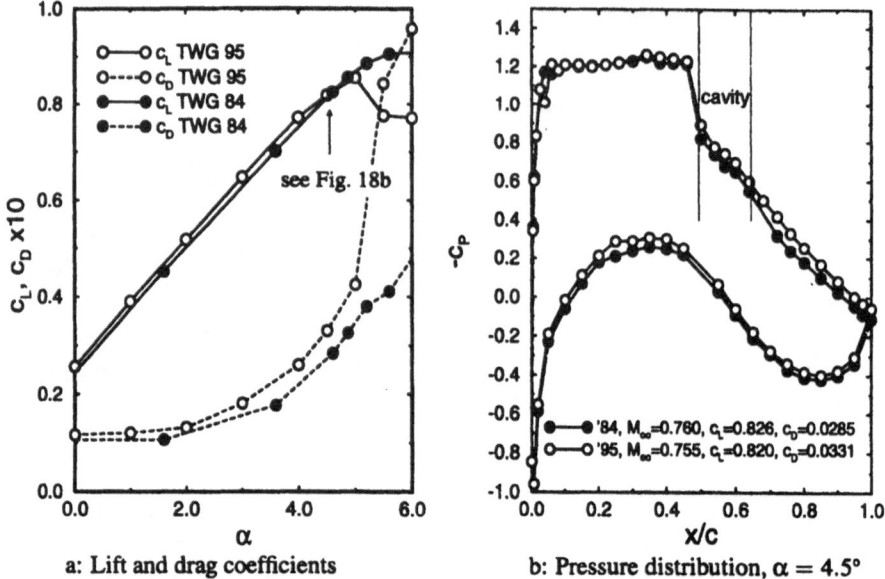

a: Lift and drag coefficients   b: Pressure distribution, $\alpha = 4.5°$

Figure 18: Comparison between 1984 and 1995 results, passive control,
$M_{\infty N} = 0.76, Re = 2.5 \times 10^6$

- at $M_{\infty N} = 0.76$ and $\alpha = 4.5°$, Fig. 18b, prior to the lift-curve break, a good agreement in the pressure distributions except for the reduced pressure recovery in the present data which is probably the cause for the higher drag, and

- at $M_{\infty N} = 0.78$ and $\alpha = 3.0°$, Fig. 19b, again a good agreement in the pressure distributions, including the cavity region where a similar spreading of the shock. hence reduction in wave drag, occurs. Particularly pronounced is here the discrepancy in the pressure recovery which is strongly reduced in the present measurements leading to the increase in total drag mainly as a result of an increase in viscous drag.

As a résumé of the comparison between the present VA-2 measurements and the earlier measurements, the following can be stated: Without control, *i.e.*, for the datum configuration, agreement is quite good except for drag which is slightly lower in correspondence with a better pressure recovery in the TWG 84-tests, and maximum lift which is somewhat higher in these tests. This trend is similar but strongly amplified in the case of passive control. It seems that especially differences in the initial boundary layer development are, at least for the datum airfoil, the cause for the discrepancies between the two data sets. The differences observed are not likely to be related to changes in the wind tunnel environment.

21.3.4.2 Effect of passive and active shock control

In spite of the reduced control effectiveness observed in the present tests, there still remains a positive effect on the flow development, including total drag, due to passive control. This is demonstrated together with the improvements due to active control in the present section where forces, surface pressure distributions and wake data are compared for the datum airfoil

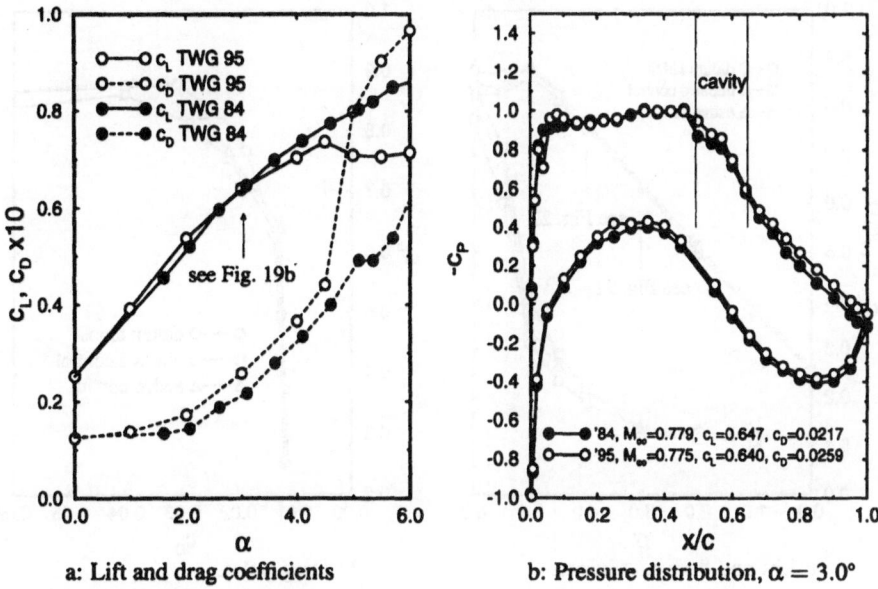

a: Lift and drag coefficients   b: Pressure distribution, $\alpha = 3.0°$

Figure 19: Comparison between 1984 and 1995 results, passive control,
$M_{\infty N} = 0.78, Re = 2.5 \times 10^6$

and the airfoil with passive and active shock control, respectively, at nominal Mach numbers of
$M_{\infty N} = 0.74, 0.76$ and 0.78. Note that the Mach numbers and angles of attack are not corrected
for wall interference; approximate corrections can, however, be obtained by assuming the wind
tunnel walls to act as an open-jet boundary.

At a Mach number of $M_{\infty} = 0.735$, Fig. 20, passive control has in the lower incidence range
$\alpha \leq 3°$, generally no effect on lift and drag while active control shows a slight increase in lift
and a small reduction in drag consistent with a better pressure recovery and a reduced wake
width, Fig. 21. At $\alpha > 3°$, passive as well as active control are associated with a pronounced
increase in total drag and generally a lower lift coefficient which again is consistent with the
reduced pressure recovery and an increased wake in the case of control, Fig. 22. The reason
for the ineffectiveness of control, especially in the higher α-range, can be seen in the pressure
distributions of Fig. 22a: shocks become stronger with a more pronounced negative influence
on the boundary layer development, but the shock locations are just upstream of the cavity out
of reach of any effective control.

At a higher free-stream Mach number, $M_{\infty} = 0.755$, Fig. 23, the effect of control on lift
and drag are in the lower angle of attack range similar to the one observed for $M_{\infty} = 0.735$.
At higher angles of attack, *i.e.*, $\alpha > 3°$, passive as well as active shock control reduce the total
drag considerably and both control modes result in an increase in maximum lift indicating the
positive effect on the buffet boundary. The corresponding pressure distributions, Fig. 24a and
25a, show that the shock location now coincides with the location of the cavity with control
noticeably reducing the pressure gradient in the shock region. This is, however, at $\alpha = 3°$ in
the case of passive control still not sufficient to reduce total drag; however, at higher incidences
conditions are such that wave drag, but especially viscous drag are reduced which is also reflected in the wake data, Fig. 24b and 25b. Pressure distributions corresponding to maximum

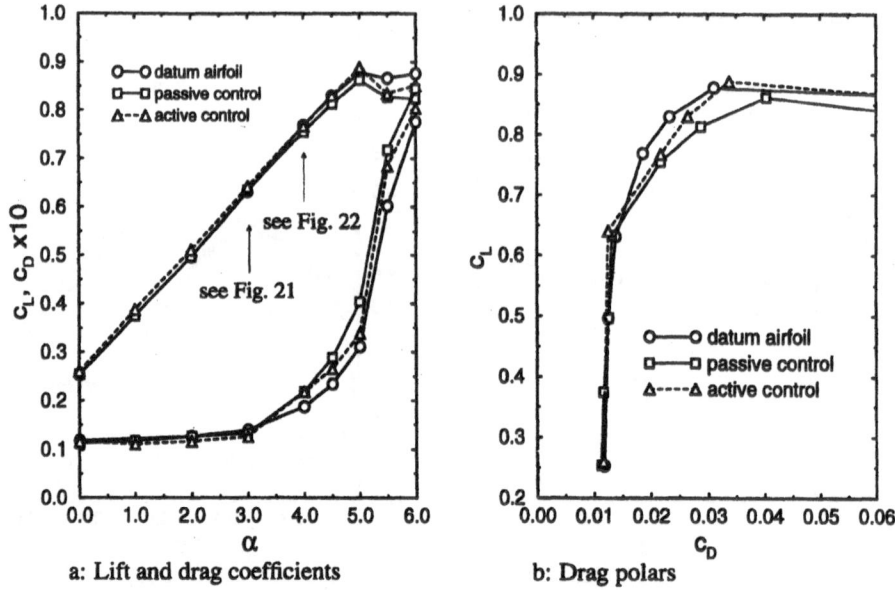

a: Lift and drag coefficients   b: Drag polars

Figure 20: Effect of passive and active control on lift and drag,
$M_\infty = 0.735, Re = 2.5 \times 10^6$

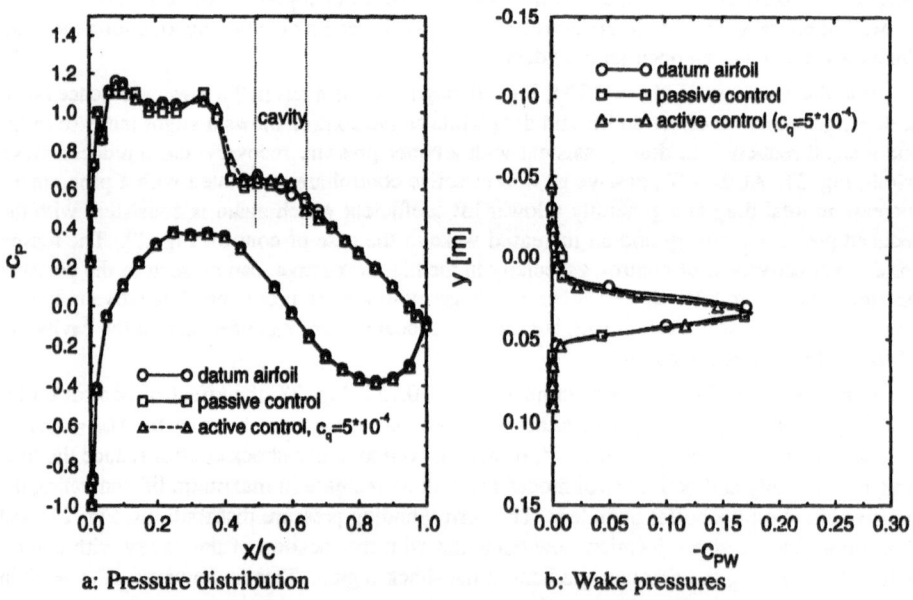

a: Pressure distribution   b: Wake pressures

Figure 21: Effect of passive and active control on pressure distributions,
$M_\infty = 0.735, Re = 2.5 \times 10^6, \alpha = 3°$

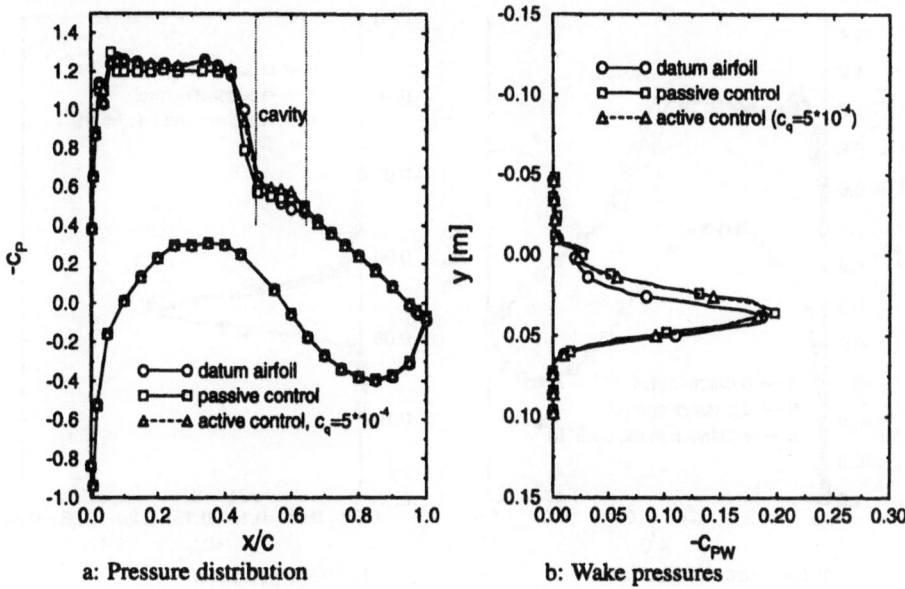

a: Pressure distribution   **b: Wake pressures**

Figure 22: Effect of passive and active control on pressure distributions,
$M_\infty = 0.735$, $Re = 2.5 \times 10^6$, $\alpha = 4°$

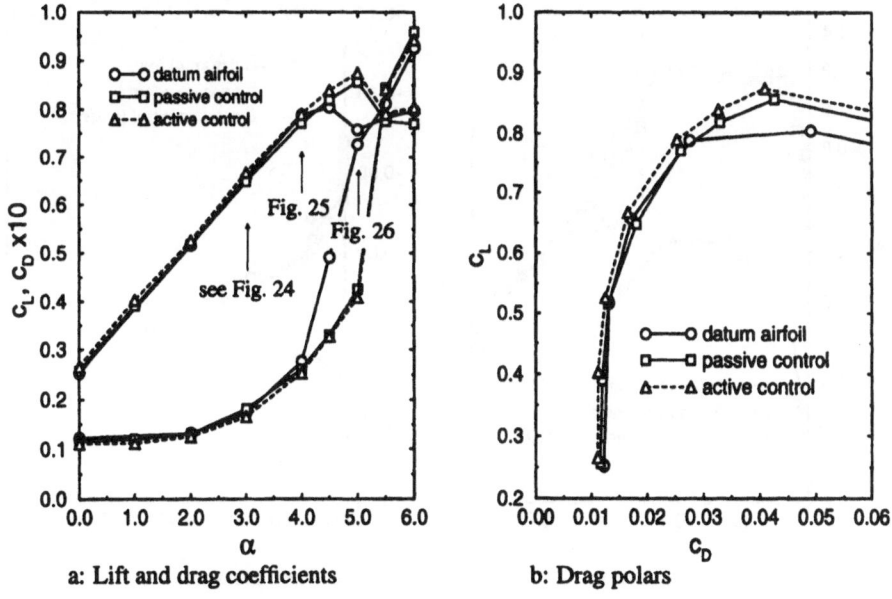

a: Lift and drag coefficients   **b: Drag polars**

Figure 23: Effect of passive and active control on lift and drag,
$M_\infty = 0.755$, $Re = 2.5 \times 10^6$

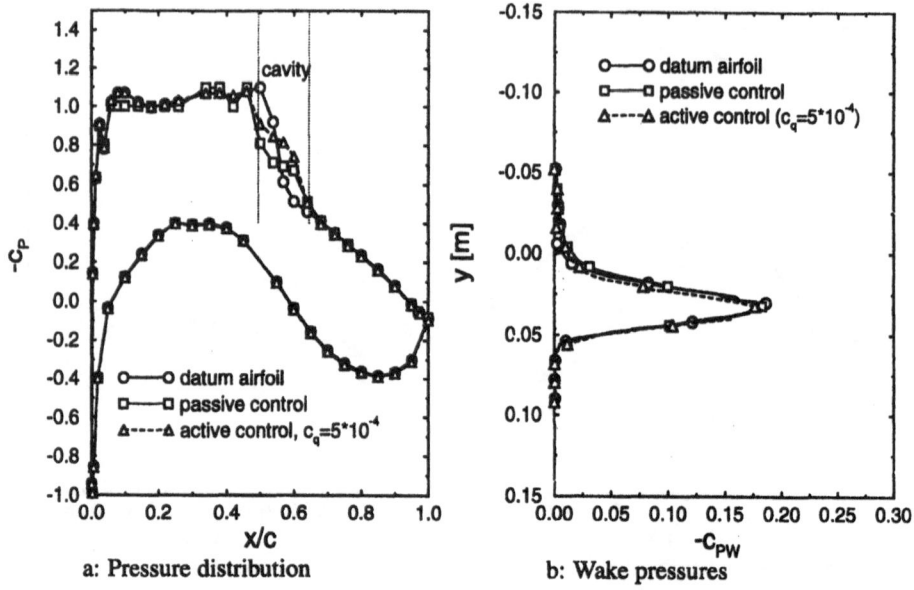

Figure 24: Effect of passive and active control on pressure distributions,
$M_\infty = 0.755, Re = 2.5 \times 10^6, \alpha = 3°$

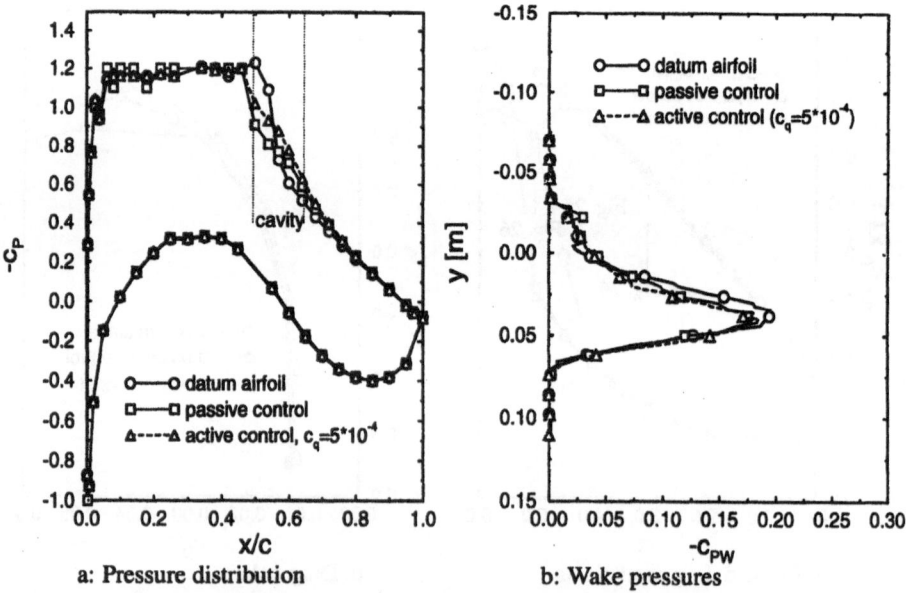

Figure 25: Effect of passive and active control on pressure distributions,
$M_\infty = 0.755, Re = 2.5 \times 10^6, \alpha = 4°$

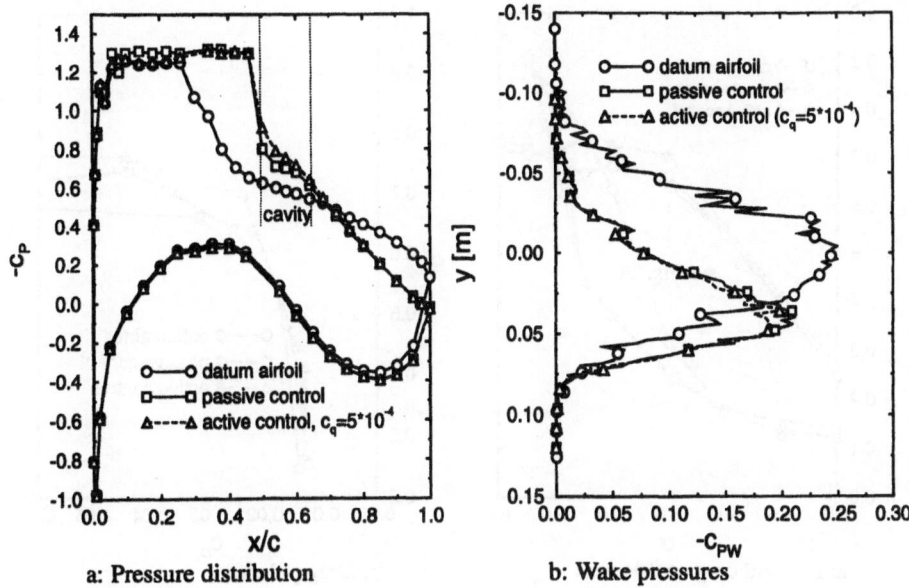

a: Pressure distribution   b: Wake pressures

Figure 26: Effect of passive and active control on pressure distributions,
$M_\infty = 0.755, Re = 2.5 \times 10^6, \alpha = 5°$

lift, Fig. 26a, demonstrate that total separation can be delayed by passive as well as active control. Note that the spread of the pressure distribution in the shock region for the datum airfoil indicates an oscillating shock wave (buffet).

At $M_\infty = 0.775$, Fig. 27, passive as well as active control reduce the total drag over the entire incidence range investigated up to complete separation. The results are otherwise very similar to the ones at $M_\infty = 0.755$ described above. Noteworthy with respect to the pressure distributions, Fig. 28, is that at certain conditions, possibly when the shock in case of the closed cavity (datum airfoil) is located in the center of the cavity, active control results in a strong pressure recovery which causes the shock to move far downstream (here to the downstream face of the cavity) which results in an appreciable increase in lift, Fig. 27, but also in an increase in wave drag, Fig. 28b. However, this process is still associated with a decrease in total drag indicating the importance of the effect of control on viscous drag which seems to dominate control effectiveness.

As a résumé of the renewed investigation of the effectiveness of passive and active shock control for the VA-2 airfoil it can be stated that passive, but more so active control, reduces total drag over a considerable range of free-stream conditions. Also maximum lift and correspondingly the buffet boundary are positively affected. Although not in the magnitude of performance increase but in trend, the TWG 84 results could thus be confirmed.

375

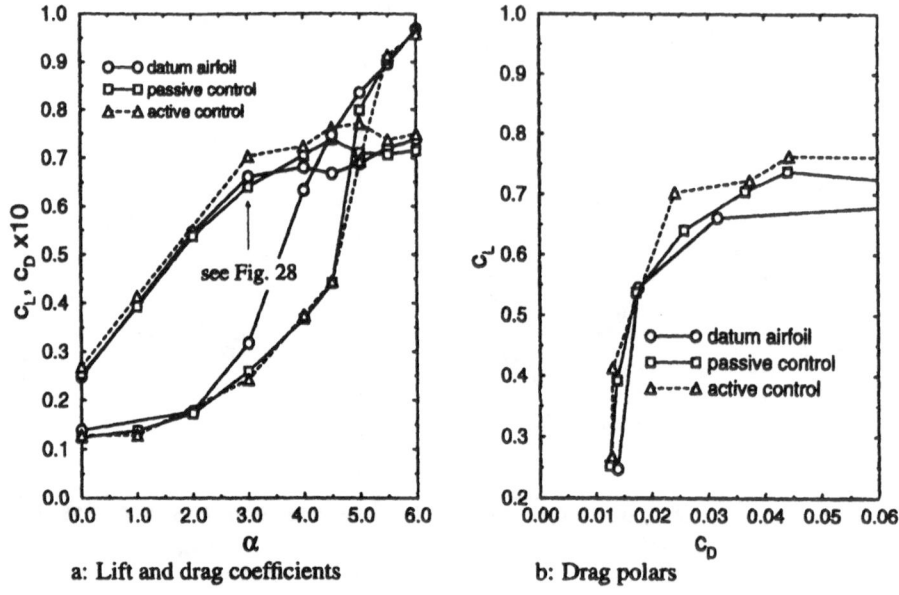

a: Lift and drag coefficients

b: Drag polars

Figure 27: Effect of passive and active control on lift and drag,
$M_\infty = 0.775, Re = 2.5 \times 10^6$

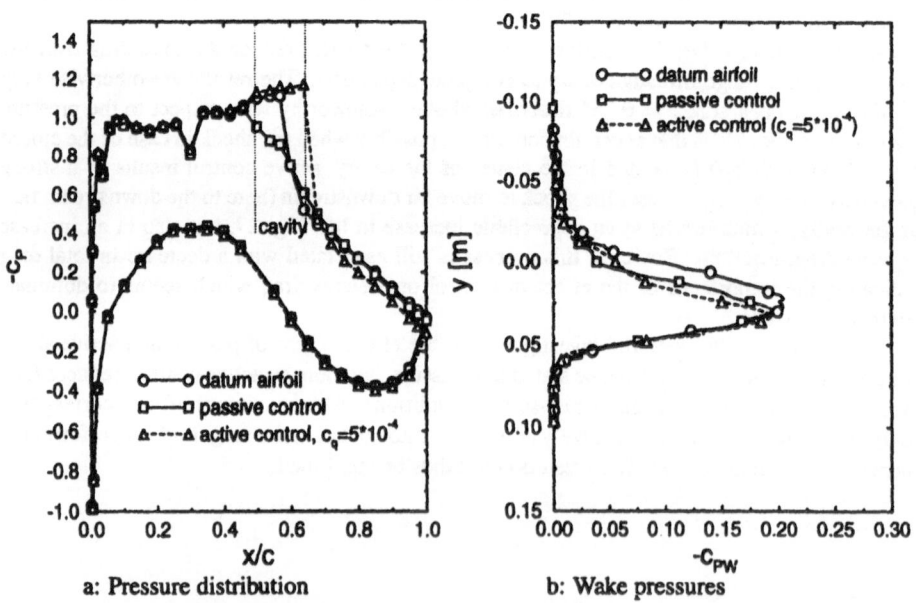

a: Pressure distribution

b: Wake pressures

Figure 28: Effect of passive and active control on pressure distributions,
$M_\infty = 0.775, Re = 2.5 \times 10^6, \alpha = 3°$

21.4 Conclusions

The experiments performed in the KRG have confirmed the results obtained earlier in the course of the EUROSHOCK program on the DRA 2303 and LVA-1Ae airfoils. Contrary to the investigation on the turbulent airfoil VA-2, where a drag reduction could be achieved over a considerable range of free-stream conditions (*cf.* Sec. 21.3), these tests have shown that on airfoils with extended laminar boundary layers passive control results in an significant increase of viscous drag outweighing the reduction of wave drag thus resulting in an increase of total drag.

To gain further insight into this different behavior, obviously depending on the state and structure of the boundary layer, tests were conducted with the LVA-1Ae model in the KRG with and without transition fixing in the Reynolds number range $4.6 \times 10^6 \leq Re \leq 12 \times 10^6$.

The results obtained here with the same model at the same Reynolds number as in the T2 tests, where experiments with emphasis on flow field surveys were carried out, agree well with the T2 data with only a slight difference in shock position of about 2%.

For the test cases with high Reynolds numbers, where turbulent boundary layers exist on the upper and lower side of the airfoil, the increase of drag due to passive control is generally considerably reduced, and at conditions with a tendency to trailing edge separation a drag reduction could be observed. This is similar to the findings for turbulent airfoils with fully developed turbulent boundary layers. However, tripping of the boundary layer a short distance upstream of the shock does not have the same effect. Therefore, it is assumed that the thickness and structure of the turbulent boundary layer in the region where passive control is applied is an important parameter determining the amount of viscous drag increase.

In order to ensure that the positive results obtained with the turbulent VA-2 airfoil during the early tests in 1984 were not spurious and caused by experimental faults, these measurements were repeated in 1995.

A comparison between the new and earlier results for the datum airfoil showed good agreement in forces and pressure distributions except for a slightly lower drag, consistent with a somewhat lesser pressure recovery on the airfoil upper surface, and a somewhat lower maximum lift coefficient in the case of the TWG 95-measurements. A similar trend was observed for the case of passive shock control, however, with the differences in the results strongly amplified. It is judged that differences in the initial boundary layer development due to deviations in surface roughness, including differences in the tripping device, might be the cause for the discrepancies in the results of the two test series.

Concerning the effectiveness of shock control determined in the present tests, it seems that passive control, but even more so active control, through a perforated surface with a cavity underneath can reduce total drag over a wide range of free-stream conditions. Also maximum lift is being increased by control shifting the buffet boundary to higher lift coefficients. The original TWG 84-results for passive control were thus confirmed, although not in the extent of the improvements but in trend. The question why there are differences in the drag behavior between the two laminar airfoils investigated and the turbulent VA-2 airfoil must, therefore, still be answered, leaving merit for further analysis and investigations.

21.5   Nomenclature

b	span
b_s	span-wise width of perforation
c	chord
c_D	drag coefficient
c_L	lift coefficient
c_p	pressure coefficient
c_{pw}	wake pressure coefficient
c_q	suction coefficient, $\dot{m}/(\rho_\infty V_\infty b_s c)$
D	drag
h	test section height
L	lift
$\dot{m}$	mass flow
$M_{\infty N}$	nominal free-stream Mach number
M_∞	free-stream Mach number (measured but uncorrected)
p	pressure
p_0	stagnation pressure
Re, Re_c	Reynolds number based on airfoil chord
t	time
T	wall temperature
V_∞	free-stream velocity
x	chord-wise coordinate
y	span-wise coordinate
α	angle of attack
ρ_∞	free-stream density

21.6   References

[1] P. Thiede, P. Krogmann, and E. Stanewsky. Active and passive shock boundary layer interaction control on supercritical airfoils. AGARD-CP-365, (Brussels, Belgium), 1984.

[2] J. Delery, J. Fulker, W. Geissler, and E. Stanewsky. European shock control investigation: Objectives, approach and previous results. In *Proceedings 2nd Community Aeronautics RTD Conference*, Napoli (Italy), 4–5 October 1993.

[3] R. D. Boehe. Transonic measurements on the airfoil VFW-F VA-2 at the DFVLR Göttingen. Report No. 8, ZKP Wing Section, November 1976. Ministry of Science and Technology Reference LFK 7511.

[4] B. Binder, L. Riethmüller, S. Tusche, and R. Wulf. Modernisierung des Transsonischen Windkanals in Göttingen (Upgrading the Transonic Wind Tunnel Göttingen). In *Jahrbuch der DGLR*, volume 1, 1992.

[5] E. Stanewsky. Shock boundary layer interaction (simulation). In *Boundary Simulation and Control in Wind Tunnels*, AGARD-AR-224, April 1988.

22 AN EXPERIMENTAL INVESTIGATION OF PASSIVE SHOCK/BOUNDARY-LAYER CONTROL ON AN AEROFOIL

J L Fulker and M J Simmons

DERA, High Speed and Weapon Aerodynamics Department
Bedford, MK41 6AE, UK

Summary: An experiment at high subsonic speeds on a two-dimensional aerofoil model is described, with a passive system for the control of shock waves via a porous skin and under-surface cavity. Passive control utilises the pressure gradient across the shock wave to induce a secondary flow through the surface and underlying cavity, causing injection into and suction from the boundary layer upstream and downstream of the shock wave, respectively. The displacement surface so formed induces compression waves which weaken the main shock wave, thereby reducing wave drag.

The measurements show that while passive control apparently reduces wave drag it also significantly increases viscous drag, consequently increasing drag overall.

22.1 Introduction

The work reported in this document is the DERA contribution to Task 3 of the EUROSHOCK project, devoted to the study of passive control of transonic shock wave/boundary-layer interactions.

The appearance of shock waves is an inevitable consequence of flight at high subsonic speeds. At these speeds shock waves and their interaction with the boundary layer define the performance boundaries of aircraft in terms of drag rise, buffet onset and buffet penetration. The potential difficulties associated with the development of shock waves has stimulated research into ways of controlling them in order to reduce wave drag and the likelihood of shock-induced separation. Several techniques exist for control of this interaction, including supercritical aerofoil technology, where, by design, only weak shock waves are allowed, but at off-design conditions strong shock waves will inevitably form. The use of vortex generators, boundary layer suction and tangential blowing have also been tried, with varying degrees of success [1,2,3].

An alternative approach, involving the passive control of shock waves, has been explored by a number of workers [4,5,6]. This technique involves replacing part of the solid surface by porous material over a shallow cavity in the region of the shock wave. The pressure gradient across the shock wave causes a secondary flow through the surface and cavity, with injection into and suction from the boundary layer upstream and downstream of the shock, respectively. The effect of the displacement surface formed by this injection and suction is to induce compression waves which weaken the shock wave, thereby reducing wave drag and the likelihood of shock-induced separation. However, there is evidence [6] that the drop in wave drag is accompanied by an increase in viscous drag. It is believed that this increase in viscous drag is due to disturbances caused by the flow through the wing surface or the surface itself.

The present Report describes an experimental investigation of passive shock control, on an aerofoil model in the 8ft x 8ft Subsonic-Supersonic Wind Tunnel at DERA Bedford. The aim of the work is to investigate the potential for drag reduction with such a system at Reynolds numbers approaching full scale, previous tests having been made at relatively low Reynolds numbers.

After the experimental techniques are described in section 2, data from the experiments are presented in section 3. Salient features of the effect of the control system on the flow in the region of the shock wave are presented in section 4. The conclusions of the study are given in section 5.

22.2 Model Details and Measurements

22.2.1 The model

The layout of the model in the Wind Tunnel is illustrated in Fig 1 and a photograph is shown in Fig 2.

The model was mounted firmly to a rotating mechanism on the starboard side of the working section and within a free rotating bearing on the port side.

The model consists of a main spar with detachable leading edge (0 - 0.17c) and trailing edge (0.7 - 1.0c) sections, manufactured from high-tensile steel. The datum aerofoil section was designed to be representative of a natural laminar-flow section with long runs of favourable pressure gradient on both the upper and lower surfaces extending to 0.5c close to the design conditions (a freestream Mach number, M, of 0.68 and a lift coefficient of 0.5). On the upper surface of the main spar there is a removable panel between 0.39c and 0.69c allowing control systems to be inserted. An insert was also manufactured to form the original profile in order to obtain a datum configuration.

The passive control insert had a perforated surface between 0.5c and 0.6c with an open area ratio of 8% based on the local area; the perforations were formed by laser drilling with a nominal diameter of 0.076mm (0.003ins). A calibration of the control surface by the method suggested by Poll et al.[7] allowed a value of the so called K modifier to be derived which in this case was found to be 1.0 for flow both into and out of the plenum, based on an extensive survey of the surface for various mass flows. Thus, although it is not possible to specify the diameter of an individual hole accurately, there is a nominal hole diameter which can be determined for a large number of holes, which in this case is identical to that specified.

Ordinates of the datum aerofoil are given in Table 1 together with ordinates for the control insert where it differs from the datum. All of these ordinates were obtained from an inspection in March 1993 and are accurate to within ±0.02mm (0.00003c).

The aerofoil and inserts are fitted with surface static pressure holes at three spanwise stations (0.25, 0.5 and 0.75 span) each having 35 orifices on the upper surface and 22 on the lower surface. A list of chordwise positions for the static holes is given in Table 2. All the static holes in the wing are of 0.5mm diameter and are drilled normal to the surface, provision is also made for measuring static pressure at 5 positions in the control cavity (plenum) when the passive control insert is in place.

22.2.2 Measurements

As well as the static pressures at the model surface, measurements of static pressures were made on the working section walls. Total and static pressures were measured on a wake rake of pitot and static tubes 1.75 chord lengths downstream of the model trailing-edge. All of these measurements were made using seven electronic pressure scanning modules having a working range of ±1.35bar.

The overall accuracy of the pressure coefficients, allowing for uncertainties in transducer calibration and dynamic and static pressures, is estimated to be ±0.002 at the test conditions.

Normal force and pitching-moment coefficients were determined by appropriate integration of the surface static pressures around the aerofoil.

Geometric angle of incidence was measured by a digital encoder attached to the half-model balance housing, supporting one end of the model, the setting for zero angle of incidence having previously been determined using an electro level meter. The estimated accuracy of the setting is ±0.005°. The model was loaded before the tests to a maximum of 740Nm about its flexural axis. No detectable movement could be recorded; hence it has been assumed that aeroelastic distortion of the model is negligible.

No corrections have been applied to the wake pitot readings for displacement effect. Sectional drag was inferred from the pitot and static pressure measurements in the wake, using the method described in [8].

22.2.3 Boundary-layer transition trips

Boundary layer transition was fixed by narrow bands of sparsely distributed ballotini cemented to the model by epoxy resin. Transition was fixed at 5% chord on both surfaces of the model, the band width in each case being 2.5mm and the diameter of ballotini used being 0.1mm - 0.13mm.

22.2.4 Test conditions

Tests with transition fixed on both surfaces were performed at freestream Mach numbers of 0.67, 0.68, 0.69 and 0.70 at nominal Reynolds numbers of $R_C = 19 \times 10^6$ and 6×10^6 for each configuration.

A limited amount of data was recorded at a Mach number of 0.68 over a range of Reynolds numbers varying from $R_C = 3 \times 10^6$ to 19×10^6 in order to check the effectiveness of the transition trips.

The 8ft x 8ft Wind Tunnel has very accurate control of total pressure and total temperature during a run. Hence variations in these parameters during the acquisition of data are expected to be small and to exert negligible effect on the measurements taken.

A correction to Mach number for the blockage effect of the model has been applied. A complete description of the method is given in [9] but, briefly, it relies on linear theory to calculate a) the effect of the model and its images beyond the tunnel walls on velocity increments at two points on the roof and corresponding points on the floor, in both cases on the tunnel centre line and close to the model centre of volume, and b) the arithmetic mean value of the blockage increment. The ratio of this increment to the arithmetic mean of the increments in a) is combined with the mean of the measured changes in static pressure at the four points relative to those in the empty tunnel to infer blockage at the model.

The correction to angle of incidence due to tunnel-wall interference is determined using linear theory [10], together with the lift coefficient inferred from the measured model pressures to determine the strength of the vortex simulating aerofoil circulation. The vortex is placed at the centre of pressure, which is also inferred from measured pressures.

Unless the 'nozzle' Mach number is adjusted, corrected Mach number increases with angle of incidence owing to changes in the blockage effect of the model.

Although the corrected Mach number was kept to within ±0.002 of the required value during testing, the sensitivity of the aerofoil forces to small changes in Mach number led to investigation of an interpolation routine to "correct" the data to the required nominal value. With the relatively fine spacing of the free stream Mach numbers tested this technique has been successful and has yielded the figures presented in this report.

22.3 Experimental Data

The model was tested over a range of angles of incidence, from a value closely corresponding to that for zero lift to an angle of incidence above that at which trailing-edge pressure decreases rapidly with increase in angle of incidence, this being deemed to be a good indicator of the onset of flow separation and hence buffet onset.

The data for M = 0.67, 0.68, 0.69 and 0.70 at Reynolds numbers of, R_C = 6 and 19 x 10^6 are shown in the figures to be discussed below. The variation of normal-force coefficient with angle of incidence is shown in Figs. 3 to 10. It can be seen that, generally, the effect on normal force of control is small at low Reynolds number, R_C = 6 x 10^6 (Figs. 3 to 6); however, the maximum value of normal-force coefficient is greater and delayed to a higher angle of incidence by control. The 'lift break' or condition where normal-force coefficient diverges from its low angle of incidence trend, can be seen to occur at a higher value of angle of incidence with control, suggesting a possible increase in the normal-force coefficient for buffet onset. For Mach numbers of 0.68 and above, the increase in normal-force curve slope for values of normal force of 0.4 and above is reduced by control.

At the higher Reynolds number, R_C = 19 x 10^6 (Figs. 7 to 10), similar phenomena are observed; however the features associated with the 'lift break' are less pronounced, whereas the reduction in non-linear lift is exaggerated.

Figs. 11 to 18 show the variation of drag coefficient with normal-force coefficient for the same conditions of Mach and Reynolds numbers mentioned above.

It can be seen that at low Reynolds number, R_C = 6 x 10^6, (Figs. 11 to 14), for all normal- force coefficient values below 'lift break' the effect of the addition of control is to increase drag. At low values of normal-force coefficient (C_N < 0.3) the increase in drag coefficient is of the order of 10 drag 'counts' ($0.0010C_D$), increasing to approximately 30 drag 'counts' ($0.0030C_D$) in some cases at higher values of normal-force coefficient (C_N = 0.7). It is only for higher values of normal-force coefficient, close to flow breakdown or buffet onset, that control can be seen to have the desired effect of reducing drag.

At high Reynolds number, R_C = 19 x 10^6, (Figs. 15 to 18), the effect of control is similar to that at low Reynolds number; however, the region close to buffet onset where control is effective in reducing drag, is less marked and in some cases, (Figs. 16 to 18), it is not clear that there is any reduction in drag.

Typical pressure distributions for a range of normal-force coefficients at $M = 0.68$ and 0.70 at both Reynolds numbers, $R_C = 6$ and 19×10^6, are shown in Figs. 19 to 34. In theses figures the pressure distributions for the configuration with passive control is compared with the corresponding pressure distribution for the datum section at essentially the same value of normal-force coefficient.

In all cases the lower surface pressure distributions and those on the upper surface ahead of the shock wave, when present, are identical. The effect of control is to weaken the shock wave by "smearing" it along the chord compressing the flow from a point just ahead of the control region ($x/c = 0.47$) to close to the downstream end ($x/c = 0.57$). The significant increase in pressure coefficient just ahead of the shock wave indicates a reduction in shock upstream Mach number and thus a reduction in wave drag as found by other workers [4,5,6].

In all cases, for normal-force coefficients above 0.5, (Figs. 25 and 26) there is an increased tendency for pressures downstream of the shock wave on the upper surface to be lower than those of the datum case. The significance of this feature will be discussed in section 4.

22.4 Discussion

It is worth studying in detail some of the passive control data with the aim of identifying the reason behind the overall increase in drag due to the addition of a control surface. Figs. 11 to 18 show that for nearly all values of normal-force coefficient the drag of the aerofoil is increased. It can be inferred from the pressure distributions in Figs. 19 to 34 that passive control has the effect of replacing the single straight shock of the datum case with a multi-shock system. It is well known that, for a given freestream Mach number and a given total pressure rise, a compression through several shocks entails the production of less entropy and thus lower wave drag than a compression through a single shock.

The increase in shock upstream pressure coefficient and hence reduction in shock upstream Mach number suggests a significant reduction in shock strength and wave drag (strictly only as long as the shock position remains unchanged). The increase in drag can therefore only arise from an increase in viscous drag due to the aerodynamic roughness of the porous surface and/or the flow through the surface causing excess thickening of the boundary layer and hence increasing drag.

The potential benefits of the reduction in wave drag, in this case, more than been off-set by an increase in viscous drag, giving an overall penalty. If the increase in drag is associated with the injection of air into the boundary layer ahead of the shock wave, the possibility of active suction in the shock region should be investigated as an alternative.

It is interesting to note from Figs. 19, 23, 27 and 31 at a normal-force coefficient of 0.1, where the upper surface flow is entirely subcritical, that there is still a pressure gradient across the control region ($x/c = 0.5$ to 0.6), with the consequence that there will, presumably, still be a secondary flow via the plenum. The existence of a secondary flow at shock free conditions at low lift is consistent with the observation from Figs. 12, 14, 16 and 18 that control causes an increase in drag at these conditions. Since shock waves are absent in these cases the increase in drag arises purely from an increase in viscous drag, suggesting that at least part of this increase is due to the secondary flow.

Further evidence for this increase in viscous drag can be gleaned from Figs. 35 to 38. Here the local contribution to drag, measured at each pitot tube in the wake rake, is plotted against vertical distance ($y/c = 0.0$ representing the central tube in the rake) for the $C_N =$

0.1 cases shown in Figs. 19 to 34. It is immediately obvious that the area under the curve (representing the overall drag) is greater for the case with control. The major contributor to the excess drag can be seen to be the upper surface for the cases of interest.

The inference from Figs. 35 to 38, is that there is a large increase in viscous drag due to the addition of the control surface. Bur [11] shows similar results from experiments in a channel flow. Measurements of the boundary layer development over a passive control surface indicate significant increases in boundary layer displacement and momentum thicknesses. Bur [11] also concluded that the increase in viscous drag could upset the overall drag balance (wave drag + viscous drag) such as to weigh against the use of passive control.

This observation is consistent with the reduction in pressure downstream of the shock wave, observed in Figs. 25. This feature is usually associated with an increase in boundary layer thickness, leading to premature separation of the flow.

Thibert et al. [6] observed that with passive control no significant overall drag reduction is achievable suggesting that the increase in viscous drag exactly matches, and thus offsets, the benefits of decreases in wave drag over almost the whole range of conditions of interest. This suggests that it may not be possible to optimise the control surface to reap the benefits of reduced wave drag whilst at the same time minimising or eliminating the increase in viscous drag to realise an overall drag reduction.

Figs. 21 to 36 also give an indication of the effectiveness of the perforated surface. The mean static pressure in the plenum chamber is shown, demonstrating that the pressure in the plenum differs only slightly from the mean of the external static pressure. This confirms the assumption that the holes are behaving as ideal, and therefore the pressure drop for flow either into or out of the plenum is the same, which is consistent with the K modifier (see Section 2.1) having a value of 1.0. The use of a laser drilled surface for passive control would therefore appear to be ideal, provided the increase in viscous drag can be overcome. Several possibilities are conceivable: a) decrease the hole size, although the current size is close to the physical limit, b) reduce the open area ratio, c) reduce the chordwise extent of the porous surface. This parametric study would best be carried out theoretically, using the methods developed within Euroshock Task 2 (whose aim is to extend available CFD methods to allow for passive control), validating them using the existing data.

Whilst the use of passive control increases overall drag at most conditions there are indications from the steady measurements described here that there is a modest increase in the normal-force coefficient for buffet onset for the aerofoil by use of control. As described in Section 3, the normal-force coefficient variations with angle of incidence (Figs 5 to 12) display a small increase in maximum normal force and delay the normal-force 'break', both indicators of buffet onset. These results are in broad agreement with the unsteady measurements conducted on the model by DLR Institute of Aeroelasticity and to reported independently.

22.5 Conclusions

The experimental study described in this report of a passive system for the control of shock waves on an aerofoil at high Reynolds numbers typical of flight has generated the following conclusions:

1   The use of a porous surface in the region of the shock wave to provide passive control results in a significant increase in drag, for the arrangement tested.

2   The increase in drag arises from a large increase in viscous drag, which
     swamps the apparent large decrease in wave drag, due to the aerodynamic
     roughness of the porous surface and/or the excess thickening of the boundary
     layer due to the flow through the surface.

3   Passive shock control reduces wave drag by compressing the flow through
     several weak shock waves, rather than by a single shock. However, with the
     increase in viscous drag appearing to be a strong function of the control
     surface geometry, it may be possible to optimise the control surface to give an
     overall reduction in drag at some points in the performance envelope.

4   The addition of passive control gives rise to a modest improvement in the
     aerofoil normal-force coefficient for buffet onset as implied by the behaviour of the
     normal-force close to its maximum value.

22.6 References

[1]   "Boundary layer and flow control" Lachmann G V, Pergammon Press, 1961.

[2]   "Control of flow separation" Chang P, Pergammon Press, 1972.

[3]   "Shock wave/turbulent boundary-layer interaction and its control" Délery J M, Progress in Aerospace
      Sciences, 22 pp209-280, 1985.

[4]   "Transonic shock-boundary layer control" Krogmann P and Stanewsky E, Proceedings of the 14th
      Congress of ICAS, Toulouse, France, Paper No. 84-2.3.2. September 1984.

[5]   "Improvement of transonic airfoil performance through passive shock/boundary-layer interaction
      control" Thiede P and Krogmann P, IUTAM Symposium, Palaiseau, France, 1985, Ed J. Délery,
      Springer, Berlin, 1986.

[6]   "ONERA activities on drag reduction" Thibert J J, Reneaux J and Schmitt V, Proceedings of the 17th
      Congress of ICAS, Stockholm, Sweden, Paper No. 90-3.6.1, September 1990.

[7]   "The aerodynamic performance of laser drilled sheets" Poll D I A, Danks M and Humphreys B E, paper
      presented to First European Symposium on Laminar Flow, Hamburg, Germany, March 1992.

[8]   "Determination of profile drag at high speeds by pitot traverse methods" Lock C N H, Hilton W F and
      Goldstein S, ARC R&M 1970, 1940.

[9]   "An improved semi-inverse version of the viscous Garabedian and Korn method"
      Ashill P R, Wood R F and Weeks D J, RAE Technical Report TR 87002, 1987.

[10] "Calculation and measurement of transonic flows over aerofoils with novel rear sections" Ashill P R,
      Proceedings of the 16th Congress of ICAS, Jerusalem, Israel, Paper No. 88-3.10.2. September 1988.

[11] "Passive control of a shock wave/turbulent boundary layer interaction in a transonic flow" Bur R, Rech
      Aérosp. No 1992-6 pp11-30, 1992.

22.7 List of symbols

c   aerofoil chord length, distance between aerofoil leading and trailing edges along aerofoil reference axis.

C_D drag coefficient.

$C_D{}'$ local contribution to drag coefficient. $C_D = \int_{WAKE} C_D{}' d(y/c)$

C_N normal-force coefficient.

C_p static pressure coefficient.

C_{pp} mean static pressure coefficient in plenum

K   (effective diameter/nominal diameter)^4, see [7].

M   freestream Mach number, ie Mach number of empty working section corrected for blockage.

R_C Reynolds number based on freestream conditions and aerofoil chord.

x   distance along aerofoil reference axis.

y   distance from centre of wake rake.

z   ordinate of aerofoil, relative to reference axis.

α   angle of incidence, angle of freestream flow vector relative to reference axis.

     Suffices

*   value corresponding to a local Mach number of unity.

Table 1 Aerofoil Ordinates

Datum Section		Passive control	Datum Section		Passive control
x/c	z/c	z/c	x/c	z/c	z/c
0.000002	0.000000		0.008895	0.014560	
0.000172	0.002156		0.012185	0.017126	
0.000902	0.004439		0.015947	0.019748	
0.002158	0.006839		0.020161	0.022390	
0.003903	0.009358		0.024867	0.025042	
0.006130	0.011956		0.030083	0.027650	

Datum Section		Passive Control
x/c	z/c	z/c
0.035743	0.030265	
0.041900	0.032870	
0.048494	0.035432	
0.055538	0.038006	
0.063033	0.040585	
0.070962	0.043154	
0.079329	0.045654	
0.088124	0.048106	
0.097331	0.050549	
0.106940	0.052912	
0.116947	0.055216	
0.127344	0.057477	
0.138107	0.059683	
0.149242	0.061855	
0.160720	0.063984	
0.160720	0.063983	
0.172541	0.066044	
0.184591	0.068031	
0.197175	0.069930	
0.209954	0.071709	
0.223027	0.073431	
0.236381	0.075094	
0.249987	0.076669	
0.263847	0.078126	
0.277959	0.079492	
0.292289	0.080762	
0.306827	0.081914	
0.321552	0.082962	
0.336466	0.083894	
0.351538	0.084710	0.084697
0.366767	0.085381	0.085369
0.382123	0.085884	0.085917
0.397589	0.086256	0.086568
0.428851	0.086612	0.086633
0.444584	0.086570	0.086561
0.460381	0.086372	0.086318
0.476208	0.085971	0.085928
0.492067	0.085395	0.085348
0.507943	0.084616	0.084510
0.523802	0.083599	0.083323
0.539631	0.082294	0.081897
0.555425	0.080677	0.080263
0.571174	0.078777	0.078447
0.586830	0.076597	0.076349

0.602421	0.074118	0.073998
Datum Section		Passive control
x/c	z/c	z/c
0.617883	0.071408	0.071341
0.633256	0.068648	0.068612
0.648467	0.065882	0.065874
0.663532	0.063129	0.063114
0.678432	0.060364	0.060390
0.693161	0.057638	0.057631
0.707699	0.055001	0.054998
0.722013	0.052415	0.052405
0.736126	0.049880	
0.749990	0.047393	
0.763598	0.044947	
0.776956	0.042540	
0.790029	0.040189	
0.802805	0.037902	
0.815280	0.035662	
0.827443	0.033486	
0.839267	0.031350	
0.850752	0.029293	
0.861886	0.027280	
0.872643	0.025347	
0.883037	0.023477	
0.893039	0.021677	
0.902646	0.019947	
0.911857	0.018290	
0.920648	0.016714	
0.929011	0.015208	
0.936947	0.013783	
0.944445	0.012441	
0.951484	0.011169	
0.958068	0.009989	
0.964210	0.008883	
0.969864	0.007869	
0.975060	0.006933	
0.979769	0.006082	
0.983988	0.005312	
0.987731	0.004616	
0.990975	0.004034	
0.993735	0.003570	
0.996005	0.003176	
0.997754	0.002862	
0.999015	0.002646	
0.999773	0.002503	
0.000002	0.000000	

0.000183	-0.002004	
0.000960	-0.003826	
Datum Section		**Passive control**
x/c	z/c	z/c
0.002218	-0.005495	
0.003960	-0.007188	
0.006217	-0.008887	
0.008989	-0.010608	
0.012247	-0.012401	
0.015997	-0.014248	
0.020237	-0.016113	
0.024927	-0.017946	
0.030128	-0.019806	
0.035780	-0.021658	
0.041927	-0.023478	
0.048518	-0.025282	
0.063061	-0.028752	
0.071000	-0.030447	
0.079359	-0.032118	
0.088145	-0.033757	
0.097358	-0.035333	
0.106966	-0.036841	
0.116968	-0.038298	
0.127365	-0.039689	
0.138123	-0.041037	
0.149255	-0.042339	
0.160735	-0.043579	
0.172562	-0.044749	
0.184719	-0.045863	
0.197193	-0.046888	
0.209965	-0.047854	
0.223037	-0.048748	
0.236390	-0.049604	
0.249996	-0.050390	
0.263854	-0.051100	
0.277965	-0.051722	
0.292296	-0.052265	
0.306833	-0.052729	
0.321558	-0.053105	
0.336472	-0.053409	
0.351543	-0.053641	
0.366771	-0.053774	
0.382126	-0.053815	
0.397591	-0.053757	
0.413181	-0.053584	
0.428851	-0.053296	

0.444583	-0.052867	
0.460378	-0.052301	
0.476205	-0.051588	
Datum Section		**Passive control**
x/c	z/c	z/c
0.492063	-0.050732	
0.507935	-0.049719	
0.523792	-0.048544	
0.539619	-0.047224	
0.555416	-0.045766	
0.571165	-0.044154	
0.586818	-0.042401	
0.602407	-0.040532	
0.617874	-0.038567	
0.633243	-0.036512	
0.648454	-0.034388	
0.663524	-0.032240	
0.678434	-0.030043	
0.693174	-0.027886	
0.707725	-0.025823	
0.722036	-0.023697	
0.736142	-0.021609	
0.750000	-0.019622	
0.763604	-0.017703	
0.776959	-0.015911	
0.790033	-0.014213	
0.802802	-0.012609	
0.815274	-0.011124	
0.827431	-0.009741	
0.839261	-0.008533	
0.850743	-0.007436	
0.861877	-0.006429	
0.872632	-0.005519	
0.883024	-0.004724	
0.893024	-0.004052	
0.902630	-0.003487	
0.911843	-0.003019	
0.920630	-0.002647	
0.928992	-0.002355	
0.936929	-0.002139	
0.944425	-0.001985	
0.951465	-0.001902	
0.958047	-0.001861	
0.964189	-0.001859	
0.969842	-0.001886	
0.975039	-0.001920	

0.979748	-0.001969
0.983968	-0.002031
0.987716	-0.002103
0.990959	-0.002179

Datum Section		Passive control
x/c	z/c	z/c
0.993714	-0.002273	

0.995980	-0.002360
0.997726	-0.002434

Datum Section		Passive control
x/c	z/c	z/c
0.998986	-0.002496	
0.999758	-0.002522	

Table 2 position of surface static pressure holes

Upper Surface	Lower Surface
0.000	
0.0025	0.0025
0.005	0.005
0.010	0.010
0.015	
0.020	0.020
0.030	
0.040	0.040
0.060	0.060
0.080	
0.100	0.100
0.150	0.150
0.200	0.200
0.250	
0.300	0.300
0.350	
0.400#	0.400
0.450#	0.450
0.475#	
0.500#	0.500
0.525#	
0.550#$	0.550*
0.575#	
0.600#$	0.600
0.625#	
0.650#	0.650
0.675#	
0.700	0.700
0.750	0.750
0.800	0.800
0.850	0.850
0.900	0.900
0.950	0.950
0.985	
1.000	

*   Centre station pressure hole blocked for the duration of test series.

\#   Pressure holes on port and starboard stations not connected for passive control configuration.

$ Centre station pressure holes blocked during passive control configuration.

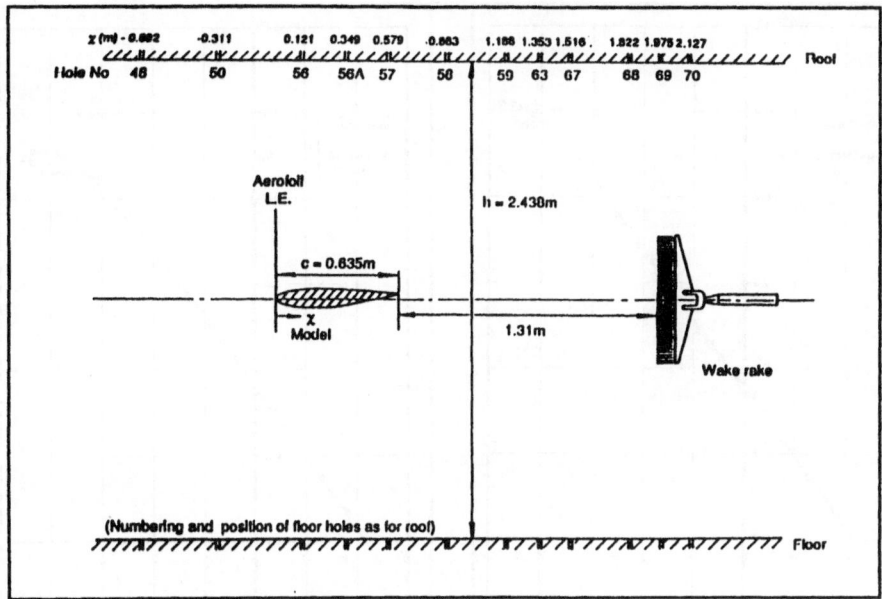

Figure 1 : Layout of the model and wake rake in the working section

Figure 2 : Photograph of model in working section

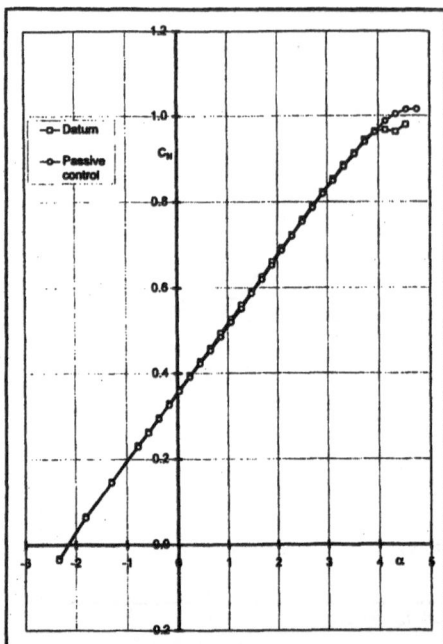

Figure 3 : Variation of normal-force coefficient with angle of incidence, $R_C = 6 \times 10^6$, M = 0.67

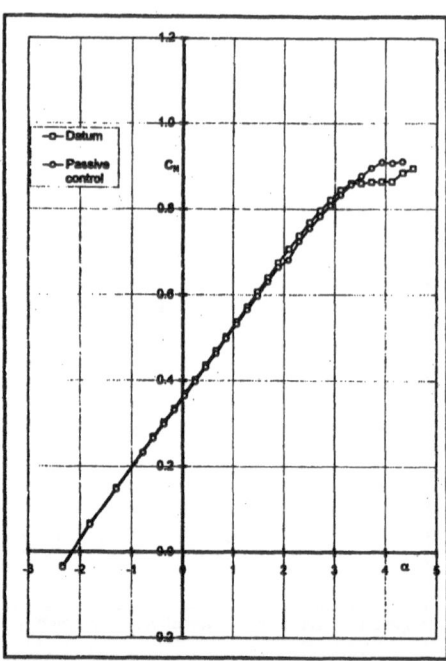

Figure 4 : Variation of normal-force coefficient with angle of incidence, $R_C = 6 \times 10^6$, M = 0.68

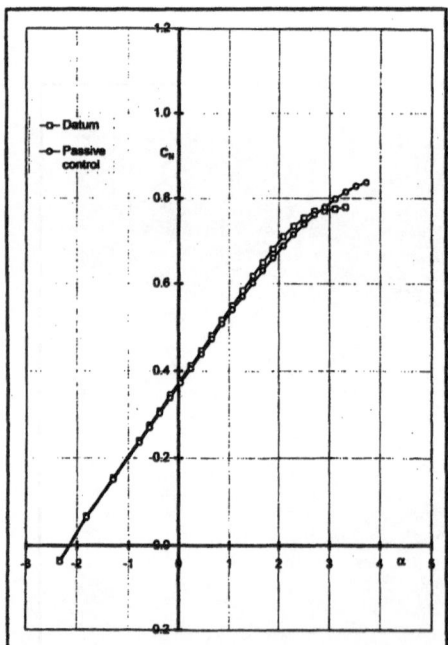

Figure 5 : Variation of normal-force coefficient with angle of incidence, $R_C = 6 \times 10^6$, M = 0.69

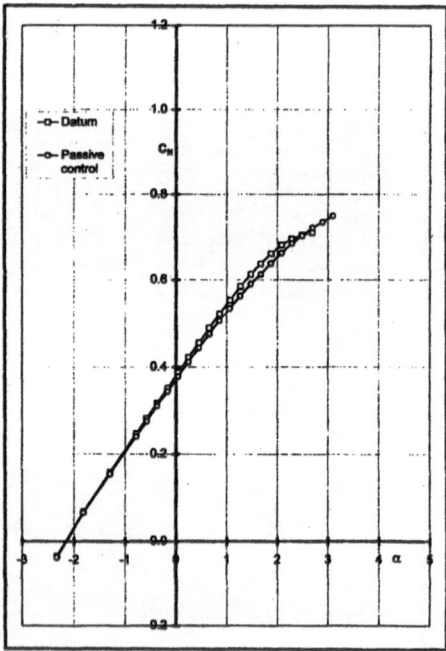

Figure 6 : Variation of normal-force coefficient with angle of incidence, $R_C = 6 \times 10^6$, M = 0.70

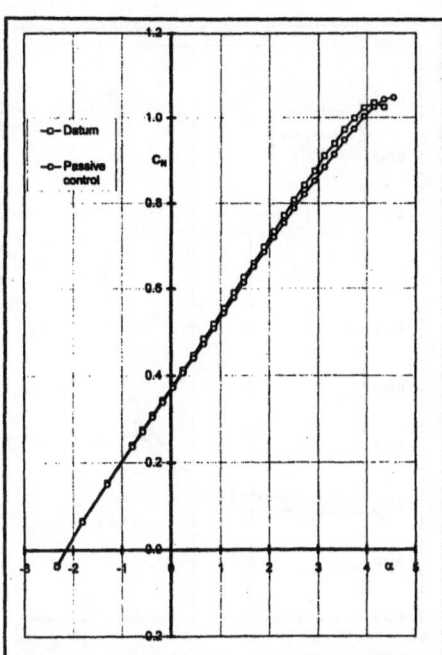

Figure 7 : Variation of normal-force coefficient with angle of incidence, R_C = 19 x 10^6, M =0.67

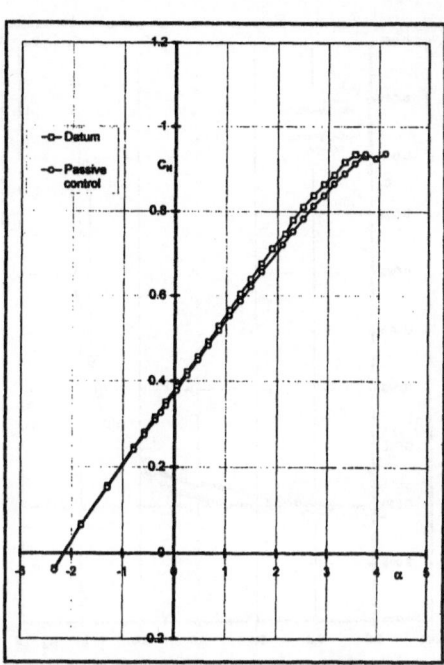

Figure 8 : Variation of normal-force coefficient with angle of incidence, R_C = 19 x 10^6, M =0.68

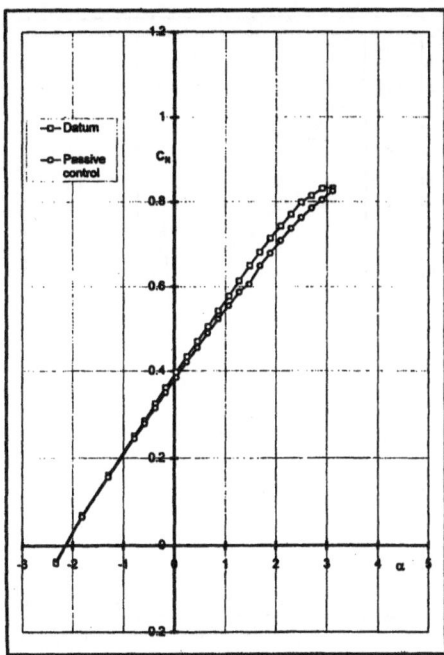

Figure 9 : Variation of normal-force coefficient with angle of incidence, R_C = 19 x 10^6, M =0.69

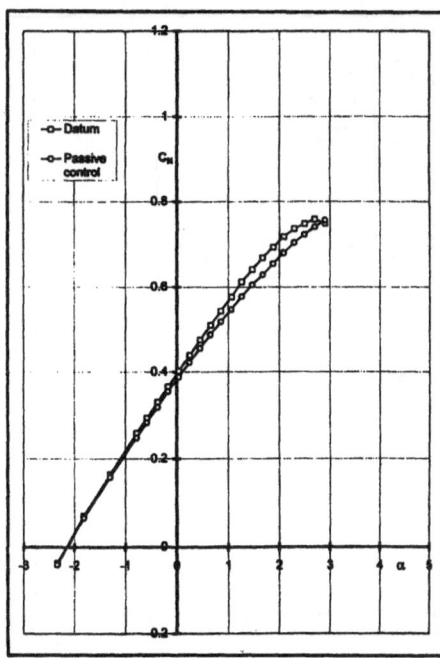

Figure 10 : Variation of normal-force coefficient with angle of incidence, R_C = 19 x 10^6, M = 0.70

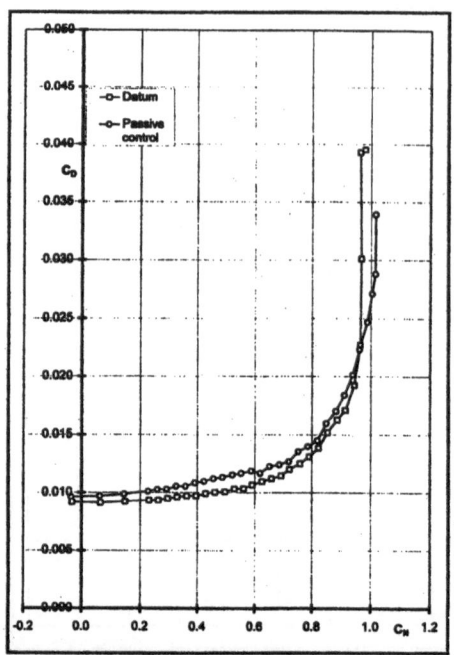

Figure 11 : Variation of drag coefficient with normal-force coefficient $R_C = 6 \times 10^6$, M = 0.67

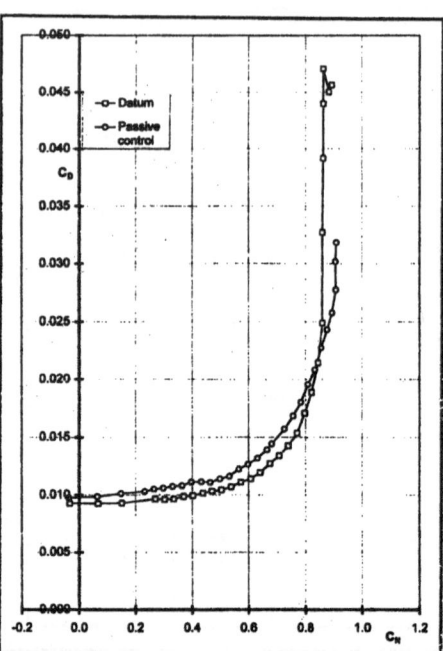

Figure 12 : Variation of drag coefficient with normal-force coefficient $R_C = 6 \times 10^6$, M = 0.68

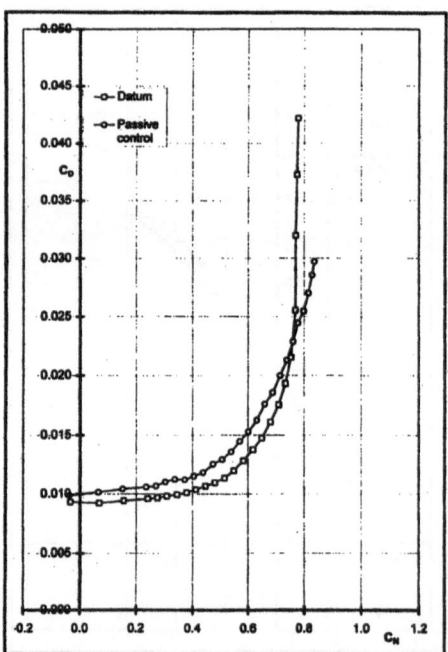

Figure 13 : Variation of drag coefficient with normal-force coefficient $R_C = 6 \times 10^6$, M = 0.69

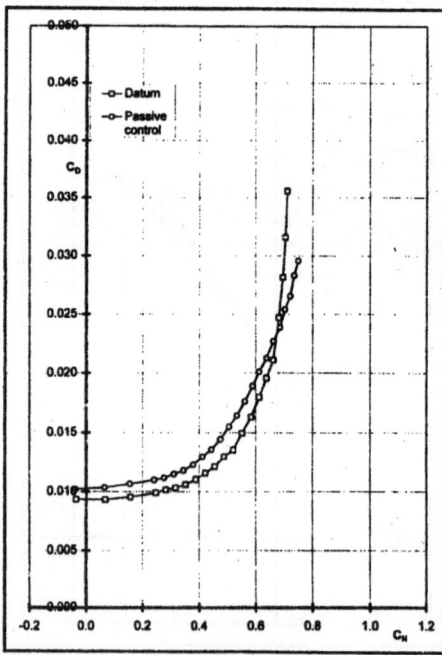

Figure 14 : Variation of drag coefficient with normal-force coefficient $R_C = 6 \times 10^6$, M = 0.70

394

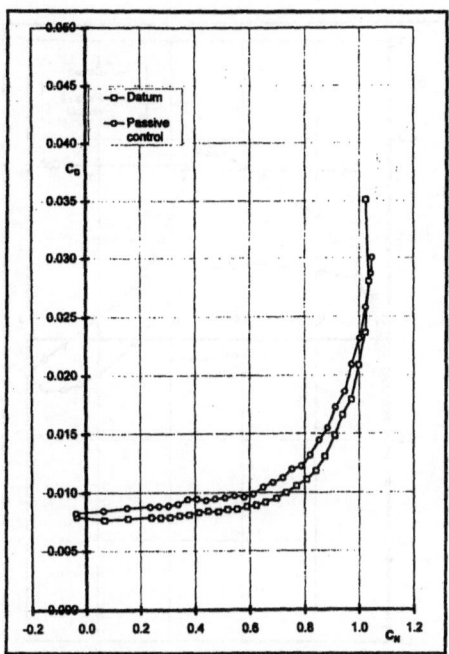

Figure 15 : Variation of drag coefficient with normal-force coefficient $R_C = 19 \times 10^6$, M = 0.67

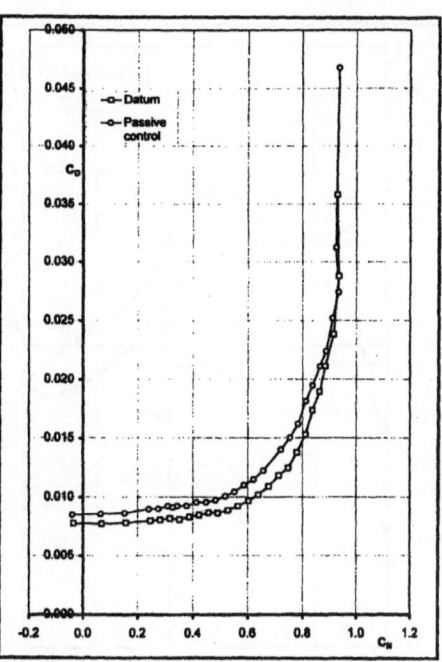

Figure 16 : Variation of drag coefficient with normal-force coefficient $R_C = 19 \times 10^6$, M = 0.68

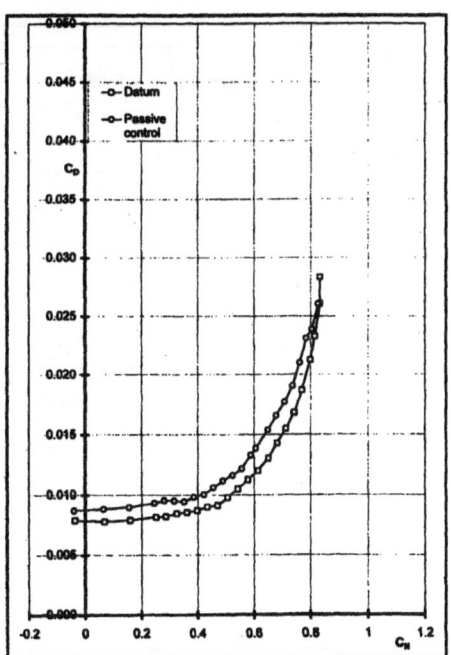

Figure 17 : Variation of drag coefficient with normal-force coefficient $R_C = 19 \times 10^6$, M = 0.69

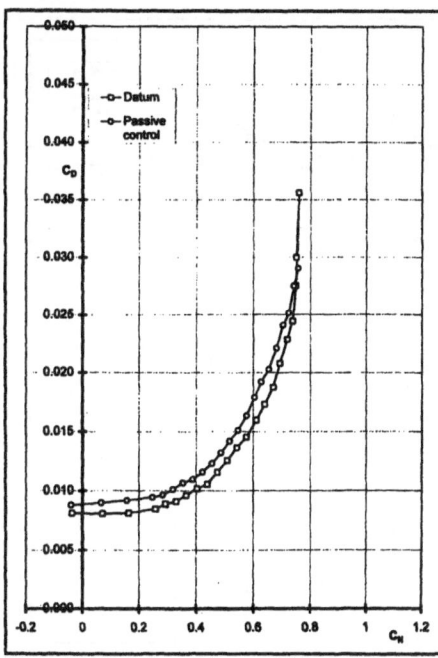

Figure 18 : Variation of drag coefficient with normal-force coefficient $R_C = 19 \times 10^6$, M = 0.70

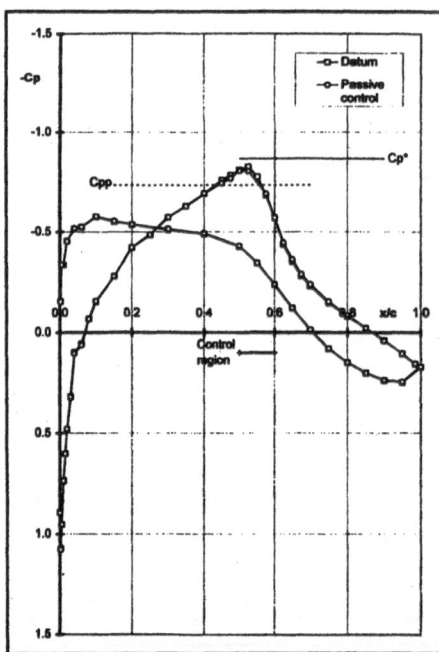

Figure 19 : Comparison of pressure distributions, $R_C = 6 \times 10^6$, $M = 0.68$, $C_N = 0.1$

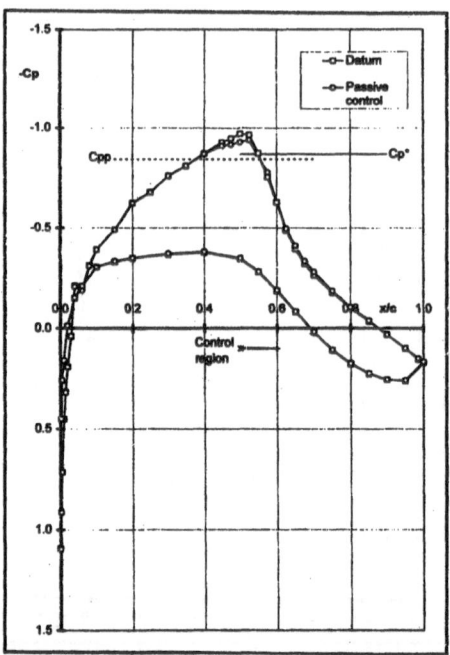

Figure 20 : Comparison of pressure distributions, $R_C = 6 \times 10^6$, $M = 0.68$, $C_N = 0.3$

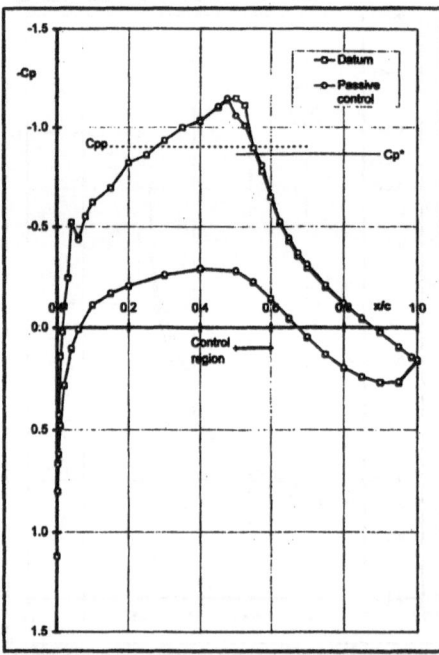

Figure 21 : Comparison of pressure distributions, $R_C = 6 \times 10^6$, $M = 0.68$, $C_N = 0.5$

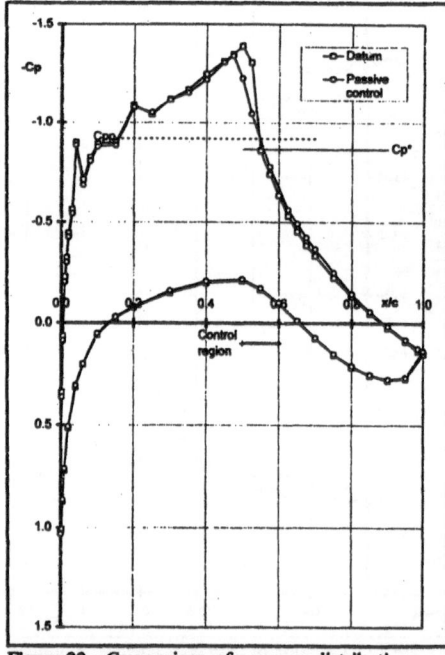

Figure 22 : Comparison of pressure distributions, $R_C = 6 \times 10^6$, $M = 0.68$, $C_N = 0.7$

396

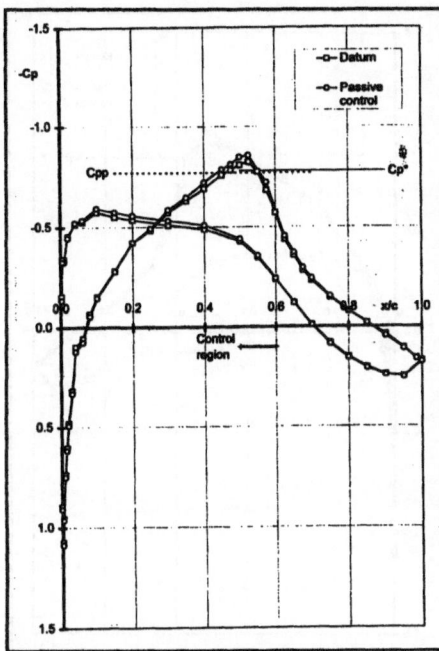

Figure 23 : Comparison of pressure distributions, $R_C = 6 \times 10^6$, $M = 0.70$, $C_N = 0.1$

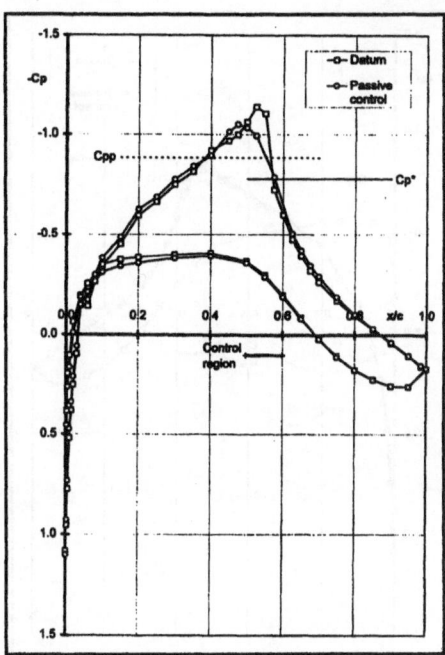

Figure 24 : Comparison of pressure distributions, $R_C = 6 \times 10^6$, $M = 0.70$, $C_N = 0.3$

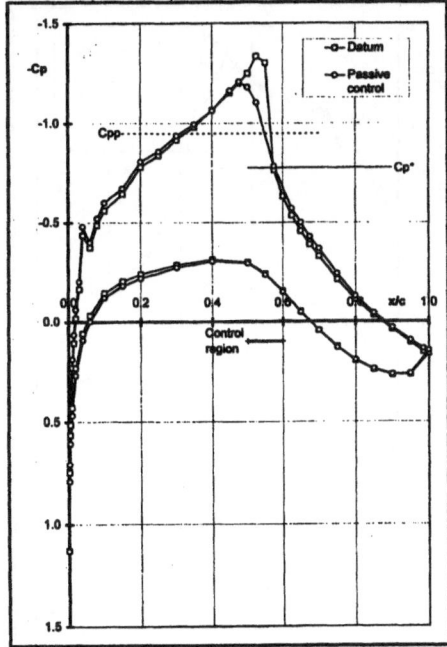

Figure 25 : Comparison of pressure distributions, $R_C = 19 \times 10^6$, $M = 0.70$, $C_N = 0.5$

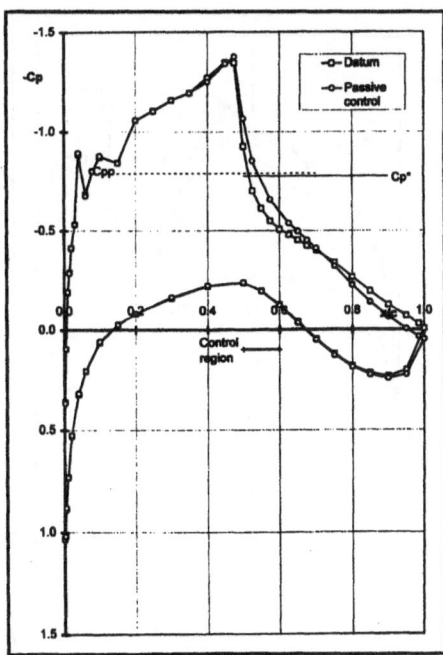

Figure 26 : Comparison of pressure distributions, $R_C = 19 \times 10^6$, $M = 0.70$, $C_N = 0.7$

397

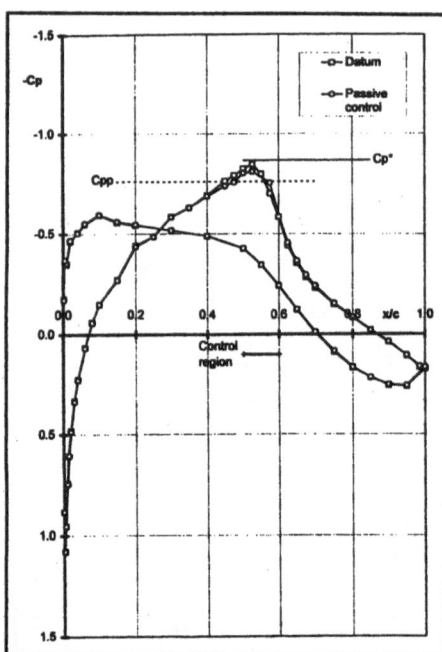

Figure 27 : Comparison of pressure distributions, $R_C = 19 \times 10^6$, M = 0.68, $C_N = 0.1$

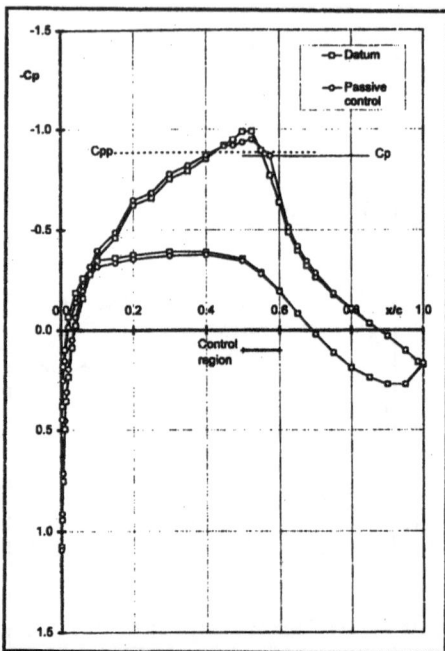

Figure 28 : Comparison of pressure distributions, $R_C = 19 \times 10^6$, M = 0.68, $C_N = 0.3$

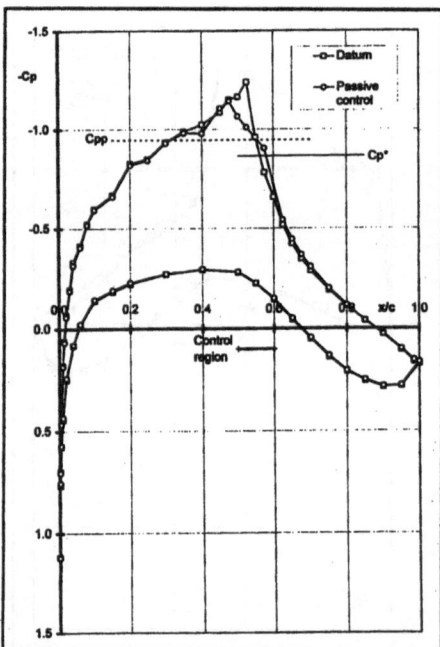

Figure 29 : Comparison of pressure distributions, $R_C = 19 \times 10^6$, M = 0.68, $C_N = 0.5$

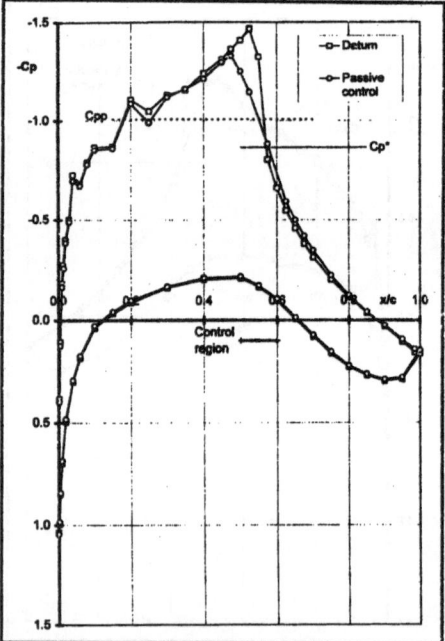

Figure 30 : Comparison of pressure distributions, $R_C = 19 \times 10^6$, M = 0.68, $C_N = 0.7$

398

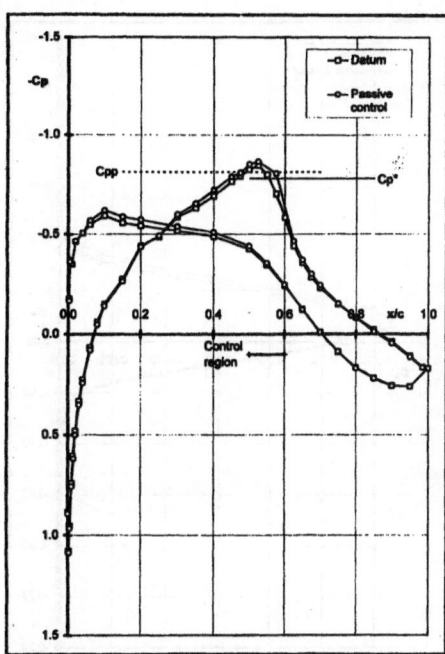

Figure 31 : Comparison of pressure distributions, $R_C = 19 \times 10^6$, M = 0.70, $C_N = 0.1$

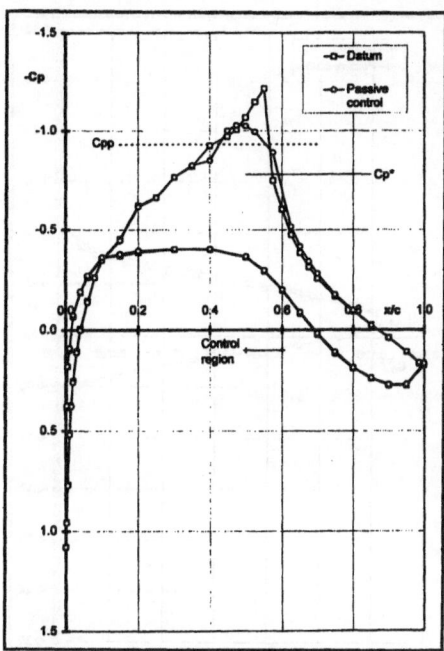

Figure 32 : Comparison of pressure distributions, $R_C = 19 \times 10^6$, M = 0.70, $C_N = 0.3$

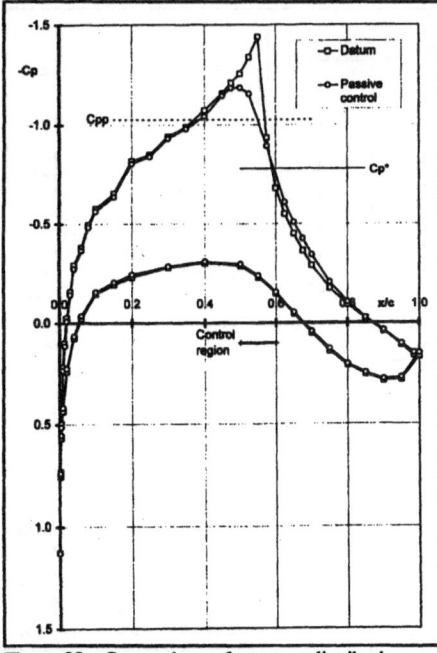

Figure 33 : Comparison of pressure distributions, $R_C = 19 \times 10^6$, M = 0.70, $C_N = 0.5$

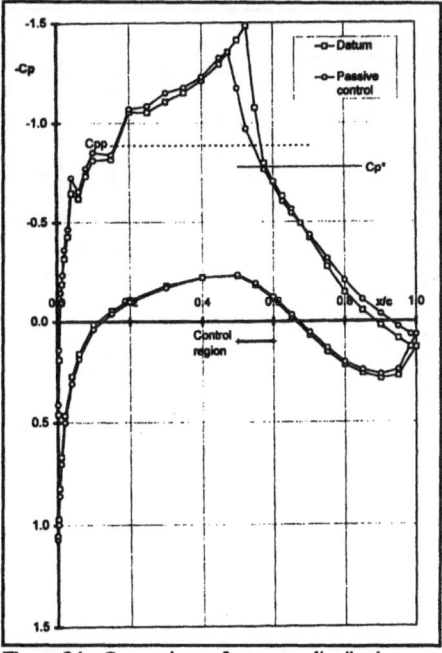

Figure 34 : Comparison of pressure distributions, $R_C = 19 \times 10^6$, M = 0.70, $C_N = 0.7$

399

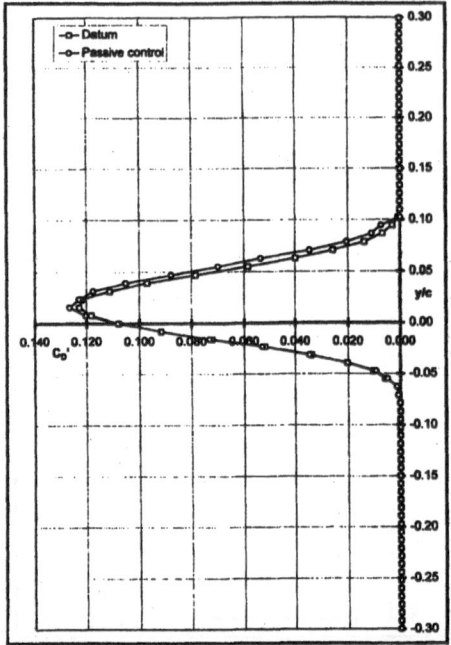

Figure 35 : Comparison of local drag variation in the wake, $R_C = 6 \times 10^6$, M = 0.68, $C_N = 0.1$

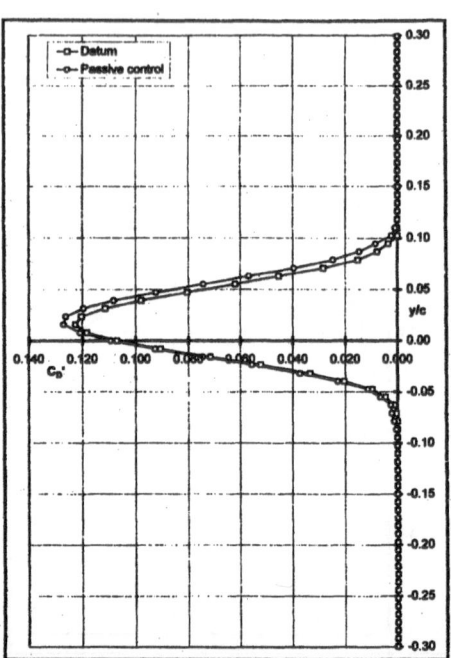

Figure 36 : Comparison of local drag variation in the wake, $R_C = 6 \times 10^6$, M = 0.70, $C_N = 0.1$

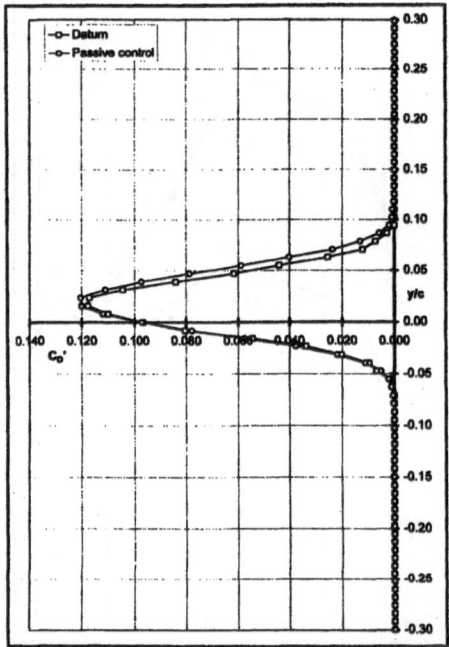

Figure 37 : Comparison of local drag variation in the wake, $R_C = 19 \times 10^6$, M = 0.68, $C_N = 0.1$

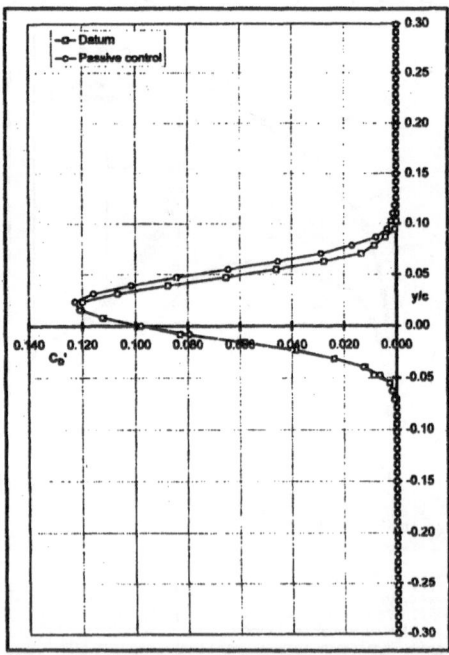

Figure 38 : Comparison of local drag variation in the wake, $R_C = 19 \times 10^6$, M = 0.70, $C_N = 0.1$

23 AN EXPERIMENTAL INVESTIGATION OF PASSIVE SHOCK/BOUNDARY LAYER INTERACTION CONTROL ON AN AIRFOIL: UNSTEADY MEASUREMENTS

W.Wagner, DLR Institut of Aeroelasticity

Bunsenstraße 10, 37073 Göttingen

Summary: During an experiment at high subsonic speeds on a two-dimensional airfoil with a passive system for the control of shock waves, unsteady pressure measurements were performed via a porous skin and an underneath cavity. Passive control, caused by the pressure gradient across the shock wave, induces a secondary flow through the surface and underlying cavity. This secondary flow causes an injection into the upstream boundary layer and a suction from the boundary layer upstream of the shock. With this, the main shock wave is weakened. The stationary measurement does not show reduced drag as expected but an increased overall drag instead. The unsteady pressures show an increase in the higher frequency fluctuations downstream of the shock, indicating a thickening of the boundary layer which is the reason for the increase in the viscous drag.

23.1 Introduction

As a part within the EUROSHOCK program to study Shock Boundary Layer Interaction Control (SBLIC), tests were performed in the 8 x 8 ft wind tunnel at DRA, Bedford during March of 1993. The task of the DLR Institute of Aeroelasticity was to perform the unsteady pressure measurements, whereby DRA carried out the stationary measurements. DRA was also responsible for the model construction and the wind tunnel test set up, described in detail by Fulker and Simmons [1]. This two-dimensional airfoil model (described in [1]) was equipped with a passive system for the control of shock waves via a porous skin and an underneath cavity. The aim of this work was to test the possibility of the drag reduction of such a system at Reynolds numbers scale from 6e+06 to 19e+06.

23.2 Measurements

The measurements were undertaken on two configurations:
The smooth (datum) airfoil and the airfoil with a perforated insert positioned between a 50% and 60% chord with an open area of 8%. The test conditions consisted of the free steam Mach numbers of 0.67, 0.68, 0.69 and 0.7 at Reynolds numbers Re = 19e+06 and 6e+06. A special selection of unsteady data were taken from these conditions. Incidence sweep runs were

performed to take a look at the buffet condition. Unsteady data were taken for selected free streem conditions. Fig. 1 shows a sketch of the model with the positions of the 19 miniature pressure sensors installed for the unsteady pressure measurements. The sensors were placed close together in the region of the insert in order to obtain the most information possible in the area of the expected shock oscillations.

23.3 Data Acquisition and Instrumentation

A scheme of the complete data acquisition system used to acquire the unsteady airfoil pressures is presented in Figs. 2 and Figs. 3. This data acquisition system is described in detail in [2]. The measuring signals were digitized and stored after proper signal conditioning (digital temperature compensation was needed) .
The signal conditions were as follows:

Number of channels:	19
Sample rate:	4096 Hz
Antialiasing filter:	at 2048 Hz with 98dB/Oct.
Frequency resolution:	2048 Hz
Block length of each channel:	200 K 16 bit words.

The steady state pressure sensitivity calibration was previously conducted by the DRA, so we were able to use this information for the evaluation of the data.

23.4 Experimental Results

Representative data are selected to demonstrate the effect of the perforated insert on the unsteady flow behavior. We first look at the results for the conditions Data point 543: incidence 2.4 degrees, Mach number 0.7, no insert. Data point 1469: incidence 2.8 degrees, Mach number 0.7, with perforated insert.
Figs. 4 and 5 show the steady values at the top of the page and the root mean square values at the bottom of the diagram (due to the spatial distribution of the pressure probes, the values are not available for the leading edge part of the profile). The steady pressures are in agreement with the results of Fulker and Simmons [1] and are discussed in detail [1]. The unsteady information at the bottom of the diagram shows the maximum *rms* values for both cases at approximately x/c = 0.5.
 In the case without shock control the value is 5 times higher than in the case with control. As the following diagram shows in more detail, this indicates strong shock oscillations at a position of about x/c=0.5 which induces buffet in the case of the datum airfoil. In the case with shock control the conditions are highly damped which means that the shock oscillations here are suppressed.

Figs. 6 and 7 show the individual time signals of the pressures. Beginning with the case without control at x/c= 0.5, an oscillation can be seen - first with a rectangular character and downstream with a more sinusoidal character. This is an indication of the shock oscillating around the position of about x/c = 0.5 with an amplitude of approximately δ x/c = 0.01 and further downstream flow oscillation as a consequence. In the case with control there is also shock located at x/c = 0.49 but the amplitude of oscillation is rather small. A slight oscillation of the downstream flow can also be detected.

Figs. 8 and 9 show on a logarithmic scale the *rms* spectra of the previously discussed pressure signals. In both cases the dominant frequency is 36 Hz. In the case without control the higher harmonics at x/c = 0.49, 0.5, 0.51, 0.52, which are related to a rectangular signal, are also seen.

In Fig. 9, the case with control, an increase of amplitude at the higher frequencies of the spectra can be detected by approaching the trailing edge. This indicates a more rapidly thickening of the boundary layer compared to the case without control . This behavior is also obvious out of the thickening of the signal trace of the time signals at x/c = 0.57 in Fig. 7 compared to Fig. 6.

Fig. 10 presents the pressures with and without control obtained during a sweep where the angle of attack was slowly altered from -2.5 degrees to +2.5 degrees. This is done within 100 seconds utilizing a constant speed of rotation to the model. The pressure variation with and without control is rather linear with angle of attack up to 1.5 degrees for the positions considered. At the angle of 0.25 degrees there is a small irregularity which may be due to the fact that, here, the mechanical moment on the model changes its sign, causing a sudden click in the mechanical bearings. For the case with control the pressure reaches at x/c = 0.503 the maximum at 1.5 degrees and then decreases. For the case without control the buffet onset is at 2.25 degrees.

Fig. 11 shows a more detailed analysis of the sweep signals. By means of the effective value of the pressure signal for the cases with and without control at different x/c positions, this figure indicates the boundary layer development. This illustration is useful since the results of the steady measurements with respect to the lower lift of the airfoil with the perforated insert can only be explained by the different behavior of the boundary layers: The effective values of the pressure variations for the case without control rise steeply to high values only for incidences higher than 2.25 degrees i.e.buffet onset. In the case of control it can be observed that at x/c = 0.517 i.e., just downstream of the leading edge for angles higher than 0 degrees that the effective value of the pressure fluctuations is higher than in the case without control.

For x/c = 0.583 and x/c = 0.9 this behavior remains unchanged. This indicates a thicker turbulent boundary layer, caused by the action of blowing and suction through the perforated insert. The thicker boundary layer changes the flow conditions, thus causing a decline in the lift especially at the trailing edge.

Further tests with a varied open area ratio of the inserts will lead to a thickening of the boundary layer, and only this will lead to an increase in lift.

23.5 References

[1] Fulker,J. L.;   An Experimental Investigation of Passive Shock/Boundary -
     Simmons, M. J.   Layer Interaction Control on an Aerofoil. DRA/AS/HWA/CR
                         95216/1 EUROSHOCK TR AER 2 - 92 - 49/3.2 .

[2] Wagener, J.;   AMIS II Anlage zur Messung zeit veränderlicher mechanischer
     Wagner, W.;   Größen in Windkanalversuchen. DLR - Nachrichten Heft 76 .

23.6 List of Symbols

α	angle of incidence,
c	airfoil chord length.
x/c, x/l	distance along airfoil chord axis.
M	free stream Mach number.
Re	Reynolds number based on free stream conditions and airfoil chord.
Peff	Root mean square value of the pressure time signal.

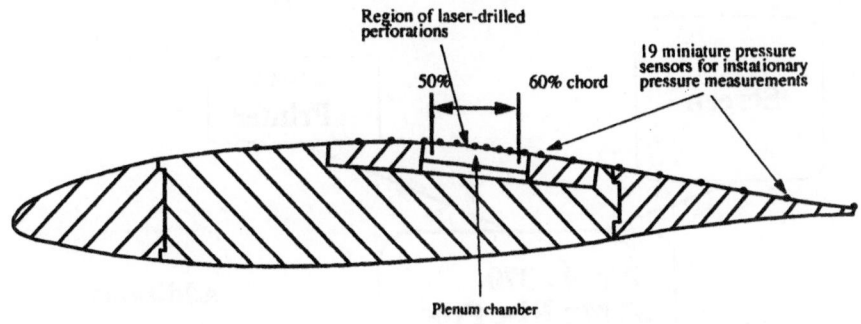

Figure 1: Airfoil model for the study of passive shock control

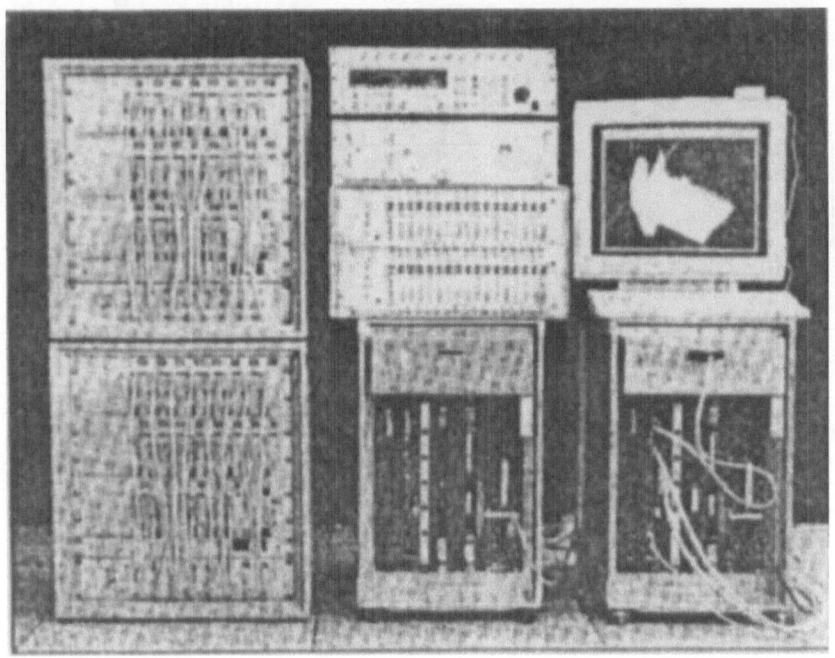

Figure 2: The AMIS II data acquisition system

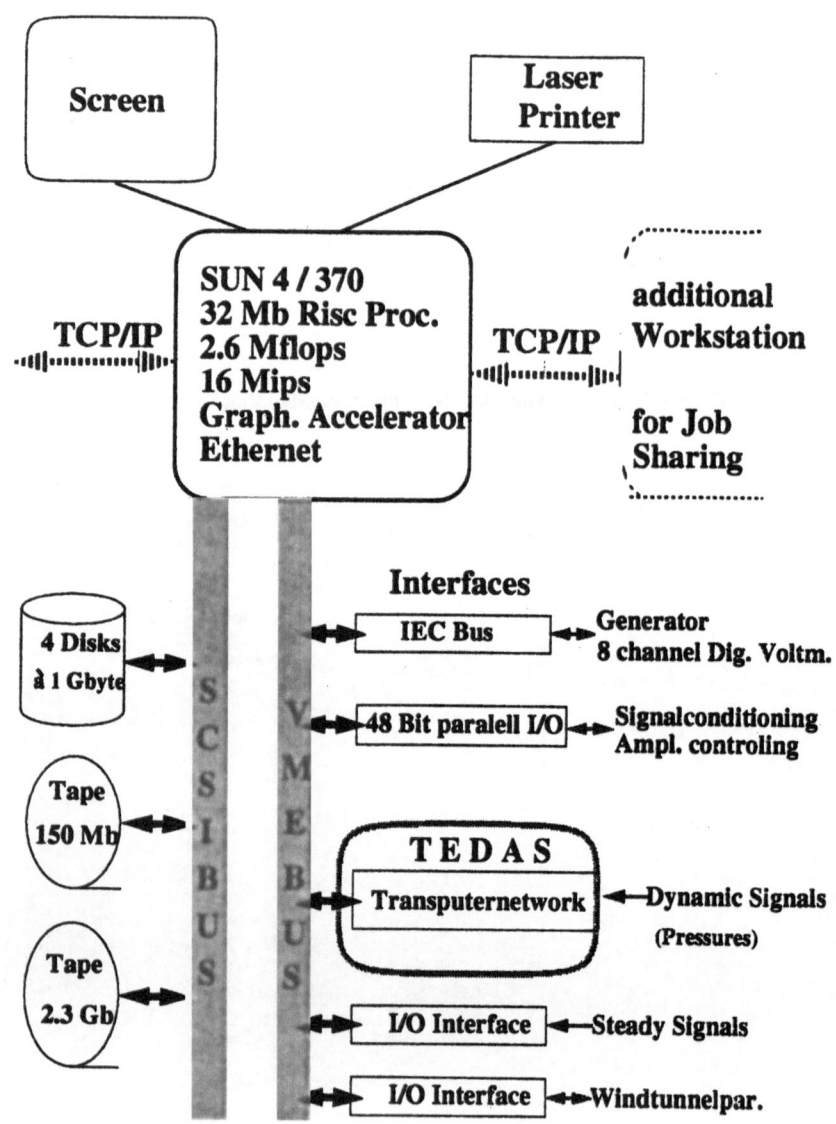

Figure 3: Hardware components of the AMIS II data acquisition system

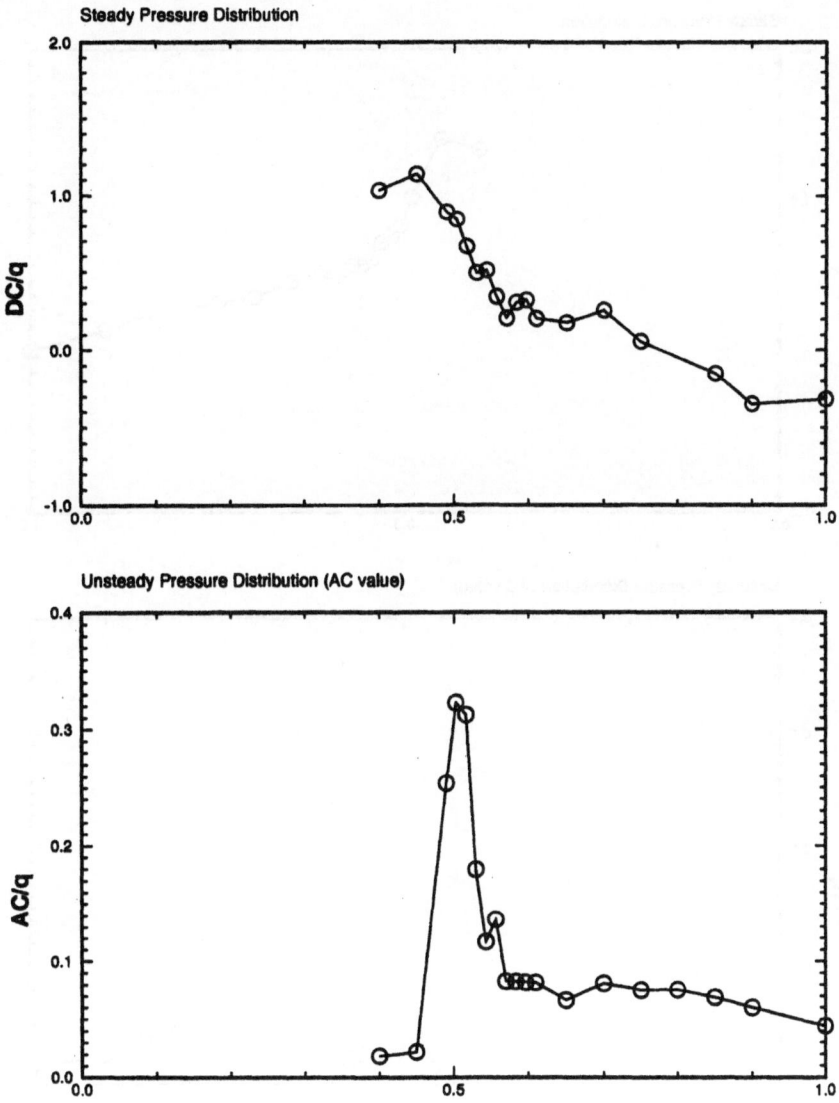

Figure 4: Wing pressures for the DRA-2303 airfoil, Data Point 543, M = 0.7022, α = 2.41°, without control

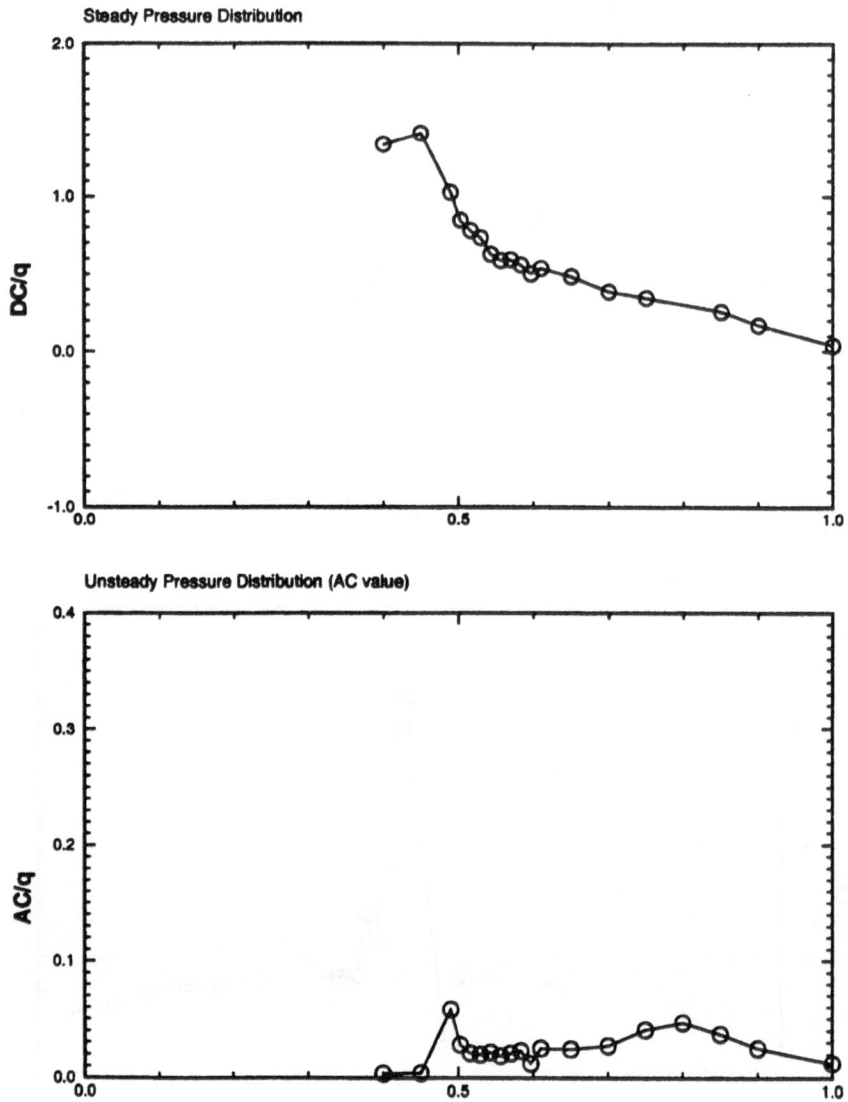

Figure 5: Wing pressures for the DRA-2303 airfoil, Data Point 1469,
M = 0.7022, α = 2.80°, with control

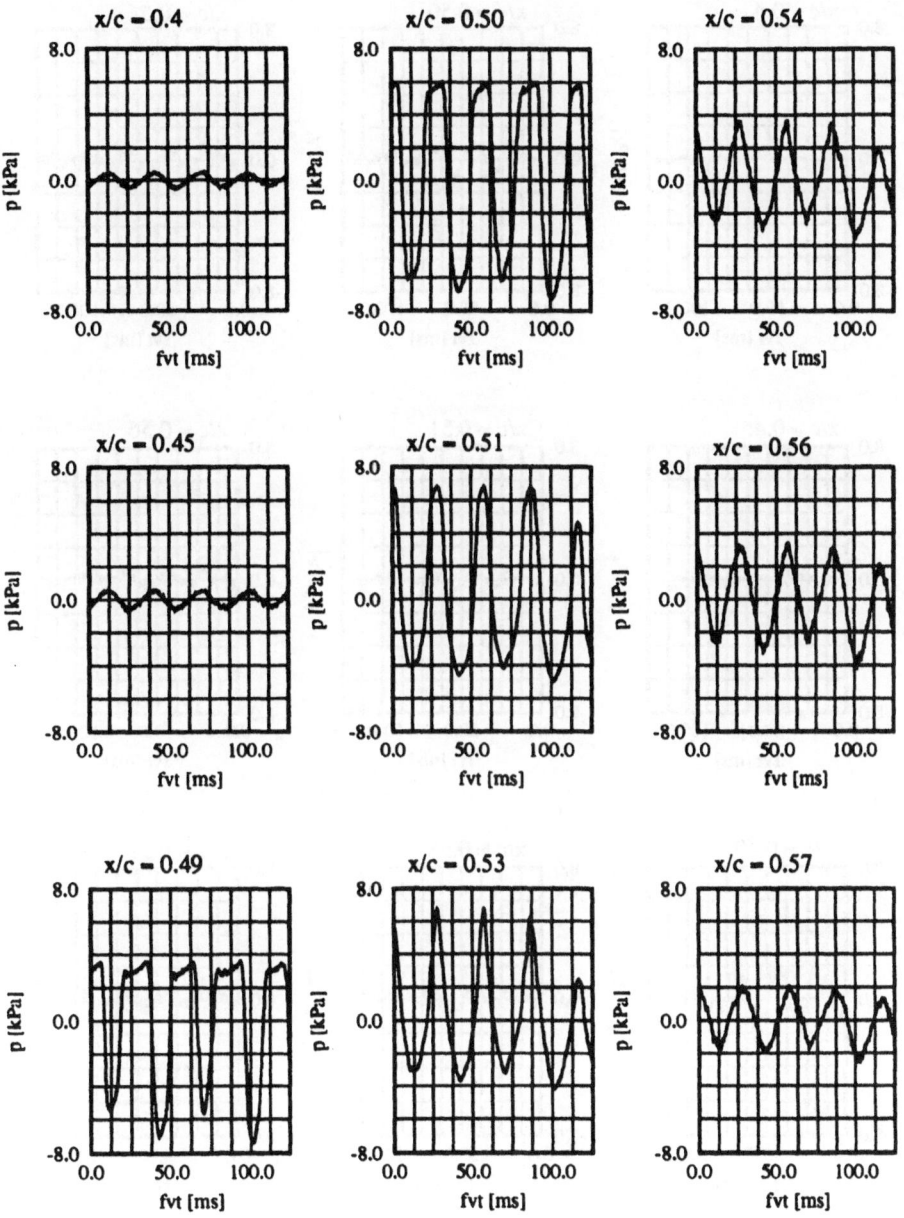

Figure 6: Instationary wing pressures at different chord positions (DRA - Airfoil M 2303)

Data Point Number	543
M =	7.022380e-01
α =	2.405000e+00
without control	

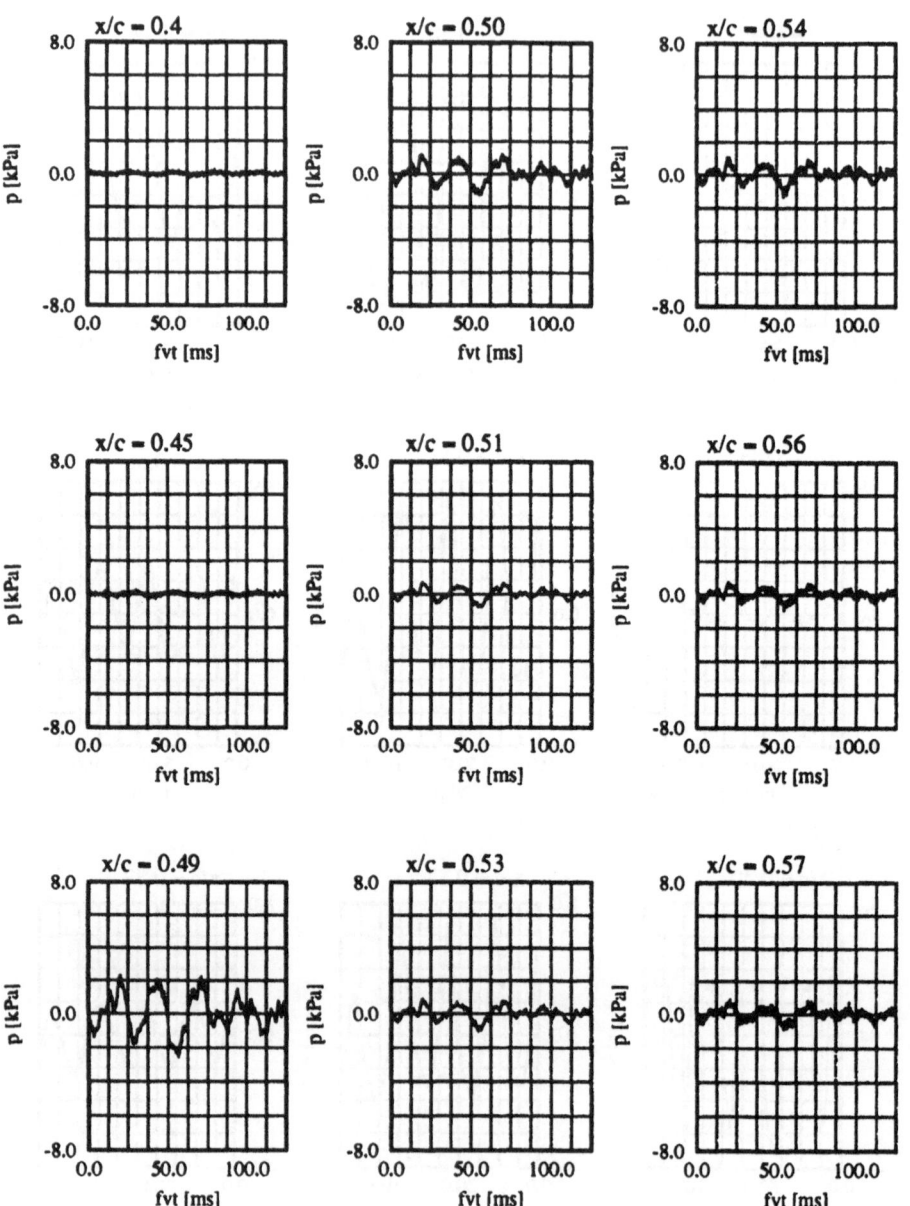

Figure 7: Instationary wing pressures at different chord positions (DRA - Airfoil M 2303)
Data Point Number   1469
M =   7.022380e-01
a =   2.805000e+00
with control

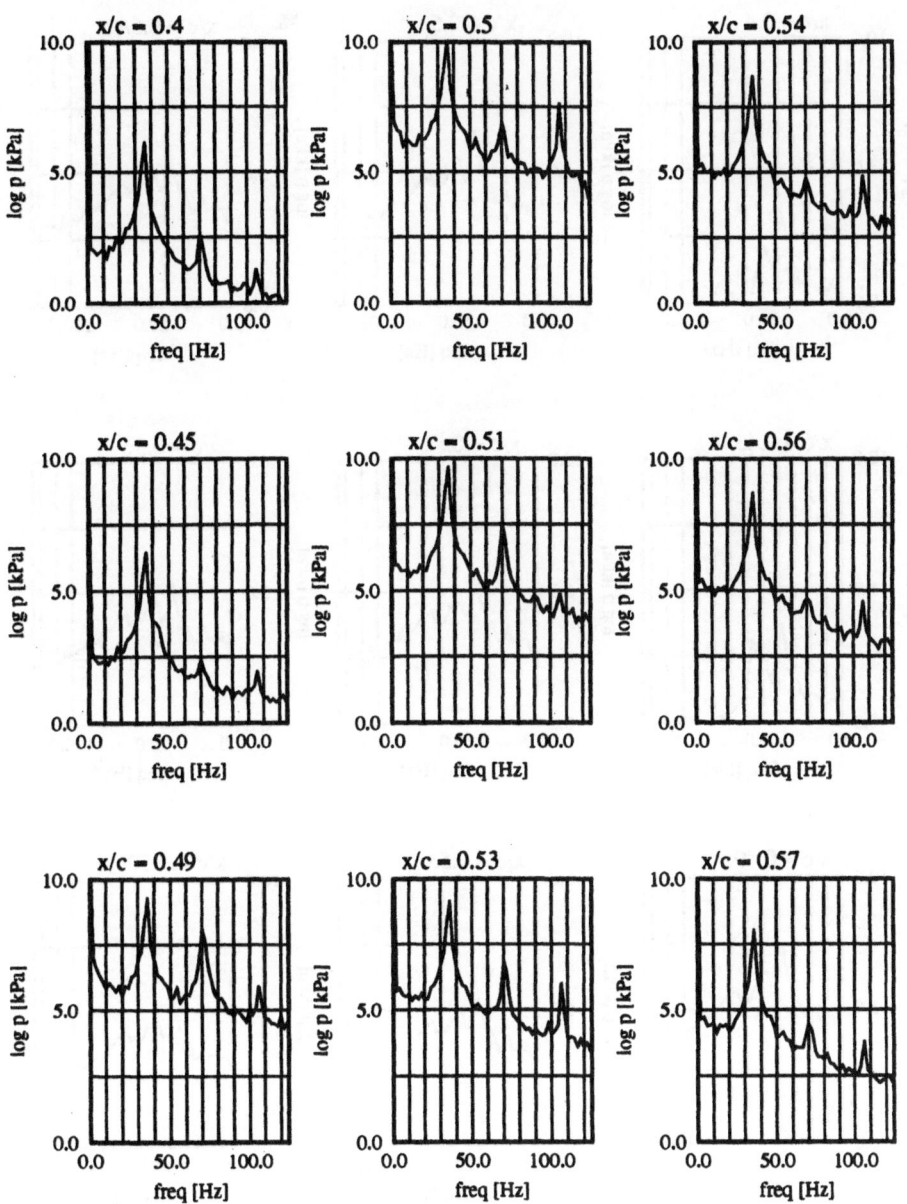

Figure 8: Ampl.Spectra of wing pressures at different chord positions (DRA - Airfoil M 2303)

Data Point Number	543
M =	7.022380e-01
a =	2.405000e+00

without control

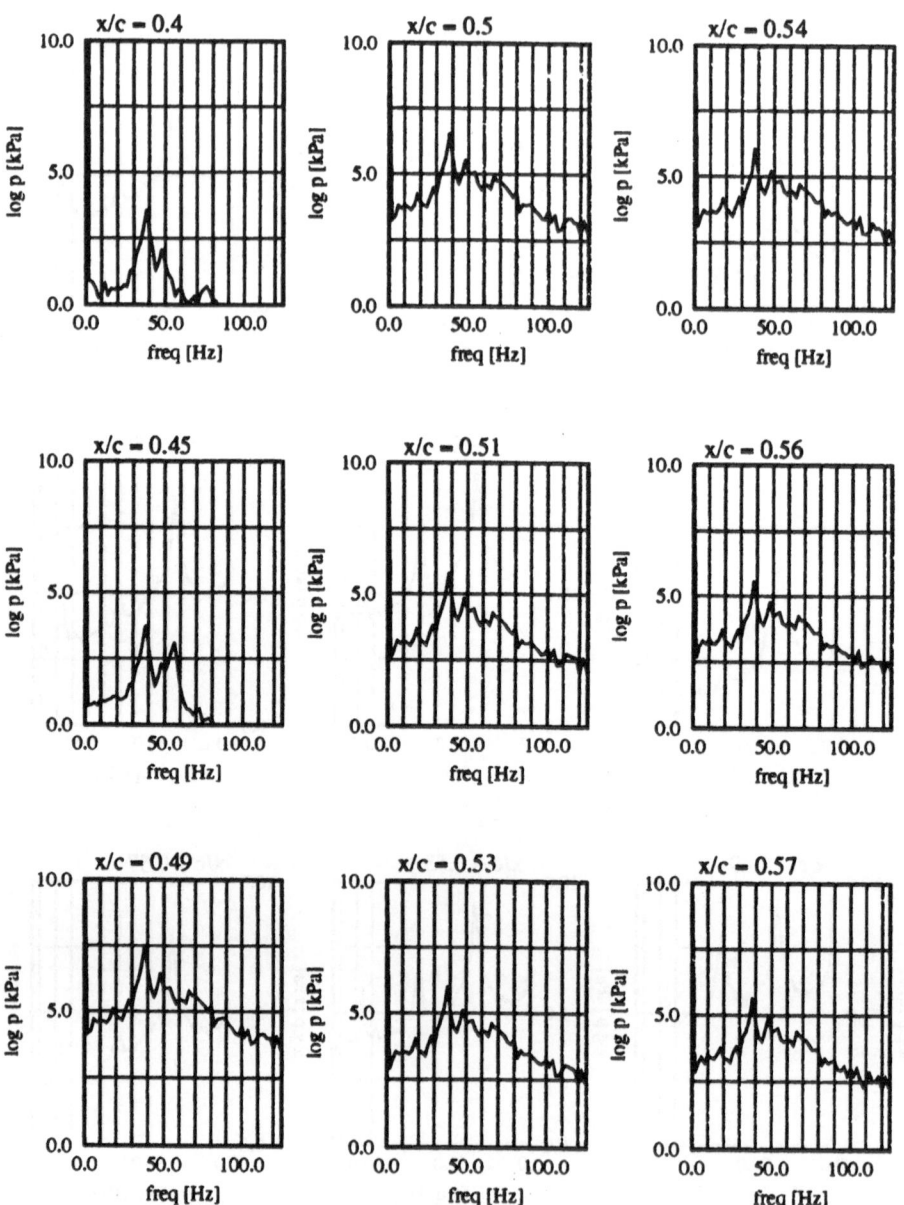

Figure 9: Ampl. Spectra of wing pressures at different chord positions (DRA - Airfoil M 2303)
Data Point Number   1469
M =   7.022380e-01
a =   2.805000e+00
with control

412

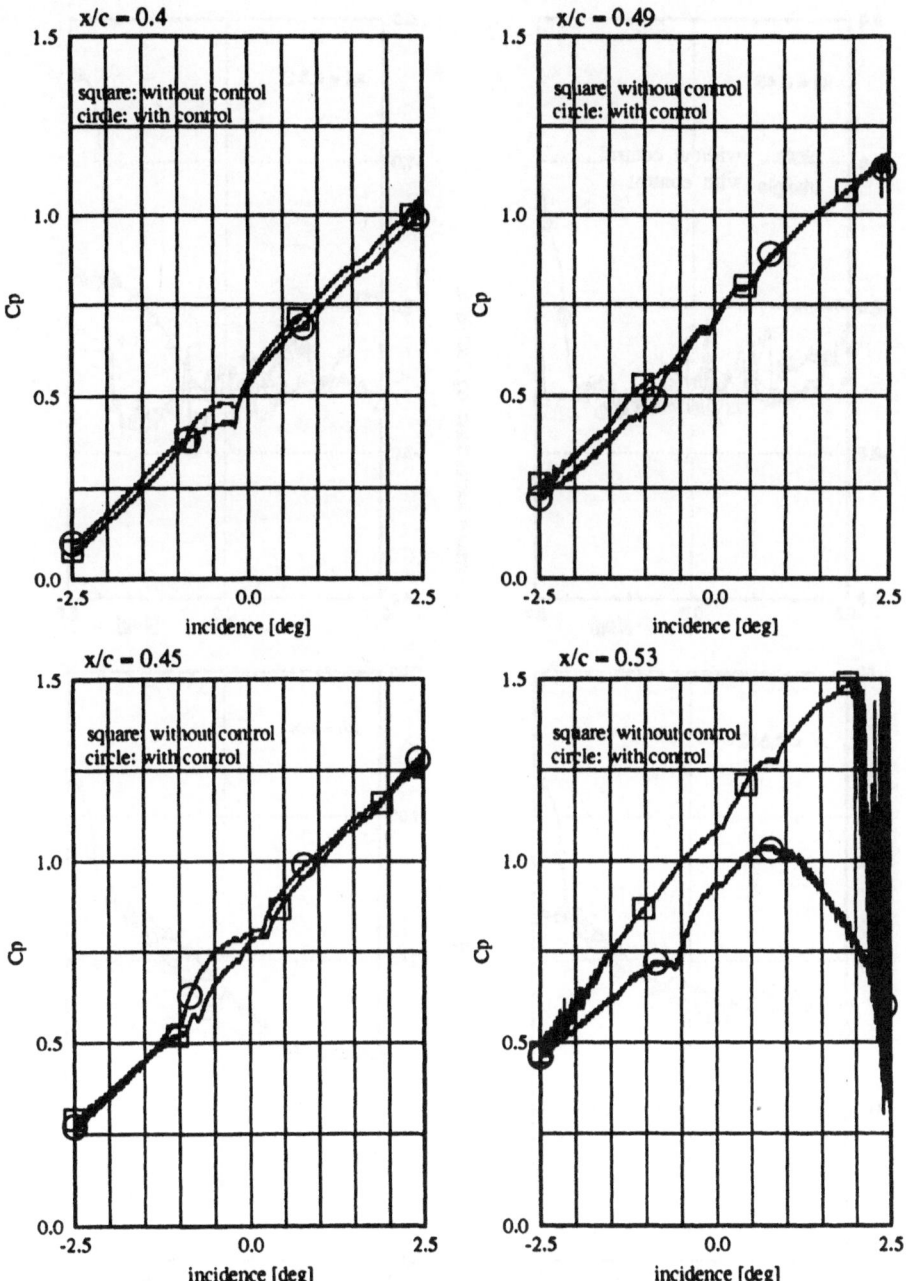

Figure 10: Instationary wing pressures during incidence sweep (DRA - Airfoil M 2303)
Data Point Numbers 1313 and 763

M =   6.99 e-01
Re =   6 .0 e+06

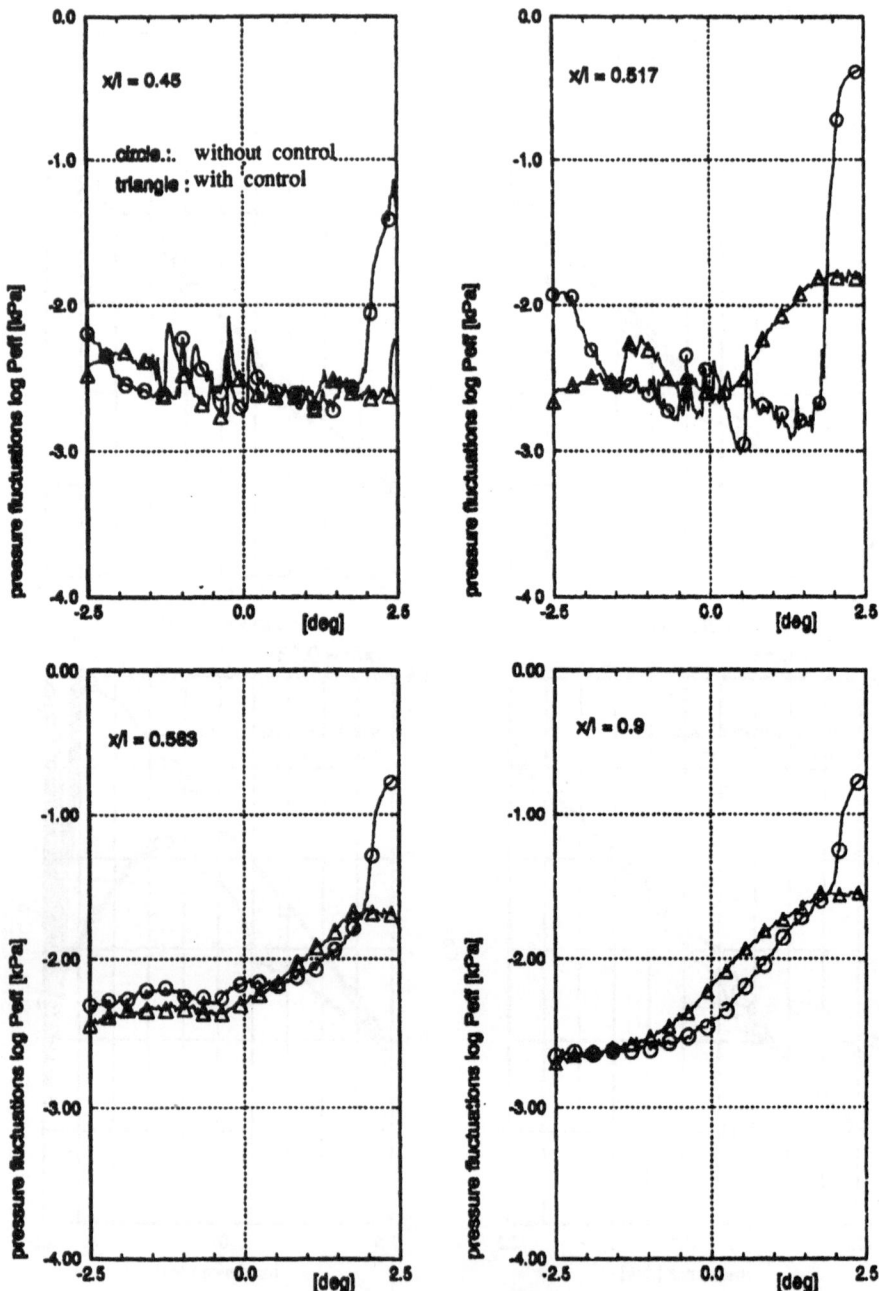

Figure 11: Pressure signal during incidence sweep, rms-values
Data Point Number   1313 and 763
M   =   7.022380e-01
Re =   6.0 e +06